T0155875

Lecture Notes in Physics

Volume 944

The Lecture Notes in Physics

The series Lecture Notes in Physics (LNP), founded in 1969, reports new developments in physics research and teaching-quickly and informally, but with a high quality and the explicit aim to summarize and communicate current knowledge in an accessible way. Books published in this series are conceived as bridging material between advanced graduate textbooks and the forefront of research and to serve three purposes:

- to be a compact and modern up-to-date source of reference on a well-defined topic
- to serve as an accessible introduction to the field to postgraduate students and nonspecialist researchers from related areas
- to be a source of advanced teaching material for specialized seminars, courses and schools

Both monographs and multi-author volumes will be considered for publication. Edited volumes should, however, consist of a very limited number of contributions only. Proceedings will not be considered for LNP.

Volumes published in LNP are disseminated both in print and in electronic formats, the electronic archive being available at springerlink.com. The series content is indexed, abstracted and referenced by many abstracting and information services, bibliographic networks, subscription agencies, library networks, and consortia.

Proposals should be sent to a member of the Editorial Board, or directly to the managing editor at Springer:

Christian Caron
Springer Heidelberg
Physics Editorial Department I
Tiergartenstrasse 17
69121 Heidelberg/Germany
christian.caron@springer.com

More information about this series at http://www.springer.com/series/5304

Cecilia Flori

A Second Course in Topos Quantum Theory

Cecilia Flori
Computing and Mathematical Sciences
The Waikato University
Hamilton, Waikato
New Zealand

ISSN 0075-8450 ISSN 1616-6361 (electronic)
Lecture Notes in Physics
ISBN 978-3-319-71107-2 ISBN 978-3-319-71108-9 (eBook)
https://doi.org/10.1007/978-3-319-71108-9

Library of Congress Control Number: 2017962000

Printed on acid-free paper

This Springer imprint is published by Springer Nature
The registered company is Springer International Publishing AG
The registered company address is: Gewerbestrasse 11, 6330 Cham, Switzerland

Vorrei dedicare questo libro ai mie Zii, Gina Ricciardi e Celestino Cruciani per la stima e l'affetto che hanno sempre mostrato verso di me ed il mio lavoro.

Contents

Notation and Terminology

For us, 'C*-algebra' always means 'unital C*-algebra'. Likewise, our $*$-homomorphisms are always assumed to be unital, unless noted otherwise (as in the proof of Theorem 13.4.1). This already applies to the following index of our notation, which lists the conventions for our most commonly used mathematical symbols:

W, X, Y, Z	Compact Hausdorff spaces
$\mathbf{1}, \ldots, \mathbf{4}$	A compact Hausdorff on the corresponding number of points, where we write e.g. $\mathbf{4} = \{0, 1, 2, 3\}$
w, x, y, z	Points in a compact Hausdorff space
f, g, h, k	Continuous functions between compact Hausdorff spaces
$\square, \bigcirc, \mathbb{T}$	Unit square, unit disk and unit circle, considered as compact subsets of $\mathbb{C}$
A, B	C*-algebras or piecewise C*-algebras (Definition 3.1.5)
M_n	the C*-algebra of $n \times n$ matrices with entries in $\mathbb{C}$
$\alpha, \beta, \gamma, \nu, \tau$	Normal elements in a C*-algebra, or (more generally) $*$-homomorphisms of the type $C(X) \to A$
ζ	A $*$-homomorphism or piecewise $*$-homomorphism of the type $A \to B$
$\mathfrak{a}, \mathfrak{b}$	Self-action of a piecewise C*-algebra (Definition 13.4.1) or a piecewise group (Definition 13.5.3)

The normal part of a C*-algebra A is

$$\mathbb{C}(A) := \{\, \alpha \in A \mid \alpha\alpha^* = \alpha^*\alpha \,\}.$$

We also think of it as the set of 'A-points' of $\mathbb{C}$. More generally, for $A \in \mathsf{C^*alg}_1$ and a closed subset $S \subset \mathbb{C}$, we also write

$$S(A) := \{\, \alpha \in \mathbb{C}(A) \mid \mathrm{sp}(\alpha) \subseteq S \,\}$$

for the set of normal elements with spectrum in S, and similarly $S(\zeta) : S(A) \to S(B)$ for the resulting action of a $*$-homomorphism $\zeta : A \to B$ on these elements. For example, $\mathbb{R}(A)$ denotes the self-adjoint part of a C*-algebra, and similarly $\mathbb{T}(A)$ is the unitary group. This sort of notation may be familiar from algebraic geometry, where the set of A-points of a scheme S (over a ring A) is denoted $S(A)$. We also use the standard notation $C(X)$ for the $\mathbb{C}$-valued continuous functions on a space X. Unfortunately, this is very similar notation despite being different in nature.

We work with the following categories:

CHaus	Compact Hausdorff spaces with continuous maps
CGHaus	Compactly generated Hausdorff spaces with continuous maps
$\mathsf{C^*alg_1}$	C*-algebras with $*$-homomorphisms
$\mathsf{cC^*alg_1}$	Commutative C*-algebras with $*$-homomorphisms
$\mathcal{V}(\mathcal{H})$	Context category (Definition 1.1.1)
$\underline{\Omega}$	Sub-object classifier (Definition 1.3.1)
S	This generally indicates a sieve (Definition 1.3.2)
$\delta^o(\hat{P})$	Outer daseinisation of projector $\hat{P}$ (Definition 2.2.2)
cHa	Complete Heyting algebra (Definition 2.1.5)
$Sub_{cl}(\underline{\Sigma})$	Set of all clopen sub-objects of $\underline{\Sigma}$ (Definition 2.2.1)
$\mathfrak{S}_V$	Isomorphism of complete Boolean algebras (Definition 2.2.1 and Eq. (3.4.3))
$\underline{\Sigma}$	Spectral presheaf over $\mathcal{V}(\mathcal{H})$ (Definition 3.1.1)
$\mathsf{pC^*alg_1}$	Piecewise C*-algebras (Definition 3.1.5) with piecewise $*$-homomorphisms (Definition 3.1.6)
Sets	Category of sets
$\mathbf{Sets}^{\mathcal{V}(\mathcal{H})^{op}}$	Topos of presheaves over $\mathcal{V}(\mathcal{H})$
$\underline{\Sigma}^{\mathcal{A}}$	Spectral presheaf over a C*-algebra $\mathcal{A}$ (Definition 3.1.1)
ucC*	Category of unital abelian C^*-algebras and unital $*$-homomorphisms
KHaus	Category of compact Hausdorff spaces and continuous maps
$\langle \Phi, \mathcal{G}_\phi \rangle$	Automorphism of spectral presheaf (Definition 3.2.1)
$F_{\hat{A}}$	Flow on the spectral presheaf (Definition 3.4.1)
$\overline{F}^{-1}(t)$	General flow (Definition 3.4.2)
$\overline{F_{\hat{A}}}^{-1}(t)$	Flows induced by unitaries (Definition 3.4.3)
μ	Measure on the state-space $\underline{\Sigma}$ (Definition 3.4.4)
$\underline{CP}$	Presheaf of classical probability measures on $\underline{\Sigma}^{\mathcal{N}}$ (Definition 3.4.5)
$\overline{F}_{\hat{A}}$	Flow on $\Gamma\underline{CP}$ induced by one-parameter group of unitaries (Definition 3.4.7)
$C_{\mathcal{N}}(F)$	Filter in $P(\mathcal{N})$ (Eq. (4.1.4))
$g_{\hat{A}}$	Antonymous function of $\hat{A}$ (Definition 4.1.2)
$f_{\hat{A}}$	Observable function of $\hat{A}$ (Definition 4.1.2)
$\delta^i(\hat{P})$	Inner daseinisation (Definition 4.1.4)
$\breve{\delta}(\hat{A})$	Physical quantity associated with $\hat{A}$ (Definition 4.1.16)
$\overline{\delta^i(\hat{A})_V}$	Gelfand transform associated with $\delta^i(\hat{A})_V$ (Corollary 4.1.1)
$\overline{\delta^o(\hat{A})_V}$	Gelfand transform associated with $\delta^o(\hat{A})_V$ (Corollary 4.1.2)

$\underline{\mathfrak{w}}^{|\psi\rangle}$ Pseudo-state (Definition 4.2.1)

$\overline{\mathbb{R}}$ Extended reals (Definition 5.1.1)

E Spectral family (Definition 5.1.1)

$o^{\hat{A}}$ q-Observable function associated with $\hat{A}$ (Definition 5.1.4)

$SA(\mathcal{N})$ Set of self-adjoint operators affiliated with a von Neumann algebra $\mathcal{N}$

$SF(\overline{\mathbb{R}}, P(\mathcal{N}))$ Set of extended, right-continuous spectral families of $P(\mathcal{N})$

$QO(P(\mathcal{N}), \overline{\mathbb{R}})$ Set of all abstract q-observable functions

$a^{\hat{A}}$ q-Antonymous function associated with $\hat{A}$ (Definition 5.3.1)

$\tilde{C}^A$ A cumulative distribution function (CDF) of a random variable A (Definition 5.5.1)

C^A An extended cumulative distribution function (ECDF) of a random variable A (Definition 5.5.1)

q^A Quantile function of A (Eq. (5.5.1))

$\overline{C}^A$ Lattice valued CDF (Definition 5.5.3)

$Sub(\mathcal{N})$ Set of all von Neumann subalgebras of $\mathcal{N}$ (Definition 6.1.1)

$AbSub(\mathcal{N})$ Set of all abelian subalgebras of $\mathcal{N}$ (Definition 6.1.1)

$FAbSub(\mathcal{N})$ Set of all abelian subalgebras of $\mathcal{N}$ containing only finitely many projections (Definition 6.1.1)

$Sub(Proj(\mathcal{N}))$ Poset of subalgebras of $Proj(\mathcal{N})$ ordered by subset inclusion (Definition 6.1.2)

$BSub(Proj(\mathcal{N}))$ Poset of Boolean subalgebras of $Proj(\mathcal{N})$ ordered by subset inclusion (Definition 6.1.2)

$FBSub(Proj(\mathcal{N}))$ Poset of finite Boolean subalgebras of $Proj(\mathcal{N})$ ordered by subset inclusion (Definition 6.1.2)

PG Set of commutative Lie subalgebras of $L(G)$ (Section 14.1)

$\underline{R}$ Presheaf of quantizations over PG (Definition 14.1.1)

$\underline{I}$ Pre-quantization presheaf (Definition 14.1.3)

$\mathcal{O}(X)$ Category of open subsets of the topological space X

J A Grothendieck topology seen as a function on a category (Definition 7.1.5)

$(\mathcal{C}, J)$ Site, consisting of a category $\mathcal{C}$ and a Grothendieck topology J

K Basis for a Grothendieck topology (Definition 7.1.11)

Loc Category of locales with continuous maps

Spaces Category of topological space with continuous maps

pt(X) Points of a local X

$\underline{A}$ Internal C^*-Algebra in a topos (Definition 9.2.1)

CStar Category of internal unital C^*-algebras, together with internal unital *-homomorphism

KRegLoc Category of compact regular locales

Σ A first order signature (Definition 10.2.1)

M A Σ-structure (Definition 10.3.1)

Σ-**Str**(τ) Category of Σ-structures and Σ-structure homomorphisms

$\vec{x}.t$ Term in context (Definition 10.3.2)

$\vec{x}.\phi$ Formula in context (Definition 10.3.3)

$\mathbb{T}$	Theory over a sequent Σ (Definition 10.2.5)
$\mathbb{T}$-$\mathbf{Mod}(\tau)$	Category of models for a given theory $\mathbb{T}$
Poset	Category of partially ordered sets and monotone functions
Topos	Category whose objects are topos and whose morphisms are geometric morphisms
$\underline{\mathcal{A}}$	Presheaf representing the internal C^*-algebra (Definition 11.1.1)
$\underline{\Sigma}_{\underline{\mathcal{A}}}$	Internal spectrum of $\underline{\mathcal{A}}$
$\Sigma_{\downarrow}$	Topological space associated with $\underline{\Sigma}_{\underline{\mathcal{A}}}$ (Definition 11.2.1)
$I : \mathcal{A}_{sa} \to \mathbb{R}$	Probability integral (Definition 11.3.3)
$\Sigma_{\uparrow}$	Alternative definition of state space (Definition 11.4.1)
$\overline{\delta}(\hat{A})^{-1}$	Covariant daseinisation map (Definition 11.4.2)
$[\hat{A} \in (p, q)]_1$	Covariant proposition (Definition 11.4.3)
$\underline{\mathbb{R}}_l$	Internal lower reals (Definition 11.5.1)
$\underline{\mathbb{R}}_u$	Internal upper reals (Definition 11.5.2)
$\underline{\mathbb{R}^{\leftrightarrow}}$	Quantity valued object (Definition 12.3.1)
$\underline{\breve{\mathbb{R}}^{\leftrightarrow}}$	Alternative quantity value object (Definition 12.4.1)
$\mathsf{aC^*alg_1}$	Almost C*-algebras (Definition 13.4.1) with almost $*$-homomorphisms (Definition 13.4.2)
Grp	Groups with group homomorphisms
pGrp	Piecewise groups (Definition 13.5.1) with piecewise group homomorphisms (Definition 13.5.2)
aGrp	Almost groups (Definition 13.5.3) with almost group homomorphisms (Definition 13.5.4)

Chapter 1
Introduction

The present book is a follow up to the first book [26] which was aimed at introducing the new promising field of Topos Quantum Theory. Since the publication of the first volume many more exciting results have been developed. The aim of this second volume is to explain these new results.

For a thorough understanding of the topics in this book the reader is advised to first read volume one, since this book builds up on the concepts explained there. Nonetheless, in the introduction we will summarise the main results dealt with in [26] so as to refresh the reader with useful concepts which will be used throughout this book.

1.1 Conceptual and Mathematical Preliminaries

The main conceptual problems inherent in quantum theory, which are mainly due to how the theory is mathematically expressed, are the following:

- Due to the Kochen-Specker theorem, quantum theory is non-realist.[1]

  Theorem 1.1.1 (Kochen-Specker Theorem) *If the dimension of the Hilbert space $\mathcal{H}$ is greater than 2, then there does not exist any valuation function $V_{\vec{\Psi}} : \mathcal{O} \to \mathbb{R}$ from the set $\mathcal{O}$ of all bounded self-adjoint operators $\hat{A}$ of $\mathcal{H}$ to the reals $\mathbb{R}$, such that for all $\hat{A} \in \mathcal{O}$ and all $f : \mathbb{R} \to \mathbb{R}$ the following holds: $V_{\vec{\Psi}}(f(\hat{A})) = f(V_{\vec{\Psi}}(\hat{A}))$.*

[1] By a 'realist' theory we mean one in which the following conditions are satisfied: (1) propositions form a Boolean algebra; (2) propositions can always be assessed to be either true or false. As it will be delineated in the following, in the topos approach to quantum theory, both conditions are relaxed, leading to what Isham and Döring called a *neo-realist* theory.

C. Flori, *A Second Course in Topos Quantum Theory*,
Lecture Notes in Physics 944, https://doi.org/10.1007/978-3-319-71108-9_1

- Notions of 'measurement' and 'external observer' pose problems when dealing with cosmology since the universe is a closed system.
- Standard quantum theory employs, in its formulation, the use of a fixed spatio-temporal structure needed to make measurements. This fixed background seems to cause problems in quantum gravity, where one is trying to make measurements of space-time properties.

All these conceptual problems lead to the idea that, maybe, a new mathematical formulation of quantum theory, which leads to a more realist interpretation might be needed. This is precisely what the topos approach aims at. In particular, in such a reformulation of quantum theory it is possible to express probabilities in terms of truth values, hence probabilities become derived concepts.

One strategy to reformulate quantum theory in a more realist way is to re-express it, such that it 'looks like' classical physics, which is the paradigmatic example of a realist theory.

This is precisely the main idea in the topos approach [24, 26].

Furthermore, this reformulation of quantum theory has the key advantages that (1) propositions can be given truth values without needing to invoke the concepts of 'measurement' or 'observer'; (2) probabilities can be expressed in terms of truth values, hence they acquire a logical interpretation; (3) the internal logic which arises is distributive.

In order to make quantum theory 'look like' classical theory we first of all need to single out the underlining structure which makes classical physics a realist theory and then, mimic, in the context of quantum theory, the way in which this structure is defined.

The mathematical building blocks which render classical theory a realist theory are:

1. The existence of a state space S.
2. Each physical quantity, A, is represented by a function $f_A : S \to \mathbb{R}$.
3. Any propositions of the form "$A \in \Delta$"[2] is represented by a subset of the state space S: $f_A^{-1}(\Delta) = \{s \in S | f_A(s) \in \Delta\}$. The collection of all such subsets forms a Boolean algebra denoted $\mathrm{Sub}(S)$.
4. States s are identified with singletons $\{s\} \subset S$.

The aim is now to define the above constructs for quantum theory in an appropriate topos.

The issue one has to face is to identify which topos is the right one to use. This is solved by noticing that, because of the Kochen-Specker theorem, the only way of obtaining quantum analogues of requirements 1, 2, 3 and 4 is by defining them with respect to commutative subalgebras (the 'contexts') of the non-commuting algebra, $\mathcal{B}(\mathcal{H})$, of all bounded operators on the quantum theory's Hilbert space.

[2]"The value of the quantity A lies in the subset $\Delta \in \mathbb{R}$".

The set of all such commuting algebras (chosen to be von Neumann algebras) forms a category,[3] $\mathcal{V}(\mathcal{H})$, called the *context category*. These contexts will represent classical 'snapshots' of reality.

Definition 1.1.1 The category $\mathcal{V}(\mathcal{H})$ of abelian von Neumann algebras has:

- Objects: abelian von Neumann algebras V_i.
- Morphisms: given two algebras V and V', there exists a map between them $i_{VV'}$: $V \rightarrow V'$ iff $V \subseteq V'$.

The category $\mathcal{V}(\mathcal{H})$ is actually a poset ordered by subset inclusion and its elements represent the contexts, with respect to which any object is defined.

Thus, in the topos approach each object will be defined as a collection of context-dependent definitions related in a coherent way. One can, intuitively, think of a topos quantum object as a collection of classical approximations, one for each abelian subalgebra V. The quantum information is then carried by the categorical structure of the collection of all these classical approximations.

1.1.1 What Is Topos Theory?

The very hand wavy definition of a topos is that of a category with extra properties, which make a topos "look like" **Sets** in the sense that, any mathematical operation which can be done in set theory, can be done in a general topos.

Of particular importance in a topos are the notions of *Heyting algebra* and *sub-object classifier* [24, 26, 29, 33].

A *Heyting algebra* is the internal logic derived from the collection of all sub-objects of any object in the topos and represents a generalisation of the Boolean algebra in **Sets**. As such it is distributive, but the law of excluded middle does not hold, i.e. $S \vee \neg S \leq 1$. An example of Heyting algebra is given by the collection of all open sets in a topological space.

The *sub-object classifier* Ω represents the generalisation of the set $\{0, 1\} \simeq \{\text{true}, \text{false}\}$ of truth-values in the category **Sets**, therefore its elements are truth values and undergo a Heyting algebra. However, differently from **Sets**, in a general topos Ω will contain many more elements than just $0, 1$, leading to a multivalued logic.

There are many different kinds of topoi, however we are looking for a topos which allows us to obtain 'classical' local descriptions of objects. Such a topos is the topos $\mathbf{Sets}^{\mathcal{V}(\mathcal{H})^{\mathrm{op}}}$ of presheaves over the category $\mathcal{V}(\mathcal{H})$ whose objects are abelian von Neumann sub-algebras of the algebra of bounded operators on $\mathcal{H}$ and morphisms are inclusions.

[3]Roughly a category is a collection of objects and relations between these objects.

The definition of a presheaf is as follows:

Definition 1.1.2 Let $\mathcal{C}, \mathcal{D}$ be categories, then a presheaf is an assignment to each $\mathcal{D}$-object A of a $\mathcal{C}$-object $X(A)$, and to each $\mathcal{D}$-arrow $f : A \rightarrow B$ a $\mathcal{C}$-arrow $X(f) : X(B) \rightarrow X(A)$, such that: i) $X(1_A) = 1_{X(A)}$ and ii) $X(f \circ g) = X(g) \circ X(f)$ for any $g : C \rightarrow A$, i.e. a presheaf is a contravariant functor.

1.2 Topos Quantum Theory

In this section we will define the topos analogues of the constructs 1, 3 and 4 of Sect. 1.1. For an analysis of construct 2 the reader should refer to [24, 26].

State Space

As a first element we consider the representation of the state space in $Sets^{\mathcal{V}(\mathcal{H})^{op}}$. This is given by the spectral presheaf:

Definition 1.2.1 The spectral presheaf, $\underline{\Sigma}$, is the contravariant functor from $\mathcal{V}(\mathcal{H})$ to **Sets**, defined by:

- Objects: given an object V in $\mathcal{V}(\mathcal{H})^{op}$, the associated set $\underline{\Sigma}(V) = \underline{\Sigma}_V$ is defined to be the Gel'fand spectrum of the (unital) commutative von Neumann subalgebra V, i.e. the set of all multiplicative linear functionals $\lambda : V \rightarrow \mathbb{C}$, such that $\lambda(\hat{1}) = 1$.
- Morphisms: given a morphism $i_{V'V} : V' \rightarrow V$ ($V' \subseteq V$) in $\mathcal{V}(\mathcal{H})$, the associated function $\underline{\Sigma}(i_{V'V}) : \underline{\Sigma}(V) \rightarrow \underline{\Sigma}(V')$ is defined for all $\lambda \in \underline{\Sigma}(V)$ to be the restriction of the functional $\lambda : V \rightarrow \mathbb{C}$ to the subalgebra $V' \subseteq V$, i.e. $\underline{\Sigma}(i_{V'V})(\lambda) := \lambda_{|V'}$.

Propositions

In topos quantum theory propositions are identified with clopen sub-objects of the spectral presheaf. A *clopen* sub-object $\underline{S} \subseteq \underline{\Sigma}$ is an object such that, for each context $V \in \mathcal{V}(\mathcal{H})$, the set $\underline{S}(V)$ is a clopen (both closed and open) subset of $\underline{\Sigma}(V)$, where the latter is equipped with the usual compact and Hausdorff spectral topology. To understand how propositions are defined we need to introduce the concept of 'daseinisation'. Roughly speaking, what daseinisation does is to approximate operators so as to 'fit' into any given context V. More precisely, the *outer daseinisation* (see Definition 2.2.2), $\delta^o(\hat{P})$, of $\hat{P}$, at each context V, is defined by

$$\delta^o(\hat{P})_V := \bigwedge\{\hat{R} \in P(V)|\hat{R} \geq \hat{P}\}. \tag{1.2.1}$$

One then assigns, to each such daseinised projection $\delta^o(\hat{P})_V$, the subset of the state space consisting of all those elements which give value 1 to the daseinised projection operator, i.e.

$$S_{\delta^o(\hat{P})_V} := \{\lambda \in \underline{\Sigma}_V|\lambda(\delta^o(\hat{P})_V) = 1\}. \tag{1.2.2}$$

This subset can be shown to be clopen [24, 26]. Moreover, the collection of subsets $S_{\delta(\hat{P})_V}$, $V \in \mathcal{V}(\mathcal{H})$ forms a sub-object of $\underline{\Sigma}$. This enables us to define the (outer) daseinisation as a mapping from the projection operators to the clopen sub-object of the spectral presheaf as:

$$\delta : P(\mathcal{H}) \to \text{Sub}_{cl}(\underline{\Sigma})\ ; \quad \hat{P} \mapsto (S_{\delta^o(\hat{P})_V})_{V\in\mathcal{V}(\mathcal{H})} =: \underline{\delta(\hat{P})}.$$

States

In classical physics a pure state, s, is identified with a singleton $\{s\} \subset S$. However, the spectral presheaf $\underline{\Sigma}$ has *no* points. Indeed, this is equivalent to the Kochen-Specker theorem! Thus, the analogue of a pure state must be identified with some other construction. There are two (ultimately equivalent) possibilities: a 'state' can be identified with (1) an element of $P(P(\underline{\Sigma}))$; or (2) an element of $P(\underline{\Sigma})$.[4] The first choice is called the *truth-object* option, the second the *pseudo-state* option. In what follows we will concentrate only on the second option. For an analysis of the first option see [24, 26].

Given a pure quantum state $\psi \in \mathcal{H}$, we define the *pseudo-state* presheaf (see Definition 4.2.1)

$$\underline{\mathfrak{w}}^{|\psi\rangle} := \underline{\delta(|\psi\rangle\langle\psi|)} \tag{1.2.3}$$

such that, for each stage V we have

$$\underline{\delta(|\psi\rangle\langle\psi|)}_V := S_{\delta^o(|\psi\rangle\langle\psi|)_V} \subseteq \underline{\Sigma}(V) \tag{1.2.4}$$

where $\delta^o(|\psi\rangle\langle\psi|)_V = \bigwedge\{\hat{\alpha} \in P(V)|\,|\psi\rangle\langle\psi| \leq \hat{\alpha}\}$.

[4] $P(\underline{\Sigma})$ indicates the set of all subsets of $\underline{\Sigma}$.

The map $|\psi\rangle \rightarrow \underline{\mathfrak{w}}^{|\psi\rangle}$ is injective [24, 26]. Thus, for each state $|\psi\rangle$, there is associated a topos pseudo-state, $\underline{\mathfrak{w}}^{|\psi\rangle}$, which is defined as the smallest sub-object of the spectral presheaf $\underline{\Sigma}$. Roughly speaking ψ, is the closest one can get to define a point in $\underline{\Sigma}$.

1.3 Sub-object Classifier and Truth Values

In the topos $\mathbf{Sets}^{\mathcal{V}(\mathcal{H})^{op}}$ the sub-object classifier $\underline{\Omega}$ is identified with the following presheaf:

Definition 1.3.1 The presheaf $\underline{\Omega} \in \mathbf{Sets}^{\mathcal{V}(\mathcal{H})^{op}}$ has

1. Objects: for any $V \in \mathcal{V}(\mathcal{H})$, the set $\underline{\Omega}(V)$ is defined as the set of all sieves (see Definition 1.3.2) on V.
2. Morphisms: given a morphism $i_{V'V} : V' \rightarrow V$ $(V' \subseteq V)$, the associated function in $\underline{\Omega}$ is $\underline{\Omega}(i_{V'V}) : \underline{\Omega}(V) \rightarrow \underline{\Omega}(V'); S \mapsto \underline{\Omega}((i_{V'V}))(S) := \{V'' \subseteq V' | V'' \in S\}$.

A sieve on a poset, in our case $\mathcal{V}(\mathcal{H})$, is defined as follows:

Definition 1.3.2 For all $V \in \mathcal{V}(\mathcal{H})$, a sieve S on V is a collection of subalgebras $(V' \subseteq V)$ such that, if $V' \in S$ and $(V'' \subseteq V')$, then $V'' \in S$. Thus S is a downward closed set.
In this case a maximal (principal) sieve on V is $\downarrow V := \{V' \in \mathcal{V}(\mathcal{H}) | V' \subseteq V\}$.

Truth values are identified with *global* elements of the presheaf $\underline{\Omega}$, i.e. a collection, for each V, of *local* elements in $\underline{\Omega}_V$, i.e. of sieves. The global element, that consists entirely of principal sieves, is interpreted as 'totally true'. Similarly, the global element that consists of empty sieves is interpreted as 'totally false'.

A very important property of sieves is that, for each V, the set $\underline{\Omega}_V$ of sieves on V has the structure of a Heyting algebra. Similarly, the collection of (global) truth values undergoes a Heyting algebra.

We can now define how truth values are assigned to propositions. Going back to classical physics, a proposition $\hat{A} \in \Delta = f_{\hat{A}}^{-1}(\Delta)$ is true for a given state s if $\{s\} \subseteq f_{\hat{A}}^{-1}(\Delta)$.

In the quantum case, a proposition of the form "$A \in \Delta$" is represented by the presheaf $\underline{\delta(\hat{E}[A \in \Delta])}$,[5] while states are represented by the presheaves $\underline{\mathfrak{w}}^{|\psi\rangle}$. Since both presheaves are sub-objects of $\underline{\Sigma}$, it is reasonable to define truthfulness in terms of an inclusion relation, as done in classical physics. In particular, we define:

$$\underline{\mathfrak{w}}^{|\psi\rangle} \subseteq \underline{\delta(\hat{P})} . \tag{1.3.1}$$

[5]Here $\hat{E}[A \in \Delta]$ represents the spectral projector for the self-adjoint operator $\hat{A}$, which projects onto the subset Δ of the spectrum of $\hat{A}$.

This equation shows that, whether or not a proposition $\underline{\delta(\hat{P})}$ is 'totally true' given a pseudo state $\underline{\mathfrak{w}}^{|\psi\rangle}$, is determined by whether or not the pseudo-state is a sub-presheaf of the presheaf $\underline{\delta(\hat{P})}$. With motivation, given the state $\underline{\mathfrak{w}}^{|\psi\rangle}$, we can now define the *truth value* of the proposition "$A \in \Delta$" as:

$$v(A \in \Delta; |\psi\rangle) = v(\underline{\mathfrak{w}}^{|\psi\rangle} \subseteq \underline{\delta(\hat{E}[A \in \Delta])}) . \tag{1.3.2}$$

However, since presheaves are defined locally, we need to evaluate the above expression at each V obtaining

$$\begin{aligned} v(A \in \Delta; |\psi\rangle)_V &= v(\underline{\mathfrak{w}}^{|\psi\rangle} \subseteq \underline{\delta(\hat{E}[A \in \Delta])})_V \\ &:= \{V' \subseteq V | (\underline{\mathfrak{w}}^{|\psi\rangle})_{V'} \subseteq \underline{\delta(\hat{E}[A \in \Delta])}_{V'}\} \\ &= \{V' \subseteq V | \langle\psi|\delta(\hat{E}[A \in \Delta])_{V'} |\psi\rangle = 1\} . \end{aligned} \tag{1.3.3}$$

The last equality is derived by the fact that the relation $(\underline{\mathfrak{w}}^{|\psi\rangle})_V \subseteq \underline{\delta(\hat{P})}_V$, at the level of projection operators, becomes $\delta^o(\hat{P})_V \geq (\mathfrak{w}^{|\psi\rangle})_V$. However, since $(\mathfrak{w}^{|\psi\rangle})_V$ is the smallest projection operator, such that $\langle\psi|(\mathfrak{w}^{|\psi\rangle})_V |\psi\rangle = 1$, then $\delta^o(\hat{P})_V \geq (\mathfrak{w}^{|\psi\rangle})_V$ implies that $\langle\psi|\delta^o(\hat{P}) |\psi\rangle = 1$.

The right hand side of Eq. (1.3.3) means that the truth value, defined at V of the proposition "$A \in \Delta$", given the state $\underline{\mathfrak{w}}^{|\psi\rangle}$, is given in terms of all those sub-contexts $V' \subseteq V$, for which the projection operator $\delta(\hat{E}[A \in \Delta]))_{V'}$ has expectation value equal to one with respect to the state $|\psi\rangle$.

Equation (1.3.3) represents a *sieve* on V and the set of all of them is a Heyting algebra, thus the set of truth values is also a Heyting algebra.

From the above discussion it emerges that in the topos formulation of quantum theory, truth values can be simultaneously assign to any set of propositions, also incompatible ones. Moreover, in [23, 26] it was shown that probabilities can be described in terms of truth values. In such a formulation, logical concepts are seen as fundamental, while probabilities become derived concepts. This approach to probability theory allows for a new type of non-instrumentalist interpretation, which does not require the problematic notions of measurement and external observer and might be particularly appropriate in those schemes which interpret probabilities as propensities.

Chapter 2
Logic of Propositions in Topos Quantum Theory

In Chap. 10 of the first series of lecture notes on topos quantum theory [26] we showed that quantum propositions were represented by clopen sub-objects of the spectral presheaf $\underline{\Sigma}$ [26, Ch.10, Sec.1]. The collection of all such clopen sub-object, which we denoted by $Sub_{cl}(\underline{\Sigma})$, was shown to form a Heyting algebra [26, Th.10.2], hence the logic of quantum theory derived from the topos approach is an intuitionistic logic. In this chapter we will explain some recent results obtained in [15] in which it is shown that $Sub_{cl}(\underline{\Sigma})$ is not only a complete Heyting algebra, but also a complete co-Heyting algebra, therefore quantum logic is represented by a complete bi-Heyting algebra where two types of implications and negations are present.

2.1 Bi-Heyting Algebras

Before explaining the results shown in [15] we will state a few definitions we will use along the way. Most of these definitions were already given in [26].

Definition 2.1.1 A lattice consists of a set, A, equipped with elements, $0, 1 \in A$, (bottom and top elements, respectively) and binary operations $\vee, \wedge : A \times A \to A$ which satisfy the following conditions:

1. $\wedge$ and $\vee$ are both associative, commutative and idempotent.
2. $x \vee 0 = x$ and $x \wedge 1 = x$, for all $x \in A$.
3. $x \wedge (x \vee y) = x \vee (x \wedge y) = x$ for all $x \in A$.

The operations $\wedge$ and $\vee$ are called *meet* and *join*, respectively.

C. Flori, *A Second Course in Topos Quantum Theory*,
Lecture Notes in Physics 944, https://doi.org/10.1007/978-3-319-71108-9_2

Definition 2.1.2 Given a lattice A, then A is said to be distributive if for all $x, y, z \in A$ the following hold:

$$x \wedge (y \vee z) = (x \wedge y) \vee (x \wedge z)$$

$$x \vee (y \wedge z) = (x \vee y) \wedge (x \vee z)$$

Definition 2.1.3 A Boolean algebra is a distributive lattice, A, equipped with an operation, $\neg : A \to A$, such that the following equalities hold for all $x \in A$

$$\neg x \wedge x = 0$$

$$\neg x \vee x = 1$$

Given any lattice A, it is possible to equip it with a partial ordering as follows: given any two elements $x, y \in A$ then $x \leq y$ iff $x \wedge y = x$ or equivalently iff $x \vee y = y$. The element $x \wedge y$ is called the *greatest lower bound* of the set $\{x, y\}$, while $x \vee y$ is the *least upper bound* of the set $\{x, y\}$. If a lattice has the lest upper bound and greatest lower bound for all sets, not only finite ones, then the lattice is said to be *complete*.

Definition 2.1.4 A Heyting algebra is a distributive lattice, A, equipped with a binary operation, $\Rightarrow: A \times A \to A$, such that for all $x, y, z \in A$ then

$$a \leq (b \Rightarrow c) \text{ iff } a \wedge b \leq c$$

From a categorical perspective what this means is that the (meet) functor $a \wedge - : A \to A$ has a right adjoint $a \Rightarrow - : A \to A$ for all $a \in A$.

Definition 2.1.5 A complete Heyting algebra is a Heyting algebra which is complete as a lattice.
Given a complete Heyting algebra (**cHa**) A, any element $b \in A$ and a family of elements $\{a_i | i \in I\}$, then the following holds

$$\bigvee_{i \in I} (b \wedge a_i) = b \wedge \bigvee_{i \in I} a_i \tag{2.1.1}$$

In fact, given any other element $c \in A$, then we have

$$\begin{aligned}
\bigvee_{i \in I} (b \wedge a_i) \leq c &\text{ iff } b \wedge a_i \leq c \text{ for all } i \\
&\text{iff } a_i \leq b \Rightarrow c \text{ for all } i \\
&\text{iff } \bigvee_{i \in I} a_i \leq (b \Rightarrow c) \\
&\text{iff } b \wedge \bigvee_{i \in I} a_i \leq c.
\end{aligned}$$

On the other hand, if A is a lattice with arbitrary suprimums and such that the identity (2.1.1) holds then we have

$$a \leq \bigvee \{d | d \wedge b \leq c\} \Rightarrow a \wedge b \leq \bigvee \{d | d \wedge b \leq c\} \wedge b$$

therefore

$$a \wedge b \leq \bigvee \{d \wedge b | d \wedge b \leq c\} \leq c.$$

However, if $a \wedge b \leq c$ then $a \in \{d | d \wedge b \leq c\}$, therefore, $a \leq \bigvee \{d | d \wedge b \leq c\}$. This shows that

$$a \leq \bigvee \{d | d \wedge b \leq c\} \text{ iff } a \wedge b \leq c.$$

Hence

$$b \Rightarrow c = \bigvee \{d | d \wedge b \leq c\}.$$

The negation in a **cHa** A is defined in terms of the implication $\Rightarrow$ relation defined above. In particular, we have

$$\neg a = (a \Rightarrow 0) = \bigvee \{b | a \wedge b \leq 0\}.$$

This represents the largest element in A such that $a \wedge \neg a = 0$. One of the main properties of Heyting algebra is that the law of excluded middle needs not hold, i.e. $a \vee \neg a \leq 1$. The canonical example of a Heyting algebra is given by the collection of all open sets of a topological space. In this case, the negation $\neg a$ is given by the interior of the complement of the open set a. Clearly $a \vee \neg a \leq 1$.

Definition 2.1.6 Given a Heyting algebra A, an element $a \in A$ is called regular if $\neg\neg a = a$.

There is also the notion of a co-Heyting algebra

Definition 2.1.7 A co-Heyting algebra A is a distributive lattice equipped with a binary operation $\Leftarrow: A \times A \to A$ such that, $a \Leftarrow b \leq c$ iff $a \leq b \vee c$.

From a categorical perspective what this means is that the (join) functor $b \vee - : A \to A$ has a right adjoint $- \Rightarrow b : A \to A$ for all $b \in A$.

Definition 2.1.8 A complete co-Heyting algebra is a co-Heyting algebra which is complete as a lattice.

Given a complete co-Heyting algebra A, any element $b \in A$ and family $\{a_i | i \in I\}$, then the following equality holds

$$b \vee \bigwedge_{i \in I} a_i = \bigwedge_{i \in I} b \vee a_i. \tag{2.1.2}$$

In fact, given any other element $c \in A$ then

$$c \leq \bigwedge_{i \in I}(b \vee a_i) \text{ iff } c \leq b \vee a_i \text{ for all } i \in I$$

$$\text{iff } c \Leftarrow b \leq a_i \text{ for all } i \in I$$

$$\text{iff } c \Leftarrow b \leq \bigwedge_{i \in I} a_i$$

$$\text{iff } c \leq b \bigwedge_{i \in I} a_i.$$

On the other hand, if A is a lattice with arbitrary meets satisfying Eq. (2.1.2), then A is a co-Heyting algebra where the co-implication is defined by $a \Rightarrow b = \bigwedge\{c|c \vee b \geq a\}$. In fact, if

$$\bigwedge\{d|d \vee b \geq a\} \leq c$$

then

$$\bigwedge\{d|d \vee b \geq a\} \vee b \leq c \vee b$$

Given (2.1.2), it follows that

$$\bigwedge\{d \vee b|d \vee b \geq a\} \leq c \vee b$$

hence

$$a \leq \bigwedge\{d \vee b|d \vee b \geq a\} \leq c \vee b$$

Moreover, if $a \leq c \vee b$, then $c \in \{d|d \vee b \geq a\}$, therefore $\bigwedge\{d|d \vee b \geq a\} \leq c$.

The co-negation operation is define by $\sim a = 1 \Leftarrow a = \bigwedge\{b|a \wedge b = 1\}$, hence it represents the smallest element in A such that $a \vee \sim a = 1$. Generally, in a co-Heyting algebra the law of contradiction does not hold, i.e. $\sim a \wedge a \geq 0$. The canonical example of a co-Heyting algebra is given by the collection of all closed sets in a topological space. In this setting $\sim a$ is given by the complement of the interior of a, then clearly $\sim a \wedge a \geq 0$.

Definition 2.1.9 Given a co-Heyting algebra A, an element $a \in A$ is called regular if $\sim\sim a = a$.

If we combine the notions of a Heyting algebra and a co-Heyting algebra then we obtain the notion of a bi-Hyeting algebra which is both a Heyting and co-Heyting algebra.

Definition 2.1.10 A bi-Heyting algebra A is a lattice which is both a Heyting algebra and a co-Heyting algebra. It is a complete bi-Heyting algebra if it is a complete Heyting algebra and a complete co-Heyting algebra.
An example of a bi-Heyting algebra is given by the collection of all open and closed sets of a topological space. Moreover, any Boolean algebra is a bi-Heyting algebra. In this case the negation and co-negation coincide with the standard Boolean negation.

2.2 Bi-Heyting Algebra in Topos Quantum Theory

In [15] it was shown that the collection of quantum propositions as expressed in terms of topos quantum theory form a bi-Heyting algebra. In order to explain this result we need to recall how a proposition is represented in terms of topos quantum theory. For an in depth analysis the reader should refer to [26]. Here we will just recall that propositions are identified with clopen sub-objects of the spectral presheaf constructed through the process of *outer daseinisation*. In particular we have:

Definition 2.2.1 A clopen sub-object $\underline{S}$ of the spectral presheaf $\underline{\Sigma}$ is a sub-object $\underline{S} \subseteq \underline{\Sigma}$ such that for each $V \in \mathcal{V}(\mathcal{H})$ the set $\underline{S}_V$ is a clopen subset of the Gelfand spectrum $\underline{\Sigma}_V$. $Sub_{cl}(\underline{\Sigma})$ denotes the set of all clopen sub-objects of $\underline{\Sigma}$.

The fact that proposition are identified with clopen sub-object is obtained though the process of outer daseinisation.

Definition 2.2.2 Given the Von Neumann algebra $\mathcal{V}(\mathcal{H})$ with lattice of projection operators $\mathcal{P}(\mathcal{V}(\mathcal{H}))$, outer daseinisation is given by the following map[1]

$$\underline{\delta}^o : \mathcal{P}(\mathcal{V}(\mathcal{H})) \rightarrow Sub_{cl}(\underline{\Sigma})$$
$$\hat{P} \mapsto \underline{\delta}^o(\hat{P}) := \left(\mathfrak{S}_V(\delta^o(\hat{P})_V)\right)_{V \in \mathcal{V}(\mathcal{H})}.$$

Here $\delta^o(\hat{P})_V := \bigwedge\{\hat{R} \in P(V) | \hat{R} \geq \hat{P}\}$, while

$$\mathfrak{S}_V : \mathcal{P}(\mathcal{V}(\mathcal{H})) \rightarrow Sub_{cl}(\underline{\Sigma})_V \tag{2.2.1}$$

is an isomorphism from the complete Boolean algebra of projection operators present in the abelian subalgebra $V \in \mathcal{V}(\mathcal{H})$ to the complete Boolean algebra of clopen subsets of the Gel'fand spectrum $\underline{\Sigma}_V$, therefore $\mathfrak{S}_V(\delta^o(\hat{P})_V) = \{\lambda \in \underline{\Sigma}_V | \lambda(\delta^o(\hat{P})_V) = 1\}$.

[1]Note that the notation $\underline{\delta}^o(\hat{P})$ and $\delta^o(\hat{P})$ are equivalent. Moreover, for notational simplicity sometimes we will denote $\underline{\delta}^o$ simply by $\underline{\delta}$ since generally when talking about daseinisation we mean outer daseinisation. If considering inner daseinisation we will always put the superscript i.

The collection of clopen sub-objects $Sub_{cl}(\underline{\Sigma})$ of $\underline{\Sigma}$ is given a partial ordering by stating that for all $\underline{S}, \underline{T} \in Sub_{cl}(\underline{\Sigma})$, then $\underline{S} \leq \underline{T}$ iff, for all $V \in \mathcal{V}(\mathcal{H})$ then $\underline{S}_V \subseteq \underline{T}_V$. $Sub_{cl}(\underline{\Sigma})$ is equipped with arbitrary joins and meets which are defined context wise as follows:
Given a family of clopen sub-objects $(\underline{S}_i)_{i \in I}$ then, for all $V \in \mathcal{V}(\mathcal{H})$,

$$(\bigwedge_{i \in I} \underline{S}_i)_V = int(\bigcap_{i \in I} \underline{S}_{i,V})$$

while

$$(\bigvee_{i \in I} \underline{S}_i)_V = cl(\bigcup_{i \in I} \underline{S}_{i,V}).$$

The need to take the interior and the closure is to guaranty that one obtains clopen subsets at each context not just closed and open respectively. This fact also implies that $Sub_{cl}\underline{\Sigma}$ is not a Heyting subalgebra of the Heyting algebra of sub-objects $Sub(\underline{\Sigma})$ of the spectral presheaf. We will now show that $Sub_{cl}\underline{\Sigma}$ is a bi-Heyting algebra [15].

Theorem 2.2.1 *The collection* $Sub_{cl}\underline{\Sigma}$ *of clopen sub-objects of the spectral presheaf forms a bi-Heyting algebra.*

Proof To prove that $Sub_{cl}\underline{\Sigma}$ is a bi-Heyting algebra we need to show that it is both a Heyting algebra and a co-Heyting algebra. The former was shown in [26], so, what remains to be shown is the latter. We know from the above discussion that $Sub_{cl}\underline{\Sigma}$ has arbitrary joins and meets, therefore we need to show that, given a family of $(\underline{S}_i)_{i \in I}$ of clopen sub-objects of $\underline{\Sigma}$ and any other clopen sub-object $\underline{S} \subseteq \underline{\Sigma}$, then

$$(\underline{S} \wedge \bigvee_{i \in I} \underline{S}_i) = \bigvee_{i \in I} (\underline{S} \wedge \underline{S}_i).$$

However, since we are in the context of presheaves, the above equation has to be defined for each context. Thus for each $V \in \mathcal{V}(\mathcal{H})$ we have:

$$(\underline{S} \wedge \bigvee_{i \in I} \underline{S}_i)_V = \bigvee_{i \in I} (\underline{S}_V \wedge \underline{S}_{i,V}).$$

However, for each context $V \in \mathcal{V}(\mathcal{H})$ then $Sub_{cl}(\underline{\Sigma})_V$ is a Boolean algebra hence

$$\begin{aligned}\bigvee_{i \in I} (\underline{S}_V \wedge \underline{S}_{i,V}) &= int\left(\bigcup_{i \in I} (\underline{S}_V \cap \underline{S}_{i,V})\right) \\ &= int\left(\underline{S}_V \cap \bigcup_{i \in I} \underline{S}_{i,V}\right)\end{aligned}$$

$$= \underline{S}_V \cap int\left(\bigcup_{i\in I} \underline{S}_{i,V}\right)$$

$$= \underline{S}_V \wedge (\bigvee_{i\in I} \underline{S}_i)_V$$

$$= (\underline{S} \wedge \bigvee_{i\in I} \underline{S}_i)_V.$$

□

The above discussion proves that $Sub_{cl}(\underline{\Sigma})_V$ is a bi-Heyting algebra where the two negations are defined as follows:

$$\sim \underline{S} := \underline{\Sigma} \Leftarrow \underline{S} = \bigwedge\{\underline{R} \in Sub_{cl}(\underline{\Sigma})|\underline{\Sigma} = \underline{R} \vee \underline{S}\} \tag{2.2.2}$$

$$\neg\underline{S} := \underline{S} \Rightarrow \underline{0} = \bigvee\{\underline{R} \in Sub_{cl}(\underline{\Sigma})|\underline{R} \wedge \underline{S} = \underline{0}\}. \tag{2.2.3}$$

Therefore, $\sim \underline{S}$ is the smallest clopen sub-object of $\underline{\Sigma}$ such that $\sim \underline{S} \vee \underline{S} = \underline{\Sigma}$, while $\neg\underline{S}$ is the biggest clopen sub-object of $\underline{\Sigma}$ such that $\neg\underline{S} \wedge \underline{S} = \underline{0}$. It then follows that:

Corollary 2.2.1 ([15]) *For all $\underline{S} \in Sub_{cl}(\underline{\Sigma})$ then $\neg\underline{S} \leq\sim \underline{S}$.*

Proof For each context $V \in \mathcal{V}(\mathcal{H})$, $\underline{\Sigma}_V/\underline{S}_V$ is the biggest subset such that $\underline{\Sigma}_V/\underline{S}_V \cap \underline{S}_V = \emptyset$. But from the definition of $\neg\underline{S}$ we know that for each $V \in \mathcal{V}(\mathcal{H})$ $(\neg\underline{S})_V \wedge \underline{S}_V = \emptyset$, therefore $(\neg\underline{S})_V \subseteq \underline{\Sigma}_V/\underline{S}_V$. On the other hand $\underline{\Sigma}_V/\underline{S}_V$ is the smallest subset such that $\underline{\Sigma}_V/\underline{S}_V \cup \underline{S}_V = \underline{\Sigma}_V$. But from the definition of $\sim \underline{S}$ we know that for each $V \in \mathcal{V}(\mathcal{H})$, $(\sim \underline{S})_V \cup \underline{S}_V = \underline{\Sigma}_V$. Therefore $\underline{\Sigma}_V/\underline{S}_V \subseteq (\sim (\underline{S})_V$. Thus we obtain that, for all $V \in \mathcal{V}(\mathcal{H})$, $(\neg\underline{S})_V \subseteq (\sim \underline{S})_V$. □

Corollary 2.2.2 $\sim \underline{S} \wedge \underline{S} \geq \underline{0}$.

Proof Since $\neg\underline{S}$ is the largest sub-object of $\underline{\Sigma}$, such that $\neg\underline{S} \wedge \underline{S} \geq \underline{0}$ and since from the above Lemma $\neg\underline{S} \leq\sim \underline{S}$, it follows that $\sim \underline{S} \wedge \underline{S} \geq \underline{0}$. □

From the above corollary it follows that the logic of topos quantum theory is a *paraconsistent logic*. This is a logic for which the principle of explosion does not hold. This principle asserts that, given a contradiction anything can be entailed. As a consequence any logic which has inconsistencies becomes trivial, since any statement becomes a theorem. On the other hand in a paraconsistent logic, contradiction does not entail truthfulness of any statement, hence it is possible to have inconsistent but non-trivial theories.

2.3 Two Types of Negations

In the previous section we have seen that the collection of quantum propositions forms a bi-Heyting algebra in which two types of negations are present. In this section we will analyse these negations and define their respective regular elements [15].

2.3.1 *Heyting Negation*

Given any clopen sub-object $\underline{S}$ we want to understand what $\neg\underline{S}$ is. From (2.2.3) we know that $\neg\underline{S}$ is the largest element in $Sub_{cl}\underline{\Sigma}$ such that $\underline{S} \wedge \neg\underline{S} = \underline{0}$. To understand how $\neg\underline{S}$ is defined context wise, we need to start with the context wise definition of a general pseudo element. For all $V \in \mathcal{V}(\mathcal{H})$ we have

$$(\underline{S} \Rightarrow \underline{R})_V = \{\lambda \in \underline{\Sigma}_V | \forall\, V' \subseteq V;\ \text{if}\ \lambda|_{V'} \in \underline{S}_{V'},\ \text{then}\ \lambda|_{V'} \in \underline{R}_{V'}\}$$

It then follows that:

$$\begin{aligned}(\underline{S} \Rightarrow \underline{0})_V &= \{\lambda \in \underline{\Sigma}_V | \forall\, V' \subseteq V;\ \text{if}\ \lambda|_{V'} \in \underline{S}_{V'},\ \text{then}\ \lambda|_{V'} \in \underline{0}_{V'}\} \\ &= \{\lambda \in \underline{\Sigma}_V | \forall\, V' \subseteq V; \lambda|_{V'} \notin \underline{S}_{V'}\}.\end{aligned} \tag{2.3.1}$$

We now would like to express $\neg\underline{S}$ in terms of projection operators. To this end we need to utilise the isomorphisms

$$\begin{aligned}\mathfrak{S}_V : \mathcal{P}(\mathcal{V}(\mathcal{H})) &\to Sub_{cl}(\underline{\Sigma})_V \\ \hat{P} &\mapsto S_{\hat{P}} := \{\lambda \in \underline{\Sigma}_V | \lambda(\hat{P}) = 1\}.\end{aligned}$$

This map associates to each projection operator in $\mathcal{P}(\mathcal{V}(\mathcal{H}))$ a clopen subset of $\underline{\Sigma}_V$. Since it is an isomorphisms, given any clopen subset $S \in Sub_{cl}(\underline{\Sigma})_V$, the associated projection operator is given by $\mathfrak{S}_V^{-1}(S) =: \hat{P}_S$. We can now re-write Eq. (2.3.1) in terms of projection operators as follows:

$$\begin{aligned}(\underline{S} \Rightarrow \underline{0})_V &= \{\lambda \in \underline{\Sigma}_V | \forall\, V' \subseteq V; \lambda|_{V'} \notin \underline{S}_{V'}\} \\ &= \{\lambda \in \underline{\Sigma}_V | \forall\, V' \subseteq V; \lambda|_{V'}(\hat{P}_{\underline{S}_{V'}}) = 0\} \\ &= \{\lambda \in \underline{\Sigma}_V | \forall\, V' \subseteq V; \lambda(\hat{P}_{\underline{S}_{V'}}) = 0\} \\ &= \{\lambda \in \underline{\Sigma}_V | \lambda(\bigvee_{V' \subseteq V} \hat{P}_{\underline{S}_{V'}}) = 0\}.\end{aligned} \tag{2.3.2}$$

We know from the definition of a sub-object of $\underline{\Sigma}$ that for $V' \subseteq V$ then $\underline{S}_{V'} \subseteq \underline{S}_V$, which implies that $\hat{P}_{\underline{S}_{V'}} \geq \hat{P}_{\underline{S}_V}$. Therefore, as the context becomes smaller the associated projections become bigger. This implies that when considering $\lambda(\bigvee_{V' \subseteq V} \hat{P}_{\underline{S}_{V'}}) = 0$ it suffices to consider only the "small" algebras $V' \subseteq V$. These "small" algebras are the so called *minimal* contexts which are generated by a single projection and the identity, i.e. $V_{\hat{P}} = \{\hat{P}, \hat{1}\}'' = \mathbb{C}\hat{Q} + \mathbb{C}\hat{1}$, with the exclusion of $V_{\hat{1}} = \{\hat{1}\}'' = \mathbb{C}\hat{1}$. The collection of minimal subalgebras for a given algebra V is identified as follows:

$$m_V := \{V' \subseteq V | V' \text{ minimal}\} = \{V_{\hat{P}} | \hat{P} \in \mathcal{P}(V)\}.$$

We can thus re-write Eq. (2.3.2) as follows:

$$\begin{aligned}(\neg \underline{S})_V &= \{\lambda \in \underline{\Sigma}_V | \lambda(\bigvee_{V' \in m_V} \hat{P}_{\underline{S}_{V'}}) = 0\} \\ &= \{\lambda \in \underline{\Sigma}_V | \lambda(\hat{1} - \bigvee_{V' \in m_V} \hat{P}_{\underline{S}_{V'}}) = 1\} \\ &= S_{\hat{1} - \bigvee_{V' \in m_V} \hat{P}_{\underline{S}_{V'}}}.\end{aligned} \tag{2.3.3}$$

This implies that

$$\hat{P}_{(\neg \underline{S})_V} = \hat{1} - \bigvee_{V' \in m_V} \hat{P}_{\underline{S}_{V'}}. \tag{2.3.4}$$

Given this result, we now want to show that:

Lemma 2.3.1 *Given any $\underline{S} \in Sub_{cl}(\underline{\Sigma})$, $\underline{S}$ is Heyting regular ($\neg\neg\underline{S} = \underline{S}$), iff for all $V \in \mathcal{V}(\mathcal{H})$*

$$\hat{P}_{\underline{S}_V} = \bigwedge_{V' \in m_V} \hat{P}_{\underline{S}_{V'}}$$

where $m_v = \{V' \subseteq V | V' \text{ minimal}\}$.

Proof For a general element $\underline{S} \in Sub_{cl}\underline{\Sigma}$ we have that $\neg\neg\underline{S} \geq \underline{S}$. In terms of projections this inequality translates as follows: from Eq. (2.3.3) we have that

$$(\neg\neg\underline{S})_V = S_{\hat{1} - \bigvee_{V' \in m_V} \hat{P}_{(\neg\underline{S})_{V'}}}.$$

Therefore, in terms of projections we have

$$\hat{P}_{(\neg\neg\underline{S})_V} = \hat{1} - \bigvee_{V' \in m_V} \hat{P}_{(\neg\underline{S})_{V'}}$$
$$= \hat{1} - \bigvee_{V' \subseteq m_V} (\hat{1} - \bigvee_{W \in m_{V'}} \hat{P}_{\underline{S}_W}).$$

Since $V' \in m_V$, we obtain that $m_{V'} = \{V'\}$, therefore

$$\hat{P}_{(\neg\neg\underline{S})_V} = \hat{1} - \bigvee_{V' \in m_V} (\hat{1} - \bigvee_{W \in m_{V'}} \hat{P}_{\underline{S}_W})$$
$$= \hat{1} - \bigvee_{V' \in m_V} (\hat{1} - \hat{P}_{\underline{S}_{V'}})$$
$$= \bigwedge_{V' \in m_V} \hat{P}_{\underline{S}_{V'}}.$$

However we know that for $V' \in m_V$ which implies $\hat{P}_{\underline{S}_{V'}} \geq \hat{P}_{\underline{S}_V}$, therefore

$$\neg\neg\underline{S} \geq \underline{S} \text{ iff } \hat{P}_{(\neg\neg\underline{S})_V} = \bigwedge_{V' \in m_V} \hat{P}_{\underline{S}_{V'}} \geq \hat{P}_{\underline{S}_V}.$$

This means that

$$\neg\neg\underline{S} = \underline{S} \text{ iff } \bigwedge_{V' \in m_V} \hat{P}_{\underline{S}_{V'}} = \hat{P}_{\underline{S}_V}.$$

□

Next we would like to relate regular Heyting elements with tight clopen sub-objects of $\underline{\Sigma}$.

Definition 2.3.1 A clopen sub-object $\underline{S} \in Sub_{cl}(\underline{\Sigma})$ is called tight if, for all $V', V \in \mathcal{V}(\mathcal{H})$ such that $V' \subseteq V$, then

$$\underline{\Sigma}(i_{V'V})(\underline{S}_V) = \underline{S}_{V'} \tag{2.3.5}$$

Note that in general we have that $\underline{\Sigma}(i_{V'V})(\underline{S}_V) \subseteq \underline{S}_{V'}$.

Lemma 2.3.2 *A clopen sub-object $\underline{S} \in Sub_{cl}(\underline{\Sigma})$ is called tight if, for all $V', V \in \mathcal{V}(\mathcal{H})$ such that $V' \subseteq V$, then*

$$\hat{P}_{\underline{S}_{V'}} = \mathcal{O}(i_{V,V'})(\hat{P}_{\underline{S}_V}).$$

where $\mathcal{O}(i_{V,V'}) : \mathcal{P}(V) \to \mathcal{P}(V')$ are the presheaf maps of the outer daseinisation presheaf Section 11.1 in [26], which are defined by $\mathcal{O}(i_{V,V'})(\hat{P}) := \delta^o(\hat{P})'_V$.

Proof In terms of projection operators equation (2.3.5) becomes

$$\hat{P}_{\underline{\Sigma}(i_{V'V})(\underline{S}_V)} = \hat{P}_{\underline{S}_{V'}}$$

so, what we need to show is that $\hat{P}_{\underline{\Sigma}(i_{V'V})(\underline{S}_V)} = \mathcal{O}(i_{V,V'})(\hat{P}_{\underline{S}_V})$ which in terms of clopen subsets becomes $\underline{\Sigma}(i_{V'V})(\underline{S}_V) = \underline{S}_{\mathcal{O}(i_{V,V'})(\hat{P}_{\underline{S}_V})}$. This was shown in Theorem 3.1 in [22], but for completeness sake we have reported the theorem and proof in Appendix A.3. □

Lemma 2.3.3 *Tight sub-objects are regular Heyting elements.*

Proof Given a tight sub-object $\underline{S} \in Sub_{cl}(\underline{\Sigma})$, by definition we have that $\hat{P}_{\underline{S}_{V'}} = \mathcal{O}(i_{V,V'})\hat{P}_{\underline{S}_V}$ for all $V, V' \in \mathcal{V}(\mathcal{H})$ such that $V' \subseteq V$, where $\mathcal{O}(i_{V,V'})\hat{P}_{\underline{S}_V} = \delta^o(\hat{P}_{\underline{S}_V})_{V'} = \bigwedge\{\hat{R} \in \mathcal{P}(V')|\hat{R} \geq \hat{P}_{\underline{S}_V}\} \geq \hat{P}_{\underline{S}_V}$. For this same sub-object $\underline{S}$ we want to show that $\bigwedge_{V' \in m_V} \hat{P}_{\underline{S}_{V'}} = \hat{P}_{\underline{S}_V}$. For each $V \in \mathcal{V}(\mathcal{H})$, the minimal subalgebra generated by $\hat{P}_{\underline{S}_V}$ is $V_{\hat{P}_{\underline{S}_V}} = \{\hat{P}_{\underline{S}_V}, \hat{1}\}''$. This belongs to m_V, therefore we obtain

$$\mathcal{O}(i_{V,V_{\hat{P}_{\underline{S}_V}}})\hat{P}_{\underline{S}_V} = \delta^o(\hat{P}_{\underline{S}_V})_{V_{\hat{P}_{\underline{S}_V}}} = \bigwedge\{\hat{R} \in \mathcal{P}(V_{\hat{P}_{\underline{S}_V}})|\hat{R} \geq \hat{P}_{\underline{S}_V}\} = \hat{P}_{\underline{S}_V} \tag{2.3.6}$$

Therefore, for each $V \in \mathcal{V}(\mathcal{H})$

$$\bigwedge_{V' \in m_V} \hat{P}_{\underline{S}_{V'}} = \hat{P}_{\underline{S}_V}.$$

□

As an immediate consequence of the above we have that

Corollary 2.3.1 *The outer daseinisation map $\underline{\delta}^o : \mathcal{P}(\mathcal{V}(\mathcal{H})) \to Sub_{cl}\underline{\Sigma}$ defined in Definition 2.2.2, maps projection operators to regular elements of the Heyting algebra $Sub_{cl}\underline{\Sigma}$ of clopen sub-object.*

Proof By definition of the outer daseinisation presheaf we have that $\mathcal{O}(i_{V,V'})(\delta^o(\hat{P})_V) = \delta^o(\hat{P})_{V'}$, therefore

$$\hat{P}_{(\underline{\delta}^o(\hat{P}))_{V'}} = \delta^o(\hat{P})_{V'} = \mathcal{O}(i_{V,V'})(\delta^o(\hat{P})_V) = \mathcal{O}(i_{V,V'})(\hat{P}_{(\underline{\delta}^o(\hat{P}))_V}).$$

Therefore clopen sub-objects of the form $\underline{\delta}^o(\hat{P})$ are tight. Application of Lemma 2.3.3 gives us the desired result. □

2.3.2 Co-Heyting Negation

Given an object $\underline{S} \in Sub_{cl}(\underline{\Sigma})$ we now would like to analyse the object $\sim \underline{S}$ and understand what it represents in terms of projection operators.

Lemma 2.3.4 *Given any sub-object $\underline{S} \in Sub_{cl}(\underline{\Sigma})$, then for any $V \in \mathcal{V}(\mathcal{H})$ we have*

$$\hat{P}_{(\sim\underline{S})_V} = \bigvee_{\bar{V} \in M_V} \mathcal{O}(i_{\bar{V},V})(\hat{1} - \hat{P}_{\underline{S}_V})$$

where $M_V = \{\bar{V} \supseteq V | \bar{V}$ is a maximal abelian subalgebra of $\mathcal{V}(\mathcal{H})\}$.

Proof Since $\sim \underline{S}$ is a sub-object of $\underline{\Sigma}$, then for $V \subseteq \bar{V}$ we have that $\underline{\Sigma}(i_{\bar{V},V})(\sim \underline{S})_{\bar{V}} \subseteq (\sim \underline{S})_V$. Moreover, we know from Lemma 2.2.1 that for all $V \in \mathcal{V}(\mathcal{H})$, $\underline{\Sigma}_V / \underline{S}_V \subseteq (\sim \underline{S})_V$. Therefore we have that $\underline{\Sigma}(i_{\bar{V},V})(\underline{\Sigma}_{\bar{V}} / \underline{S}_{\bar{V}}) \subseteq \underline{\Sigma}(i_{\bar{V},V})(\sim \underline{S})_{\bar{V}} \subseteq (\sim \underline{S})_V$. In terms of projection operators this translates into $\hat{P}_{(\sim\underline{S})_V} \geq \mathcal{O}(i_{\bar{V},V})(\hat{1} - \hat{P}_{\underline{S}_{\bar{V}}})$. Since the set M_V contains all maximal sub-algebras (maximal contexts) containing V, it follows that

$$\hat{P}_{(\sim\underline{S})_V} \geq \bigvee_{\bar{V} \in M_V} \mathcal{O}(i_{\bar{V},V})(\hat{1} - \hat{P}_{\underline{S}_{\bar{V}}}).$$

From the definition of $\sim \underline{S}$ we know that $\sim \underline{S}$ is the smallest sub-object such that $\sim \underline{S} \vee \underline{S} = \underline{\Sigma}$. Context-wise we have that for all $V \in \mathcal{V}(\mathcal{H})$, $(\sim \underline{S})_V$ is the smallest subset such that $(\sim \underline{S})_V \cup \underline{S}_V = \underline{\Sigma}_V$. In terms of projection operators this means that $\hat{P}_{(\sim\underline{S})_V}$ is the smallest projection operator such that $\hat{P}_{(\sim\underline{S})_V} \vee \hat{P}_{\underline{S}_V} = \hat{1}$. Therefore, if we show that $\bigvee_{\bar{V} \in M_V} \mathcal{O}(i_{\bar{V},V})(\hat{1} - \hat{P}_{\underline{S}_{\bar{V}}}) \vee \hat{P}_{\underline{S}_V} = \hat{1}$ for all $V \in \mathcal{V}(\mathcal{H})$ then it follows that $\hat{P}_{(\sim\underline{S})_V} = \bigvee_{\bar{V} \in M_V} \mathcal{O}(i_{\bar{V},V})(\hat{1} - \hat{P}_{\underline{S}_{\bar{V}}})$. To show this we subdivide our analysis in two case:

i) If V is maximal, then $M_V = \{V\}$ and $\bigvee_{\bar{V} \in M_V} \mathcal{O}(i_{\bar{V},V})(\hat{1} - \hat{P}_{\underline{S}_{\bar{V}}}) = \hat{1} - \hat{P}_{\underline{S}_V}$, therefore $\hat{P}_{\underline{S}_V} \vee (\hat{1} - \hat{P}_{\underline{S}_V}) = \hat{1}$ and $\hat{P}_{(\sim\underline{S})_V} = \bigvee_{\bar{V} \in M_V} \mathcal{O}(i_{\bar{V},V})(\hat{1} - \hat{P}_{\underline{S}_{\bar{V}}})$.
ii) If V is non-maximal, then for any $\bar{V} \in m_V$ we have that $\bigvee_{\bar{V} \in M_V} \mathcal{O}(i_{\bar{V},V})(\hat{1} - \hat{P}_{\underline{S}_{\bar{V}}}) \geq \bigvee_{\bar{V} \in M_{\bar{V}}} \mathcal{O}(i_{\bar{V},\bar{V}})(\hat{1} - \hat{P}_{\underline{S}_{\bar{V}}}) = \hat{1} - \hat{P}_{\underline{S}_{\bar{V}}}$. Since $\hat{P}_{\underline{S}_V} \geq \hat{P}_{\underline{S}_{\bar{V}}}$, then $\bigvee_{\bar{V} \in M_V} \mathcal{O}(i_{\bar{V},V})(\hat{1} - \hat{P}_{\underline{S}_{\bar{V}}}) \vee \hat{P}_{\underline{S}_V} \geq (\hat{1} - \hat{P}_{\underline{S}_{\bar{V}}}) \vee \hat{P}_{\underline{S}_{\bar{V}}} = \hat{1}$. It follows that $\hat{P}_{(\sim\underline{S})_V} = \bigvee_{\bar{V} \in M_V} \mathcal{O}(i_{\bar{V},V})(\hat{1} - \hat{P}_{\underline{S}_{\bar{V}}})$.

□

Now that we have defined the co-Heyting negation in terms of projection operators, we want to identify the regular elements [15].

Lemma 2.3.5 *A sub-object $\underline{S} \subseteq \underline{\Sigma}$ is co-Heyting regular ($\sim\sim \underline{S} = \underline{S}$) iff for all $V \in \mathcal{V}(\mathcal{H})$, then*

$$\hat{P}_{\underline{S}} = \bigvee_{\bar{V} \in M_V} \mathcal{O}(i_{\bar{V},V})\hat{P}_{\underline{S}_V} = \bigvee_{\bar{V} \in M_V} \delta^o(\hat{P}_{\underline{S}_{\bar{V}}})_V$$

Proof For each $V \in \mathcal{V}(\mathcal{H})$ the condition of $\underline{S} \in Sub_{cl}(\underline{\Sigma})$ of being co-Heyting regular can be expressed in terms of projection operators as follows: $\hat{P}_{(\sim\sim\underline{S})_V} = \hat{P}_{\underline{S}_V}$. By applying Lemma 2.3.4 we obtain

$$\begin{aligned}
\hat{P}_{(\sim\sim\underline{S})_V} &= \bigvee_{\bar{V}\in M_V} \mathcal{O}(i_{\bar{V},V})(\hat{1} - \hat{P}_{(\sim\underline{S})_{\bar{V}}}) \\
&= \bigvee_{\bar{V}\in M_V} \mathcal{O}(i_{\bar{V},V})(\hat{1} - \bigvee_{W\in M_{\bar{V}}} \mathcal{O}(i_{\bar{V},V})(\hat{1} - \hat{P}_{(\sim\underline{S})_W})) \\
&\overset{M_{\bar{V}}=\{\bar{V}\}}{=} \bigvee_{\bar{V}\in M_V} \mathcal{O}(i_{\bar{V},V})(\hat{1} - (\hat{1} - \hat{P}_{(\sim\underline{S})_{\bar{V}}})) \\
&= \bigvee_{\bar{V}\in M_V} \mathcal{O}(i_{\bar{V},V})(\hat{P}_{(\sim\underline{S})_{\bar{V}}}) \\
&= \bigvee_{\bar{V}\in M_V} \delta^o(\hat{P}_{\underline{S}_{\bar{V}}})_V.
\end{aligned}$$

□

Lemma 2.3.6 *Tight sub-objects of $\underline{\Sigma}$ are co-Heyting regular.*

Proof By definition, if $\underline{S}$ is tight then for all $V, V' \in \mathcal{V}(\mathcal{H})$ such that $V' \subseteq V$, $\underline{\Sigma}(i_{V,V'})\underline{S}_V = \underline{S}_{V'}$ which in terms of projection operators becomes $\mathcal{O}(i_{V,V'})(\hat{P}_{\underline{S}_V}) = \hat{P}_{\underline{S}_{V'}}$. Now consider the case in which $\bar{V} \in M_V$, then we obtain $\mathcal{O}(i_{\bar{V},V})(\hat{P}_{\underline{S}_{\bar{V}}}) = \hat{P}_{\underline{S}_V}$. Since this holds for all $\bar{V} \in M_V$ it follows that $\bigvee_{\bar{V}\in M_V} \mathcal{O}(i_{\bar{V},V})(\hat{P}_{\underline{S}_{\bar{V}}}) = \hat{P}_{\underline{S}_V}$. □

An immediate consequence of this is the following:

Corollary 2.3.2 *The outer daseinisation map $\underline{\delta}^o : \mathcal{P}(\mathcal{V}(\mathcal{H})) \to Sub_{cl}\underline{\Sigma}$ defined in Definition 2.2.2, maps projection operators to regular elements of the co-Heyting algebra $Sub_{cl}\underline{\Sigma}$ of clopen sub-object.*

Proof As shown in the proof of Corollary 2.3.1 clopen sub-objects of the form $\underline{\delta}^o(\hat{P})$ are tight. Applying Lemma 2.3.6 proves our result. □

2.4 Examples of the Two Negations

We will now give several examples for both the Heyting and co-Heyting negation of a quantum proposition. To this end let us consider a four dimensional Hilbert space $\mathcal{H} = \mathbb{C}^4$ with orthonormal basis $(\psi_1, \psi_2, \psi_3, \psi_4)$ and projection operators $(\hat{P}_1, \hat{P}_2, \hat{P}_3, \hat{P}_4)$, such that each $\hat{P}_i$ projects onto one dimensional subspace $\mathbb{C}\psi_i$. One possible abelian von Neumann algebra is $V = lin_{\mathbb{C}}(\hat{P}_1, \hat{P}_2, \hat{P}_3, \hat{P}_4)$. The spectral presheaf, evaluated at this context, is given by the Gelfand spectrum of V: $\underline{\Sigma}_V = \{\lambda_1, \lambda_2, \lambda_3, \lambda_4\}$, where $\lambda_i(\hat{P}_j) = \delta_{ij}$. We will now give examples of both negations for two distinct quantum propositions.

2.4.1 First Example

For the first example we choose a clopen sub-object $\underline{S} \subseteq \underline{\Sigma}$, such that at the context V we have $\underline{S}_V = \{\lambda_1, \lambda_4\}$. This means that at the context V, $\underline{S}_V$ represents the proposition $\hat{P}_1 + \hat{P}_4$ given by $\underline{\mathfrak{S}}_V^{-1}(S)$. Next consider the contexts $V_{1,2} = lin_{\mathbb{C}}(\hat{P}_1.\hat{P}_2, \hat{P}_3 + \hat{P}_4\}$. The Gelfand spectrum of $V_{1,2}$ is $\underline{\Sigma}_{V_{1,2}} = \{\lambda'_1, \lambda'_2, \lambda'_{3+4}\}$, where $\lambda'_i(\hat{P}_j = \delta_{ij}$ and $\lambda'_{kl}(\hat{P}_k + \hat{P}_l) = 1$. For an in-depth analysis of the spectral presheaf the reader is referred to Section 9.2.1 of [26]. Since the restriction maps $\underline{\Sigma}(i_{V_{1,2},V}) : \underline{\Sigma}_V \to \underline{\Sigma}_{V_{1,2}}$ are such that $\underline{\Sigma}(i_{V,V_{1,2}})\underline{S}_V \subseteq \underline{S}_{V_{1,2}}$, then $\underline{S}_{V_{1,2}} = \{\lambda'_1, \lambda'_{3+4}\}$, with associated projection operator $\hat{P}_1 + \hat{P}_3 + \hat{P}_4$. We are now interested in defining

$$\hat{P}_{(\sim\underline{S})V_{1,2}} = \bigvee_{\bar{V} \in M_{V_{1,2}}} \mathcal{O}(i_{\bar{V},V_{1,2}})(\hat{1} - \hat{P}_{\underline{S}_{\bar{V}}}).$$

First of all we need to define the set $M_{V_{1,2}} = \{V, V_{1,2}\}$, then we obtain

$$\mathcal{O}(i_{V,V_{1,2}})(\hat{1} - \hat{P}_{\underline{S}_V}) = \mathcal{O}(i_{V,V_{1,2}})(\hat{P}_2 + \hat{P}_3) = \hat{P}_2 + \hat{P}_3 + \hat{P}_4$$
$$\mathcal{O}(i_{V_{1,2},V_{1,2}})(\hat{1} - \hat{P}_{\underline{S}_{V_{1,2}}}) = \hat{1} - \hat{P}_{\underline{S}_{V_{1,2}}} = \hat{P}_2.$$

Putting these results together we obtain

$$\hat{P}_{(\sim\underline{S})V_{1,2}} = \hat{P}_2 + \hat{P}_3 + \hat{P}_4$$

Then, clearly

$$\hat{P}_{(\sim\underline{S})V_{1,2}} \geq \hat{1} - \hat{P}_{\underline{S}_{V_{1,2}}} = \hat{P}_2.$$

This confirms that the projection $\hat{P}_{(\sim\underline{S})V}$ is always greater than or equal to $\hat{1} - \hat{P}_{\underline{S}_V}$.

Given the definition the co-Heyting negation in Lemma 2.3.4, $\hat{P}_{(\sim\underline{S})V}$ represents the disjunction of all the coarse-grainings of complements of (finer) local propositions at "bigger" contexts $\bar{V} \supseteq V$. Therefore, for each context V, the propositions associated to $\underline{S}_V$ and $(\sim \underline{S})_V$ are not mutually exclusive in general.

We now compute the other negation

$$\hat{P}_{(\neg\underline{S})_{V_{1,2}}} = \hat{1} - \bigvee_{V' \in m_{V_{1,2}}} \hat{P}_{\underline{S}'_V}. \tag{2.4.1}$$

As a first step we identify the set $m_{V_{1,2}} = \{V_1, V_2, V_{1,2}\}$ where $V_i = \mathbb{C}\hat{1} + \mathbb{C}\hat{P}_i$. The Gelfand spectrum for contests V_1 and V_2 are $\underline{\Sigma}_{V_1} = \{\lambda'_1, \lambda'_{234}\}$ and $\underline{\Sigma}_{V_2} = \{\lambda'_2, \lambda'_{134}\}$, respectively. Therefore we obtain $\underline{S}_{V_1} = \{\lambda'_1, \lambda'_{234}\}$ and $\underline{S}_{V_2} = \{\lambda'_2, \lambda'_{134}\}$. The associate projection operator is $\hat{P}_1 + \hat{P}_2 + \hat{P}_3 + \hat{P}_4$ in both cases. We now plug in all these results into (2.4.1), obtaining

$$\hat{P}_{(\neg\underline{S})_{V_{1,2}}} = \hat{1} - (\hat{P}_1 + \hat{P}_2 + \hat{P}_3 + \hat{P}_4) = \hat{0}.$$

Clearly

$$\hat{P}_{(\neg\underline{S})_{V_{1,2}}} \leq (\hat{1} - \hat{P}_{\underline{S}_{V_{1,2}}}) = \hat{P}_2.$$

Confirming the fact that the projection $\hat{P}_{(\neg\underline{S})_V}$ is always smaller than or equal to $\hat{1} - \hat{P}_{\underline{S}_V}$, since $\hat{P}_{\underline{S}_{V'}} \geq \hat{P}_{\underline{S}_V}$ for all $V' \in M_V$.

Given the definition of the Heyting negation in Eq. (2.3.4) it follows that, the projection $\hat{P}_{\neg\underline{S}_V}$ is determined at each stage V as the complement of the join of all the coarse-grainings $\hat{P}_{\underline{S}_{V'}}$ of $\hat{P}_{\underline{S}_V}$, where $V' \in m_V$.

2.4.2 *Second Example*

For our second example we choose the sub-object $\underline{S}_V = \{\lambda_3\}$ with associated projection operator *hatP*$_3$. We would like to evaluate $\hat{P}_{(\neg\underline{S}_{V_{1,3}})}$ for the context $V_{1,3} = lin_{\mathbb{C}}(\hat{P}_1, \hat{P}_3, \hat{P}_2 + \hat{P}_4)$. The Gelfand spectrum for $V_{1,3}$ is $\underline{\Sigma}_{V_{1,3}} = \{\lambda'_1, \lambda'_3, \lambda'_{24}\}$, while $m_{V_{1,3}} = \{V_{1,3}, V_1, V_3\}$. Next we need to identify the sub-object $\underline{S}$ as defined for each of the contexts V' in $m_{V_{1,3}}$. These are subject to the condition $\underline{\Sigma}(i_{V,V'})\underline{S}_V \subseteq \underline{S}_{V'}$. Therefore for the context $V_{1,3}$ we can choose $\underline{S}_{V_{1,3}}$ to be $\{\lambda'_3\}$, $\{\lambda'_3, \lambda'_1\}$ or $\{\lambda'_3, \lambda'_1, \lambda'_{34}\}$. We choose $\underline{S}_{V_{1,3}} = \{\lambda'_3\}$ whose associated projection is $\hat{P}_3$. Next for V_1 we choose $\underline{S}_{V_1} = \{\lambda'_{234}\}$ with associated projection $\hat{P}_2 + \hat{P}_3 + \hat{P}_4$ and for V_3 we choose $\underline{S}_{V_3} = \{\lambda'_3\}$ with associated projection $\hat{P}_3$. We then obtain

$$\begin{aligned} \hat{P}_{(\neg\underline{S})_{V_{1,3}}} &= \hat{1} - \bigvee_{V' \in m_{V_{1,3}}} \hat{P}_{\underline{S}'_V} \\ &= \hat{1} - (\hat{P}_3 + \hat{P}_2 + \hat{P}_4) = \hat{P}_1. \end{aligned}$$

Clearly

$$\hat{P}_{(\neg\underline{S})_{V_{1,3}}} \leq \hat{1} - \hat{P}_{\underline{S}_{V_{1,3}}} = \hat{P}_1 + \hat{P}_2 + \hat{P}_4.$$

Next we compute

$$\hat{P}_{(\sim\underline{S})_{V_{1,3}}} = \bigvee_{\bar{V} \in M_{V_{1,3}}} \mathcal{O}(i_{\bar{V},V_{1,3}})(\hat{1} - \hat{P}_{\underline{S}_{\bar{V}}}).$$

Here the set $M_{V_{1,3}}$ is $M_{V_{1,3}} = \{V, V_{1,3}\}$, therefore we obtain

$$\hat{P}_{(\sim\underline{S})_{V_{1,3}}} = \hat{P}_1 + \hat{P}_2 + \hat{P}_4.$$

In this case $\hat{P}_{(\sim\underline{S})_{V_{1,3}}} = \hat{1} - \hat{P}_{\underline{S}_{V_{1,3}}}$.

2.4.3 *Interpretation*

As discussed above, the Heyting complement of the sub-object $\underline{S}$ at each stage V is the complement of all the coarse graining of the complement of $\underline{S}$ at each context $V' \subseteq V$.

On the other hand the co-Heyting negation of the sub-object $\underline{S}$ at each stage V is the disjunction of all the coarse grainings of the complement of $\underline{S}$ for all contexts $\bar{V} \supset V$.

Given the above, then for each context V we have

$$\hat{P}_{\neg\underline{S}_V} \leq \hat{1} - \hat{P}_{\underline{S}_V} \leq \hat{P}_{(\sim\underline{S})V}$$

When V is a minimal context we have that $\hat{P}_{\neg\underline{S}_V} = \hat{1} - \hat{P}_{\underline{S}_V}$, while if V is a maximal context (maximal sub-algebra) we have that $\hat{1} - \hat{P}_{\underline{S}_V} = \hat{P}_{(\sim\underline{S})V}$. For any other contexts V we seem to be getting a lower and a higher bound for $\hat{1} - \hat{P}_{\underline{S}_V}$.

Chapter 3
Alternative Group Action in Topos Quantum Theory

In this Chapter we will explain an alternative way of describing group actions in topos quantum theory. The definition of group and group action in topos quantum theory was first introduced in [27]. Later, an alternative definition was put forward in [13]. In the following chapter we will explain this new definition which rests on the idea of *flows* in the spectral presheaf.

3.1 Maps Between Spectral Presheaves

In order to understand how to define flows of the spectral presheaf we, first of all, need to introduce the notion of maps between spectral presheaves [14]. These are maps between two distinct spectral presheaves associated to two different algebras. In particular, given two C^*-algebras $\mathcal{A}$ and $\mathcal{B}$, then a map between the spectral presheaf $\underline{\Sigma}^{\mathcal{A}}$ and $\underline{\Sigma}^{\mathcal{B}}$ is uniquely determined by a unital $*$-homomorphism $\phi : \mathcal{A} \rightarrow \mathcal{B}$. Clearly when the two algebras coincide then we get the notion of an automorphisms on the spectral presheaf. Such automorphisms will be utilised to define flows on the spectral presheaf.

Before going into the details of how these flows on the spectral presheaf are defined, it would be useful to remind ourselves what the spectral presheaf $\underline{\Sigma}^{\mathcal{A}}$ associated to a C^*-algebra A is. In [26] we defined the spectral presheaf associated to the category $\mathcal{V}(\mathcal{H})$ of abelian von Neumann subalgebras of the von Neumann algebra $\mathcal{N} \subseteq \mathcal{B}(\mathcal{H})$.[1] Clearly, such a definition can be easily extended to any unital C^*-algebra obtaining the following:

[1] $\mathcal{B}(\mathcal{H})$ indicates the algebra of bounded operators on the Hilbert space.

C. Flori, *A Second Course in Topos Quantum Theory*,
Lecture Notes in Physics 944, https://doi.org/10.1007/978-3-319-71108-9_3

Definition 3.1.1 Consider a unital C^*-algebra $\mathcal{A}$ with associated category $\mathcal{C}(\mathcal{A})$ of unital abelian C^*-subalgebras of $\mathcal{A}$ which forms a poset under algebra inclusion. The spectral presheaf $\underline{\Sigma}^{\mathcal{A}} \in \mathbf{Sets}^{\mathcal{C}(\mathcal{A})^{\text{op}}}$ associated to $\mathcal{A}$ is defined on

- Objects: given any $C \in \mathcal{C}(\mathcal{A})$, then $\underline{\Sigma}^{\mathcal{A}}_C$ is the set of multiplicative states $\lambda : C \to \mathbb{C}$ equipped with the Gelfand topology.
- Morphisms: given an inclusion map $i_{C'C} : C' \hookrightarrow C$, the corresponding presheaf map is defined in terms of restriction as follows:

$$\begin{aligned} \underline{\Sigma}(i_{C'C}) : \underline{\Sigma}^{\mathcal{A}}_C &\to \underline{\Sigma}_{C'} \\ \lambda &\mapsto \lambda|_{C'} \end{aligned} \tag{3.1.1}$$

This map is surjective and continuous with respect to the Gelfand topology.

This definition implies that the spectral presheaf associated to a unital C^*-algebra consists of the collection of the Gelfand spectrum of all of the unital abelian C^*-algebra glued together by the presheaf maps.

Equipped with this definition we will define a map between two spectral presheaves associated to two distinct C^*-algebras $\mathcal{A}$ and $\mathcal{B}$, respectively. To this end we note that these two algebras are objects in the category $\mathbf{uC}^*$ whose objects are unital C^*-algebras and whose arrows are $*$-homomorphisms. Therefore, given $\mathcal{A}, \mathcal{B} \in \mathbf{uC}^*$, the map $\phi : \mathcal{A} \to \mathcal{B} \in \mathbf{uC}^*(\mathcal{A}, \mathcal{B})$ is a unital *-homomorphisms. The claim maid in [14] is that any such unital *-homomorphism induces a map $\underline{\Sigma}^{\mathcal{B}} \to \underline{\Sigma}^{\mathcal{A}}$. In particular, ϕ, induces a map at the level of the category of unital abelian C^*-subalgebras[2] as follows:

$$\begin{aligned} \tilde{\phi} : \mathcal{C}(\mathcal{A}) &\to \mathcal{C}(\mathcal{B}) \\ C &\mapsto \phi|_C(C). \end{aligned}$$

This map is called the *base map*. Given a map between the base categories $\mathcal{C}(\mathcal{A})$ and $\mathcal{C}(\mathcal{B})$, this induced an essential geometric morphisms [55, Theorem 2, VII]

$$\Phi : \mathbf{Sets}^{\mathcal{C}(\mathcal{A})} \to \mathbf{Sets}^{\mathcal{C}(\mathcal{B})}$$

whose inverse image is defined as follows:

$$\begin{aligned} \Phi^* : \mathbf{Sets}^{\mathcal{C}(\mathcal{B})^{\text{op}}} &\to \mathbf{Sets}^{\mathcal{C}(\mathcal{A})^{\text{op}}} \\ \underline{Q} &\mapsto \Phi^*(\underline{Q}) := \underline{Q} \circ \tilde{\phi};. \end{aligned}$$

For the sake of completeness we will recall the definition of a geometric morphism.

[2] Note that since ϕ is a *-homomorphisms, then the restriction $\phi|_C$ is norm-closed and hence $\phi|_C(C)$ is a C^*-algebra.

Definition 3.1.2 A *geometric morphism* [55, 73] $\phi : \tau_1 \to \tau_2$ between topoi τ_1 and τ_2 is defined to be a pair of functors $\phi_* : \tau_1 \to \tau_2$ and $\phi^* : \tau_2 \to \tau_1$, called respectively the *inverse image* and the *direct image* part of the geometric morphism, such that

1. $\phi^* \dashv \phi_*$ i.e., ϕ^* is the left adjoint of ϕ_*,
2. ϕ^* is left exact, i.e., it preserves all finite limits.

A geometric morphism f is said to be essential if the inverse image functor f^* has both a right adjoint f_* and a left adjoint $f_!$.

Since Φ is essential, there exists $\Phi_! : \mathbf{Sets}^{\mathcal{C}(\mathcal{A})^{\text{op}}} \to \mathbf{Sets}^{\mathcal{C}(\mathcal{B})^{\text{op}}}$ such that $\Phi_! \dashv \Phi^* \dashv \Phi_*$. Applying ϕ^* to the spectral presheaf $\underline{\Sigma}^{\mathcal{B}}$ associated to the algebra $\mathcal{B}$ we obtain that, for each $C \in \mathcal{C}(\mathcal{A})$, $\Phi^*(\underline{\Sigma}^{\mathcal{B}})_C = \underline{\Sigma}^{\mathcal{B}}_{\tilde{\phi}(C)}$. Moreover, given an inclusion map $i_{C'C} : C' \hookrightarrow C$, then

$$\Phi^*(\underline{\Sigma}^{\mathcal{B}})(i_{C'C}) : \Phi^*(\underline{\Sigma}^{\mathcal{B}})_C \to \Phi^*(\underline{\Sigma}^{\mathcal{B}})_{C'} \tag{3.1.2}$$
$$\lambda \mapsto \lambda|_{\phi(C')}.$$

Therefore Φ^* allows us to map the object $\underline{\Sigma}^{\mathcal{B}} \in \mathbf{Sets}^{\mathcal{C}(\mathcal{B})^{\text{op}}}$ to the object $\Phi^*(\underline{\Sigma}^{\mathcal{B}}) \in \mathbf{Sets}^{\mathcal{C}(\mathcal{A})^{\text{op}}}$. The next step is to define a map $\Phi^*(\underline{\Sigma}^{\mathcal{B}}) \to \underline{\Sigma}^{\mathcal{A}}$ in the topos $\mathbf{Sets}^{\mathcal{C}(\mathcal{A})^{\text{op}}}$. To this end we will utilise the existence of a duality between the category of commutative C^*-algebras and the category of locally compact Hausdorff spaces. Such a duality is called *Gelfand duality*. In our case we only consider the sub-category of unital abelian C^*-algebras which, by Gelfand duality is related to the category of compact Hausdorff spaces. In particular, let $\mathbf{ucC^*}$ be the category of unital abelian C^*-algebras and unital *-homomorphisms and **KHaus** the category of compact Hausdorff spaces and continuous maps, the Gelfand duality is expressed by the existence of the following adjunction:

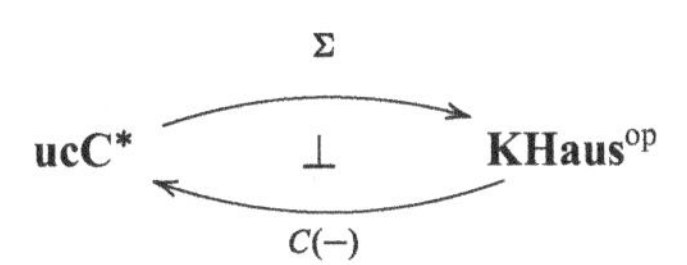

The action of the left adjoint Σ is to associate to each $A \in \mathbf{ucC^*}$ the set $\Sigma(A)$ of characters of A equipped with the topology of pointwise convergence, and to each morphism (unital *-homomorphism) $\phi : A \to B$, the map $\Sigma(\phi) : \Sigma(B) \to \Sigma(A)$, defined by $\Sigma(\phi)(\lambda) := \lambda \circ \phi$. Clearly $\Sigma(A)$ represents the spectral presheaf of the algebra A. In particular, going back to the definition of the spectral presheaf for an algebra $\mathcal{A}$ as in Definition 3.1.1 we note that for each $C \in \mathcal{C}(\mathcal{A})$ then $\underline{\Sigma}^{\mathcal{A}}_C := \Sigma(C)$.

On the other hand the right adjoint $C(-)$ assigns to each object $X \in \mathbf{KHaus}$ the set $C(X)$ of continuous complex valued functions on X equipped with the supremum norm. This is a commutative C^*-algebra under the pointwise algebraic operations. Give a continuous function $f : X \to Y$, then $C(f) : C(Y) \to C(X)$ is defined by $C(f)(g) := g \circ f \in C(X)$.

For the case at hand, we have the morphisms $\phi|_C : C \to \phi(C)$ between two commutative unital C^*-algebras which by Gelfand duality induces a continuous map

$$\mathcal{G}_{\phi;C} : \Sigma(\phi(C)) \to \Sigma(C) \tag{3.1.3}$$

$$\lambda \mapsto \lambda \circ \phi|_C. \tag{3.1.4}$$

Since $\phi : \mathcal{A} \to \mathcal{B}$, then $\Sigma(\phi(C)) = \underline{\Sigma}^{\mathcal{B}}_{\tilde{\phi}(C)} = \Phi^*(\underline{\Sigma}^{\mathcal{B}})_C$ and $\Sigma(C) = \underline{\Sigma}^{\mathcal{A}}_C$, therefore, for each $C \in \mathcal{C}(\mathcal{A})$, the above map translates to

$$\mathcal{G}_{\phi;C} : \Phi^*(\underline{\Sigma}^{\mathcal{B}})_C \to \underline{\Sigma}^{\mathcal{A}}_C.$$

We now would like to show that, for each $C \in \mathcal{C}(\mathcal{A})$, $\mathcal{G}_{\phi;C}$ are the components of the natural transformation $\mathcal{G}_\phi : \Phi^*(\underline{\Sigma}^{\mathcal{B}}) \to \underline{\Sigma}^{\mathcal{A}}$. In particular we need to show that, given an inclusion map $i_{C'C} : C' \hookrightarrow C$, the following diagram commutes

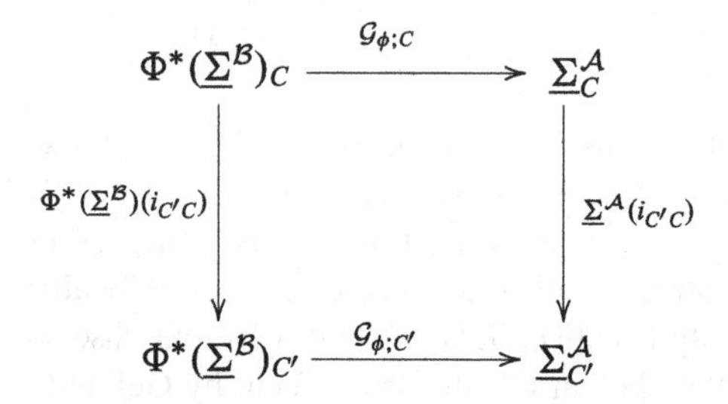

Chasing the diagram clockwise, we obtain, for $\lambda \in \Phi^*(\underline{\Sigma}^{\mathcal{B}})_C$

$$\begin{aligned}
\underline{\Sigma}^{\mathcal{A}}(i_{C'C})(\mathcal{G}_{\phi,C}(\lambda)) &\overset{(3.1.3)}{=} \underline{\Sigma}^{\mathcal{A}}(i_{C'C})(\lambda \circ \phi|_C) \\
&\overset{(3.1.1)}{=} (\lambda \circ \phi|_C)|_{C'} \\
&= \lambda|_{\phi|_{C'}} \circ \phi|_{C'} \\
&\overset{(3.1.3)}{=} \mathcal{G}_{\phi,C'}(\lambda|_{\phi|_{C'}}) \\
&\overset{(3.1.2)}{=} \mathcal{G}_{\phi,C'}(\Phi^*(\underline{\Sigma}^{\mathcal{B}})(i_{C'C})(\lambda)).
\end{aligned}$$

This proves that the map $\mathcal{G}_\phi : \Phi^*(\underline{\Sigma}^{\mathcal{B}}) \to \underline{\Sigma}^{\mathcal{A}}$ is a natural transformation in $\mathbf{Sets}^{\mathcal{C}(\mathcal{A})^{\text{op}}}$. Combining the two newly constructed maps Φ^* and $\mathcal{G}_\phi$ we obtain the desired map

$$\underline{\Sigma}^{\mathcal{B}} \xrightarrow{\Phi^*} \Phi^*(\underline{\Sigma}^{\mathcal{B}}) \xrightarrow{\mathcal{G}_\phi} \underline{\Sigma}^{\mathcal{A}}.$$

The above discussion uncovers the fact that to each unital *-homomorphism $\phi : \mathcal{A} \to \mathcal{B}$ there is associated a map $\mathcal{G}_\phi \circ \Phi^* : \underline{\Sigma}^{\mathcal{B}} \to \underline{\Sigma}^{\mathcal{A}}$ going in the opposite direction, i.e. we have a contravariant association. This finding is summarised in the following Lemma [14]:

Lemma 3.1.1 *Given two unital C^*-algebras $\mathcal{A}, \mathcal{B} \in$ **ucC*** with a unital *-homomorphism $\phi : \mathcal{A} \to \mathcal{B}$, then there exists a map*

$$\langle \Phi, \mathcal{G}_\phi \rangle = \mathcal{G}_\phi \circ \Phi^* : \underline{\Sigma}^{\mathcal{B}} \to \underline{\Sigma}^{\mathcal{A}}$$

between the respective presheaves going in the opposite direction.

In this Chapter we are interested in explaining the concept of *flow of the spectral presheaf* put forward in [13]. To this end we need to define maps from a spectral presheaf to itself. So far we have been able to define maps between two distinct spectral presheaves the next step is to understand under what conditions such maps are isomorphisms. The definition of an isomorphism between spectral presheaves associated to two distinct algebras was given in [14].

Definition 3.1.3 Consider two unital C^*-algebras $\mathcal{A}, \mathcal{B} \in$ **ucC*** with corresponding spectral presheaves $\underline{\Sigma}^{\mathcal{A}}$ and $\underline{\Sigma}^{\mathcal{B}}$, respectively. An isomorphism between $\underline{\Sigma}^{\mathcal{A}}$ and $\underline{\Sigma}^{\mathcal{B}}$ consists of a pair $\langle \Phi, \mathcal{G}_\phi \rangle$ where $\Phi : \mathbf{Sets}^{\mathcal{C}(\mathcal{A})} \to \mathbf{Sets}^{\mathcal{C}(\mathcal{B})}$ is the essential geometric isomorphism induced by the order-isomorphisms[3] $\phi : \mathcal{C}(\mathcal{A}) \to \mathcal{C}(\mathcal{B})$ (base map) and $\mathcal{G}_\phi : \Phi^*(\underline{\Sigma}^{\mathcal{B}}) \to \underline{\Sigma}^{\mathcal{A}}$ is a natural isomorphism, i.e. each component $(\mathcal{G}_\phi)_C : (\Phi^*(\underline{\Sigma}^{\mathcal{B}}))_C \to \underline{\Sigma}^{\mathcal{A}}_C$ is a homeomorphism for all $C \in \mathcal{C}(\mathcal{A})$. As in Lemma 3.1.1 the action of $\langle \Phi, \mathcal{G}_\phi \rangle$ is defined by

$$\mathcal{G}_\phi \circ \Phi^* : \underline{\Sigma}^{\mathcal{B}} \to \underline{\Sigma}^{\mathcal{A}}.$$

If $\mathcal{A} = \mathcal{B}$ then an isomorphism $\langle \Phi, \mathcal{G}_\phi \rangle : \underline{\Sigma}^{\mathcal{A}} \to \underline{\Sigma}^{\mathcal{A}}$ is called an automorphism.
An *order isomorphism* is essentially an isomorphism of partially ordered sets, where the isomorphism is extended to the partial order as well (recall that the categories $\mathcal{C}(\mathcal{A})$ and $\mathcal{C}(\mathcal{B})$ are posets). The formal definition is as follows

Definition 3.1.4 Given two posets P and Q, $f : P \to Q$ is an order isomorphism iff f is a bijection such that, for every $x, y \in P$, then $x \leq_P y$ if and only if $f(x) \leq_Q f(y)$.
We would now like to characterise a way of obtaining isomorphisms of spectral presheaves as defined in Definition 3.1.3. This is given by the following theorem [14]:

Theorem 3.1.1 *Given two unital C^*-algebras $\mathcal{A}, \mathcal{B} \in ucC^*$ with associated spectral presheaves $\underline{\Sigma}^{\mathcal{A}}$ and $\underline{\Sigma}^{\mathcal{B}}$ respectively, then there exists an injective map*

$$Iso(\mathcal{A}, \mathcal{B}) \to Iso(\underline{\Sigma}^{\mathcal{B}}, \underline{\Sigma}^{\mathcal{A}})$$

[3]Note that from [50, 4.2.7] we know that when the base map ϕ is surjective on objects then the induced essential geometric morphisms is surjective. Moreover, from [50, 4.2.12] we know that when the base map ϕ is full and faithful, then the induced essential geometric morphisms is an inclusion. It then follows that in our case the essential geometric morphism induced by the isomorphism ϕ is itself an isomorphism.

where $Iso(\mathcal{A}, \mathcal{B})$ *denotes all isomorphisms from* $\mathcal{A}$ *to* $\mathcal{B}$ *and* $Iso(\underline{\Sigma}^{\mathcal{B}}, \underline{\Sigma}^{\mathcal{A}})$ *denotes all isomorphisms from* $\underline{\Sigma}^{\mathcal{B}}$ *to* $\underline{\Sigma}^{\mathcal{A}}$.

In order to prove this theorem we will need to introduce the notion of a partial C^*-algebra also sometimes denoted as a piecewise C^*-algebra. The following definition was inspired by Kochen and Specker's consideration of partial algebras [51].[4]

Definition 3.1.5 ([72]) A *piecewise C*-algebra* is a set A equipped with the following pieces of structure:

1. a reflexive and symmetric relation $⫫ \subseteq A \times A$. If $\alpha ⫫ \beta$, we say that α and β *commute*;
2. binary operations $+, \cdot : ⫫ \rightarrow A$;
3. a scalar multiplication $\cdot : \mathbb{C} \times A \rightarrow A$;
4. distinguished elements $0, 1 \in A$;
5. an involution $* : A \rightarrow A$;
6. a norm $|| - || : A \rightarrow \mathbb{R}$;

such that every subset $C \subseteq A$ of pairwise commuting elements is contained in some subset $\bar{C} \subseteq A$ of pairwise commuting elements which is a commutative C*-algebra with respect to the data above.

The piecewise C*-algebras in which the relation $⫫$ is total are precisely the commutative C*-algebras. Our choice of the symbol "$⫫$" is explained by the special case of rank one projections, which commute if and only if they are either orthogonal ($\perp$) or parallel ($\|$).

Definition 3.1.6 ([72]) Given piecewise C^*-algebras A and B, a *piecewise* $*$-*homomorphism* is a function $\zeta : A \rightarrow B$ such that

1. If $\alpha ⫫ \beta$ in A, then

$$\zeta(\alpha) ⫫ \zeta(\beta), \qquad \zeta(\alpha\beta) = \zeta(\alpha)\zeta(\beta), \qquad \zeta(\alpha+\beta) = \zeta(\alpha)+\zeta(\beta). \tag{3.1.5}$$

2. $\zeta(z\alpha) = z\zeta(\alpha)$ for all $a \in A$ and $z \in \mathbb{C}$,
3. $\zeta(\alpha^*) = \zeta(\alpha^*)$ for all $\alpha \in A$.
4. $\zeta(1) = 1$.

Example 3.1.1 It is well-known that there is no $*$-homomorphism $M_n \rightarrow \mathbb{C}$ for $n \geq 2$. The Kochen-Specker theorem [51] states that for $n \geq 3$ not even a piecewise $*$-homomorphism $M_n \rightarrow \mathbb{C}$ exist.

The collection of piecewise C^*-algebras and piecewise $*$-homomorphisms form a category which we denote by $\mathsf{pC^*alg_1}$. Still following [72], there is a forgetful

[4] For this reason van den Berg and Heunen introduced their definition as *partial C*-algebras*, but the term was subsequently changed to *piecewise C*-algebra* [41].

functor[5] $\mathbb{C}(-) : \mathsf{C^*alg}_1 \to \mathsf{pC^*alg}_1$ sending every C*-algebra $\mathcal{A}$ to its normal part,

$$\mathbb{C}(\mathcal{A}) = \{ \alpha \in A \mid \alpha\alpha^* = \alpha^*\alpha \}. \tag{3.1.6}$$

This set forms a piecewise C*-algebra by postulating that $\alpha \perp\!\!\!\perp \beta$ holds whenever α and β commute. $\mathbb{C}(-)$ is easily seen to be a faithful functor that reflects isomorphisms. In the language of *property, structure and stuff* [59], this means that it forgets at most the structure. So we may think of a C*-algebra as a piecewise C*-algebra together with additional structure, namely the specifications of sums and products of noncommuting elements.

Example 3.1.2 For $A, B \in \mathsf{C^*alg}_1$, any Jordan homomorphism $\mathbb{R}(A) \to \mathbb{R}(B)$ extends linearly to a piecewise $*$-homomorphism $\mathbb{C}(A) \to \mathbb{C}(B)$.

Given the category of partial C^*-algebras it was shown in [14] that there exists a bijective correspondence between isomorphisms of partial C^*-algebras and isomorphisms of the respective spectral presheaves.

Theorem 3.1.2 *Let $\mathcal{A}$ and $\mathcal{B}$ be unital C^*-algebras whose spectral presheaves are $\underline{\Sigma}^{\mathcal{A}}$ and $\underline{\Sigma}^{\mathcal{B}}$ respectively. There exists a bijective correspondence between isomorphisms $\langle \Phi, \mathcal{G}_\phi \rangle : \underline{\Sigma}^{\mathcal{B}} \to \underline{\Sigma}^{\mathcal{A}}$ and isomorphisms $T : \mathbb{C}(\mathcal{A}) \to \mathbb{C}(\mathcal{B})$ of the associated partial C^*-algebras. This means that $\mathbb{C}(\mathcal{A}) \simeq \mathbb{C}(\mathcal{B})$ iff $\underline{\Sigma}^{\mathcal{A}} \simeq \underline{\Sigma}^{\mathcal{B}}$.*

The way to prove this theorem is to show that every isomorphism $\langle \Phi, \mathcal{G}_\phi \rangle : \underline{\Sigma}^{\mathcal{B}} \to \underline{\Sigma}^{\mathcal{A}}$ induces an isomorphism $T : \mathbb{C}(\mathcal{A}) \to \mathbb{C}(\mathcal{B})$ and vice versa.

Lemma 3.1.2 ([14]) *Let $\mathcal{A}$ and $\mathcal{B}$ be unital C^*-algebras whose spectral presheaves are $\underline{\Sigma}^{\mathcal{A}}$ and $\underline{\Sigma}^{\mathcal{B}}$ respectively, then each isomorphisms $\langle \Phi, \mathcal{G}_\phi \rangle : \underline{\Sigma}^{\mathcal{B}} \to \underline{\Sigma}^{\mathcal{A}}$ induces an isomorphisms $T : \mathbb{C}(\mathcal{A}) \to \mathbb{C}(\mathcal{B})$.*

Proof Since $\langle \Phi, \mathcal{G}_\phi \rangle : \underline{\Sigma}^{\mathcal{B}} \to \underline{\Sigma}^{\mathcal{A}}$ is an isomorphism, the associated base map $\tilde{\phi} : \mathcal{C}(\mathcal{A}) \to \mathcal{C}(\mathcal{B})$ is an order-isomorphism. Therefore, for each $C' \in \mathcal{C}(\mathcal{A})$ the associate map

$$\mathcal{G}_{\phi;C'} : \Phi^*(\underline{\Sigma}^{\mathcal{B}})_{C'} = \underline{\Sigma}^{\mathcal{B}}_{\gamma(C')} \to \underline{\Sigma}^{\mathcal{A}}_{C'}$$

is a homeomorphism. Gelfand duality then determines a unique unital isomorphism

$$\begin{aligned} k_{C'} : C(\underline{\Sigma}^{\mathcal{A}}_{C'}) &\to C(\underline{\Sigma}^{\mathcal{B}}_{C'}) \\ f &\mapsto f \circ \mathcal{G}_{\phi;C'}. \end{aligned}$$

The unit of the adjunction for Gelfand duality is given by

$$\eta : id_{\mathbf{ucC^*}} \to C(-) \circ \underline{\Sigma}$$

[5] We recall that a forgetful functor 'forgets' or drops some or all of the input's structure or properties 'before' mapping to the output.

such that for each $C' \in \mathcal{C}(\mathcal{A})$ we get the (components) isomorphisms

$$\eta_{C'} : C' \to C(\underline{\Sigma}^{\mathcal{A}}_{C'})$$
$$\eta^{-1}_{\phi(C')} : C(\underline{\Sigma}^{\mathcal{B}}_{\phi(C')}) \to \phi(C').$$

These two maps together with the map $k_{C'}$ can be combined to define, for each $C' \in \mathcal{C}(\mathcal{A})$ an isomorphism

$$\alpha_{C'} : \eta^{-1}_{\phi(C')} \circ k_{C'} \circ \eta_{C'} : C' \to \phi(C').$$

We will use such a map to define the desired isomorphism between the partial C^*-algebras as follows

$$T : \mathbb{C}(\mathcal{A}) \to \mathbb{C}(\mathcal{B})$$
$$\hat{A} \mapsto \alpha_C(\hat{A})$$

where $C \in \mathcal{C}(\mathcal{A})$ is any context such that $\hat{A} \in C$. We now need to show that this map is well defined, i.e. that it does not depend on the context C containing $\hat{A}$. This can be easily seen from the fact that the following diagram commutes

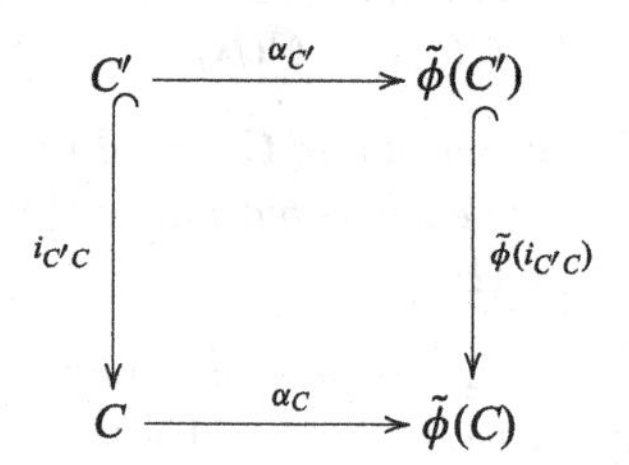

where $C, C' \in \mathcal{C}(\mathcal{A})$ such that $\hat{A} \in C$ and $\hat{A} \in C'$, and $i_{C'C} : C' \hookrightarrow C$ in $\mathcal{C}(\mathcal{A})$. □

On the other hand, we also have the "reverse" of the above lemma, namely

Lemma 3.1.3 ([14]) *Given the unital partial C^*-algebras $\mathbb{C}(\mathcal{A})$ and $\mathbb{C}(\mathcal{B})$, every isomorphism $T : \mathbb{C}(\mathcal{A}) \to \mathbb{C}(\mathcal{B})$ induces an isomorphism $\langle \Phi, \mathcal{G}_\phi \rangle : \underline{\Sigma}^{\mathcal{B}} \to \underline{\Sigma}^{\mathcal{A}}$.*

Proof Given an isomorphism $T : \mathbb{C}(\mathcal{A}) \to \mathbb{C}(\mathcal{B})$, then, for every $C \in \mathcal{C}(\mathcal{A})$, the restriction $T|_C : C \to \mathbb{C}(\mathcal{B})$ is a unital *-homomorphism which implies that $T|_C(C)$ is norm-closed and hence a unital abelian C^*-subalgebra of $\mathbb{C}(\mathcal{B})$, i.e. $T|_C(C) \in \mathcal{C}(\mathcal{B})$. Since for $C' \subseteq C$ then $T|_{C'}(C') \subseteq T|_C(C)$ and it is possible to construct an order preserving map

$$\tilde{\phi} : \mathcal{C}(\mathcal{A}) \to \mathcal{C}(\mathcal{B})$$
$$C \mapsto T|_C(C)$$

whose inverse is

$$\begin{aligned}\tilde{\phi}^{-1} : \mathcal{C}(\mathcal{B}) &\rightarrow \mathcal{C}(\mathcal{A}) \\ C &\mapsto T^{-1}|_C(C)\end{aligned}$$

where T^{-1} is the unital partial isomorphism inverse of T. The fact that $\tilde{\phi}$ has an inverse, also defined in terms of an isomorphism T^{-1}, implies that $\tilde{\phi}$ is an order isomorphism and hence the desired base map. This induces the essential geometric isomorphism $\Phi : \mathbf{Sets}^{\mathcal{C}(\mathcal{A})} \rightarrow \mathbf{Sets}^{\mathcal{C}(\mathcal{B})}$. We now need to define the natural isomorphism $\mathcal{G}_\phi$. This will be done in terms of the map T as follows: for each $C \in \mathcal{C}(\mathcal{A})$

$$\begin{aligned}\mathcal{G}_{\phi;C} : (\Phi^*(\underline{\Sigma}^{\mathcal{B}}))_C = \underline{\Sigma}^{\mathcal{B}}_{\tilde{\phi}(C)} &\rightarrow \underline{\Sigma}^{\mathcal{B}}_C \\ \lambda &\mapsto \lambda \circ T|_C.\end{aligned}$$

Clearly this map is an isomorphism by construction. Since it is defined for each $C \in \mathcal{C}(\mathcal{A})$, these are the components of the natural isomorphism

$$\mathcal{G}_\phi : \Phi^*(\underline{\Sigma}^{\mathcal{B}}) \rightarrow \underline{\Sigma}^{\mathcal{B}}.$$

□

Putting together Lemmas 3.1.2 and 3.1.3 we obtain a proof for Theorem 3.1.2.

In [36] is was shown that there exists a correspondence between order isomorphisms $\phi : \mathcal{C}(\mathcal{A}) \rightarrow \mathcal{C}(\mathcal{B})$ and unital quasi-Jordan isomorphisms. Before stating and proving this theorem we will need to define what a quasi-Jordan isomorphism is.

Definition 3.1.7 Given a unital C^*-algebra A, the set $\mathcal{A}_{sa}$ of self-adjoint operators in $\mathcal{A}$ forms a real unital Jordan algebra with product

$$\forall\, \hat{A}, \hat{B} \in \mathcal{A}_{sa};\ \ \hat{A} \cdot \hat{B} := \frac{1}{2}(\hat{A}\hat{B} + \hat{B}\hat{A}).$$

Definition 3.1.8 Given two real unital Jordan algebras $\mathcal{A}_{sa}$ and $\mathcal{B}_{sa}$, then a quasi-Jordan homomorphism is a unital map

$$Q : \mathcal{A}_{sa} \rightarrow \mathcal{B}_{sa}$$

such that for all $C \in \mathcal{C}(\mathcal{A})$, then the restriction

$$Q|_{C_{sa}} : C_{sa} \rightarrow \mathcal{B}_{sa}$$

is a unital Jordan homomorphism and thus Q is required to be linear only on commuting self-adjoint operators, i.e. a quasi-linear map. Clearly Q preserves the Jordan product on commuting operators since if $\hat{A}\hat{B} = \hat{B}\hat{A}$, then $\hat{A} \cdot \hat{B} = \hat{A}\hat{B}$.

If Q is a bijection with inverse Q^{-1}, which is itself a quasi-Jordan homomorphism, then Q is a quasi-Jordan isomorphism.

From the above definition it is clear that there exists an intimate connection between the Jordan algebra $\mathcal{A}_{sa}$ associated to a unital C^*-algebra $\mathcal{A}$ and the partial algebra $\mathbb{C}(\mathcal{A})$ associated with $\mathcal{A}$. In particular, $\mathcal{A}_{sa}$ is the self-adjoint part of $\mathbb{C}(\mathcal{A})$, therefore any partial *-isomorphism $T : \mathbb{C}(\mathcal{A}) \to \mathbb{C}(\mathcal{B})$ restricts to an isomorphism $T|_{sa} : \mathcal{A}_{sa} \to \mathcal{B}_{sa}$ and any unital quasi-Jordan isomorphisms $Q : \mathcal{A}_{sa} \to \mathcal{B}_{sa}$ extends linearly to a partial *-isomorphism.

Theorem 3.1.3 ([36]) *Given two unital C^*-algebras $\mathcal{A}$ and $\mathcal{B}$ such that neither are isomorphism to $\mathbb{C}^2$ or $\mathcal{B}(\mathbb{C})$, then*

$$Iso(\mathcal{C}(\mathcal{A}), \mathcal{C}(\mathcal{B})) \simeq Iso(\mathcal{A}_{sa}, \mathcal{B}_{sa}) \simeq Iso(\mathbb{C}(\mathcal{A}), \mathbb{C}(\mathcal{B}))$$

*where $Iso(\mathcal{C}(\mathcal{A}), \mathcal{C}(\mathcal{B}))$ is the set of all order-isomorphisms, $Iso(\mathcal{A}_{sa}, \mathcal{B}_{sa})$ is the set of all unital quasi-Jordan isomorphisms and $Iso(\mathbb{C}\mathcal{A}, \mathbb{C}(\mathcal{B}))$ is the set of all partial *-isomorphism.*

An immediate consequence of Theorems 3.1.3 and 3.1.2 is the following:

Theorem 3.1.4 *Given two unital C^*-algebras $\mathcal{A}$ and $\mathcal{B}$, such that neither are isomorphism to $\mathbb{C}^2$ or $\mathcal{B}(\mathbb{C})$ then*

$$Iso(\underline{\Sigma}^{\mathcal{B}}, \underline{\Sigma}^{\mathcal{A}}) \simeq Iso(\mathcal{A}_{sa}, \mathcal{B}_{sa})$$

where $Iso(\underline{\Sigma}^{\mathcal{B}}, \underline{\Sigma}^{\mathcal{A}})$ denotes the set of isomorphisms between the presheaves $\underline{\Sigma}^{\mathcal{B}}$ and $\underline{\Sigma}^{\mathcal{A}}$, while $Iso(\mathcal{A}_{sa}, \mathcal{B}_{sa})$ denotes the set of unital quasi-Jordan isomorphisms between the real unital Jordan algebras $\mathcal{A}_{sa}$ and $\mathcal{B}_{sa}$.

3.2 Group Action as Flows on the Spectral Presheaf

In this section we will construct flows on the spectral presheaf as defined in [13]. These are given in terms of maps between the spectral presheaf and itself. Therefore, we need to adapt what said in the previous section, to the case in which $\mathcal{A} = \mathcal{B}$, i.e. the two unital C^*-algebras coincide. As a first step we modify Definition 3.1.3 to obtain:

Definition 3.2.1 Given a unital C^*-algebras $\mathcal{A} \in$ **ucC*** with corresponding spectral presheaves $\underline{\Sigma}^{\mathcal{A}}$. An automorphism on $\underline{\Sigma}^{\mathcal{A}}$ consists of a pair $\langle \Phi, \mathcal{G}_\phi \rangle$ where $\Phi : \mathbf{Sets}^{\mathcal{C}(\mathcal{A})} \to \mathbf{Sets}^{\mathcal{C}(\mathcal{A})}$ is the essential geometric isomorphism induced by the order-isomorphisms $\phi : \mathcal{C}(\mathcal{A}) \to \mathcal{C}(\mathcal{A})$ (base map) and $\mathcal{G}_\phi : \Phi^*(\underline{\Sigma}^{\mathcal{A}}) \to \underline{\Sigma}^{\mathcal{A}}$ is a natural isomorphism, i.e. each component $(\mathcal{G}_\phi)_C : (\Phi^*(\underline{\Sigma}^{\mathcal{A}}))_C \to \underline{\Sigma}^{\mathcal{A}}_C$ is a homeomorphism for all $C \in \mathcal{C}(\mathcal{A})$. As in Lemma 3.1.1 the action of $\langle \Phi, \mathcal{G}_\phi \rangle$ is defined by

$$\mathcal{G}_\phi \circ \Phi^* : \underline{\Sigma}^{\mathcal{A}} \to \underline{\Sigma}^{\mathcal{A}}.$$

The set of all automorphisms on $\underline{\Sigma}^{\mathcal{A}}$ is denoted by $Aut(\underline{\Sigma}^{\mathcal{A}})$ and it forms a group under the operation

$$Aut(\underline{\Sigma}^{\mathcal{A}}) \times Aut(\underline{\Sigma}^{\mathcal{A}}) \to Aut(\underline{\Sigma}^{\mathcal{A}})$$
$$(\langle \Phi, \mathcal{G}_\phi \rangle, \langle \Phi', \mathcal{G}'_{\phi'} \rangle) \mapsto \langle \Phi' \circ \Phi, \mathcal{G}_\phi \circ \mathcal{G}'_{\phi'} \rangle$$

where $\Phi' \circ \Phi : \mathbf{Sets}^{\mathcal{C}(\mathcal{A})} \to \mathbf{Sets}^{\mathcal{C}(\mathcal{A})}$ is the essential geometric morphism induced by $\phi' \circ \phi : \mathcal{C}(\mathcal{A}) \to \mathcal{C}(\mathcal{A})$ and $\mathcal{G}_\phi \circ \mathcal{G}'_{\phi'}$ is the natural isomorphism such that, for each $C \in \mathcal{C}(\mathcal{A})$ we have

$$(\mathcal{G}_\phi \circ \mathcal{G}'_{\phi'})_C = \mathcal{G}_{\phi;C} \circ \mathcal{G}'_{\phi';C} : ((\Phi' \circ \Phi)^* \underline{\Sigma}^{\mathcal{A}})_C = \underline{\Sigma}^{\mathcal{A}}_{\tilde{\phi}'(\tilde{\phi}(C))} \to \underline{\Sigma}^{\mathcal{A}}_{\tilde{\phi}(C)} \to \underline{\Sigma}^{\mathcal{A}}_C .$$

We will now show that indeed $Aut(\underline{\Sigma}^{\mathcal{A}})$ forms a group. As a first step we will prove associativity, namely that $(\langle \Phi, \mathcal{G}_\phi \rangle \circ \langle \Phi', \mathcal{G}'_{\phi'} \rangle) \circ \langle \Phi'', \mathcal{G}''_{\phi''} \rangle = \langle \Phi, \mathcal{G}_\phi \rangle \circ (\langle \Phi', \mathcal{G}'_{\phi'} \rangle \circ \langle \Phi'', \mathcal{G}''_{\phi''} \rangle)$. This is equivalent to showing that

$$\langle \Phi'' \circ (\Phi' \circ \Phi), (\mathcal{G}_\phi \circ \mathcal{G}_{\phi'}) \circ \mathcal{G}_{\phi''} \rangle = \langle (\Phi'' \circ \Phi') \circ \Phi, \mathcal{G}_\phi \circ (\mathcal{G}_{\phi'} \circ \mathcal{G}_{\phi''}) \rangle .$$

To prove the above equality we need to show that $\Phi'' \circ (\Phi' \circ \Phi) = (\Phi'' \circ \Phi') \circ \Phi$ and $(\mathcal{G}_\phi \circ \mathcal{G}_{\phi'}) \circ \mathcal{G}_{\phi''} = \mathcal{G}_\phi \circ (\mathcal{G}_{\phi'} \circ \mathcal{G}_{\phi''})$. We know that the essential geometric morphism $\Phi'' \circ (\Phi' \circ \Phi)$ is induced by the base map $\tilde{\phi}'' \circ (\tilde{\phi}' \circ \tilde{\phi}) : \mathcal{C}(\mathcal{A}) \to \mathcal{C}(\mathcal{A})$ while $(\Phi'' \circ \Phi') \circ \Phi$ is induced by the base map $(\tilde{\phi}'' \circ \tilde{\phi}') \circ \tilde{\phi} : \mathcal{C}(\mathcal{A}) \to \mathcal{C}(\mathcal{A})$. However, for each $C \in \mathcal{C}(\mathcal{A})$ we have

$$\tilde{\phi}'' \circ (\tilde{\phi}' \circ \tilde{\phi})(C) = \tilde{\phi}'' \circ (\tilde{\phi}'(\tilde{\phi}(C))) = \tilde{\phi}''(\tilde{\phi}'(\tilde{\phi}(C)))$$

and

$$(\tilde{\phi}'' \circ \tilde{\phi}') \circ \tilde{\phi}(C) = (\tilde{\phi}'' \circ \tilde{\phi}')(\tilde{\phi}(C)) = \tilde{\phi}''(\tilde{\phi}'(\tilde{\phi}(C))).$$

Therefore $\Phi'' \circ (\Phi' \circ \Phi) = (\Phi'' \circ \Phi') \circ \Phi$.

To prove that $(\mathcal{G}_\phi \circ \mathcal{G}_{\phi'}) \circ \mathcal{G}_{\phi''} = \mathcal{G}_\phi \circ (\mathcal{G}_{\phi'} \circ \mathcal{G}_{\phi''})$ we note that

$$(\mathcal{G}_\phi \circ \mathcal{G}_{\phi'}) \circ \mathcal{G}_{\phi''} : ((\Phi'' \circ (\Phi' \circ \Phi))^* \underline{\Sigma}^{\mathcal{A}})_C = \underline{\Sigma}^{\mathcal{A}}_{\tilde{\phi}''(\tilde{\phi}'(\tilde{\phi}(C)))} \to \underline{\Sigma}^{\mathcal{A}}_{\tilde{\phi}'(\tilde{\phi}(C))} \to \underline{\Sigma}^{\mathcal{A}}_{\tilde{\phi}(C)} \to \underline{\Sigma}^{\mathcal{A}}_C$$

and

$$\mathcal{G}_\phi \circ (\mathcal{G}_{\phi'} \circ \mathcal{G}_{\phi''}) : ((\Phi'' \circ \Phi') \circ \Phi))^* \underline{\Sigma}^{\mathcal{A}})_C = \underline{\Sigma}^{\mathcal{A}}_{\tilde{\phi}''(\tilde{\phi}'(\tilde{\phi}(C)))} \to \underline{\Sigma}^{\mathcal{A}}_{\tilde{\phi}'(\tilde{\phi}(C))} \to \underline{\Sigma}^{\mathcal{A}}_{\tilde{\phi}(C)} \to \underline{\Sigma}^{\mathcal{A}}_C .$$

From the previous result it then follows that indeed $(\mathcal{G}_\phi \circ \mathcal{G}_{\phi'}) \circ \mathcal{G}_{\phi''} = \mathcal{G}_\phi \circ (\mathcal{G}_{\phi'} \circ \mathcal{G}_{\phi''})$.

Next we will show that the automorphism $\langle Id, id\rangle$ is the identity element. In fact

$$\langle Id, id\rangle \circ \langle \Phi, \mathcal{G}_\phi\rangle = \langle \Phi \circ Id, id \circ \mathcal{G}_\phi\rangle = \langle \Phi, \phi\rangle = \langle Id \circ \Phi, \mathcal{G}_\phi \circ id\rangle = \langle \Phi, \mathcal{G}_\phi\rangle \circ \langle Id, id\rangle.$$

Finally, given any element $\langle \Phi, \mathcal{G}_\phi\rangle$ with underlying base map $\tilde{\phi}$, then the element $\langle \Phi^{-1}, \mathcal{G}^{-1}_{\phi^{-1}}\rangle$ with underlying base map $\tilde{\phi}^{-1}$ is its inverse. In fact,

$$\begin{aligned}\langle \Phi^{-1}, \mathcal{G}^{-1}_{\phi^{-1}}\rangle \circ \langle \Phi, \mathcal{G}_\phi\rangle = \langle \Phi \circ \Phi^{-1}, \mathcal{G}^{-1}_{\phi^{-1}} \circ \mathcal{G}_\phi\rangle = \langle Id, id\rangle &= \langle \Phi^{-1} \circ \Phi, \mathcal{G}_\phi \circ \mathcal{G}^{-1}_{\phi^{-1}}\rangle \\ &= \langle \Phi, \mathcal{G}_\phi\rangle \circ \langle \Phi^{-1}, \mathcal{G}^{-1}_{\phi^{-1}}\rangle.\end{aligned}$$

We would now like to prove the following theorem [14]

Theorem 3.2.1 *Given a unital C^*-algebra $\mathcal{A}$ with associated spectral presheaf $\underline{\Sigma}^{\mathcal{A}}$, then there exists an injective group homomorphism*

$$\begin{aligned} Aut(\mathcal{A}) &\to Aut(\underline{\Sigma}^{\mathcal{A}})^{\mathrm{op}} \\ \phi &\mapsto \langle \Phi, \mathcal{G}_\phi\rangle = \mathcal{G}_\phi \circ \Phi^*. \end{aligned} \tag{3.2.1}$$

In order to prove this theorem we need to consider Theorem 3.1.2 for the special case in which $\mathcal{A} = \mathcal{B}$. This theorem, then, tells us that there exists a (contravariant) group isomorphisms

$$Aut(\underline{\Sigma}^{\mathcal{A}})^{\mathrm{op}} \xrightarrow{\simeq} Aut_{part}(\mathbb{C}(\mathcal{A})) \tag{3.2.2}$$

between the group of automorphisms on $\underline{\Sigma}^{\mathcal{A}}$ and the group of unital partial *-automorphism on the partial C^*-algebra $\mathbb{C}(\mathcal{A})$.

The fact that the above map preserves the group structure can be seen from the fact that, given two automorphisms $\langle \Phi, \mathcal{G}_\phi\rangle$ and $\langle \Phi', \mathcal{G}'_{\phi'}\rangle$ in $Aut(\underline{\Sigma}^{\mathcal{A}})$ with corresponding automorphisms T and T' in $Aut_{part}(\mathbb{C}(\mathcal{A}))$, then the composite automorphism $\langle \Phi, \mathcal{G}_\phi\rangle \circ \langle \Phi', \mathcal{G}'_{\phi'}\rangle$ gets mapped to the composite $T' \circ T$.

We can now prove Theorem 3.2.1.

Proof Given an automorphism $\phi : \mathcal{A} \to \mathcal{A}$ on the C^*-algebra $\mathcal{A}$, this gives rise to an automorphism $\phi|_{\mathbb{C}(\mathcal{A})} : \mathbb{C}(\mathcal{A}) \to \mathbb{C}(\mathcal{A})$ by restriction. From the above discussion this corresponds contravariantly to an automorphism on $\underline{\Sigma}^{\mathcal{A}}$. Therefore we obtain a contravariant group homomorphism from $Aut(\mathcal{A})$ to $Aut(\underline{\Sigma}^{\mathcal{A}})$. We now need to show that this homomorphism is injective. In particular given two automorphisms $\alpha, \beta : \mathcal{A} \to \mathcal{A}$ if they are distinct then there will exist an $\hat{A} \in \mathbb{C}(\mathcal{A})$ such that $\alpha(\hat{A}) \neq \beta(\hat{A})$, therefore $\alpha|_{\mathbb{C}(\mathcal{A})} \neq \beta|_{\mathbb{C}(\mathcal{B})}$ which implies by Theorem 3.1.2 that these corresponds to two distinct automorphisms on $\underline{\Sigma}^{\mathcal{A}}$. □

The reason why the map of Eq. (3.3.9) is not surjective is because the group $Aut_{part}(\mathbb{C}(\mathcal{A}))$ contains more elements then the group $Aut(\mathcal{A})$. In fact, given any automorphism $\alpha \in Aut(\mathcal{A})$, this induces an automorphism $\alpha|_{\mathbb{C}(\mathcal{A})} \in Aut_{part}(\mathbb{C}(\mathcal{A}))$. However the converse is not true, i.e. not every unital partial *-automorphisms on $\mathbb{C}(\mathcal{A})$ gives rise to an automorphism on $\mathcal{A}$.

If we consider Theorem 3.1.3 for the case in which $\mathcal{A} = \mathcal{B}$, we obtain the following group isomorphisms

$$Aut_{ord}(\mathcal{C}(\mathcal{A}))) \simeq Aut(\mathbb{C}(\mathcal{A}))) \simeq Aut(\mathcal{A}_{sa}).$$

This together with Theorem 3.2.1 proves that

Theorem 3.2.2 *Given a unital C^*-algebra $\mathcal{A}$ which is neither isomorphic to $\mathbb{C}^2$ not $\mathcal{B}(\mathbb{C}^2)$, then*

$$Aut(\mathcal{C}(\mathcal{A})) \simeq Aut(\underline{\Sigma}^{\mathcal{A}})^{op}.$$

We now would like to consider a special case of automorphisms on a unital C^*-algebra $\mathcal{A}$, i.e. inner automorphisms.

Definition 3.2.2 Given a unital C^*-algebra $\mathcal{A}$, then each element $\hat{U} \in \mathcal{U}(\mathcal{A})$ of the unitary group $\mathcal{U}(\mathcal{A})$ of $\mathcal{A}$ induces an inner automorphism

$$\begin{aligned} \phi_{\hat{U}} : \mathcal{A} &\to \mathcal{A} \\ \hat{A} &\mapsto \hat{U}\hat{A}\hat{U}^*. \end{aligned}$$

As a corollary of Theorem 3.2.1 we have:

Corollary 3.2.1 *Given a unital C^*-algebra $\mathcal{A}$ with associated spectral presheaf $\underline{\Sigma}^{\mathcal{A}}$, then there exists a group homomorphism*

$$\begin{aligned} \mathcal{U}(\mathcal{A}) &\to Aut(\underline{\Sigma}^{\mathcal{A}})^{\mathrm{op}} \\ \hat{U} &\mapsto \langle \Phi_{\hat{U}}, \mathcal{G}_{\hat{U}} \rangle = \mathcal{G}_{\hat{U}} \circ \Phi^*_{\hat{U}}. \end{aligned} \tag{3.2.3}$$

To understand how each automorphism $\langle \Phi_{\hat{U}}, \mathcal{G}_{\hat{U}} \rangle$ is constructed we recall that each inner automorphism $\phi_{\hat{U}}$ induced an order automorphism on $\mathcal{C}(\mathcal{A})$

$$\begin{aligned} \tilde{\phi}_{\hat{U}} : \mathcal{C}(\mathcal{A}) &\to \mathcal{C}(\mathcal{A}) \\ C &\mapsto \hat{U}C\hat{U}* \end{aligned}$$

with the associated essential geometric morphism $\Phi_{\hat{U}} : \mathbf{Sets}^{\mathcal{C}(\mathcal{A})^{\mathrm{op}}} \to \mathbf{Sets}^{\mathcal{C}(\mathcal{A})^{\mathrm{op}}}$, whose inverse image part is

$$\begin{aligned} \Phi^*_{\hat{U}} : \mathbf{Sets}^{\mathcal{C}(\mathcal{A})^{\mathrm{op}}} &\to \mathbf{Sets}^{\mathcal{C}(\mathcal{A})^{\mathrm{op}}} \\ \underline{\Sigma}^{\mathcal{A}} &\mapsto \Phi^*_{\hat{U}}(\underline{\Sigma}^{\mathcal{A}}). \end{aligned}$$

As described in Sect. 3.1, to each such essential geometric morphism there is associated a natural isomorphism $\mathcal{G}_{\hat{U}} : \Phi^*_{\hat{U}}(\underline{\Sigma}^{\mathcal{A}}) \rightarrow \underline{\Sigma}^{\mathcal{A}}$, such that for each $C \in \mathcal{C}(\mathcal{A})$ we have

$$\begin{aligned} \mathcal{G}_{\hat{U};C} : \Phi^*_{\hat{U}}(\underline{\Sigma}^{\mathcal{A}})_C &\rightarrow \underline{\Sigma}^{\mathcal{A}}_C \\ \lambda &\mapsto \lambda \circ \phi_{\hat{U}}|_C. \end{aligned}$$

Similarly, as in Sect. 3.1, we then obtain an automorphism of the spectral presheaf

$$\langle \Phi_{\hat{U}}, \mathcal{G}_{\hat{U}} \rangle := \mathcal{G}_{\hat{U}} \circ \Phi^*_{\hat{U}} : \underline{\Sigma}^{\mathcal{A}} \rightarrow \underline{\Sigma}^{\mathcal{A}}.$$

Clearly the inverse $\langle \Phi_{\hat{U}}, \mathcal{G}_{\hat{U}} \rangle^{-1} = \langle \Phi_{\hat{U}^*}, \mathcal{G}_{\hat{U}^*} \rangle$ is the automorphism induced by $\hat{U}^*$, while the identity element $\hat{1}$ induces the identity automorphism $\langle \Phi_{\hat{1}}, \mathcal{G}_{\hat{1}} \rangle = Id_{\underline{\Sigma}^{\mathcal{A}}}$.

It is important to note that the map (3.2.3), differently from the map (3.3.9), is not injective. This is because each element $\hat{U}$, which belongs to the centre $\mathcal{Z}(\mathcal{A})$ of $\mathcal{A}$, induces the identity automorphism $\langle Id, id \rangle$ on $\underline{\Sigma}^{\mathcal{A}}$. To fix this problem it was suggested in [13] to consider the quotient[6]

$$\mathcal{U}(\mathcal{A})_{proper} := \mathcal{U}(\mathcal{A})/\mathcal{U}(\mathcal{A})_0$$

where $\mathcal{U}(\mathcal{A})_0 = \mathcal{U}(\mathcal{A}) \cap \mathcal{Z}(\mathcal{A})$. Having done this we obtain:

Corollary 3.2.2 *Given a unital C^*-algebra $\mathcal{A}$ with associated spectral presheaf $\underline{\Sigma}^{\mathcal{A}}$, then there exists an injective group homomorphism*

$$\begin{aligned} \mathcal{U}(\mathcal{A})_{proper} &\rightarrow Aut(\underline{\Sigma}^{\mathcal{A}})^{\text{op}} \\ [\hat{U}] &\mapsto \langle \Phi_{\hat{U}}, \mathcal{G}_{\hat{U}} \rangle = \mathcal{G}_{\hat{U}} \circ \Phi^*_{\hat{U}} \end{aligned} \tag{3.2.4}$$

for $\hat{U} \in [\hat{U}]$.

Now that we have defined automorphisms on the spectral presheaf we would like to understand how these are defined in the context of topos quantum theory. Therefore, instead of considering general unital C^*-algebras, we will focus on von Neumann algebras and how, in that case, automorphisms on the spectral presheaf are defined. This will be the topic of the next section.

[6]As stated in [13], this quotient is unproblematic when considering single systems and their unitary evolution.

3.3 From C^*-Algebras to von Neumann Algebras

Most of the definitions and theorems developed in the previous sections can be easily adapted to the case of von Neumann algebra however, for the sake of completeness we will, nonetheless, report them here. The reader is also referred to [14].

In [26] Chapter 9 we introduced the topos analogue of the state-space. This was given by the spectral presheaf $\underline{\Sigma}^{\mathcal{N}}$ of a von Neumann algebra $\mathcal{N}$, which was defined as follows:

Definition 3.3.1 Given a von Neumann algebra $\mathcal{N}$ we denote the category of its abelian subalgebras (which share the unit element) by $\mathcal{V}(\mathcal{N})$. This is a poset ordered by inclusion, often referred to as the *context category* of $\mathcal{N}$. $\underline{\Sigma}^{\mathcal{N}} \in \mathbf{Sets}^{\mathcal{V}(\mathcal{N})^{\text{op}}}$ is the spectral presheaf associated with $\mathcal{N}$, which is defined on

- objects: for all $V \in \mathcal{V}(\mathcal{N})$, $\underline{\Sigma}^{\mathcal{N}}_V := \Sigma(V)$ is the Gelfand spectrum of V.
- Morphisms: for each morphism $i_{V'V} : V' \hookrightarrow V$ in $\mathcal{V}(\mathcal{N})$, the associated presheaf map is

$$\underline{\Sigma}^{\mathcal{N}}(i_{V'V}) : \underline{\Sigma}^{\mathcal{N}}_V \rightarrow: \underline{\Sigma}^{\mathcal{N}}_{V'} \tag{3.3.1}$$

$$\lambda \mapsto \lambda|_{V'}. \tag{3.3.2}$$

$\underline{\Sigma}^{\mathcal{N}}(i_{V'V})$ is surjective, continuous, open and closed with respect to the Gelfand topology.

We would now like to understand how maps between different state-spaces associated to different quantum systems are defined. This is essentially an adaptation of Definition 3.1.3, as applied to von Neumann algebras [14]. In particular we have:

Definition 3.3.2 Given two von Neumann algebras $\mathcal{N}$ and $\mathcal{M}$ with corresponding spectral presheaves $\underline{\Sigma}^{\mathcal{N}}$ and $\underline{\Sigma}^{\mathcal{M}}$ respectively, an isomorphism between $\underline{\Sigma}^{\mathcal{N}}$ and $\underline{\Sigma}^{\mathcal{M}}$ consists of a pair $\langle \Phi, \mathcal{G}_\phi \rangle$ where $\Phi : \mathbf{Sets}^{\mathcal{C}(\mathcal{A})} \rightarrow \mathbf{Sets}^{\mathcal{C}(\mathcal{B})}$ is the essential geometric isomorphism induced by the order-isomorphism $\phi : \mathcal{V}(\mathcal{N}) \rightarrow \mathcal{V}(\mathcal{M})$ (base map) and $\mathcal{G}_\phi : \Phi^*(\underline{\Sigma}^{\mathcal{M}}) \rightarrow \underline{\Sigma}^{\mathcal{N}}$ is a natural isomorphism, i.e. each component $(\mathcal{G}_\phi)_C : (\Phi^*(\underline{\Sigma}^{\mathcal{M}}))_C \rightarrow \underline{\Sigma}^{\mathcal{N}}_C$ is a homeomorphism for all $C \in \mathcal{V}(\mathcal{N})$. As in Lemma 3.1.1 the action of $\langle \Phi, \mathcal{G}_\phi \rangle$ is defined by

$$\mathcal{G}_\phi \circ \Phi^* : \underline{\Sigma}^{\mathcal{M}} \rightarrow \underline{\Sigma}^{\mathcal{N}}.$$

If $\mathcal{N} = \mathcal{M}$ then an isomorphism $\langle \Phi, \mathcal{G}_\phi \rangle : \underline{\Sigma}^{\mathcal{N}} \rightarrow \underline{\Sigma}^{\mathcal{N}}$ is called an automorphism. Similarly, as for the C^*-algebra case, the collection $Aut(\underline{\Sigma}^{\mathcal{N}})$ of automorphisms on $\underline{\Sigma}^{\mathcal{N}}$ forms a group.

We now would like to prove that there exists a bijective correspondence between order-isomorphisms $\phi : \mathcal{V}(\mathcal{N}) \rightarrow \mathcal{V}(\mathcal{M})$ and spectral presheaf isomorphisms $\langle \Phi, \mathcal{G}_\phi \rangle : \underline{\Sigma}^{\mathcal{N}} \rightarrow \underline{\Sigma}^{\mathcal{N}}$. This correspondence is given by the following theorem [14, Thm. 5.14]:

Theorem 3.3.1 *Consider two von Neumann algebras $\mathcal{N}$ and $\mathcal{M}$ without type I_2 summands, whose associated spectral presheaves are $\underline{\Sigma}^{\mathcal{N}}$ and $\underline{\Sigma}^{\mathcal{M}}$, respectively. Assume also that $\tilde{\mathcal{N}}$ and $\tilde{\mathcal{M}}$ are the Jordan algebras[7] of $\mathcal{N}$ and $\mathcal{M}$ respectively. Then to every isomorphism $\underline{\Sigma}^{\mathcal{N}} \to \underline{\Sigma}^{\mathcal{M}}$ there corresponds a unique Jordan *-isomorphism $T : \tilde{\mathcal{M}} \to \tilde{\mathcal{N}}$ and vice versa, to each Jordan *-isomorphism $T : \tilde{\mathcal{M}} \to \tilde{\mathcal{N}}$ there corresponds a unique isomorphisms $\langle \Phi, \mathcal{G}_\phi \rangle : \underline{\Sigma}^{\mathcal{N}} \to \underline{\Sigma}^{\mathcal{M}}$.*

Before being able to prove this theorem we need a few more results. First of all we need the analogue of Theorem 3.1.2 for von Neumann algebras. In particular:

Theorem 3.3.2 *Let $\mathcal{N}$ and $\mathcal{M}$ be unital C^*-algebras whose spectral presheaves are $\underline{\Sigma}^{\mathcal{N}}$ and $\underline{\Sigma}^{\mathcal{M}}$, respectively. There exists a bijective correspondence between isomorphisms $\langle \Phi, \mathcal{G}_\phi \rangle : \underline{\Sigma}^{\mathcal{M}} \to \underline{\Sigma}^{\mathcal{N}}$ and isomorphisms $T : \mathbb{C}(\mathcal{N}) \to \mathbb{C}(\mathcal{M})$ of the associated partial von Neumann algebras. This means that $\mathbb{C}(\mathcal{N}) \simeq \mathbb{C}(\mathcal{M})$ iff $\underline{\Sigma}^{\mathcal{N}} \simeq \underline{\Sigma}^{\mathcal{M}}$.*

The proof is a straightforward consequence of Theorem 3.1.2.

In Definition 3.1.5 we introduced the notion of a partial/piecewise C^*-algebra. This definition also applies for partial/piecewise von Neumann algebras. With respect to the latter, we obtain the following theorem [14]:

Theorem 3.3.3 *Given two von Neumann algebras $\mathcal{N}$ and $\mathcal{M}$ with associated partial von Neumann algebras $\tilde{\mathrm{N}}$ and $\tilde{\mathrm{M}}$ respectively, then*

$$Iso(\mathcal{P}(\mathcal{M}), \mathcal{P}(\mathcal{N})) \simeq Iso(\mathbb{C}(\mathcal{M}, \mathbb{C}(\mathcal{N}))$$

where $Iso(\mathcal{P}(\mathcal{M}), \mathcal{P}(\mathcal{N}))$ is the collection of all isomorphisms between the complete orthomodular lattices $\mathcal{P}(\mathcal{M})$ and $\mathcal{P}(\mathcal{N})$, while $Iso(\mathbb{C}(\mathcal{M}, \mathbb{C}(\mathcal{N})$ is the collection of all isomorphisms between the partial von Neumann algebras $\mathbb{C}(\mathcal{M})$ and $\mathbb{C}(\mathcal{N})$. Therefore $\mathbb{C}(\mathcal{M}) \simeq \mathbb{C}(\mathcal{N})$ iff $\mathcal{P}(\mathcal{M}) \simeq \mathcal{P}(\mathcal{N})$.

Proof In order to prove this theorem we will consider, on the one hand an isomorphism $T' : \mathbb{C}(\mathcal{M}) \to \mathbb{C}(\mathcal{N})$ and construct the associated isomorphism $T : \mathcal{P}(\mathcal{M}) \to \mathcal{P}(\mathcal{N})$, on the other hand we will consider an isomorphism $T : \mathcal{P}(\mathcal{M}) \to \mathcal{P}(\mathcal{N})$ and construct the corresponding isomorphism $T' : \mathbb{C}(\mathcal{M}) \to \mathbb{C}(\mathcal{N})$.

Consider now an isomorphism $T' : \mathbb{C}(\mathcal{M}) \to \mathbb{C}(\mathcal{N})$. Since it is an isomorphism it preserves multiplication and involution, therefore, given any $\hat{P} \in \mathcal{P}(\mathcal{M}) \subseteq \mathbb{C}(\mathcal{M})$ we obtain

$$T'(\hat{P}) = T'(\hat{P}^2) = T'(\hat{P})^2, \quad \text{and} \quad T'(\hat{P}) = T'(\hat{P}^*) = T'(\hat{P})^*.$$

Hence $T'(\hat{P}) \in \mathcal{P}(\mathcal{N})$ and $T := T'|_{\mathcal{P}(\mathcal{M})} : \mathcal{P}(\mathcal{M}) \to \mathcal{P}(\mathcal{N})$ is a well defined bijection. Next we need to show that T, as defined above, preserves the orthomodular

[7]Note that in this context the Jordan algebras $\tilde{\mathcal{N}}$and $\tilde{\mathcal{M}}$ associated to the von Neumann algebras $\mathcal{N}$ and $\mathcal{M}$ respectively, are JBW-algebras, hence, their ground field is $\mathbb{C}$.

lattice structure. In particular, we need to show that it preserves complements, all joins, all meets and that it preserves and reflects the order $\leq$.

To this end consider two elements $\hat{P}$ and $\hat{Q}$ in $\mathcal{P}(\mathcal{M})$, such that $\hat{P} \leq \hat{Q}$ (i.e. $\hat{P}\hat{Q} = \hat{P}$), then,

$$T'(\hat{P}\hat{Q}) = T'(\hat{P}) = T'(\hat{P})T'(\hat{Q}).$$

Therefore $T'(\hat{P}) \leq T'(\hat{Q})$. To show that T' reverses the order we consider its inverse T'^{-1}, which is also an isomorphism of partial von Neumann algebras. Then given $\hat{S}, \hat{T} \in \mathcal{P}(\mathcal{N})$, such that $\hat{S} \leq \hat{T}$, we have that

$$T'(\hat{s}\hat{T}) = T'(\hat{S}) = T'(\hat{R})T'(\hat{T}).$$

Hence $T'^{-1}(\hat{S}) \leq T'^{-1}(\hat{T})$ and T' reflects the order.

Next, consider a projection $\hat{P} \in \mathcal{P}(\mathcal{M})$ and its complement $\hat{1} - \hat{P}$. Applying T' we obtain

$$T'(\hat{1} - \hat{P}) = T'(\hat{1}) - T'(\hat{P}) = \hat{1} - T'(\hat{P}).$$

To show that T' preserves all joins we consider a family of projections $(\hat{P}_i)_{i\in I} \subseteq \mathcal{P}(\mathcal{M})$, not necessarily commuting. Applying the map T' we obtain that, for all $i \in I$,

$$T'(\hat{P}_i) \leq T'(\bigvee_{i\in I} \hat{P}_i) \tag{3.3.3}$$

which implies that

$$\bigvee_{i\in I} T'(\hat{P}_i) \leq T'(\bigvee_{i\in I} \hat{P}_i).$$

Since T'^{-1} is the inverse of T', it preserves and reverses the order, hence,

$$T'^{-1}(\bigvee_{i\in I} T'(\hat{P}_i)) \leq T'^{-1}(T'(\bigvee_{i\in I} \hat{P}_i)) = \bigvee_{i\in I} \hat{P}_i). \tag{3.3.4}$$

Moreover, for all $i \in I$ applying T'^{-1} to (3.3.3) we obtain

$$\hat{P}_i = T'^{-1}T'(\hat{P}_i) \leq T'^{-1}(\bigvee_{i\in I} T'\hat{P}_i)$$

therefore

$$\bigvee_{i\in I} \hat{P}_i \leq T'^{-1}(\bigvee_{i\in I} T'\hat{P}_i). \tag{3.3.5}$$

Combining (3.3.4) and (3.3.5) we obtain that

$$\bigvee_{i\in I}\hat{P}_i = T'^{-1}(\bigvee_{i\in I}T'\hat{P}_i)$$

which is equivalent to

$$T'(\bigvee_{i\in I}\hat{P}_i) = \bigvee_{i\in I}T'(\hat{P}_i).$$

To show that T' preserves all meets we proceed in a similar fashion. In particular, for all $i \in I$, we have that

$$T'(\bigwedge_{i\in I}\hat{P}_i) \leq T'(\hat{P}_i) \tag{3.3.6}$$

therefore

$$T'(\bigwedge_{i\in I}\hat{P}_i) \leq \bigwedge_{i\in I}T'(\hat{P}_i). \tag{3.3.7}$$

Using the properties of T'^{-1} we obtain that

$$T'^{-1}(T'(\bigwedge_{i\in I}\hat{P}_i) = \bigwedge_{i\in I}\hat{P}_i \leq T'^{-1}(\bigwedge_{i\in I}T'(\hat{P}_i)).$$

Moreover, for all $i \in I$, we have that

$$T'(\hat{P}_i) \geq \bigwedge_{i\in I}T'(\hat{P}_i)$$

applying T'^{-1} we obtain that

$$T'^{-1}T'(\hat{P}_i) = \hat{P}_i \geq T'^{-1}\bigwedge_{i\in I}T'^{-1}(\hat{P}_i).$$

Therefore

$$\bigwedge_{i\in I}\hat{P}_i \geq T'^{-1}\bigwedge_{i\in I}T'(\hat{P}_i)$$

which is equivalent to

$$\bigwedge_{i\in I}T'(\hat{P}) \leq T(\bigwedge_{i\in I}\hat{P}_i). \tag{3.3.8}$$

Combining (3.3.7) and (3.3.8) we obtain that

$$\bigwedge_{i\in I} T'(\hat{P}) = T(\bigwedge_{i\in I} \hat{P}_i).$$

Therefore $T := T'|_(\mathcal{P}(\mathcal{M}))$ defines an isomorphism from $\mathcal{P}(\mathcal{M})$ to $\mathcal{P}(\mathcal{N})$.

On the other hand, given an isomorphism $T : \mathcal{P}(\mathcal{M}) \to \mathcal{P}(\mathcal{N})$, we construct an isomorphism $T' : \mathbb{C}(\mathcal{M}) \to \mathbb{C}(\mathcal{N})$ in the following recursive way:

1. Consider self-adjoint operators, these are finite real-linear combinations of projections, therefore we define

$$T'(\hat{A}) = T'(\sum_{i=1}^{n} a_i\hat{P}_i) := \sum_{i=1}^{n} a_i T(\hat{P}_i)$$

   where $\hat{P}_i\hat{P}_j = \delta_{ij}$ and $\sum \hat{P}_i = \hat{1}$.
2. Arbitrary self-adjoint operators, on the other hand, can be approximate in norm by a family of real self-adjoint operators which are finite real-linear combinations of projections. Therefore we define

$$T'(\hat{A}) := \lim_{i\to\infty} T(\hat{A}_i)$$

   where the limit is taken in the norm topology.
3. Non-self-adjoint normal operators can be decomposed into a sum of self-adjoint operators, therefore we define

$$T(\hat{B}) := T'(\hat{A}_1) + i\hat{A}_2.$$

4. For the unit element $\hat{1}$ we define

$$T'(\hat{1}) := T(\hat{1}) = \hat{1}.$$

This shows that T' is a well defined map, but it remains to show that it is, indeed, an isomorphism, i.e. that it preserves the algebraic structure. The preservation of involution can be shown from the fact that $T(\hat{P}^*) := T'(\hat{P}^*) = (T'(\hat{P}))^* = (T(\hat{P}))^*$, together with the definitions (a)–(d) above. To show that T' preserves sums and multiplications we first consider self-adjoint operators which can be expressed as finite real-linear combinations of projections.

In particular, consider two self-adjoint operators $\hat{A}$ and $\hat{B}$ such that $\hat{A} = \sum_{i=1}^{n} a_i\hat{P}_i$ and $\hat{B} = \sum_{i=1}^{n} b_i\hat{P}_i$, then $\hat{A} + \hat{B} = \sum_{i=1}^{n}(a_i + b_i)\hat{P}_i$ and $\hat{A}\hat{B} =$

$\sum_{i=1}^{n} a_i b_i \hat{P}_i$. Therefore

$$\begin{aligned} T'(\hat{A}+\hat{B}) &= \sum_{i=1}^{n}(a_i+b_i)T(\hat{P}_i) \\ &= \sum_{i=1}^{n} a_i T(\hat{P}_i) + \sum_{i=1}^{n} b_i T(\hat{P}_i) \\ &= T'(\hat{A}) + T'(\hat{B}). \end{aligned}$$

For multiplication we have instead that

$$\begin{aligned} T'(\hat{A})T'(\hat{B}) &= \sum_{i=1}^{n} a_i T(\hat{P}_i) \sum_{j=1}^{n} b_j T(\hat{P}_j) \\ &= \sum_{i=1}^{n} a_i T(\hat{P}_i)(\sum_{i=1}^{n} b_j T(\hat{P}_j)) \\ &= \sum_{i=1}^{n} a_i \delta_{ij} b_j T(\hat{P}_j) \\ &= \sum_{i=1}^{n} a_i b_i T(\hat{P}_i) \end{aligned}$$

while

$$T'(\hat{A}\hat{B}) = \sum_{i=1}^{n} a_i b_i T(\hat{P}_i).$$

Therefore

$$T'(\hat{A}\hat{B}) = T'(\hat{A})T'(\hat{B})$$

Hence also multiplication is preserved. By continuity in the norm topology, the action of T can be extended to all self-adjoint operators and, by linearity, to all normal operators. □

We will now state a result proven in [25]

Theorem 3.3.4 *Given two von Neumann algebras $\mathcal{N}$ and $\mathcal{M}$ without type I_2 summands, there exists a bijective correspondence between isomorphisms $T : \mathcal{P}(\mathcal{M}) \rightarrow \mathcal{P}(\mathcal{N})$ of complete orthomodular lattices of projection operators and Jordan *-isomorphisms $\tilde{T} : \tilde{\mathcal{M}} \rightarrow \tilde{\mathcal{N}}$. This correspondence is given by*

$$\begin{aligned} &T \mapsto \tilde{T}, \quad s.t. \;\; \tilde{T}(\hat{P}) := T(\hat{P}) \;\; \forall\, \hat{P} \in \mathcal{P}(\mathcal{M}) \\ &\tilde{T} \mapsto T, \quad s.t. \;\; T := \tilde{T}|_{\mathcal{P}(\mathcal{M})}. \end{aligned}$$

The last result needed to prove Theorem 3.3.1 is the following [13]:

Theorem 3.3.5 *Given two von Neumann algebras $\mathcal{N}$ and $\mathcal{M}$ without type I_2 summands and not isomorphic to $\mathbb{C}^2$, there exists a bijective correspondence between order-isomorphisms $\tilde{\phi}: \mathcal{V}(\mathcal{M}) \rightarrow \mathcal{V}(\mathcal{N})$ and Jordan *-isomorphisms $T : \tilde{\mathcal{M}} \rightarrow \tilde{\mathcal{N}}$.*

The proof of this Theorem is given in Sect. 6.2.

We have now all the tools to prove Theorem 3.3.1.

Proof

$$\begin{aligned} Iso(\underline{\Sigma}^{\mathcal{N}}, \underline{\Sigma}^{\mathcal{N}}) &\overset{\text{Theorem 5.1.6}}{\simeq} Iso(\mathbb{C}(\mathcal{M}), \mathbb{C}(\mathcal{N})) \\ &\overset{\text{Theorem 3.3.3}}{\simeq} Iso(\mathcal{P}(\mathcal{M}), \mathcal{P}(\mathcal{N})) \\ &\overset{\text{Theorem 3.3.4}}{\simeq} Iso(\tilde{\mathcal{M}}, \tilde{\mathcal{N}}). \end{aligned}$$

□

As an immediate consequence of the above prove we also obtain the following isomorphisms

$$Iso(\underline{\Sigma}^{\mathcal{N}}, \underline{\Sigma}^{\mathcal{N}}) \simeq Iso(\tilde{\mathcal{M}}, \tilde{\mathcal{N}})$$
$$Iso(\mathbb{C}(\mathcal{M}), \mathbb{C}(\mathcal{N})) \simeq Iso(\tilde{\mathcal{M}}, \tilde{\mathcal{N}}).$$

If $\mathcal{M} \nsim \mathbb{C}^2$, then we also obtain

$$\begin{aligned} Iso(\mathcal{P}(\mathcal{M}), \mathcal{P}(\mathcal{N})) &\overset{\text{Theorem 3.3.4}}{\simeq} Iso(\tilde{\mathcal{M}}, \tilde{\mathcal{N}}) \\ &\overset{\text{Theorem 3.3.5}}{\simeq} Iso(\mathcal{V}(\mathcal{M}), \mathcal{V}(\mathcal{M}) \end{aligned}$$

where each $\tilde{\phi} \in Iso(\mathcal{V}(\mathcal{M}), \mathcal{V}(\mathcal{M})$ is the base map underlying an essential geometric morphism $\langle \Phi, \mathcal{G}_{\phi} \rangle$.

Having defined maps between state-spaces of different quantum systems, we can now apply this definition to maps between the state-space of the same quantum system. These maps will be then used to describe time evolution of the system. To define such maps we simply apply the definitions given in the previous section in terms of general C^*-algebra to the case of von Neumann algebras. All proves will then be a straightforward consequence of the proves given above. In particular, from Theorem 3.1.2 we obtain:

Corollary 3.3.1 *Given a von Neumann algebra $\mathcal{N}$, we have the following group isomorphism:*

$$Aut(\underline{\Sigma}^{\mathcal{N}})^{\text{op}} \xrightarrow{\simeq} Aut_{part}(\mathbb{C}(\mathcal{N})).$$

This together with Theorem 3.2.1 give us:

Theorem 3.3.6 *Given a von Neumann algebra $\mathcal{N}$, with associated spectral presheaf $\underline{\Sigma}^{\mathcal{N}}$, then there exists an injective group homomorphism*

$$Aut(\mathcal{N}) \to Aut(\underline{\Sigma}^{\mathcal{N}})^{\mathrm{op}} \tag{3.3.9}$$
$$\phi \mapsto \langle \Phi, \mathcal{G}_{\phi} \rangle = \mathcal{G}_{\phi} \circ \Phi^{*}.$$

Moreover, as a consequence of Theorem 3.3.3 we obtain

Corollary 3.3.2 *Given a von Neumann algebra $\mathcal{N}$, there exists a group isomorphism*

$$Aut(\mathcal{P}(\mathcal{N})) \simeq Aut_{part}(\mathbb{C}(\mathcal{N}))$$

where $\mathcal{P}(\mathcal{N})$ is the complete orthomodular lattice of projections in $\mathcal{N}$ and $\mathbb{C}(\mathcal{N})$ is the partial von Neumann algebra associated to $\mathcal{N}$.
On the other hand, as a consequence of Theorem 3.3.4 we obtain:

Corollary 3.3.3 *Given a von Neumann algebra $\mathcal{N}$ without summands of type I_2, there exists a group isomorphism*[8]

$$Aut(\mathcal{P}(\mathcal{N})) \simeq Aut(\tilde{\mathcal{N}})$$

where $\mathcal{P}(\mathcal{N})$ is the complete orthomodular lattice of projections in $\mathcal{N}$ and $\tilde{\mathcal{N}}$ is the Jordan algebra associated to $\mathcal{N}$.
Finally from Theorem 3.3.5 we obtain:

Corollary 3.3.4 *Given a von Neumann algebra $\mathcal{N}$ without type I_2 summands and not isomorphic to $\mathbb{C}^2$, then there exists a group isomorphism*

$$Aut_{ord}(\mathcal{V}(\mathcal{N})) \simeq Aut(\tilde{\mathcal{N}})$$

*between the group of order automorphisms $\phi : (\mathcal{V}(\mathcal{N}) \to (\mathcal{V}(\mathcal{N})$ and the group of Jordan *-automorphisms $T : \tilde{\mathcal{N}} \to \tilde{\mathcal{N}}$.*
The above discussion uncovers the fact that we effectively obtain six group isomorphisms as follows:

$$Aut_{ord}(\mathcal{V}(\mathcal{N}) \simeq Aut(\tilde{\mathcal{N}}) \simeq Aut(\mathcal{P}(\mathcal{N})) \simeq Aut_{part}(\mathbb{C}(\mathcal{N})). \tag{3.3.10}$$

[8]Here $Aut(\tilde{\mathcal{N}})$ represents the group of Jordan *-automorphisms associated to $\mathcal{N}$.

3.4 Time Evolution of Quantum Systems

In this section we will develop a way of defining time evolution of quantum systems in the context of topos quantum theory. This was put forwards in [13].

Given a von Neumann algebra $\mathcal{N}$, Stone's theorem tells us that any self-adjoint operator $\hat{A} \in \mathcal{N}$ induces a strongly continuous one-parameter group of unitary operators in $\mathcal{N}$, as a representation of $\mathbb{R}$ in $\mathcal{U}(\mathcal{N})$:

$$\begin{aligned} U : \mathbb{R} &\to \mathcal{U}(\mathcal{N}) \\ t &\mapsto U(t) := \hat{U}_t = e^{it\hat{A}} \end{aligned}$$

Each $\hat{U}_t$ induces an inner automorphism as follows:

$$\begin{aligned} \phi_{\hat{U}_t} : \mathcal{N} &\to \mathcal{N} \\ \hat{A} &\mapsto \hat{U}_t \hat{A} \hat{U}_t^*. \end{aligned}$$

We then obtain a representation of $\mathbb{R}$ in $Aut(\mathcal{N})$:

$$\begin{aligned} \mathbb{R} &\to Aut(\mathcal{N}) \\ t &\mapsto \phi_{\hat{U}_t}. \end{aligned}$$

Since each $\phi_{\hat{U}_t}$ (which we also denote by ϕ_t for simplicity) induces an automorphism on $Aut(\underline{\Sigma}^{\mathcal{N}})$, we effectively get a representation of $\mathbb{R}$ on $Aut(\underline{\Sigma}^{\mathcal{N}})$. Such a representation is called a *flow on the spectral presheaf*.

Definition 3.4.1 Given a von Neumann algebra $\mathcal{N}$, a flow on the spectral presheaf $\underline{\Sigma}^{\mathcal{N}}$ associated to an operator $\hat{A}$, is a representation

$$\begin{aligned} F_{\hat{A}} : \mathbb{R} &\to Aut(\underline{\Sigma}^{\mathcal{N}}) \\ t &\mapsto \langle \Phi_t, \mathcal{G}_{\phi_t} \rangle. \end{aligned}$$

These flows on the spectral presheaf will allow us to define time evolution in topos quantum theory. We recall that in canonical quantum theory time evolution is represented in two distinct, but related ways, depending on whether one chooses the Heisenberg picture or the Schrodinger picture. In particular, in the Heisenberg picture physical quantities change in time, while states remain fixed. On the other hand, in the Schrodinger picture physical quantities remain fixed while, states change in time. In [13] it was shown that both "types" of time evolution can be represented in topos quantum theory utilising the flows on the spectral presheaf.

3.4.1 Heisenberg Picture

In the Heisenberg picture, as mentioned above, time evolution is relegated to physical quantities while states remain fixed. In topos quantum theory, as explained in [26], physical quantities are represented as maps $\breve{\delta}(\hat{A}) : \underline{\Sigma}^{\mathcal{N}} \to \underline{\mathbb{R}^{\leftrightarrow}}$ from the state-space to the quantity value object. Mostly, when doing physics, we are interested in determining the values of physical quantities, hence we are interested in establishing the truthfulness or falsehood of propositions regarding values of physical quantities. Such propositions are generally of the form "$A \in \Delta$", which states that the quantity A takes values in the interval $\Delta \subseteq \mathbb{R}$. In topos quantum theory propositions of this type are identified with clopen sub-objects of the spectral presheaf [26]. Since such propositions pertain physical quantities they will undergo time evolution. It is precisely the time evolution of quantum propositions that we will analyse in this section. In topos quantum theory propositions are represented by clopen sub-objects of the spectral presheaf, therefore time evolution of propositions will be defined through the flow of the spectral presheaf as defined in Definition 3.4.1.

In what follows we will firstly define the action of the flow of the spectral presheaf on the bi-Heyting algebra $Sub_{cl}(\underline{\Sigma}^{\mathcal{N}})$ (see Chap. 2) of all clopen sub-objects of the spectral presheaf and then restrict it to only quantum propositions.

As a first step we need to show that automorphisms $\langle \Phi, \mathcal{G}_\phi \rangle$ of the spectral presheaf map clopen sub-objects to clopen sub-objects. Recall that:

$$\underline{\Sigma}^{\mathcal{N}} \xrightarrow{\Phi^*} \Phi^*(\underline{\Sigma}^{\mathcal{N}}) \xrightarrow{\mathcal{G}_\phi} \underline{\Sigma}^{\mathcal{N}}.$$

Given any clopen sub-object $\underline{S} \in Sub_{cl}\underline{\Sigma}^{\mathcal{N}}$ we want to show that $\langle \Phi, \mathcal{G}_\phi \rangle$ maps it to another clopen sub-object in $\underline{\Sigma}^{\mathcal{N}}$. In particular, for each $V \in \mathcal{V}(\mathcal{N})$ we have

$$(\Phi^*(\underline{S}))_V = \underline{S}_{\phi(V)} \subseteq \underline{\Sigma}^{\mathcal{N}}_{\phi(V)}.$$

If we then restrict the action of $\mathcal{G}_{\phi,V}$ to the sub-object $(\Phi^*(\underline{S}))_V$, we obtain

$$\mathcal{G}_{\phi,V}((\Phi^*(\underline{S}))_V) \subseteq \underline{\Sigma}^{\mathcal{N}}_V.$$

To show that the collection $\mathcal{G}_{\phi,V}((\Phi^*(\underline{S}))_V)$ for all $V \in \mathcal{V}(\mathcal{N})$ forms a sub-object of $\underline{\Sigma}^{\mathcal{N}}$, we need to show that given an inclusion maps $i_{V'V} : V' \hookrightarrow V$ then $\mathcal{G}_{\phi,V'}((\Phi^*(\underline{S}))_{V'}) \subseteq \mathcal{G}_{\phi,V}((\Phi^*(\underline{S}))_V)$. From the fact that $\underline{S} \in Sub_{cl}(\underline{\Sigma}^{\mathcal{N}})$ we have that

$$\mathcal{G}_{\phi,V'}((\Phi^*(\underline{S}))_{V'}) = \underline{S}_{\phi(V')} \subseteq \underline{S}_{\phi(V)}.$$

From the fact that $\mathcal{G}_\phi$ is a natural isomorphism we then have that

$$\mathcal{G}_{\phi,V'}((\Phi^*(\underline{S}))_{V'}) \subseteq \mathcal{G}_{\phi,V}((\Phi^*(\underline{S}))_V)$$

This shows that indeed the automorphisms $\langle \Phi, \mathcal{G}_\phi \rangle$ of the spectral presheaf map clopen sub-objects to clopen sub-objects. Hence each $\langle \Phi, \mathcal{G}_\phi \rangle$ induces a bijection

$$\begin{aligned} \overline{\phi} : Sub_{cl}(\underline{\Sigma}^{\mathcal{N}}) &\to Sub_{cl}(\underline{\Sigma}^{\mathcal{N}}) \\ \underline{S} &\mapsto \mathcal{G}_\phi(\Phi^*(\underline{S})). \end{aligned}$$

We now need to analyse whether the bi-Heyting structure of $Sub_{cl}(\underline{\Sigma}^{\mathcal{N}})$ is preserved by $\overline{\phi}$. In particular we need to show that given two sub-objects $\underline{S}_1, \underline{S}_2 \in Sub_{cl}(\underline{\Sigma}^{\mathcal{N}})$, such that $\underline{S}_1 \leq \underline{S}_2$, then $\overline{\phi}(\underline{S}_1) \leq \overline{\phi}(\underline{S}_2)$. Applying the various definitions we obtain that, for each $V \in \mathcal{V}(\mathcal{N})$,

$$\begin{aligned} \overline{\phi}(\underline{S}_1)_V &= \mathcal{G}_{\phi,V}((\Phi^*(\underline{S}_1))_V) \\ &= \mathcal{G}_{\phi,V}(\underline{S}_{1,\phi(V)}) \\ &\subseteq \mathcal{G}_{\phi,V}(\underline{S}_{2,\phi(V)}) \\ &= \overline{\phi}(\underline{S}_2)_V \end{aligned}$$

where the $\subseteq$ is given by the fact that $\mathcal{G}_{\phi,V}$ is a natural isomorphism and $\underline{S}_{1,\phi(V)} \subseteq \underline{S}_{2,\phi(V)}$.

The collection of automorphisms on the spectral presheaf forms a group $Aut(\underline{\Sigma}^{\mathcal{N}})$, hence the map $\overline{\phi}$ has an inverse $\overline{\phi}^{-1}$ which is order reversing. This implies that $\overline{\phi}$ preserves the bi-Heyting structure of $Sub_{cl}(\underline{\Sigma}^{\mathcal{N}})$.

In [13] the author decided to work with the inverse $\overline{\phi}^{-1}$ to define the action on clopen sub-objects. This was motivated by the fact that the automorphisms $\langle \Phi, \mathcal{G}_\phi \rangle$ on the spectral presheaf are interpreted as measurable functions while the collection of clopen sub-object $Sub_{cl}(\underline{\Sigma}^{\mathcal{N}})$ is interpreted as the algebra of measurable subsets. In terms of $\overline{\phi}^{-1}$ we obtain the following:

Lemma 3.4.1 *Given a von Neumann algebra $\mathcal{N}$ there exists an injective group homomorphism*

$$\begin{aligned} Aut(\underline{\Sigma}^{\mathcal{N}}) &\to Aut_{biHeyt}(Sub_{cl}(\underline{\Sigma}^{\mathcal{N}})^{\mathrm{op}} \\ \langle \Phi, \mathcal{G}_\phi \rangle &\mapsto \overline{\phi}^{-1} \end{aligned}$$

where

$$\begin{aligned} \overline{\phi}^{-1} : Sub_{cl}(\underline{\Sigma}^{\mathcal{N}}) &\to Sub_{cl}(\underline{\Sigma}^{\mathcal{N}}) \\ \underline{S} &\mapsto \mathcal{G}_\phi^{-1}((\Phi^{-1})^*(\underline{S})). \end{aligned} \tag{3.4.1}$$

In order to understand how $\overline{\phi}^{-1}$ is defined we consider the inverse of $\langle \Phi, \mathcal{G}_\phi \rangle$, which is $\langle \Phi, \mathcal{G}_\phi \rangle^{-1} = \langle \Phi^{-1}, \mathcal{G}_\phi^{-1} \rangle$, where $\Phi^{-1} : \mathbf{Sets}^{\mathcal{V}(\mathcal{N})} \to \mathbf{Sets}^{\mathcal{V}(\mathcal{N})}$ is the essential geometric automorphism induced by $\phi^{-1} : \mathcal{V}(\mathcal{N}) \to \mathcal{V}(\mathcal{N})$ and $\mathcal{G}_{\phi}^{-1} : (\Phi^{-1})^{*}(\underline{\Sigma}^{\mathcal{N}}) \to \underline{\Sigma}^{\mathcal{N}}$ is the natural isomorphism with components, for each $V \in \mathcal{V}(\mathcal{N})$, defined by

$$\begin{aligned} \mathcal{G}_{\phi,V}^{-1} : ((\Phi^{-1})^{*}(\underline{\Sigma}^{\mathcal{N}}))_V = \underline{\Sigma}^{\mathcal{N}}_{\phi^{-1}(V)} &\to \underline{\Sigma}^{\mathcal{N}} \\ \lambda &\mapsto \lambda \circ T^{-1}|_V. \end{aligned} \tag{3.4.2}$$

The map $T^{-1} : \tilde{\mathcal{N}} \to \tilde{\mathcal{N}}$ is the Jordan *-automorphism associated to ϕ^{-1} via Eq. (3.3.10), hence $T^{-1}|_V : V \to \phi^{-1}(V)$.

We have seen in Lemma 3.4.1 that the automorphism $\overline{\phi}^{-1} : Sub_{cl}(\underline{\Sigma}^{\mathcal{N}}) \to Sub_{cl}(\underline{\Sigma}^{\mathcal{N}})$ is defined on clopen sub-objects as $\overline{\phi}^{-1}(\underline{S}) = \mathcal{G}_{\phi}^{-1}((\Phi^{-1})^{*}(\underline{S}))$. Each clopen sub-object corresponds to a projection operator via the isomorphism defined for each $V \in \mathcal{V}(\mathcal{N})$ by

$$\begin{aligned} \mathfrak{S}_V : \mathcal{P}(\mathcal{V}(\mathcal{N})) &\to Sub_{cl}(\underline{\Sigma}^{\mathcal{N}})_V \\ \hat{P} &\mapsto \{\lambda \in \underline{\Sigma}^{\mathcal{N}}_V | \lambda(\hat{P}) = 1\} \end{aligned} \tag{3.4.3}$$

hence we are interested in analysing how $\overline{\phi}^{-1}$ "acts" on the corresponding projections. In particular it was show in [13] that the projection operators (one for each $V \in \mathcal{V}(\mathcal{N})$) corresponding to the clopen sub-object $\overline{\phi}^{-1}(\underline{S})$ are given by the projection operators corresponding to the sub-object $\underline{S}$, shifted by the Jordan *-automorphisms T which corresponds (contravariantly), via Eqs. (3.2.2) and (3.3.10), to the automorphism $\langle \Phi, \mathcal{G}_{\phi} \rangle$.

Lemma 3.4.2 *Consider a von Neumann algebra $\mathcal{N}$ with no type I_2 summands whose spectral presheaf if us$^{\mathcal{N}}$ and an automorphism $\langle \Phi, \mathcal{G}_{\phi} \rangle : \underline{\Sigma}^{\mathcal{N}} \to \underline{\Sigma}^{\mathcal{N}}$ with associated automorphism $\overline{\phi}^{-1} : Sub_{cl}(\underline{\Sigma}^{\mathcal{N}}) \to Sub_{cl}(\underline{\Sigma}^{\mathcal{N}})$. For each $\underline{S} \in Sub_{cl}(\underline{\Sigma}^{\mathcal{N}})$ we define, for all $V \in \mathcal{V}(\mathcal{N})$*

$$\hat{P}_{\overline{\phi}^{-1}(\underline{S})_V} := \mathfrak{S}_V^{-1}(\overline{\phi}^{-1}(\underline{S})_V);$$

$$\hat{P}_{\underline{S}_{\phi^{-1}(V)}} := \mathfrak{S}^{-1}_{\phi^{-1}(V)}(\underline{S}_{\phi^{-1}(V)}) = \mathfrak{S}^{-1}_{\phi^{-1}(V)}(((\Phi^{-1})^{*}(\underline{S}))_V).$$

Then, for all $V \in \mathcal{V}(\mathcal{N})$

$$\hat{P}_{\overline{\phi}^{-1}(\underline{S})_V} = T|_{\phi^{-1}(V)}(\hat{P}_{\underline{S}_{\phi^{-1}(V)}}). \tag{3.4.4}$$

Proof From Eq. (3.4.3), we know that, given any sub-object $\underline{S} \in Sub_{cl}(\underline{\Sigma}^{\mathcal{N}})$, then for each $V \in \mathcal{V}(\mathcal{N})$, $\lambda \in \underline{S}_V$ iff $\lambda(\hat{P}_{\underline{S}_V}) = 1$. Therefore we have

$$\begin{aligned} \lambda \in \underline{S}_{\phi^{-1}(V)} &\iff \lambda(\hat{P}_{\underline{S}_{\phi^{-1}(V)}}) = 1 \\ &\overset{(T^{-1}|_V \circ T|_{\phi^{-1}(V)} = id)}{\iff} \lambda \circ T^{-1}|_V \circ T|_{\phi^{-1}(V)})(\hat{P}_{\underline{S}_{\phi^{-1}(V)}}) = 1 \end{aligned}$$

$$\iff (\lambda \circ T^{-1}|_V)(T|_{\phi^{-1}(V)})(\hat{P}_{\underline{S}_{\phi^{-1}(V)}})) = 1$$

$$\overset{(3.4.2)}{\iff} \phi_V^{-1}(\lambda)(T|_{\phi^{-1}(V)})(\hat{P}_{\underline{S}_{\phi^{-1}(V)}})) = 1.$$

But

$$\lambda \in \underline{S}_{\phi^{-1}(V)} = ((\Phi^{-1})^*(\underline{S}))_V \iff \phi_V^{-1}(\lambda) \in \phi_V^{-1}(((\Phi^{-1})^*(\underline{S}))_V)$$

$$\overset{(3.4.1)}{\iff} \phi_V^{-1}(\lambda) \in (\overline{\phi}^{-1}(\underline{S})_V)$$

$$\iff \phi_V^{-1}(\lambda)(\hat{P}_{\overline{\phi}^{-1}(\underline{S})_V}) = 1.$$

This implies that

$$T|_{\phi^{-1}(V)}(\hat{P}_{\underline{S}_{\phi^{-1}(V)}}) = \hat{P}_{\overline{\phi}^{-1}(\underline{S})_V}.$$

□

We now consider Corollary 3.2.2. When applied to von Neumann algebras we obtain:

Corollary 3.4.1 *Given a von Neumann algebra $\mathcal{N}$ with associated spectral presheaf $\underline{\Sigma}^{\mathcal{N}}$, then there exists an injective group homomorphism*

$$\mathcal{U}(\mathcal{N})_{proper} \to Aut(\underline{\Sigma}^{\mathcal{N}})^{\text{op}} \tag{3.4.5}$$

$$[\hat{U}] \mapsto \langle \Phi_{\hat{U}}, \mathcal{G}_{\hat{U}} \rangle = \mathcal{G}_{\hat{U}} \circ \Phi_{\hat{U}}^*$$

for $\hat{U} \in [\hat{U}]$.

This Corollary, together with Lemma 3.4.1, proves the following:

Corollary 3.4.2 *Given a von Neumann algebra $\mathcal{N}$ with no type I_2 summand and not isomorphic to $\mathbb{C}^2$, then there is an injective group homomorphism*

$$\mathcal{U}(\mathcal{N})_{proper} \to Aut_{biHeyt}(Sub_{cl}(\underline{\Sigma}^{\mathcal{N}}))^{\text{op}}$$

$$[\hat{U}] \mapsto \overline{\phi_{\hat{U}}}^{-1} = \overline{\phi}_{\hat{U}*}$$

for $\hat{U} \in [\hat{U}]$, and

$$\overline{\phi}_{\hat{U}*} : Sub_{cl}(\underline{\Sigma}^{\mathcal{N}}) \to Sub_{cl}(\underline{\Sigma}^{\mathcal{N}})$$

$$\underline{S} \mapsto \mathcal{G}_{\hat{U}*}((\Phi_{\hat{U}*})^*(\underline{S}))$$

where $\overline{\phi}_{\hat{U}*}$ is the automorphism induced by $\phi_{\hat{U}}^{-1} = \phi_{\hat{U}*}$.

We would now like to apply Lemma 3.4.2 for the case in which the automorphism on the spectral presheaf is induced by an element $U \in \mathcal{U}(\mathcal{N})$. In particular, given a von Neumann algebra $\mathcal{N}$ with no type I_2 summand, to each $U \in \mathcal{U}(\mathcal{N})$ there corresponds an automorphism $\overline{\phi}_{\hat{U}*} : Sub_{cl}\underline{\Sigma}^{\mathcal{N}} \rightarrow Sub_{cl}\underline{\Sigma}^{\mathcal{N}}$ therefore, for each $V \in \mathcal{V}(\mathcal{N})$, the equivalent of Eq. (3.4.4) is

$$\hat{P}_{\overline{\phi}_{\hat{U}*}(\underline{S}_V)} = \hat{U}\hat{P}_{\underline{S}_{\hat{U}*V\hat{U}}}\hat{U}^* \tag{3.4.6}$$

where $\hat{P}_{\overline{\phi}_{\hat{U}*}(\underline{S}_V} = \mathfrak{S}_V^{-1}(\overline{\phi}_{\hat{U}*}(\underline{S})_V)$ and $\hat{P}_{\underline{S}_{\hat{U}*V\hat{U}}} = \mathfrak{S}^{-1}_{\hat{U}*V\hat{U}}(\underline{S}_{\hat{U}*V\hat{U}})$.

In order to understand why this is the case, we note that each element $U \in \mathcal{U}(\mathcal{N})$ induces an inner algebra automorphism $\phi_{\hat{U}} : \mathcal{N} \rightarrow \mathcal{N}; \hat{A} \mapsto \hat{U}\hat{A}\hat{U}^*$. Being an algebra automorphism it gives a Jordan *-automorphism, hence we can interpret $\phi_{\hat{U}}$ as a Jordan *-automorphism. Hence we obtain:

$$\hat{P}_{\overline{\phi}^{-1}(\underline{S})_V} \rightarrow \hat{P}_{\overline{\phi}_{\hat{U}*}(\underline{S}_V)}$$

$$T|_{\phi^{-1}(V)}(\hat{P}_{\underline{S}_{\phi^{-1}(V)}}) \rightarrow \phi_{\hat{U}}\hat{P}_{\underline{S}_{\phi_{\hat{U}*}}(V)} = \hat{U}\hat{P}_{\underline{S}_{\hat{U}*V\hat{U}}}\hat{U}^*.$$

Now that we have described how automorphisms on the bi-Hyeting algebra of clopen sub-objects act, we can extend the definition of flow from flows on the spectral presheaf to flows on the bi-Hyeting algebra $Sub_{cl}(\underline{\Sigma}^{\mathcal{N}})$ [13].

Definition 3.4.2 (General Flows) Given a flow $F : \mathbb{R} \rightarrow Aut(\underline{\Sigma}^{\mathcal{N}})$ on the spectral presheaf $\underline{\Sigma}^{\mathcal{N}}$ the associated flow $\overline{F}^{-1} : \mathbb{R} \rightarrow Aut(Sub_{cl}(\underline{\Sigma}^{\mathcal{N}}))$ on clopen sub-objects is defined as the one-parameter group $(\overline{F}^{-1}(t))_{t\in\mathbb{R}} : Sub_{cl}(\underline{\Sigma}^{\mathcal{N}}) \rightarrow Sub_{cl}(\underline{\Sigma}^{\mathcal{N}})$ such that, given any $t \in \mathbb{R}$, if $F(t) = \langle \Phi_t, \mathcal{G}_t \rangle$ then

$$\overline{F}^{-1}(t) := \overline{\phi}_t^{-1} : Sub_{cl}(\underline{\Sigma}^{\mathcal{N}}) \rightarrow Sub_{cl}(\underline{\Sigma}^{\mathcal{N}})$$

$$\underline{S} \mapsto \mathcal{G}_t^{-1}((\Phi_t^{-1})^*(\underline{S})).$$

Definition 3.4.3 (Flows Induced by Unitaries) Given a self-adjoint operator $\hat{A}$ affiliated with $\mathcal{N}$, this induces a one-parameter group of unitaries in $\mathcal{N}$, namely $(\hat{U}_t)_{t\in\mathbb{R}} = (e^{it\hat{A}})_{t\in\mathbb{R}}$. Utilising Definition 3.4.2, the flow $(\overline{F_{\hat{A}}}^{-1}(t))_{t\in\mathbb{R}}$ on $Sub_{cl}(\underline{\Sigma}^{\mathcal{N}})$ associated to this one parameter group of unitaries is defined for all $t \in \mathbb{R}$ by

$$\overline{F_{\hat{A}}}^{-1}(t) := \overline{\phi_{\hat{U}_t}}^{-1} : Sub_{cl}(\underline{\Sigma}^{\mathcal{N}}) \rightarrow Sub_{cl}(\underline{\Sigma}^{\mathcal{N}})$$

$$\underline{S} \mapsto \mathcal{G}_{\hat{U}_t^{-1}}((\Phi_{\hat{U}_t^{-1}})^*(\underline{S})).$$

Given an automorphism $\langle \Phi_t, \mathcal{G}_t \rangle$ its inverse $\langle \Phi_t, \mathcal{G}_t \rangle^{-1}$ is equivalent to the automorphism induced by $-t$, i.e. $\langle \Phi_t, \mathcal{G}_t \rangle^{-1} = \langle \Phi_t^{-1}, \mathcal{G}_t^{-1} = \rangle \langle \Phi_{-t}, \mathcal{G}_{-t} \rangle$. This implies that $\overline{\phi}_t^{-1} = \overline{\phi}_{-t}$, therefore for all $t \in \mathbb{R}$ we have

$$\overline{F}^{-1}(t) = \overline{\phi}_{-t} : Sub_{cl}(\underline{\Sigma}^{\mathcal{N}}) \rightarrow Sub_{cl}(\underline{\Sigma}^{\mathcal{N}}).$$

Similarly, if we are considering the flow induced by one-parameter group of unitaries, since $\hat{U}_t^{-1} = \hat{U}_{-t}$, then for all $t \in \mathbb{R}$, we have

$$\overline{F}_{\hat{A}}^{-1}(t) = \overline{\phi}_{\hat{U}_{-t}} : Sub_{cl}(\underline{\Sigma}^{\mathcal{N}}) \rightarrow Sub_{cl}(\underline{\Sigma}^{\mathcal{N}}).$$

The above discussion elucidates how flows on the bi-Heyting algebra $Sub_{cl}(\underline{\Sigma}^{\mathcal{N}})$ act on general clopen sub-objects $\underline{S} \in Sub_{cl}(\underline{\Sigma}^{\mathcal{N}})$. We now would like to restrict our attention to those clopen sub-objects which represent quantum propositions, i.e. those of the form $\underline{\delta}^o(\hat{P})$ which were defined in Definition 2.2.2. Since we are interested in defining time transformations we will label each proposition by a time label $t \in \mathbb{R}$. For example, the proposition $A \in \Delta; t$ signifies that at time t the quantity A has values in Δ. The projection corresponding to such a proposition would then be $\hat{P}_t$. The topos quantum theory analogue of such a proposition is the clopen sub-object $\underline{\delta}^o(\hat{P}_t)$ of $\underline{\Sigma}^{\mathcal{N}}$. We now would like to analyse how time transformations act on these time dependent propositions. To this end we first of all analyse how time transformations apply in standard quantum theory. As a convention we will denote the initial time by $t_0 = 0$. Thus, consider a proposition $A \in \Delta; 0$ at the initial time 0 which is represented by the projection operator $\hat{P}_0$. At a later time t such a proposition will have evolved to the proposition $A \in \Delta; t$. Such a time evolution in quantum theory is mathematically defined by a one-parameter group $(\hat{U}_t^*)_{t\in\mathbb{R}}$, such that the projection operator $\hat{P}_t$, representing the proposition $A \in \Delta; t$, is given by $\hat{P}_t := \hat{U}_{-t}\hat{P}_0\hat{U}_t$.

The topos theory analogue of the proposition $\hat{P}_0$ is given by the clopen sub-object $\underline{\delta}^o(\hat{P}_0)$, while the proposition associated to $\hat{P}_t := \hat{U}_{-t}\hat{P}_0\hat{U}_t$ is $\underline{\delta}^o\hat{P}_t := \hat{U}_{-t}\hat{P}_0\hat{U}_t$. It was shown in [13] that the action of the flow on $Sub_{cl}(\underline{\Sigma}^{\mathcal{N}})$, induced by the one parameter group[9] $(\hat{U}_{-t})_{t\in\mathbb{R}}$, transforms $\underline{\delta}^o(\hat{P}_0)$ into $\underline{\delta}^o(\hat{P}_t)$.

Theorem 3.4.1 ([13]) *Consider the clopen sub-objects* $\underline{\delta}^o(\hat{P}_0), \underline{\delta}^o(\hat{P}_t) \in Sub_{cl}(\underline{\Sigma}^{\mathcal{N}})$ *which represent the same proposition at different time, and the flow* $(\overline{F}_{\hat{H}}^{-1}(t))_{t\in\mathbb{R}}$ *on* $Sub_{cl}(\underline{\Sigma}^{\mathcal{N}})$ *induced by the one parameter group of unitaries* $(\hat{U}_{-t})_{t\in\mathbb{R}}$ *in* $\mathcal{N}$ *(such that* $\hat{U}_{-t} = e^{it\hat{A}}$ *for some operator* $\hat{A}$ *affiliated to* $\mathcal{N}$*), then, for all* $t \in \mathbb{R}$

$$\overline{F}_{\hat{H}}^{-1}(t)(\underline{\delta}^o(\hat{P}_0)) = \underline{\delta}^o(\hat{P}_t).$$

Proof From the correspondence between projection operators and clopen sub-objects of the spectral presheaf we have that for all $V \in \mathcal{V}(\mathcal{N})$

$$\mathfrak{S}_V^{-1}(\overline{F}_{\hat{H}}^{-1}(t)(\underline{\delta}^o(\hat{P}_0)))_V = \hat{P}_{\overline{F}_{\hat{H}}^{-1}(t)(\underline{\delta}^o(\hat{P}_0))_V} = \hat{P}_{\overline{\phi}_{\hat{U}_t}(\underline{\delta}^o(\hat{P}_0))_V} = \hat{U}_{-t}\hat{P}_{\underline{\delta}^o(\hat{P}_o)_{\hat{U}_t V \hat{U}_{-t}}}\hat{U}_t$$

[9] Similarly as done in [13, 14], from now on, we will define the one-parameter group by $(\hat{U}_{-t})_{t\in\mathbb{R}}$ rather than by $(\hat{U}_t)_{t\in\mathbb{R}}$. This is in accordance with the fact that, as seen above, in canonical quantum theory time evolution is given by conjugating by $\hat{U}_t^* = \hat{U}_{-t}$ instead of $\hat{U}_t$.

where the last equality follows from Eq. (3.4.6). If we now apply the definition of outer daseinisation [26][10] to $\hat{U}_{-t}\hat{P}_{\underline{\delta}^o(\hat{P}_o)_{\hat{U}_tV\hat{U}_{-t}}}$ we obtain for all $V \in \mathcal{V}(\mathcal{N})$ that:

$$\begin{aligned}
\hat{U}_{-t}\hat{P}_{\underline{\delta}^o(\hat{P}_o)_{\hat{U}_tV\hat{U}_{-t}}}\hat{U}_t &= \hat{U}_{-t}\bigwedge\{\hat{U}_t\hat{Q}\hat{U}_{-t} \in \mathcal{P}(\hat{U}_tV\hat{U}_{-t})|\hat{U}_t\hat{Q}\hat{U}_{-t} \geq \hat{P}_0\}\hat{U}_t \\
&= \bigwedge\{\hat{Q} \in \mathcal{P}(V)|\hat{U}_t\hat{Q}\hat{U}_{-t} \geq \hat{P}_0\} \\
&= \bigwedge\{\hat{Q} \in \mathcal{P}(V)|\hat{Q} \geq \hat{U}_{-t}\hat{P}_0\hat{U}_t\} \\
&= \delta^o(\hat{U}_{-t}\hat{P}_0\hat{U}_t)_V \\
&= \hat{P}_{\underline{\delta}^o(\hat{U}_{-t}\hat{P}_0\hat{U}_t)_V} \\
&= \hat{P}_{\underline{\delta}^o(\hat{P}_t)}.
\end{aligned}$$

Since the map $\mathfrak{S}_V$ is an isomorphism for each $V \in \mathcal{V}(\mathcal{N})$, it follows that

$$\overline{F}_{\hat{H}}^{-1}(t)(\underline{\delta}^o(\hat{P}_0)) = \underline{\delta}^o(\hat{P}_t).$$

□

Generalising the above discussion, given any clopen sub-object $\underline{S}_0 \in Sub_{cl}(\underline{\Sigma}^{\mathcal{N}})$, representing a proposition at time $t_0 = 0$, this gets mapped (in the Heisenberg picture) to a proposition at a later time by the flow $(\overline{F}^{-1}(t))_{t\in\mathbb{R}}$ on $Sub_{cl}(\underline{\Sigma}^{\mathcal{N}})$ as follows:

$$\underline{S}_t := \overline{F}^{-1}(t)(\underline{S}_0) = \overline{\phi}_{\hat{U}_t}(\underline{S}_0). \tag{3.4.7}$$

3.4.2 Schrodinger Picture

We will now analyse how time evolution can be implemented in topos quantum theory in terms of the Schrodinger picture, i.e. states evolve in time while operators stay fixed. In canonical quantum theory, states are identified with states of the von Neumann algebra $\mathcal{N}$ associated with the quantum system. In topos quantum theory, as explained in [26, Ch. 15], states are identifies with probability measures on $\underline{\Sigma}^{\mathcal{N}}$. The definition of a probability measure on $\underline{\Sigma}^{\mathcal{N}}$ is as follows:

Definition 3.4.4 A measure μ on the state-space $\underline{\Sigma}$ is a map

$$\mu : Sub_{cl}(\underline{\Sigma}) \to \Gamma\underline{[0,1]^{\succeq}} \tag{3.4.8}$$

$$\underline{S} = (\underline{S}_V)_{V\in\mathcal{V}(\mathcal{H})} \mapsto (\mu(\underline{S}_V))_{V\in\mathcal{V}(\mathcal{H})} := (\rho(\hat{P}_{\underline{S}_V}))_{V\in\mathcal{V}(\mathcal{N}}$$

[10]Recall that the definition of outer daseinisation is given by $\delta^o(\hat{P}) = \bigwedge\{\hat{Q} \in \mathcal{P}(V)|\hat{Q} \geq \hat{P}\}$.

such that, the following conditions holds:

1. $\mu(\underline{\Sigma}) = 1_{V(\mathcal{H})}$.
2. for all $\underline{S}$ and $\underline{T}$ in $Sub_{cl}(\underline{\Sigma})$ then $\mu(\underline{S} \vee \underline{T}) + \mu(\underline{S} \wedge \underline{T}) = \mu(\underline{S}) + \mu(\underline{T})$.

All operations of addition, meet and join are defined context-wise at each $V \in \mathcal{V}(\mathcal{N})$.
The fact that to each such measure there is associated a state ρ is given by the following theorem [26, Th. 15.2]:

Theorem 3.4.2 *Given a measure μ, as defined above, then there exist a unique state ρ "associated" to that measure.*

In [13] the author gave an alternative but equivalent definition of a probability measure on $\underline{\Sigma}^{\mathcal{N}}$. The reason being that such an alternative definition better lends itself to describe time evolution in terms of the action of one-parameter group of unitaries on the set of measures. In particular in [13] probability measures on $\underline{\Sigma}^{\mathcal{N}}$ are defined in terms of global sections of the presheaf of classical probabilities. This is defined as follows:

Definition 3.4.5 Given a von Neumann algebra $\mathcal{N}$ with associated spectral presheaf $\underline{\Sigma}^{\mathcal{N}}$, the presheaf $\underline{CP}$ of classical probability measures on $\underline{\Sigma}^{\mathcal{N}}$ is defined on:

- objects: for each $V \in \mathcal{V}(\mathcal{N})$

$$\underline{CP}_V := \{m_V : Sub_{cl}(\underline{\Sigma}^{\mathcal{N}})_V \to [0, 1]\}$$

  where $m_V(\underline{\Sigma}^{\mathcal{N}}_V) = 1$. Given $S_1, S_2 \in Sub_{cl}(\underline{\Sigma}^{\mathcal{N}})_V$, then $m_V(S_1 \cup S_2) + m_V(S_1 \cap S_2) = m_V(S_1) + m_V(S_2)$ therefore, m_V is a finitely additive probability measure.
- Morphisms: for each map $i_{V'V} : V' \hookrightarrow V$ the associated presheaf morphism is given in terms of pushforwards as follows[11]:

$$\begin{aligned} \underline{CP}(i_{V'V}) : \underline{CP}_{V'} &\to \underline{CP}_V \\ m_{V'} &\mapsto m_V \circ \underline{\Sigma}^{\mathcal{N}}(i_{V'V})^{-1}. \end{aligned}$$

We now will show the correspondence between global elements of $\underline{CP}$ and states on $\mathcal{N}$.

Theorem 3.4.3 ([13]) *Given a von Neumann algebra $\mathcal{N}$ with no type I_2 summands, then there exists the following isomorphism of convex sets:*

$$\Gamma \underline{CP} \simeq \mathcal{S}(\mathcal{N})$$

[11]Recall that, for each $V' \subseteq V$, the morphisms $\underline{\Sigma}^{\mathcal{N}}(i_{V'V})$ are continuous and hence measurable. This implies that the inverse $\underline{\Sigma}^{\mathcal{N}}(i_{V'V})^{-1}$ maps measurable sets to measurable sets.

where $\Gamma\underline{CP}$ is the set of global section of $\underline{CP}$ while $\mathcal{S}(\mathcal{N})$ is the set of states of $\mathcal{N}$.

Proof Given a state ρ on the von Neumann algebra $\mathcal{N}$, then for each $V \in \mathcal{V}(\mathcal{N})$ the map

$$\begin{aligned}\rho|_V : \mathcal{P}(V) &\to [0, 1]\\ \hat{P} &\mapsto \rho(\hat{P})\end{aligned}$$

is a finitely additive probability measure. In fact, for all $\hat{P}, \hat{Q} \in \mathcal{P}(V)$ such that $\hat{P}\hat{Q} = 0$, then

$$\rho|_V(\hat{P} + \hat{Q}) = \rho|_V(\hat{P}) + \rho|_V(\hat{Q})$$

while

$$\rho|_V(\hat{1}) = \hat{1}.$$

For each $V \in \mathcal{V}(\mathcal{N})$, $\rho|_V$ can be "extended" to a finitely additive probability measure on $\underline{\Sigma}^{\mathcal{N}}$ via the isomorphism $\mathfrak{S}_V : \mathcal{P}(V) \to Sub_{cl}(\underline{\Sigma}^{\mathcal{N}})_V; \hat{P} \mapsto \{\lambda \in \underline{\Sigma}^{\mathcal{N}}_V | \lambda(\hat{P}) = 1\}$ as follows:

$$\rho|_V \circ \mathfrak{S}_V^{-1} : Sub_{cl}(\underline{\Sigma}^{\mathcal{N}})_V \to [0, 1].$$

Therefore, for each $V \in \mathcal{V}(\mathcal{N})$ it is possible to assign a finitely additive probability measure $\rho|_V \circ \mathfrak{S}_V^{-1}$ on $\underline{\Sigma}^{\mathcal{N}}$. Clearly the family $(\rho|_V \circ \mathfrak{S}_V^{-1})_{V \in \mathcal{V}(\mathcal{N})}$ is a global element of $\underline{CP}$.

On the other hand, given a global element $m = (m_V)_{V \in \mathcal{V}(\mathcal{N})}$ of $\underline{CP}$, then we can define a finitely additive probability measure by

$$\begin{aligned}\mu : \mathcal{P}(\mathcal{N}) &\to [0, 1]\\ \hat{P} &\mapsto \mu(\hat{P}) := (m_V \circ \alpha_V)(\hat{P})\end{aligned}$$

where $\hat{P} \in V \subseteq \mathcal{N}$. Here $m_V \circ \alpha_V : \mathcal{P}(V) \to [0, 1]$ is a finitely additive probability measure constructed from composing the finitely additive probability measure $m_V : Sub_{cl}(\underline{\Sigma}^{\mathcal{N}})_V \to [0, 1]$ with the isomorphism α_V. We now need to check that μ is well defined, that is, given $V' \subseteq V$ then the measure $\mu|_{\mathcal{P}(V)}$ restricts to a measure on $\mathcal{P}(V')$. This is indeed the case since $(m_V)_{V \in \mathcal{V}(\mathcal{N})}$ is a global section of $\underline{CP}$ hence, for all $V', V \in \mathcal{N}$, such that $V' \subseteq V$, then

$$(m_V \circ \alpha_V)|_{\mathcal{P}(V')} = (m_{V'} \circ \alpha_{V'}).$$

Therefore, given a global section of $\underline{CP}$ we can construct a finitely additive probability measure μ on $\mathcal{P}(\mathcal{N})$. By Gleason's theorem each μ corresponds uniquely to a state ρ_μ of $\mathcal{N}$, such that $\rho_\mu|_{\mathcal{P}(\mathcal{N}} = \mu$. □

From the above isomorphism we have that, given a state ρ, the associated global element of $\underline{CP}$ will be $m_\rho = (m_{\rho,V})_{V\in\mathcal{V}(\mathcal{N})}$ where, for each $V \in \mathcal{V}(\mathcal{N})$, we have

$$m_{\rho,V} = \rho|_V \circ \mathfrak{S}_V^{-1} : Sub_{cl}(\underline{\Sigma}_V^{\mathcal{N}}) \to [0, 1]. \tag{3.4.9}$$

As mentioned at the beginning of this section, in [13] the author gives an alternative definition of measures on $\underline{\Sigma}^{\mathcal{N}}$ in terms of global section of the presheaf $\underline{CP}$. This definition is given by the following Lemma:

Lemma 3.4.3 *The convex set $\Gamma\underline{CP}$ of global elements of $\underline{CP}$ is isomorphic to the convex set $\mathcal{M}(\underline{\Sigma}^{\mathcal{N}})$ of probability measures on $\underline{\Sigma}^{\mathcal{N}}$ as defined in Definition 3.4.4.*

Proof To prove the above lemma we will construct two maps $\alpha : \Gamma\underline{CP} \to \mathcal{M}(\underline{\Sigma}^{\mathcal{N}})$ and $\beta : \mathcal{M}(\underline{\Sigma}^{\mathcal{N}}) \to \Gamma\underline{CP}$ and show that they are inverse of each other. We start by defining α as follows:

$$\alpha : \Gamma\underline{CP} \to \mathcal{M}(\underline{\Sigma}^{\mathcal{N}})$$
$$m = (m_V)_{V\in\mathcal{V}(\mathcal{N})} \mapsto \tilde{m}$$

where

$$\tilde{m} : Sub_{cl}(\underline{\Sigma}) \to \Gamma\underline{[0, 1]^{\geq}}$$
$$\underline{S} = (\underline{S}_V)_{V\in\mathcal{V}(\mathcal{H})} \mapsto (m_V(\underline{S}_V))_{V\in\mathcal{V}(\mathcal{H})}.$$

We now need to show that $\tilde{m}$, as defined above, is indeed a probability measure on $\underline{\Sigma}^{\mathcal{N}}$, i.e. we need to show that conditions 1–2 in Definition 3.4.4 hold. Clearly $\tilde{m}(\underline{\Sigma}^{\mathcal{N}}) = (m_V(\underline{\Sigma}_V^{\mathcal{N}}))_{V\in\mathcal{V}(\mathcal{H})} = 1_{\mathcal{V}(\mathcal{N})}$. Next, consider two sub-objects $\underline{S}_1, \underline{S}_2 \in Sub_{cl}(\underline{\Sigma}^{\mathcal{N}})$, then for all $V \in \mathcal{V}(\mathcal{N})$, we have

$$\begin{aligned}(\tilde{m}(\underline{S}_1) + \tilde{m}(\underline{S}_2))_V &= m_V(\underline{S}_{1;V}) + m_V(\underline{S}_{2;V}) \\ &= m_V(\underline{S}_{1;V} \cup \underline{S}_{2;V}) + m_V(\underline{S}_{2;V} \cap \underline{S}_{1;V}) \\ &= (\tilde{m}(\underline{S}_1 \vee \underline{S}_2)_V + \tilde{m}(\underline{S}_1 \wedge \underline{S}_2))_V.\end{aligned}$$

This shows that indeed $\tilde{m}$ is a probability measure on $\underline{\Sigma}^{\mathcal{N}}$. On the other hand we define

$$\beta : \mathcal{M}(\underline{\Sigma}^{\mathcal{N}}) \to \Gamma\underline{CP}$$
$$\mu \mapsto \tilde{\mu} := (\mu_V)_{V\in\mathcal{V}(\mathcal{N})}$$

where each μ_V is defined as follows: given a clopen subset $S \in Sub_{cl}(\underline{\Sigma}^{\mathcal{N}}_V)$, then there exists a clopen sub-object $\underline{S} \in Sub_{cl}(\underline{\Sigma}^{\mathcal{N}})$ such that $\underline{S}_V = S$, then

$$\begin{aligned}\mu_V : Sub_{cl}(\underline{\Sigma}^{\mathcal{N}}_V) &\to [0,1]\\ S &\mapsto \mu((\underline{S}))(V).\end{aligned}$$

Clearly, for each $V \in \mathcal{V}(\mathcal{N})$ then $\mu_V(\underline{\Sigma}^{\mathcal{N}}_V) = \mu(\underline{\Sigma}^{\mathcal{N}})(V) = 1$. Moreover, for any two disjoint subsets $S, T \in Sub_{cl}(\underline{\Sigma}^{\mathcal{N}}_V)$, there correspond clopen sub-objects $\underline{T}, \underline{S} \in Sub_{cl}(\underline{\Sigma}^{\mathcal{N}})$, such that $\underline{T}_V = T$ and $\underline{S}_V = S$. It then follows that, for all $V \in \mathcal{V}(\mathcal{N})$

$$\begin{aligned}\mu_V(S \cup T) &= \mu_V((\underline{S} \vee \underline{T})_V)\\ &= \mu(\underline{S} \vee \underline{T})(V)\\ &= (\mu(\underline{S}) + \mu(\underline{T}) - \mu(\underline{S} \wedge \underline{T}))(V)\\ &= \mu(\underline{S})(V) + \mu(\underline{T})(V)\\ &= \mu_V(S) + \mu_V(T).\end{aligned}$$

This shows that, indeed, μ_V is a finitely additive probability measure on $Sub_{cl}(\underline{\Sigma}^{\mathcal{N}}_V)$, therefore $\tilde{\mu} := (\mu_V)_{V \in \mathcal{V}(\mathcal{N})} \in \Gamma\underline{CP}$. By construction the two maps α and β are inverse of each other. □

We are now interested in defining time evolution in terms of the Schrodinger picture, that is, we want to define how probability measure on the state-space $\underline{\Sigma}^{\mathcal{N}}$ evolve in time. To this end we first of all recall how unitaries act on states in canonical quantum theory. In this context, given a normal state $\rho : \mathcal{N} \to \mathbb{C}$ with associated density matrix $\tilde{\rho}$, such that $\rho = tr(\tilde{\rho}-)$, then the action of a unitary $\hat{U}$ is given by

$$\hat{U} \cdot \rho = \hat{U}tr(\tilde{\rho}-) := tr(\hat{U}\tilde{\rho}\hat{U}^*-) = tr(\tilde{\rho}\hat{U}^* - \hat{U}) = \rho \circ \phi_{\hat{U}^*}$$

where $\phi_{\hat{U}^*} : \mathcal{N} \to \mathcal{N}$ is the automorphism induced by $\hat{U}^*$. In general, such a definition of the action of a unitary can be applied to any state ρ, not necessarily normal, obtaining $\hat{U} \cdot \rho := \rho \circ \phi_{\hat{U}^*}$.

In [13] it was shown that a similar definition applies to the action of unitaries on probability measures on the state-space $\underline{\Sigma}^{\mathcal{N}}$. In particular, since to each state ρ there is associated a probability measure m_ρ, then to the time evolved state $\hat{U} \cdot \rho$ there will be associated the time evolved probability measure $\hat{U} \cdot m_\rho = m_{\hat{U} \cdot \rho}$. Before proving this result, however, we need to define what $\hat{U} \cdot m_\rho$ actually is.

Definition 3.4.6 Given a state $\rho : \mathcal{N} \to \mathbb{C}$ with associated probability measure $m_\rho \in \Gamma\underline{CP}$, then the action of a unitary $\hat{U} \in \mathcal{U}(\mathcal{N})$ on m_ρ is defined, for each $V \in \mathcal{V}(\mathcal{N})$ by

$$(\hat{U} \cdot m_\rho)_V := m_{\rho, \hat{U}^* V \hat{U}} \circ \mathfrak{S}_{\hat{U}^* V \hat{U}} \circ \phi|_{\hat{U}^*} \circ \mathfrak{S}_V^{-1} : Sub_{cl}(\underline{\Sigma}^{\mathcal{N}}_V) \to [0,1]. \quad (3.4.10)$$

Given this definition we are now ready to show that the evolved state $\hat{U} \cdot \rho$ corresponds to the evolved probability measure $\hat{U} \cdot m_\rho = m_{\hat{U} \cdot \rho}$.

Lemma 3.4.4 *The state $\hat{U} \cdot \rho : \mathcal{N} \to \mathbb{C}$ corresponds to the probability measure $\hat{U} \cdot m_\rho = m_{\hat{U} \cdot \rho} = m_{\rho \circ \phi_{\hat{U}*}} \in \Gamma \underline{CP}$.*

Proof For all $V \in \mathcal{V}(\mathcal{N})$ we obtain

$$\begin{aligned}
(\hat{U} \cdot m_\rho)_V &\overset{(3.4.10)}{=} m_{\rho, \hat{U}^* V \hat{U}} \circ \mathfrak{S}_{\hat{U}^* V \hat{U}} \circ \phi|_{\hat{U}^*} \circ \mathfrak{S}_V^{-1} \\
&\overset{(3.4.9)}{=} \rho|_{\hat{U}^* V \hat{U}} \circ \mathfrak{S}^{-1}_{\hat{U}^* V \hat{U}} \circ \mathfrak{S}_{\hat{U}^* V \hat{U}} \circ \phi|_{\hat{U}^*} \circ \mathfrak{S}_V^{-1} \\
&= \rho|_{\hat{U}^* V \hat{U}} \circ \phi|_{\hat{U}^*} \circ \mathfrak{S}_V^{-1} \\
&= (\rho \circ \phi_{\hat{U}^*})_V \circ \mathfrak{S}_V^{-1} \\
&= m_{\hat{U} \cdot \rho, V}.
\end{aligned}$$

□

We can now define the action of the flow induced by one-parameter group of unitaries on the global sections of $\underline{CP}$.

Definition 3.4.7 Given a one-parameter group of unitaries $(\hat{U}_t)_{t \in \mathbb{R}}$ in $\mathcal{N}$ induced by some operator $\hat{A}$, then the flow on $\Gamma \underline{CP}$ induced by $(\hat{U}_t)_{t \in \mathbb{R}}$ is

$$\begin{aligned}
\overline{F}_{\hat{A}} : \mathbb{R} &\to Aut(\Gamma \underline{CP}) \\
t &\mapsto m_{\rho_t}
\end{aligned}$$

where $\rho_t = \hat{U}_t \rho_0 = \rho_0 \circ \phi_{\hat{U}_t^*} = \rho_0 \circ \phi_{\hat{U}_{-t}}$ represents the state ρ_0 at time t, and $m_{\rho_t} = \hat{U}_t \cdot m_{\rho_0} = m_{\hat{U}_t \cdot \rho_0}$ represents the corresponding probability measure on $\underline{\Sigma}^{\mathcal{N}}$. Since $\Gamma \underline{CP} \simeq \mathcal{M}(\underline{\Sigma}^{\mathcal{N}})$, then the flow $\overline{F}_{\hat{A}} : \mathbb{R} \to Aut(\Gamma \underline{CP})$ is equivalent to a flow $\overline{F}_{\hat{A}} : \mathbb{R} \to Aut(\mathcal{M}(\underline{\Sigma}^{\mathcal{N}}))$.

From the discussion in this section it is clear that given a state and a clopen sub-object of the state-space $\underline{\Sigma}^{\mathcal{N}}$ one can obtain an element of $\Gamma \underline{[0,1]^{\succeq}}$.

Definition 3.4.8 We define a state-proposition paring as the following map:

$$\begin{aligned}
p : \Gamma \underline{CP} \times Sub_{cl}(\underline{\Sigma}^{\mathcal{N}}) &\to \Gamma \underline{[0,1]^{\succeq}} \\
\langle m_\rho, \underline{S} \rangle &\mapsto m_\rho(\underline{S})
\end{aligned} \tag{3.4.11}$$

such that, for all $V \in \mathcal{V}(\mathcal{N})$, we have $m_\rho(\underline{S})_V := m_{\rho;V}(\underline{S}_V)$.
From Lemma 3.4.3 it is straightforward to see that for all $V \in \mathcal{V}(\mathcal{N})$ then $(m_\rho(\underline{S}))_V = \mu_{\rho,V}(\underline{S}_V) = (\mu_\rho(\underline{S}))_V$ hence

$$m_\rho(\underline{S}) = \mu_\rho(\underline{S}). \tag{3.4.12}$$

3.5 Relation Between the Heisenberg Picture and the Schrodinger Picture

In the previous section we defined time evolution in topos quantum theory both in terms of the Heisenberg picture, where propositions evolved in time while states were constant, and the Shrodineger picture, where states evolved in time while propositions remained constant. In this section we will investigate the relation between the two and check whether or not it resembles the relation in canonical quantum theory. In particular, in canonical quantum theory, the Heisenberg and the Schrodinger picture are completely equivalent, i.e. the two are physically indistinguishable. Mathematically this equivalence is expressed by the following equation:

$$\rho_t(\hat{P}_0) = \rho_0(\hat{P}_t) \tag{3.5.1}$$

where $\hat{P}_0$ represents the proposition "$A \in \Delta$" at time t_0. This equation indicates that the expectation value of the proposition "$A \in \Delta$" at time t_0, with respect to the evolved state ρ_t, is the same as the expectation value of the time evolved proposition "$A \in \Delta$ at time t (represented by $\hat{P}_t$) with respect to the 'initial' state ρ_0.

The fact that the above equation holds can be easily verified by applying the various definitions. In fact, since $\rho_t = \hat{U}_t \cdot \rho_0 = \rho_0 \circ \phi_{\hat{U}_{-t}}$ and $\hat{P}_t = \hat{U}_{-t}\hat{P}_0\hat{U}_t$ then

$$\rho_t(\hat{P}_0) = \rho_0 \circ \phi_{\hat{U}_{-t}}(\hat{P}_0) = \rho_0(\hat{U}_{-t}\hat{P}_0\hat{U}_t) = \rho_0(\hat{P}_t).$$

In [13] the author showed that it is possible to define the topos equivalent of Eq. (3.5.1) when defining time evolution in terms of flows on $Sub_{cl}(\underline{\Sigma}^{\mathcal{N}})$ (Heisenberg) and flows on $\Gamma\underline{CP}$ (Schrodinger). Since in topos quantum theory everything is defined context-wise, we need to express Eq. (3.5.1) for each context $V \in \mathcal{V}(\mathcal{N})$. To do so we first consider $\rho_t(\hat{P}_0)$. To express it context-wise we need to pick a context $V \in \mathcal{V}(\mathcal{N})$ which contains the proposition $\hat{P}_0$. Given such a V, then clearly the context $\hat{U}_{-t}V\hat{U}_t$ will contain the proposition $\hat{P}_t$. Hence the context dependent version of (3.5.1) is

$$\rho_t|_V(\hat{P}_0) = \rho_0|_{\hat{U}_{-t}V\hat{U}_t}(\hat{P}_t).$$

We then have the following result [13]:

Theorem 3.5.1 *Given a one-parameter group of unitaries $(\hat{U}_t)_{t\in\mathbb{R}}$ in $\mathcal{N}$, then for all $V \in \mathcal{V}(\mathcal{N})$ and $t \in \mathbb{R}$*

$$(m_{\rho_t}(\underline{S}_0))_V = m_{\rho_0}(\underline{S}_t))_{\hat{U}_{-t}V\hat{U}}$$

where $\underline{S}_0 \in Sub_{cl}(\underline{\Sigma}^{\mathcal{N}})$, $\underline{S}_t = \overline{\phi}_{\hat{U}_t}(\underline{S}_0)$, $\rho_0 \in \mathcal{S}(\mathcal{N})$ and $\rho_t = \hat{U}_t \cdot \rho_0 = \rho_0 \circ \phi_{\hat{U}_{-t}}$.

Proof We know from Definition 3.4.8 that $m_{\rho_t}(\underline{S}) \in \underline{\Gamma[0,1]^{\geq}}$, hence for all $V \in \mathcal{V}(\mathcal{N})$ we have

$$\begin{aligned}
(m_{\rho_t}(\underline{S}_0))_V &\overset{(3.4.11)}{=} m_{\rho;V}(\underline{S}_{0;V}) \\
&\overset{(3.4.4)}{=} (\hat{U}_t \cdot m_{\rho_0})_V(\underline{S}_{0,V}) \\
&\overset{(3.4.10)}{=} m_{\rho_0;\hat{U}_{-t}V\hat{U}_t} \circ \mathfrak{S}_{\hat{U}_{-t}V\hat{U}_t} \circ \phi_{\hat{U}_{-t}} \circ \mathfrak{S}_V^{-1}(\underline{S}_{0,V}) \\
&= m_{\rho_0;\hat{U}_{-t}V\hat{U}_t} \circ \mathfrak{S}_{\hat{U}_{-t}V\hat{U}_t}(\hat{U}_{-t}\hat{P}_{\underline{S}_{0,V}}\hat{U}_t) \\
&\overset{(3.4.6)}{=} m_{\rho_0;\hat{U}_{-t}V\hat{U}_t} \circ \mathfrak{S}_{\hat{U}_{-t}V\hat{U}_t}(\hat{P}_{\overline{\phi}_{\hat{U}_t}(\underline{S}_0)_{\hat{U}_{-t}V\hat{U}_t}}) \\
&= m_{\rho_0;\hat{U}_{-t}V\hat{U}_t}(\overline{\phi}_{\hat{U}_t}(\underline{S}_0)_{\hat{U}_{-t}V\hat{U}_t}) \\
&\overset{(3.4.7)}{=} m_{\rho_0;\hat{U}_{-t}V\hat{U}_t}(\underline{S}_{t;\hat{U}_{-t}V\hat{U}_t}) \\
&\overset{(3.4.11)}{=} (m_{\rho_0}(\underline{S}_t))_{\hat{U}_{-t}V\hat{U}}.
\end{aligned}$$

□

In canonical quantum theory the compatibility of the Heisenberg picture and the Schrodinger picture is given by the equation

$$\rho_t|_V(\hat{P}_0) = \rho_0|_{\hat{U}_{-t}V\hat{U}_t}(\hat{P}_t)$$

which is valid *only* for the context $V \in \mathcal{V}(\mathcal{N})$, such that $\hat{P}_0 \in V$. On the other hand, in topos quantum theory the compatibility between the Heisenberg picture and the Schrodinger picture is given by the equation

$$(m_{\rho_t}(\underline{S}_0))_V = (m_{\rho_0}(\underline{S}_t))_{\hat{U}_{-t}V\hat{U}}$$

which is valid for *all* contexts $V \in \mathcal{V}(\mathcal{N})$ simultaneously.

We have seen above that in standard quantum theory, given a proposition "$A \in \Delta$" represented by the projection operator $\hat{P}$, then, given a state $\rho \in \mathcal{S}(\mathcal{N})$, the probability of $\hat{P}$ being true given ρ is

$$Prob(\text{“}A \in \Delta\text{”}; \rho) = \rho(\hat{P}) \in [0, 1].$$

Clearly if $\rho(\hat{P}) = 0$ then the proposition is false given ρ, while if $\rho(\hat{P}) = 1$ the proposition is true given ρ. However for $0 < \rho(\hat{P}) < 1$ the proposition is neither true nor false.

We would now like to understand how the *Born rule* can be defined in topos quantum theory. To this end let us consider a general proposition $\underline{\delta}(\hat{P})$ and a state $\rho \in \mathcal{S}(\mathcal{N})$. We know that to each such state there corresponds a probability measure $\mu_\rho \in \mathcal{M}(\underline{\Sigma}^{\mathcal{N}})$ on the state-space $\underline{\Sigma}^{\mathcal{N}}$, such that $\mu_\rho(\underline{\delta}(\hat{P})) : \mathcal{V}(\mathcal{N}) \to [0, 1]$. From

Eq. (3.4.8) it then follows that for all $V \in \mathcal{V}(\mathcal{N})$ such that $\hat{P} \in V$, we have that $\mu_\rho(\underline{\delta}(\hat{P}))(V) = \rho(\hat{P})$. On the other hand for all $V \in \mathcal{V}(\mathcal{N})$ such that $\hat{P} \notin V$, then $\mu_\rho(\underline{\delta}(\hat{P}))(V) > \rho(\hat{P})$ since $\delta^o(\hat{P})_V > \hat{P}$. It follows that the expectation value $Prob(\text{“}A \in \Delta\text{”}; \rho) = \rho(\hat{P})$ is given by the minimum of $\mu_\rho(\underline{\delta}(\hat{P}))$, i.e.

$$Prob(\text{“}A \in \Delta\text{”}; \rho) = min_{v\in\mathcal{V}(\mathcal{N})}\mu_\rho(\underline{\delta}(\hat{P}))_V \overset{(3.4.12)}{=} min_{v\in\mathcal{V}(\mathcal{N})}m_\rho(\underline{\delta}(\hat{P}))_V.$$

From Theorem 3.5.1 it then follows that for all $V \in \mathcal{V}(\mathcal{N})$

$$min_{v\in\mathcal{V}(\mathcal{N})}m_{\rho_t}(\underline{\delta}(\hat{P}_0))_V = min_{v\in\mathcal{V}(\mathcal{N})}m_{\rho_0}(\underline{\delta}(\hat{P}_t))_V$$

therefore the expectation values in topos quantum theory coincide for the Heisenberg picture and the Schrodinger picture.

An alternative way of expressing the compatibility between the Heisenberg picture and the Schrodinger picture in canonical quantum theory is through what it is known as *covariance*. This property states that if one considers a state ρ_0 at time t_0 and a physical quantity A_0 at time t_0, any physical prediction obtained would be the same as if we had considered the state ρ_t at time t and a physical quantity A_t at time t. In particular, given a normal state ρ_o with associated density matrix $\tilde{\rho}_0$ then

$$\rho_0(\hat{A}_0) = tr(\tilde{\rho}\hat{A}_0) = tr(\hat{U}_t\tilde{\rho}\hat{U}_{-t}\hat{U}_t\hat{A}_0\hat{U}_{-t}) = tr(\tilde{\rho}_t\hat{A}_t) = \rho_t(\hat{A}_t).$$

In terms of projection operators which represent quantum propositions, the above covariance becomes

$$\rho_0(\hat{P}_0) = \rho_t(\hat{P}_{-t}). \tag{3.5.2}$$

We would now like to express such a covariance in topos quantum theory. As it was done above, we first need to express Eq. (3.5.2) context wise. In particular, given a context $V \in \mathcal{V}(\mathcal{N})$ such that $\hat{P}_0 \in V$, then $\hat{P}_{-t} = \hat{U}_t\hat{P}_0\hat{U}_{-t} \in \hat{U}_tV\hat{U}_{-t}$, therefore we obtain

$$\rho_0|_V(\hat{P}_0) = \rho_t|_{\hat{U}_tV\hat{U}_{-t}}(\hat{P}_{-t}) \tag{3.5.3}$$

Our aim is to find a topos quantum theory analogue of Eq. (3.5.3). This was done in [13].

Theorem 3.5.2 *Given a one-parameter group of unitaries $(\hat{U}_t)_{t\in\mathbb{R}}$ in $\mathcal{N}$ and a clopen sub-object $\underline{S}_0 \in Sub_{cl}(\underline{\Sigma}^{\mathcal{N}})$ representing a proposition $\underline{\delta}(\hat{P}_0)$, then for all $V \in \mathcal{V}(\mathcal{N})$*

$$(m_{\rho_o}(\underline{S}_0))_V = (m_{\rho_t}(\underline{S}_{-t}))_{\hat{U}_tV\hat{U}_{-t}}$$

where $\underline{S}_{-t} = \overline{\phi}_{\hat{U}_{-t}}(\underline{S}_0)$, $\rho_0 \in \mathcal{S}(\mathcal{N})$ *and* $\rho_t = \hat{U}_t \cdot \rho_0 = \rho_0 \circ \phi_{\hat{U}_{-t}}$.

Proof To prove this theorem we simply apply the various results obtained in this Chapter. In particular, for all $V \in \mathcal{V}(\mathcal{N})$ we have

$$
\begin{aligned}
(m_{\rho_t}(\underline{S}_{-t}))_{\hat{U}_t V \hat{U}_{-t}} &\overset{(3.4.11)}{=} m_{\rho_t;\hat{U}_t V \hat{U}_{-t}}(\underline{S}_{\hat{U}_t V \hat{U}_{-t}}) \\
&= m_{\rho_0 \circ \phi_{\hat{U}_{-t}};\hat{U}_t V \hat{U}_{-t}}(\underline{S}_{\hat{U}_t V \hat{U}_{-t}}) \\
&\overset{(3.4.9)}{=} (\rho_0 \circ \phi_{\hat{U}_{-t}})|_{\hat{U}_t V \hat{U}_{-t}} \circ \mathfrak{S}^{-1}_{\hat{U}_t V \hat{U}_{-t}}(\underline{S}_{\hat{U}_t V \hat{U}_{-t}}) \\
&= (\rho_0 \circ \phi_{\hat{U}_{-t}})|_{\hat{U}_t V \hat{U}_{-t}}(\hat{P}_{\underline{S}_{\hat{U}_t V \hat{U}_{-t}}}) \\
&= \rho_0(\hat{U}_{-t}\hat{P}_{\underline{S}_{\hat{U}_t V \hat{U}_{-t}}}\hat{U}_t) \\
&\overset{(3.4.6)}{=} \rho_0(\hat{P}_{\overline{\phi}_{\hat{U}_t}(\underline{S}_{-t})_V}) \\
&= \rho_0 \circ \mathfrak{S}^{-1}_V(\overline{\phi}_{\hat{U}_t}(\underline{S}_{-t})_V) \\
&= \rho_0 \circ \mathfrak{S}^{-1}_V(\underline{S}_{0,V}) \\
&= m_{\rho_0}(\underline{S}_0)_V.
\end{aligned}
$$

□

Chapter 4
Observables in Terms of Antonymous and Observable Functions

In this chapter we will introduce the notions of *antonymous* and *observable functions* put forward in [18]. These are utilised to express the physical quantise $\breve{\delta}(\hat{A})$ corresponding to the self-adjoint operators $\hat{A}$ in a more efficient way, which does not relay on calculating the approximations of $\breve{\delta}(\hat{A})$ for each context $V \in \mathcal{V}$ as it was done in [26].

4.1 Observables Functions and Antonymous Functions

In this section we will describe how quantum observables can be described using special types of functions called *antonymous functions*. This was first shown in [18]. A central ingredient in the definition of antonymous functions is that of a *maximal filter* on the lattice $P(\mathcal{N})$ of projection operators of a von Neumann algebra $\mathcal{N}$.

Definition 4.1.1 Given a lattice L with zero element 0, a *(proper) filter* (o *(proper) dual ideal*) consists of a subset $F \subseteq L$ such that:

1. $0 \notin F$.
2. If $a, b \in F$, then $a \wedge b \in F$.
3. If $a \in F$ and $b \geq a$, then $b \in F$.

The set of all (proper) filters of L is denoted by $\mathcal{F}(L)$.
A filter F is *maximal* if it is proper and there is no proper filter that is strictly greater. The set of maximal filters will be denoted by $\mathcal{Q}(L)$. If the lattice L is complemented and distributive, a maximal filter is called an *ultra filter* and it is such that, for each element $a \in L$, it either contains a or it contains its complement $\neg a$.

In the case at hand, given a von Neumann algebra $\mathcal{N}$, each abelian subalgebra $V \subseteq \mathcal{N}$ gives rise to a complemented distributive lattice $P(V)$, whose elements are projection operators in V. In [18] it was shown that, to each element $\lambda \in \underline{\Sigma}_V$ of

C. Flori, *A Second Course in Topos Quantum Theory*,
Lecture Notes in Physics 944, https://doi.org/10.1007/978-3-319-71108-9_4

the Gelfand spectrum of an abelian subalgebra $V \subseteq \mathcal{N}$, there is associated to it the (maximal) filter given by

$$F_\lambda := \{\hat{P} \in P(V) | \lambda(\hat{P}) = 1\}. \tag{4.1.1}$$

We will now show that, as defined in (4.1.1), F_λ is indeed a maximal filter. In particular, we will first show that F_λ satisfies conditions 1–3 in Definition 4.1.1 and then we will show that either $\hat{P}$ or its complement $\neg\hat{P}$ belong to $F(V)$.

Proof We recall that each multiplicative linear functional $\lambda \in \underline{\Sigma}_V$ is such that, for each $\hat{P} \in P(V)$, then $\lambda(\hat{P}) \in \{0, 1\}$. Clearly $\hat{0} \notin F_\lambda$. Next we need to show that if $\hat{P}, \hat{Q} \in F_\lambda$, then $\hat{P} \wedge \hat{Q} \in F_\lambda$. Assuming that indeed $\hat{P}, \hat{Q} \in F_\lambda$ we obtain

$$\begin{aligned} \lambda(\hat{P} \wedge \hat{Q}) &= \lambda(\hat{P}\hat{Q}) \\ &\overset{\text{multiplicative property}}{=} \lambda(\hat{P})\lambda(\hat{Q}) \\ &\overset{\hat{P},\hat{Q} \in F_\lambda}{=} 1 \end{aligned}$$

For condition 3 in Definition 4.1.1, let us assume that $\hat{P} \in F_\lambda$ and $\hat{Q} \geq \hat{P}$. Since $\lambda(\hat{P}) = 1$ and $\lambda(\hat{Q}) \in \{0, 1\}$ it follows that $\lambda(\hat{Q}) = 1$ and $\hat{Q} \in F_\lambda$.

So far we have proved that F_λ is a (proper) filter. What remains to be shown is that is it also a maximal filter. To this end we recall that, for each $\hat{P} \in P(V)$ its complement is given by $\hat{1} - \hat{P}$. We thus obtain

$$1 = \lambda(\hat{1}) = \lambda(\hat{P} + \hat{1} - \hat{P}) = \lambda(\hat{P}) + \lambda(\hat{1} - \hat{P}),$$

hence it follows that either $\lambda(\hat{P}) = 1$ and $\hat{P} \in F_\lambda$, or $\lambda(\hat{1} - \hat{P}) = 1$ and $(\hat{1} - \hat{P}) \in F_\lambda$. □

The above discussion uncovers the fact that to each $\lambda \in \underline{\Sigma}_V$ ($V \subseteq \mathcal{N}$) one assigns a maximal filter F_λ. Denoting the set of maximal filters on $P(V)$ by $\mathcal{Q}(V)$ we claim that the assignment

$$\beta : \underline{\Sigma}_V \to \mathcal{Q}(V) \tag{4.1.2}$$

$$\lambda \mapsto F_\lambda \tag{4.1.3}$$

is injective.

Proof We assume that $F_{\lambda_1} = F_{\lambda_2}$. This implies that, for all $\hat{P} \in P(V)$ such that $\lambda_1(\hat{P}) = 1$, then $\lambda_2(\hat{P}) = 1$, and vice versa. It then, trivially, follows that $\lambda_1 = \lambda_2$. □

Each filter F in $P(V)$ can be extended to a filter in $P(\mathcal{N})$ ($V \subseteq \mathcal{N}$) as follows:

$$C_{\mathcal{N}}(F) :=\uparrow F = \{\hat{Q} \in P(\mathcal{N}) | \exists \hat{P} \in F \text{ s.t. } \hat{P} \leq \hat{Q}\}. \tag{4.1.4}$$

This represents the smallest filter in $P(\mathcal{N})$ which contains F. An important Lemma that will be used in subsequent section is:

Lemma 4.1.1 ([18]) *Consider two von Neumann algebras $\mathcal{N}$, $\mathcal{M}$ such that $\mathcal{M} \subseteq \mathcal{N}$ and $1_{\mathcal{N}} = 1_{\mathcal{M}}$. Defining the map*

$$\begin{aligned} \delta^i_{\mathcal{M}} &: P(\mathcal{N}) \to P(\mathcal{N}) \\ \hat{P} &\mapsto \delta^i_{\mathcal{M}}(\hat{P}) := \bigvee\{\hat{Q} \in P(\mathcal{M}) | \hat{Q} \leq \hat{P}\}, \end{aligned}$$

then, for all $F \in F(\mathcal{M})$ we have that

$$(\delta^i_{\mathcal{M}})^{-1}(F) = C_{\mathcal{N}}(F).$$

Proof ([18]) Let us assume that $\hat{P} \in C_{\mathcal{N}}(F)$. This implies that there exists a $\hat{Q} \in F \subseteq P(\mathcal{N})$ such that $\hat{Q} \leq \hat{P}$. It then follows that $\delta^i_{\mathcal{M}}(\hat{P}) \geq \hat{Q}$. Since F is a filter, this means that $\delta^i_{\mathcal{M}}(\hat{P}) \in F$. Considering that for each $\hat{Q} \in F$, $(\delta^i_{\mathcal{M}})^{-1}(\hat{Q}) = \{\hat{P} \in P(\mathcal{N}) | \delta^i_{\mathcal{M}}(\hat{P}) = \hat{Q}\}$, we obtain that $C_{\mathcal{N}}(F) \subseteq (\delta^i_{\mathcal{M}})^{-1}(F)$.

On the other hand, assume that for $\hat{P} \in P(\mathcal{N})$, $\hat{P} \in (C_{\mathcal{N}}(F))^c$, i.e. there does not exist a $\hat{Q} \in F$ such that $\hat{Q} \leq \hat{P}$. Since $\delta^i_{\mathcal{M}}(\hat{P}) \leq \hat{P}$, then also $\delta^i_{\mathcal{M}}(\hat{P}) \in (C_{\mathcal{N}}(F))^c$, i.e. there does not exist a $\hat{Q} \in F$ such that $\hat{Q} \leq \delta^i_{\mathcal{M}}(\hat{P})$, therefore $\hat{P} \notin (\delta^i_{\mathcal{M}})^{-1}(F)$. This implies that $(C_{\mathcal{N}}(F))^c \subseteq (\delta^i_{\mathcal{M}})^{-1}(F))^c$ or, equivalently, $(\delta^i_{\mathcal{M}})^{-1}(F) \subseteq C_{\mathcal{N}}(F)$. □

We are now ready to define the notions of *antonymous* and *observable* functions.

Definition 4.1.2 Given a von Neumann algebra $\mathcal{N}$ and a self-adjoint operator $\hat{A} \in \mathcal{N}$ with spectral family $(\hat{E}^{\hat{A}}_r)_{r \in \mathbb{R}} = \hat{E}^{\hat{A}}$, then the *antonymous* function of $\hat{A}$ is defined as:

$$\begin{aligned} g_{\hat{A}} &: \mathcal{F}(\mathcal{N}) \to sp(\hat{A}) \\ F &\mapsto \sup\{r \in \mathbb{R} | \hat{1} - \hat{E}^{\hat{A}}_r \in F\}. \end{aligned}$$

The *observable* function of $\hat{A}$ is defined as:

$$\begin{aligned} f_{\hat{A}} &: \mathcal{F}(\mathcal{N}) \to sp(\hat{A}) \\ F &\mapsto \inf\{r \in \mathbb{R} | \hat{E}^{\hat{A}}_r \in F\}. \end{aligned}$$

Now that we have defined the notion of antonymous and observable functions we will explain how these are used to describe the Gelfand spectrum of inner and outer daseinised self-adjoint operators, respectively. To this end we need to recall a few facts about inner and outer daseinisation of self-adjoint operators. First of all we need to recall the notion of spectral order.

Definition 4.1.3 ([24, 26]) Consider two self-adjoint operators $\hat{A}, \hat{B} \in \mathcal{N}_{sa}$ in a von Neumann algebra $\mathcal{N}$[1] with spectral families given by $(\hat{E}_r^{\hat{A}})_{r\in\mathbb{R}}$ and $(\hat{E}_r^{\hat{B}})_{r\in\mathbb{R}}$, respectively. The *spectral order* is defined as follows:

$$\hat{A} \leq_s \hat{B} \qquad \text{iff} \qquad \forall\, r \in \mathbb{R} : \hat{E}_r^{\hat{A}} \geq \hat{E}_r^{\hat{B}}.$$

With respect to the *spectral order*, $\mathcal{N}_{sa}$ becomes a boundedly complete lattice.

Both inner and outer daseinisation of self-adjoint operators are defined in terms of the spectral order. In particular, given any self-adjoint operator $\hat{A} \in \mathcal{N}_{sa}$, then for each context[2] $V \in \mathcal{C}(\mathcal{N})$ outer daseinisation of $\hat{A}$ is given by

$$\delta^o(\hat{A})_V := \bigwedge \{\hat{B} \in V_{sa} | \hat{B} \geq_s \hat{A}\},$$

while inner daseinisation of $\hat{A}$ is given by

$$\delta^i(\hat{A})_V := \bigvee \{\hat{B} \in V_{sa} | \hat{B} \leq_s \hat{A}\}.$$

Clearly these two processes give rise to the following mappings:

$$\delta^o_V : \mathcal{N}_{sa} \to V_{sa} \tag{4.1.5}$$

$$\hat{A} \mapsto \delta^o(\hat{A})_V \tag{4.1.6}$$

and

$$\delta^i_V : \mathcal{N}_{sa} \to V_{sa} \tag{4.1.7}$$

$$\hat{A} \mapsto \delta^i(\hat{A})_V. \tag{4.1.8}$$

The spectral order defined above is also utilised to define the spectral families of inner and outer daseinised self-adjoint operators. In particular, for each context $V \in \mathcal{C}(\mathcal{N})$ we define [24, 26]

$$\forall r \in \mathbb{R} : \hat{E}_r^{\delta^i(\hat{A})_V} = \bigwedge_{s>r} \delta^o(\hat{E}_s^{\hat{A}})_V \tag{4.1.9}$$

and

$$\forall r \in \mathbb{R} : \hat{E}_r^{\delta^o(\hat{A})_V} = \delta^i(\hat{E}_r^{\hat{A}})_V. \tag{4.1.10}$$

[1] Here $\mathcal{N}_{sa}$ denotes the set of all self-adjoint operators in $\mathcal{N}$.

[2] Here $\mathcal{C}(\mathcal{N})$ indicates the category of abelian von Neumann subalgebras of $\mathcal{N}$.

In the last equation we utilised the process of inner daseinisation on projection operators. This process was extensively defined in [26, Sec 13.2] but, for the sake of completeness, we will briefly recall it below.

Definition 4.1.4 Given a projection operator $\hat{P}$, for each context $V \in C(\mathcal{N})$, inner daseinisation is defined as:

$$\delta^i(\hat{P})_V = \bigvee\{\hat{\alpha} \in P(V)|\hat{\alpha} \leq \hat{P}\}. \tag{4.1.11}$$

It follows that $\delta^i(\hat{P})_V$ is the best approximation in V of $\hat{P}$ obtained by taking the 'largest' projection operator in V, which implies $\hat{P}$. From the definition, given $V' \subseteq V$ then

$$\delta^i(\delta^i(\hat{P}_V))_{V'} = \delta^i(\hat{P})_{V'} \leq \delta^i(\hat{P})_V \text{ and } \delta^i(\hat{P})_V \leq \hat{P}. \tag{4.1.12}$$

Since $\hat{P} \leq \delta^o(\hat{P})_V$ we then have

$$\delta^i(\hat{P})_V \leq \delta^o(\hat{P})_V. \tag{4.1.13}$$

While outer daseinisation would pick the smallest projection operator implied by the original projection operators (hence approximation from above), inner daseinisation picks the biggest projection operators which implies the original one (hence approximation from below).

The interesting feature of the mappings defined in (4.1.5) and (4.1.7) is that they can be generalised also to cases in which the subalgebra $W \subseteq \mathcal{N}$ is not abelian, as long as the unit elements in W and $\mathcal{N}$ coincide. Therefore, given any subalgebra $W \subseteq \mathcal{N}$, we obtain the following mappings:

$$\begin{aligned} \delta^o_W : \mathcal{N}_{sa} &\to W_{sa} \\ \hat{A} &\mapsto \delta^o(\hat{A})_W = \bigwedge\{\hat{B} \in W_{sa}|\hat{B} \geq_s \hat{A}\} \end{aligned}$$

and

$$\begin{aligned} \delta^i_W : \mathcal{N}_{sa} &\to W_{sa} \\ \hat{A} &\mapsto \delta^i(\hat{A})_W = \bigvee\{\hat{B} \in W_{sa}|\hat{B} \leq_s \hat{A}\}. \end{aligned}$$

These generalisations allow us to prove the following theorems [18]:

Theorem 4.1.1 *Consider a self-adjoint operator $\hat{A} \in \mathcal{N}_{sa}$ and any subalgebra $\mathcal{M} \subseteq \mathcal{N}$ such that $1_{\mathcal{N}} = 1_{\mathcal{M}}$. For all filters $F \in \mathcal{F}(\mathcal{M})$ we have that*

$$g_{\delta^i(\hat{A})_{\mathcal{M}}}(F) = g_{\hat{A}}(C_{\mathcal{N}}(F)).$$

Proof Applying the definitions we obtain [18]

$$
\begin{aligned}
g_{\hat{A}}(C_{\mathcal{N}}(F)) &= \sup\{r \in \mathbb{R} | \hat{1} - \hat{E}_r^{\hat{A}} \in C_{\mathcal{N}}(F)\} \\
&\overset{\text{Lemma 4.1.1}}{=} \sup\{r \in \mathbb{R} | \hat{1} - \hat{E}_r^{\hat{A}} \in (\delta^i_{\mathcal{M}})^{-1}(F)\} \\
&= \sup\{r \in \mathbb{R} | \delta^i(\hat{1} - \hat{E}_r^{\hat{A}})_{\mathcal{M}} \in F\} \\
&= \sup\{r \in \mathbb{R} | \hat{1} - \delta^o(\hat{E}_r^{\hat{A}})_{\mathcal{M}} \in F\} \\
&= \sup\{r \in \mathbb{R} | \hat{1} - \bigwedge_{s>r} \delta^o(\hat{E}_s^{\hat{A}})_{\mathcal{M}} \in F\} \\
&\overset{(4.1.9)}{=} \sup\{r \in \mathbb{R} | \hat{1} - \hat{E}_r^{\delta^i(\hat{A})_{\mathcal{M}}} \in F\} \\
&= g_{\delta^i(\hat{A})_{\mathcal{M}}}(F).
\end{aligned}
$$

□

Theorem 4.1.2 *Consider a self-adjoint operator $\hat{A} \in \mathcal{N}_{sa}$ and any subalgebra $\mathcal{M} \subseteq \mathcal{N}$ such that $1_{\mathcal{N}} = 1_{\mathcal{M}}$. For all filters $F \in \mathcal{F}(\mathcal{M})$ we have that*

$$f_{\delta^o(\hat{A})_{\mathcal{M}}}(F) = f_{\hat{A}}(C_{\mathcal{N}}(F)).$$

Proof Applying the definitions we obtain [18]

$$
\begin{aligned}
f_{\delta^o(\hat{A})_{\mathcal{M}}}(F) &= \inf\{r \in \mathbb{R} | \hat{E}_r^{\delta^o(\hat{A})_{\mathcal{M}}} \in F\} \\
&\overset{(4.1.10)}{=} \inf\{r \in \mathbb{R} | \delta^i(\hat{E}_r^{\hat{A}})_{\mathcal{M}} \in F\} \\
&= \inf\{r \in \mathbb{R} | \hat{E}_r^{\hat{A}} \in (\delta^i_{\mathcal{M}})^{-1} F\} \\
&\overset{\text{Lemma 4.1.1}}{=} f_{\hat{A}}(C_{\mathcal{N}}(F)).
\end{aligned}
$$

□

The above two theorems are very important, since they show that both the antonymous and observable functions for daseinised self-adjoint operators can be completely derived from the antonymous and observable functions of the original self-adjoint operators. In particular, Theorem 4.1.1 tells us that, given a self-adjoint operator $\hat{A}$, then the antonymous functions $g_{\delta^i(\hat{A})_{\mathcal{M}}} : \mathcal{F}(\mathcal{M}) \to sp(\delta^i(\hat{A})_{\mathcal{M}})$ of the approximated self-adjoint operator $\delta^i(\hat{A})_{\mathcal{M}}$, can be completely derived from the antonymous functions $g_{\hat{A}} : \mathcal{F}(\mathcal{N}) \to sp(\hat{A})$ of the (original) self-operator $\hat{A}$.

Similarly, Theorem 4.1.2 tells us that, given a self-adjoint operator $\hat{A}$, then the observable functions $f_{\delta^o(\hat{A})_{\mathcal{M}}} : \mathcal{F}(\mathcal{M}) \to sp(\delta^o(\hat{A})_{\mathcal{M}})$ of the approximated self-adjoint operator $\delta^o(\hat{A})_{\mathcal{M}}$ can be completely derived from the antonymous functions

$f_{\hat{A}} : \mathcal{F}(\mathcal{N}) \rightarrow sp(\hat{A})$ of the (original) self-operator $\hat{A}$. When $\mathcal{M}$ is an abelian subalgebra of $\mathcal{N}$ we obtain the following important results [18]:

Corollary 4.1.1 *Consider a von Neumann subalgebra $V \subseteq \mathcal{N}$. We know that there is an injective map* (4.1.2) $\beta : \underline{\Sigma}_V \rightarrow \mathcal{Q}(V)$; $\lambda \mapsto F_\lambda$ *which assigns to each element of the spectrum of V a maximal filter F_λ. We can then identify the Gelfand transform* $\overline{\delta^i(\hat{A})_V} : \underline{\Sigma}_V \rightarrow sp(\delta^i(\hat{A})_V)$; $\lambda \mapsto \overline{\delta^i(\hat{A})_V} = \lambda(\delta^i(\hat{A})_V)$ *with* $g_{\delta^i(\hat{A})_{\mathcal{M}}}|_{\mathcal{Q}(V)}$, *obtaining*

$$\lambda(\delta^i(\hat{A})_V) = g_{\delta^i(\hat{A})_{\mathcal{M}}}(F_\lambda) = g_{\hat{A}}(C_{\mathcal{N}}(F_\lambda)) . \tag{4.1.14}$$

Proof Upon identifying $\overline{\delta^i(\hat{A})_V} : \underline{\Sigma}_V \rightarrow sp(\delta^i(\hat{A})_V)$; $\lambda \mapsto \overline{\delta^i(\hat{A})_V} = \lambda(\delta^i(\hat{A})_V)$ with $g_{\delta^i(\hat{A})_{\mathcal{M}}}|_{\mathcal{Q}(V)}$, Eq. (4.1.14) is a direct consequence of Theorem 4.1.1. □

A similar result holds for observable functions.

Corollary 4.1.2 *Consider a von Neumann subalgebra $V \subseteq \mathcal{N}$. We know that there is an injective map* (4.1.2) $\beta : \underline{\Sigma}_V \rightarrow \mathcal{Q}(V)$; $\lambda \mapsto F_\lambda$ *which assigns to each element of the spectrum of V a maximal filter F_λ. We can then identify the Gelfand transform* $\overline{\delta^o(\hat{A})_V} : \underline{\Sigma}_V \rightarrow sp(\delta^o(\hat{A})_V)$; $\lambda \mapsto \overline{\delta^o(\hat{A})_V} = \lambda(\delta^o(\hat{A})_V)$ *with* $f_{\delta^o(\hat{A})_{\mathcal{M}}}|_{\mathcal{Q}(V)}$, *obtaining*

$$\lambda(\delta^o(\hat{A})_V) = f_{\delta^o(\hat{A})_{\mathcal{M}}}(F_\lambda) = f_{\hat{A}}(C_{\mathcal{N}}(F_\lambda)) \tag{4.1.15}$$

Proof Upon identifying $\overline{\delta^o(\hat{A})_V} : \underline{\Sigma}_V \rightarrow sp(\delta^o(\hat{A})_V)$; $\lambda \mapsto \overline{\delta^o(\hat{A})_V} = \lambda(\delta^o(\hat{A})_V)$ with $g_{\delta^o(\hat{A})_{\mathcal{M}}}|_{\mathcal{Q}(V)}$, Eq. (4.1.15) is a direct consequence of Theorem 4.1.2. □

We recall from [26, Section 13.5] that physical quantities in topos quantum theory are represented as maps from the state-space to the quantity value object. In particular, a physical quantity A with associated self-adjoint operator $\hat{A}$ is represented by the map

$$\breve{\delta}(\hat{A}) : \underline{\Sigma} \rightarrow \underline{\mathbb{R}^{\leftrightarrow}} \tag{4.1.16}$$

which, at each context V, is defined as

$$\begin{aligned} \breve{\delta}(\hat{A})_V : \underline{\Sigma}_V &\rightarrow \underline{\mathbb{R}^{\leftrightarrow}}_V \\ \lambda &\mapsto \breve{\delta}(\hat{A})_V(\lambda) := \left(\breve{\delta}^i(\hat{A})_V(\lambda), \breve{\delta}^o(\hat{A})_V(\lambda)\right), \end{aligned} \tag{4.1.17}$$

where $\breve{\delta}^o(\hat{A})_V$ is the order reversing function defined by

$$\breve{\delta}^o(\hat{A})_V(\lambda) : \downarrow V \rightarrow sp(\hat{A}) \tag{4.1.18}$$

such that

$$\begin{aligned}\big(\breve{\delta}^{o}(\hat{A})_V(\lambda)\big)(V') &:= \overline{\delta^{o}(\hat{A})_{V'}}(\underline{\Sigma}(i_{V'V})(\lambda)) \\ &= \overline{\delta^{o}(\hat{A})_{V'}}(\lambda_{|V'}) \\ &= \langle \lambda_{|V'}, \delta^{o}(\hat{A})_{V'} \rangle \\ &= \langle \lambda, \delta^{o}(\hat{A})_{V'} \rangle \\ &= \lambda(\delta^{o}(\hat{A})_{V'}) .\end{aligned} \tag{4.1.19}$$

Here $\underline{\Sigma}(i_{V'V})$ are the spectral presheaf maps defined in (3.3.1) and $\overline{\delta^{o}(\hat{A})_V} : \underline{\Sigma}_V \to \mathbb{R}$ represents the Gelfand transform of $\delta^{o}(\hat{A})_V$.

The choice of order reversing functions was determined by the fact that, for all $V' \subseteq V$, since $\delta^{o}(\hat{A})_{V'} \geq \delta^{o}(\hat{A})_V$, then

$$\overline{\delta^{o}(\hat{A})_{V'}}(\lambda_{|V'}) = \overline{\delta^{o}(\hat{A})_{V'}}(\underline{\Sigma}(i_{V'V})(\lambda)) \geq \overline{\delta^{o}(\hat{A})_V}(\lambda) . \tag{4.1.20}$$

On the other hand, the order preserving function is defined by

$$\breve{\delta}^{i}(\hat{A})_V(\lambda) : \downarrow V \to sp(\hat{A}) \tag{4.1.21}$$

such that

$$\begin{aligned}\big(\breve{\delta}^{i}(\hat{A})_V(\lambda)\big)(V') &:= \overline{\delta^{i}(\hat{A})_{V'}}(\underline{\Sigma}(i_{V'V})(\lambda)) \\ &= \overline{\delta^{i}(\hat{A})_{V'}}(\lambda_{|V'}) \\ &= \langle \lambda_{|V'}, \delta^{i}(\hat{A})_{V'} \rangle \\ &= \langle \lambda, \delta^{i}(\hat{A})_{V'} \rangle \\ &= \lambda(\delta^{i}(\hat{A})_{V'}) .\end{aligned} \tag{4.1.22}$$

In this case the appropriate Gelfand transform to use is $\overline{\delta^{i}(\hat{A})_{V'}} : \underline{\Sigma}_V \to \mathbb{R}$. The choice of order preserving function was determined by the fact that, for $i : V' \subseteq V$, since $\delta^{i}(\hat{A})_{V'} \leq \delta^{i}(\hat{A})_V$, then

$$\overline{\delta^{i}(\hat{A})_{V'}}(\lambda_{|V'}) = \overline{\delta^{i}(\hat{A})_{V'}}(\underline{\Sigma}(i_{V'V})(\lambda)) \leq \overline{\delta^{i}(\hat{A})_V}(\lambda) . \tag{4.1.23}$$

Given Eqs. (4.1.14) and (4.1.15) the above order preserving and order reversing functions can be written as

$$\big(\breve{\delta}^{i}(\hat{A})_V(\lambda)\big)(V') = g_{\hat{A}}(C_{\mathcal{N}}(F_\lambda))$$

and

$$\left(\breve{\delta}^o(\hat{A})_V(\lambda)\right)(V') = f_{\hat{A}}(C_{\mathcal{N}}(F_\lambda)),$$

respectively.

4.2 Example

We will now give an example on how to use the antonymous and observable functions to compute the value of an observable, given a state.

Example 4.2.1 Let us consider a 2 spin system in $\mathbb{C}^4$ whose algebra of bounded operators is given by $\mathcal{B}(\mathbb{C}^4)$. The self-adjoint operator representing the spin in the z direction is given by

$$\hat{S}_z = \begin{pmatrix} 2 & 0 & 0 & 0 \\ 0 & 0 & 0 & 0 \\ 0 & 0 & 0 & 0 \\ 0 & 0 & 0 & -2 \end{pmatrix}$$

while the spectral family of $\hat{S}_z$ is

$$\hat{E}_\lambda^{\hat{S}_z} = \begin{cases} \hat{0} \text{ if } \lambda < -2 \\ \hat{P}_4 \text{ if } -2 \leq \lambda < 0 \\ \hat{P}_4 + \hat{P}_3 + \hat{P}_2 \text{ if } 0 \leq \lambda < 2 \\ \hat{P}_4 + \hat{P}_3 + \hat{P}_2 + \hat{P}_1 \text{ if } 2 \leq \lambda\,. \end{cases} \tag{4.2.1}$$

The only maximal abelian subalgebra containing $\hat{S}_z$ is $V = lin_{\mathbb{C}}(\hat{P}_1, \hat{P}_2, \hat{P}_3, \hat{P}_4)$ where $\hat{P}_1 = (1,0,0,0)$, $\hat{P}_2 = (0,1,0,0)$, $\hat{P}_3 = (0,0,1,0)$, $\hat{P}_4 = (0,0,0,1)$. The Gelfand spectrum of V is $\underline{\Sigma}_V = \{\lambda_1, \lambda_2, \lambda_3, \lambda_4\}$, where $\lambda_i(\hat{P}_j) = \delta_{ij}$.

Next, let us consider the subalgebra $V_{\hat{P}_1,\hat{P}_2} = lin_{\mathbb{C}}(\hat{P}_1, \hat{P}_2, \hat{P}_3 + \hat{P}_4)$ whose spectrum is $\underline{\Sigma}_{V_{\hat{P}_1,\hat{P}_2}} = \{\lambda'_1, \lambda'_2, \lambda'_3\}$, where the λ'_i are such that the only non zero values are given by $\lambda'_1(\hat{P}_1) = 1$, $\lambda'_2(\hat{P}_2) = 1$ and $\lambda'_3(\hat{P}_3 + \hat{P}_4) = 1$. Clearly, the spectral presheaf map $\underline{\Sigma}(i_{V_{\hat{P}_1,\hat{P}_2},V}) : \underline{\Sigma}_V \to \underline{\Sigma}_{V_{\hat{P}_1,\hat{P}_2}}$ is given by

$$\begin{aligned} \underline{\Sigma}(i_{V_{\hat{P}_1,\hat{P}_2},V})(\lambda_1) &= \lambda'_1 \\ \underline{\Sigma}(i_{V_{\hat{P}_1,\hat{P}_2},V})(\lambda_2) &= \lambda'_2 \\ \underline{\Sigma}(i_{V_{\hat{P}_1,\hat{P}_2},V})(\lambda_3) &= \lambda'_3 \\ \underline{\Sigma}(i_{V_{\hat{P}_1,\hat{P}_2},V})(\lambda_4) &= \lambda'_3. \end{aligned}$$

Each $\lambda_i \in \underline{\Sigma}_{V_{\hat{P}_1,\hat{P}_2}}$ gives rise to a maximal filter

$$F_{\lambda_i} = \{\hat{R} \in P(V_{\hat{P}_1,\hat{P}_2})|\lambda_i(\hat{R}) = 1\}.$$

By renaming $\hat{Q}_1 = \hat{P}_1$, $\hat{Q}_2 = \hat{P}_2$ and $\hat{Q}_3 := \hat{P}_3 + \hat{P}_4$, since $\lambda'_i(\hat{Q}_j) = \delta_{ij}$, we can write

$$F_{\lambda_i} = \{\hat{R} \in P(V_{\hat{P}_1,\hat{P}_2})|\hat{R} \geq \hat{Q}_i\}.$$

We will now compute F_{λ_i} for λ'_1, λ'_2 and λ'_3. In particular we obtain

$$F_{\lambda'_1} = \{\hat{P}_1, \hat{P}_1 + \hat{P}_2, \hat{P}_1 + \hat{P}_3 + \hat{P}_4, \hat{1}\},$$

$$F_{\lambda'_2} = \{\hat{P}_2, \hat{P}_1 + \hat{P}_2, \hat{P}_2 + \hat{P}_3 + \hat{P}_4, \hat{1}\}$$

and

$$F_{\lambda'_3} = \{\hat{P}_3 + \hat{P}_4, \hat{P}_1 + \hat{P}_3 + \hat{P}_4, \hat{P}_2 + \hat{P}_3 + \hat{P}_4, \hat{1}\},$$

respectively.

We would now like to compute the Gelfand transforms $\overline{\delta^i(\hat{S}_Z)_{V_{\hat{P}_1,\hat{P}_2}}}$ and $\overline{\delta^o(\hat{S}_Z)_{V_{\hat{P}_1,\hat{P}_2}}}$ in terms of $g_{\hat{S}_Z}$ and $f_{\hat{S}_Z}$, respectively. We recall (see (4.1.14)) that, for a given $\lambda'_i \in \underline{\Sigma}_{V_{\hat{P}_1,\hat{P}_2}}$ then

$$\overline{\delta^i(\hat{S}_z)_{V_{\hat{P}_1,\hat{P}_2}}}(\lambda'_i) = g_{\hat{S}_z}(C_{\mathcal{B}(\mathbb{C}^4)})(F_{\lambda_i})$$

while (see (4.1.15))

$$\overline{\delta^o(\hat{S}_z)_{V_{\hat{P}_1,\hat{P}_2}}}(\lambda'_i) = f_{\hat{S}_z}(C_{\mathcal{B}(\mathbb{C}^4)})(F_{\lambda_i}).$$

In both cases, $C_{\mathcal{B}(\mathbb{C}^4)}(F_{\lambda_i})$ is the cone over the filter F_{λ_i}, i.e.

$$\begin{aligned} C_{\mathcal{B}(\mathbb{C}^4)}(F_{\lambda_i}) &= \{\hat{Q} \in P(\mathcal{B}(\mathbb{C}^4))|\exists\hat{R} \in F_\lambda : \hat{Q} \geq \hat{R}\} \\ &= \{\hat{Q} \in P(\mathcal{B}(\mathbb{C}^4))|\hat{Q} \geq \hat{P}_i\}. \end{aligned}$$

We will first consider the case for λ'_1, obtaining

$$\begin{aligned} \overline{\delta^i(\hat{S}_z)_{V_{\hat{P}_1,\hat{P}_2}}}(\lambda'_1) &= \sup\{r \in \mathbb{R}|\hat{1} - \hat{E}^{\hat{S}_z}_r \geq \hat{P}_1\} \\ &= \sup\{r \in \mathbb{R}|r < 2\} \\ &= 2\,, \end{aligned}$$

while

$$\begin{aligned}\overline{\delta^o(\hat{S}_z)_{V_{\hat{P}_1,\hat{P}_2}}}(\lambda'_1) &= \inf\{r \in \mathbb{R} | \hat{E}_r^{\hat{S}_z} \geq \hat{P}_1\} \\ &= \inf\{r \in \mathbb{R} | 2 \leq r\} \\ &= 2\,.\end{aligned}$$

If we consider λ'_2 we obtain

$$\begin{aligned}\overline{\delta^i(\hat{S}_z)_{V_{\hat{P}_1,\hat{P}_2}}}(\lambda'_2) &= \sup\{r \in \mathbb{R} | \hat{1} - \hat{E}_r^{\hat{S}_z} \geq \hat{P}_2\} \\ &= \sup\{r \in \mathbb{R} | r < 0\} \\ &= 0\,,\end{aligned}$$

while

$$\begin{aligned}\overline{\delta^o(\hat{S}_z)_{V_{\hat{P}_1,\hat{P}_2}}}(\lambda'_2) &= \inf\{r \in \mathbb{R} | \hat{E}_r^{\hat{S}_z} \geq \hat{P}_2\} \\ &= \inf\{r \in \mathbb{R} | 0 \leq r\} \\ &= 0\,.\end{aligned}$$

If we consider λ'_3 we obtain

$$\begin{aligned}\overline{\delta^i(\hat{S}_z)_{V_{\hat{P}_1,\hat{P}_2}}}(\lambda'_1) &= \sup\{r \in \mathbb{R} | \hat{1} - \hat{E}_r^{\hat{S}_z} \geq \hat{P}_3 + \hat{P}_4\} \\ &= \sup\{r \in \mathbb{R} | r < -2\} \\ &= -2\,,\end{aligned}$$

while

$$\begin{aligned}\overline{\delta^o(\hat{S}_z)_{V_{\hat{P}_1,\hat{P}_2}}}(\lambda'_3) &= \inf\{r \in \mathbb{R} | \hat{E}_r^{\hat{S}_z} \geq \hat{P}_3 + \hat{P}_4\} \\ &= \inf\{r \in \mathbb{R} | 0 \leq r\} \\ &= 0\,.\end{aligned}$$

As expected, in all above cases the values of $\overline{\delta^i(\hat{S}_z)_{V_{\hat{P}_1,\hat{P}_2}}}(\lambda'_i)$ and $\overline{\delta^o(\hat{S}_z)_{V_{\hat{P}_1,\hat{P}_2}}}(\lambda'_i)$ lie in the spectrum of $\hat{S}_z$. Moreover we have that $\overline{\delta^i(\hat{S}_z)_{V_{\hat{P}_1,\hat{P}_2}}}(\lambda'_i) \leq \overline{\delta^o(\hat{S}_z)_{V_{\hat{P}_1,\hat{P}_2}}}(\lambda'_i)$, therefore we can think of the pair of values as an interval $[\overline{\delta^i(\hat{S}_z)_{V_{\hat{P}_1,\hat{P}_2}}}(\lambda'_i),$ $\overline{\delta^o(\hat{S}_z)_{V_{\hat{P}_1,\hat{P}_2}}}(\lambda'_i)]$ representing the possible values of the operator $\hat{S}_z$ when coarse-grained to the context $V_{\hat{P}_1,\hat{P}_2}$ (see (4.1.17)). For a detailed explanation the reader should refer to [26, Section 13.5].

In the present case the physical quantity representing the spin in the z direction is represented by the map

$$\breve{\delta}(\hat{S}_z)_V : \underline{\Sigma}_V \to \underline{\mathbb{R}^{\leftrightarrow}}_V \qquad (4.2.2)$$
$$\lambda \mapsto \breve{\delta}(\hat{S}_z)_V(\lambda) := \left(\breve{\delta}^i(\hat{S}_z)_V(\lambda), \breve{\delta}^o(\hat{S}_z)_V(\lambda)\right).$$

If we consider the context $V_{\hat{P}_1,\hat{P}_2} \subseteq V$ and the element $\lambda_1 \in \underline{\Sigma}_V$ we then obtain

$$\left(\breve{\delta}^i(\hat{S}_z)_V(\lambda_1), \breve{\delta}^o(\hat{S}_z)_V(\lambda_1)\right) = \left(\overline{\delta^i(\hat{S}_z)_{V_{\hat{P}_1,\hat{P}_2}}}(\lambda_1'), \overline{\delta^o(\hat{S}_z)_{V_{\hat{P}_1,\hat{P}_2}}}(\lambda_1')\right) = (2, 2).$$

Instead, if we consider $\lambda_2 \in \underline{\Sigma}_V$ we obtain

$$\left(\breve{\delta}^i(\hat{S}_z)_V(\lambda_2), \breve{\delta}^o(\hat{S}_z)_V(\lambda_2)\right) = \left(\overline{\delta^i(\hat{S}_z)_{V_{\hat{P}_1,\hat{P}_2}}}(\lambda_2'), \overline{\delta^o(\hat{S}_z)_{V_{\hat{P}_1,\hat{P}_2}}}(\lambda_2')\right) = (0, 0).$$

Finally if we consider either $\lambda_3 \in \underline{\Sigma}_V$ (or $\lambda_4 \in \underline{\Sigma}_V$) we obtain

$$\left(\breve{\delta}^i(\hat{S}_z)_V(\lambda_3), \breve{\delta}^o(\hat{S}_z)_V(\lambda_3)\right) = \left(\overline{\delta^i(\hat{S}_z)_{V_{\hat{P}_1,\hat{P}_2}}}(\lambda_3'), \overline{\delta^o(\hat{S}_z)_{V_{\hat{P}_1,\hat{P}_2}}}(\lambda_3')\right) = (-2, 0).$$

We will now utilise the above definition of the physical quantity $\breve{\delta}^i(\hat{S}_z)$ to compute its value given a state $|\psi\rangle = (0, 0, 1, 0)$. As a first step we need to recall how states are defined in topos quantum theory. In this instance we will utilise the pseudo-state object [26, Section 11.4] that is, the object in our topos which most resembles the notion of a point state, since it represents the smallest sub-object of the state-space $\underline{\Sigma}$. Being $\underline{\Sigma}$ a presheaf, its sub-objects will be themselves presheaves, thus the *pseudo-state* is a presheaf, i.e. an object in $\mathbf{Sets}^{\mathcal{V}(\mathcal{H})^{\mathrm{op}}}$.

Specifically, given a pure quantum state $\psi \in \mathcal{H}$, we define the presheaf

$$\underline{\mathfrak{w}}^{|\psi\rangle} := \underline{\delta(|\psi\rangle\langle\psi|)} \qquad (4.2.3)$$

such that for each context V we have

$$\underline{\delta(|\psi\rangle\langle\psi|)}_V := \mathfrak{S}(\bigwedge\{\hat{\alpha} \in P(V) |\, |\psi\rangle\langle\psi| \leq \hat{\alpha}\}) = \mathfrak{S}(\delta^o(|\psi\rangle\langle\psi|)) \subseteq \underline{\Sigma}(V). \qquad (4.2.4)$$

The map $\mathfrak{S}$ is defined as

$$\mathfrak{S} : P(V) \to \mathrm{Sub}_{cl}(\underline{\Sigma})_V, \qquad (4.2.5)$$

such that

$$\delta^o(|\psi\rangle\langle\psi|)_V \mapsto \mathfrak{S}(\delta^o(|\psi\rangle\langle\psi|)_V) := S_{\delta^o(|\psi\rangle\langle\psi|)_V}. \qquad (4.2.6)$$

Thus, for each context $V \in \mathcal{V}(\mathcal{H})$, the projection operator $\underline{\delta(|\psi\rangle\langle\psi|)}_V$ is the smallest projection operator implied by $|\psi\rangle\langle\psi|$. Since $|\psi\rangle\langle\psi|$ projects on a 1-dimensional sub-space of the Hilbert space, i.e. it projects on a state, $\underline{\delta(|\psi\rangle\langle\psi|)}_V$ identifies the smallest sub-space of $\mathcal{H}$ equal or bigger than the one dimensional sub-space $|\psi\rangle$.

The collection $(\underline{\delta(|\psi\rangle\langle\psi|)}_V)_{V\in\mathcal{V}(\mathcal{H})} =: (\underline{\mathfrak{w}}_V)_{V\in\mathcal{V}(\mathcal{H})}$ forms a sub-presheaf of $\underline{\Sigma}$ which is defined as follows:

Definition 4.2.1 For each state $|\psi\rangle \in \mathcal{H}$ we obtain the pseudo-state $\underline{\mathfrak{w}}^{|\psi\rangle} \in$ **Sets**$^{\mathcal{V}(\mathcal{H})^{\text{op}}}$ which is defined on:

- Objects: for each context $V \in \mathcal{V}(\mathcal{H})$ we obtain

$$\underline{\delta(|\psi\rangle\langle\psi|)}_V := \{\lambda \in \underline{\Sigma}_V | \lambda(\delta^o(|\psi\rangle\langle\psi|)_V) = 1\}. \tag{4.2.7}$$

- Morphisms: for each $i_{V'V} : V' \subseteq V$ the corresponding map is simply the spectral presheaf map restricted to $\underline{\mathfrak{w}}^{|\psi\rangle}$, i.e.

$$\begin{aligned} \underline{\mathfrak{w}}^{|\psi\rangle}(i_{V'V}) : \underline{\mathfrak{w}}^{|\psi\rangle}_V &\to \underline{\mathfrak{w}}^{|\psi\rangle}_{V'} \\ \lambda &\mapsto \lambda_{|V'}. \end{aligned} \tag{4.2.8}$$

In the case at hand, since $\psi = (0, 0, 1, 0)$, then $|\psi\rangle\langle\psi| = \hat{P}_3$ and for the context $V_{\hat{P}_1,\hat{P}_2}$ we obtain

$$\underline{\delta(|\psi\rangle\langle\psi|)}_{V_{\hat{P}_1,\hat{P}_2}} = \{\lambda'_3\}.$$

Therefore, given the state $\psi = (0, 0, 1, 0)$, the value of the spin in the z direction in the context $V_{\hat{P}_1,\hat{P}_2}$ is given by (for an in depth analysis on how this is done the reader should refer to [26, Section 13.7]):

$$\begin{aligned} (\breve{\delta}(\hat{S}_z)(\underline{\mathfrak{w}}^{|\psi\rangle}))_{V_{\hat{P}_1,\hat{P}_2}} &= (\breve{\delta}(\hat{S}_z)_{V_{\hat{P}_1,\hat{P}_2}}(\underline{\mathfrak{w}}^{|\psi\rangle}))_{V_{\hat{P}_1,\hat{P}_2}} \\ &= \{(\breve{\delta}(\hat{S}_z)_{V_{\hat{P}_1,\hat{P}_2}}(\lambda) | \lambda \in \underline{\mathfrak{w}}^{|\psi\rangle}_{V_{\hat{P}_1,\hat{P}_2}}\} \\ &= \{(-2, 0)\}. \end{aligned}$$

We now consider a smaller context $V_{\hat{P}_2} = lin_{\mathbb{C}}(\hat{P}_2, \hat{P}_1 + \hat{P}_3 + \hat{P}_4)$ such that $V_{\hat{P}_2} \subseteq V_{\hat{P}_1,\hat{P}_2}$. The state-space $\underline{\Sigma}_{V_{\hat{P}_2}}$ is given by $\underline{\Sigma}_{V_{\hat{P}_2}} = \{\lambda''_1, \lambda''_2\}$ such that $\lambda''_1(\hat{P}_2) = 1$ and $\lambda''_2(\hat{P}_1 + \hat{P}_3 + \hat{P}_4) = 1$. The presheaf map $\underline{\Sigma}(i_{V_{\hat{P}_1,\hat{P}_2},V_{\hat{P}_2}}) : \underline{\Sigma}_{V_{\hat{P}_1,\hat{P}_2}} \to \underline{\Sigma}_{\hat{P}_2}$ is such that

$$\begin{aligned} \lambda'_2 &\mapsto \lambda''_1 \\ \lambda'_1, \lambda'_3 &\mapsto \lambda''_2. \end{aligned}$$

For this context the pseudo-state is

$$\underline{\delta(|\psi\rangle\langle\psi|)}_{V_{\hat{P}_2}} = \{\lambda_3'\}\,.$$

If we now compute the value of the spin in the z direction for the context $V_{\hat{P}_2}$, given the state $\psi = (0, 0, 1, 0)$, we obtain

$$\begin{aligned}
(\breve{\delta}(\hat{S}_z)(\underline{\mathfrak{w}}^{|\psi\rangle}))_{V_{\hat{P}_2}} &= (\breve{\delta}(\hat{S}_z)_{\hat{P}_2}(\underline{\mathfrak{w}}^{|\psi\rangle}))_{V_{\hat{P}_2}} \\
&= \{(\breve{\delta}(\hat{S}_z)_{V_{\hat{P}_2}}(\lambda)|\lambda \in \underline{\mathfrak{w}}^{|\psi\rangle}_{V_{\hat{P}_2}}\} \\
&= \{(-2, 2)\}\,.
\end{aligned}$$

As we can see from this example, if we go to smaller contexts which contain less information, the interval of possible values for a physical quantity becomes bigger, i.e. we have less precise measurements.

Chapter 5
Interpreting Self-Adjoint Operators as q-Functions

In [20, 21] the authors show how it is possible to interpret self-adjoint operators affiliated with a von Neumann algebra $\mathcal{N}$, as real-valued functions on the projection lattice $P(\mathcal{N})$ of the algebra. These functions are called q-observable functions. The method of utilising real-valued function on $P(\mathcal{N})$ to define self-adjoint operators was first introduced in [12] and, independently, in [8]. However, the novelty of the approach defined [20, 21] consists in the fact that these real valued functions are related to both the daseinisation map, central to topos quantum theory, and to quantum probabilities.

5.1 q-Observable Functions

In order to define q-observable functions we first of all need to introduce some mathematical background. In particular, we need to introduce the notions of a *complete lattice*.

A *meet-semilattice* is a poset P such that, given any two elements $a, b \in P$, then the *meet* (*greatest lower bound*) $a \wedge b$ is in P, i.e.

$$\forall\, c \in P : \; c \leq a, b \iff c \leq a \wedge b\,.$$

If every family $(a_i)_{i \in I}$ of elements in P has a meet $\bigwedge_{i \in I} a_i$ in P, then the meet-semilattice is called complete.

Dually we also have the notion of a *join-semilattice*. In particular, a *join-semilattice* is a poset P such that, given any two elements $a, b \in P$, then the *join* (*least upper bound*) $a \vee b$ is in P, i.e.

$$\forall\, a, b \in P : \; a, b \leq c \iff a \vee b \leq c\,.$$

C. Flori, *A Second Course in Topos Quantum Theory*,
Lecture Notes in Physics 944, https://doi.org/10.1007/978-3-319-71108-9_5

If every family $(a_i)_{i\in I}$ of elements in P has a join $\bigvee_{i\in I} a_i$ in P, then P is a *complete join-semilattice*.

If P is both a meet-semilattice and a join-semilattice then it is called a *lattice*. Furthermore, if it is complete as a meet-semilattice and join-semilattice then it is said to be complete as a lattice.

In what follows we will use the *extended reals* which are defined by

$$\overline{\mathbb{R}} = \{-\infty\} \cup \mathbb{R} \cup \{\infty\}\,. \tag{5.1.1}$$

This forms a complete lattice which will be used to define q-observable functions. To this end we also need to introduce the notion of *spectral family*. This was already introduced in [26] but we will report it below for the sake of completeness.

Definition 5.1.1 Given a complete Hilbert space $\mathcal{H}$ with lattice of projections given by $P(\mathcal{H})$, then a spectral family is a map

$$E : \mathbb{R} \to P(\mathcal{H})$$
$$r \mapsto \hat{E}_r$$

such that:

1. for all $r, s \in \mathbb{R}$, if $r < s$ then $\hat{E}_r \leq \hat{E}_s$.
2. $\bigvee_{r\in\mathbb{R}} \hat{E}_e = \hat{1}$.
3. $\bigwedge_{r\in\mathbb{R}} \hat{E}_r = \hat{0}$.

A spectral family can be either *left-continuous* or *right-continuous*. In particular it is called *right-continuous* if

$$\forall\, r \in \mathbb{R}\,, \qquad \bigwedge_{s>r} \hat{E}_s = \hat{E}_r\,, \tag{5.1.2}$$

while it is called *left continuous* if

$$\forall\, r \in \mathbb{R}\,, \qquad \bigvee_{s>r} \hat{E}_s = \hat{E}_r\,. \tag{5.1.3}$$

Given a von Neumann algebra $\mathcal{N}$ on a Hilbert space $\mathcal{H}$, if the image of the spectral family E is in $P(\mathcal{N})$ (projection lattice of $\mathcal{N}$), then we say that E is in $\mathcal{N}$. A spectral family $E : \mathbb{R} \to P(\mathcal{N})$ can be seen as a monotone function.[1]

Self-adjoint operators can be expressed in terms of spectral families through the spectral decomposition theorem

[1]Given two posets $(P, \leq_p)$ and $(Q, \leq_q)$, then a map $f : P \to Q$ is called monotone (order-preserving) if for all $a, b \in P$, when $a \leq_p b$ then $f(a) \leq_q f(b)$.

Theorem 5.1.1 *Given a self-adjoint operator $\hat{A}$ on $\mathcal{N}$, then there exists a unique right-continuous spectral family $E : \mathbb{R} \to P(\mathcal{N})$, such that*

$$\hat{A} = \int_{-\infty}^{+\infty} r\, d\hat{E}_r ,$$

and vice versa: each right-continuous spectral family determines a unique self-adjoint operator. Such a right-continuous spectral family will be denoted by $E^{\hat{A}}$.

We recall that a self-adjoint operator $\hat{A}$ is contained in $B(\mathcal{H})$ iff $\hat{A}$ is bounded i.e. if its spectral family $E^{\hat{A}}$ is bounded. Moreover, given a von Neumann algebra $\mathcal{N}$, if $\hat{A} \in \mathcal{N} \subseteq B(\mathcal{H})$ then $E^{\hat{A}} \in P(\mathcal{N})$ and $\hat{A}$ is bounded. However it could be the case that $E^{\hat{A}} \in P(\mathcal{N})$ but $\hat{A}$ is not bounded. If this happens we say that $\hat{A}$ is *affiliated* with $\mathcal{N}$. The set of self-adjoint operators affiliated with a von Neumann algebra $\mathcal{N}$ is denoted by $SA(\mathcal{N})$, while the set of self-adjoint operators in $\mathcal{N}$ is denoted by $\mathcal{N}_{sa}$. Clearly $\mathcal{N}_{sa} \subseteq SA(\mathcal{N})$.

In Definition 4.1.3 we defined the spectral order on the set $\mathcal{N}_{sa}$, this definition can easily be enlarged to the set $SA(\mathcal{N})$ as follows:

Definition 5.1.2 Given a von Neumann algebra $\mathcal{N}$, the spectral order on the set $SA(\mathcal{N})$ is given by:

$$\forall\, \hat{A}, \hat{B} \in SA(\mathcal{N}) : \hat{A} \leq_s \hat{B} \iff \forall\, r \in \mathbb{R} : \hat{E}_r^{\hat{A}} \geq \hat{E}_r^{\hat{B}}$$

where the order on the right hand side is the usual order of projections and $E^{\hat{A}} = (\hat{E}_r^{\hat{A}})_{r \in \mathbb{R}}$ and, $E^{\hat{B}} = (\hat{E}_r^{\hat{B}})_{r \in \mathbb{R}}$ are the right-continuous spectral families of $\hat{A}$ and $\hat{B}$, respectively.

If we replace the reals $\mathbb{R}$ with the extended reals $\overline{\mathbb{R}}$ in the definition of a spectral family we obtain the notion of an *extended spectral family* as follows:

Definition 5.1.3 Given a von Neumann algebra $\mathcal{N}$, then an extended spectral family is a map

$$E : \overline{\mathbb{R}} \to P(\mathcal{N})$$
$$r \mapsto \hat{E}_r$$

such that

1. $\hat{E}_{-\infty} = \hat{O}$.
2. $\hat{E}_{+\infty} = \hat{1}$.
3. $E|_{\mathbb{R}}$ is a spectral family.

Right- or left-continuity will depend on the right- or left-continuity of $E|_{\mathbb{R}}$. Clearly every spectral family $E : \mathbb{R} \to P(\mathcal{N})$ determines a unique extended spectral family $E : \overline{\mathbb{R}} \to P(\mathcal{N})$ and vice versa. Moreover, the spectral theorem implies that there exists a bijective correspondence between the set $SA(\mathcal{N})$ of self-adjoint operators affiliated with $\mathcal{N}$ and the set $SF(\overline{\mathbb{R}}, P(\mathcal{N}))$ of extended, right-continuous spectral families of $P(\mathcal{N})$, i.e.

$$SA(\mathcal{N}) \simeq SF(\overline{R}, P(\mathcal{N}))$$

$$\hat{A} \mapsto E^{\hat{A}}$$

$$\int_{-\infty}^{+\infty} r\, d\hat{E}_r^{\hat{A}} \leftarrow\!\shortmid E^{\hat{A}} .$$

Similarly as it was the case for the spectral family, also the extended spectral family is a monotone function, moreover, since it is defined on the extended reals, it also preserves all meets. This implies that it is a morphism of complete meet-semilattices. This result is encoded in the following Lemma [20]:

Lemma 5.1.1 *Seen as a monotone function, the extended right-continuous spectral family $E : \overline{\mathbb{R}} \to P(\mathcal{N})$ is a morphism of complete meet-semilattices. Conversely, any meet preserving map $E : \overline{\mathbb{R}} \to P(\mathcal{N})$ such that the conditions*

i. $E(-\infty) = \hat{0}$

ii. $\bigvee_{r\in\mathbb{R}} E(r) = \hat{1}$

hold, determines an extended right-continuous spectral family.

Proof Let us assume that $E : \overline{\mathbb{R}} \to P(\mathcal{N})$ is an extended right-continuous spectral family, then, given an arbitrary family $(r_i)_{i\in I}$ of elements in $\overline{\mathbb{R}}$, we obtain

$$\begin{aligned} E(\inf_{i\in I} r_i) &\overset{\text{Definition}}{=} \hat{E}_{\inf_{i\in I} r_i} \\ &\overset{\text{right-continuity}}{=} \bigwedge_{s>\inf_{i\in I} r_i} \hat{E}_s \\ &\overset{\text{monotonicity}}{=} \bigwedge_{i\in I} \hat{E}_{r_i} . \end{aligned}$$

Therefore E preserves all meets and hence it is a morphism of complete meet-semilattices.

On the other hand, let us assume that $E : \overline{\mathbb{R}} \to P(\mathcal{N})$ is a meet-preserving map satisfying condition (i) and (ii) above, then clearly it is monotone. This follows from the fact that if E is meet-preserving then, given $r, s \in \overline{\mathbb{R}}$ we obtain

$$r \leq s \Longrightarrow r = r \wedge s \Longrightarrow E(r) = E(r) \wedge E(s) \Longrightarrow E(r) \leq E(s) .$$

To show that it is right-continuous we note that for all $r \in \mathbb{R}$ then

$$\hat{E}_r = \hat{E}_{\inf_{s>r}} = \bigwedge_{s>r} \hat{E}_s ,$$

hence E is right-continuous. To show that E is an extended spectral family we need to show that

1) $E(+\infty) = \hat{1}$;
2) $E(-\infty) = \hat{0}$, which is simply assumption (i);
3) $\bigvee_{r\in\mathbb{R}} E(r) = \hat{1}$ which is simply assumption (ii) and
4) $\bigwedge_{r\in\mathbb{R}} E(r) = \hat{0}$.

Utilising the meet-preservation properties of E we obtain

$$\bigwedge_{r\in\mathbb{R}} E(r) = \hat{E}_{\inf_{r\in\mathbb{R}} r} = E(-\infty) = \hat{0}$$

and

$$E(+\infty) = E(\inf(\emptyset)) = \bigwedge \emptyset = \hat{1}\,.$$

□

Now that we have shown that E is a morphism of complete meet-semilattices we can apply the adjoint functor theorem for poset to construct the left adjoint of E. In particular the adjoint functor theorem for poset is as follows:

Theorem 5.1.2 *Consider two complete meet-semilattices* $(P, \leq_p)$ *and* $(Q, \leq_q)$*. If* $f : P \to Q$ *is a monotone map, then* f *has a left adjoint* $g : Q \to P$ *iff* f *preserves all meets. The left adjoint* g *is monotone, it preserves all joins and it is defined by*

$$\begin{aligned} g &: Q \to P \\ x &\mapsto \bigwedge\{a \in P | x \leq_q f(a)\}\,. \end{aligned}$$

Proof Let us assume that $f : P \to Q$ has a left adjoint $g : Q \to P$. If we then consider an arbitrary family $(a_i)_{i\in I} \subseteq P$ we obtain, for all $i \in I$, that

$$\begin{aligned} \bigwedge_{i\in I} a_i \leq_p a_i &\Longrightarrow f\left(\bigwedge_{i\in I} a_i\right) \leq_q f(a_i) \\ &\Longrightarrow f\left(\bigwedge_{i\in I} a_i\right) \leq_q \bigwedge_{i\in I} f(a_i)\,. \end{aligned}$$

On the other hand, for all $i \in I$ we have

$$\begin{aligned}\bigwedge_{i\in I} f(a_i) \leq_q f(a_i) &\Longrightarrow g\left(\bigwedge_{i\in I} f(a_i)\right) \leq_p a_i \\ &\Longrightarrow g\left(\bigwedge_{i\in I} f(a_i)\right) \leq_p \bigwedge_{i\in I} a_i \\ &\Longrightarrow \bigwedge_{i\in I} f(a_i) \leq_q f\left(\bigwedge_{i\in I} a_i\right).\end{aligned}$$

It follows that $\bigwedge_{i\in I} f(a_i) = f\left(\bigwedge_{i\in I} a_i\right)$ for any family $(a_i)_{i\in I} \subseteq P$.

Conversely, let us assume that $f : P \to Q$ preserves all meets and that $g : Q \to P$ is defined by $g(x) = \bigwedge\{a \in P|x \leq_q f(a)\}$ for all $x \in Q$. We now want to show that $x \leq_q f(a) \Leftrightarrow g(x) \leq_p x$.

$\Rightarrow$ Assume that $a_1 \in P$ is such that $x \leq_q f(a_1)$ for a given $x \in Q$, then $a_1 \in \{a \in P|x \leq_q f(a)\}$ and $g(x) = \bigwedge\{a \in P|x \leq_q f(a)\} \leq_p a_1$.

$\Leftarrow$ Assume that $g(x) \leq_p a_1$ then, since f preserves meets and is monotone, we have that

$$\begin{aligned}f(a_1) \geq_q f(g(x)) &= f\left(\bigwedge\{a \in P|x \leq_q f(a)\}\right) \\ &= \bigwedge\{f(a) \in Q|x \leq_q f(a)\} \geq_q x.\end{aligned}$$

□

An analogous theorem holds for join-semilattices.

Theorem 5.1.3 *Consider two complete join-semilattices $(P, \leq_p)$ and $(Q, \leq_q)$. If $f : P \to Q$ is a monotone map, then f has a right adjoint $g : Q \to P$ iff f preserves all joins. The right adjoint g is monotone, preserves all meets and is defined by*

$$\begin{aligned}g : Q &\to P \\ x &\mapsto \bigvee\{a \in P|f(x) \leq_q a\}.\end{aligned}$$

The proof is very similar to the one given above so we will omit it.

Theorem 5.1.2 together with Lemma 5.1.1 imply that $E : \overline{\mathbb{R}} \to P(\mathcal{N})$ has a left adjoint

$$o^E : P(\mathcal{N}) \to \overline{\mathbb{R}}$$

which preserves arbitrary joins, i.e.

$$o^E\left(\bigvee_{i\in I}\hat{P}_i\right) = \sup_{i\in I} o^E(\hat{P}_i)\,.$$

From Theorem 5.1.2 we know how to explicitly construct o^E, in particular, for all $\hat{P} \in P(\mathcal{N})$

$$o^E(\hat{P}) = \inf\{r \in \overline{\mathbb{R}}|\hat{P} \leq \hat{E}_r\}\,.$$

For the case in which $E = E^{\hat{A}}$ we obtain:

$$o^{\hat{A}}(\hat{P}) = \inf\{r \in \overline{\mathbb{R}}|\hat{P} \leq \hat{E}^{\hat{A}}_r\}\,. \tag{5.1.4}$$

We are now ready to define the notion of *q-observable* functions:

Definition 5.1.4 ([20]) Consider a self-adjoint operator $\hat{A}$ affiliated with a von Neumann algebra $\mathcal{N}$ and whose extended right-continuous spectral family is given by $E^{\hat{A}} = (\hat{E}^{\hat{A}}_r)_{r\in\overline{\mathbb{R}}}$. The left adjoint $o^{\hat{A}} : P(\mathcal{N}) \to \overline{\mathbb{R}}$ of $E^{\hat{A}}$ defined in (5.1.4) is called the q-observable function associated to $\hat{A}$.

In [20] the authors gave an abstract characterization of q-observable functions in terms of the adjunction $o^E \dashv E$. To this end they introduced the notions of *weak q-observables* and *abstract q-observables*.

Definition 5.1.5 A weak q-observable is a join-preserving function $o : P(\mathcal{N}) \to \overline{\mathbb{R}}$, such that

$$o(\hat{P}) > -\infty \qquad \forall\, \hat{P} > \hat{0}\,. \tag{5.1.5}$$

An abstract q-observable function is a weak q-observable function with the extra property that there exists a family $(\hat{P}_i)_{i\in I} \subseteq P(\mathcal{N})$ with $\bigvee_{i\in I}\hat{P}_i = \hat{1}$, such that

$$o(\hat{P}_i) \lneq +\infty \qquad \forall\, i \in I\,. \tag{5.1.6}$$

The set of all abstract q-observable functions is denoted by $QO(P(\mathcal{N}), \overline{\mathbb{R}})$.

Theorem 5.1.4 ([20]) *Given a von Neumann algebra $\mathcal{N}$, there exists a bijective correspondence between the set $QO(P(\mathcal{N}), \overline{\mathbb{R}})$ and the set $SF(\overline{\mathbb{R}}, P(\mathcal{N}))$, i.e.*

$$\begin{aligned} SF(\overline{\mathbb{R}}, P(\mathcal{N})) &\simeq QO(P(\mathcal{N}), \overline{\mathbb{R}}) \\ E &\mapsto o^E \\ E^o &\leftarrow\!\shortmid o\,. \end{aligned}$$

Proof

$\Rightarrow$ Let us assume that $E : \overline{\mathbb{R}} \to P(\mathcal{N})$ is an extended right-continuous spectral family. We then have to show that its left adjoint o^E satisfies the conditions in Definition 5.1.5. In particular, given any $\hat{P} > \hat{0}$, since o^E preserves all joins, we have that

$$o^E(\hat{P}) = \inf\{r \in \overline{\mathbb{R}} | \hat{P} \leq E(r)\} > -\infty .$$

Next consider that family of projections $E|_{\mathbb{R}} = (E(r))_{r \in \mathbb{R}}$, since it is a spectral family we know that $\bigvee_{r \in \mathbb{R}} E(r) = \hat{1}$ holds. Moreover given the adjunction $o^E \dashv E$, then for all $r \in \overline{\mathbb{R}}$

$$E(r) \leq E(r) \Longrightarrow o^E(E(r)) \leq r .$$

Therefore, given the above family $E|_{\mathbb{R}} = (E(r))_{r \in \mathbb{R}}$, we obtain that

$$o^E(E(r)) \leq r \lneqq +\infty , \qquad \forall r \in \mathbb{R} .$$

$\Leftarrow$ Let us assume that $o : P(\mathcal{N}) \to \overline{\mathbb{R}}$ is an abstract q-observable function, we need to show that its right adjoint E^o is a right-continuous extended spectral family. Since E^o preserves all meets we have that for all $r \in \overline{\mathbb{R}}$

$$E^o(r) = E^o(\inf\{s \in \overline{\mathbb{R}} | r < s\}) = \bigwedge_{s > r} E^o(s) .$$

This implies that E^o is right-continuous. Next we need to show that it satisfies the requirements of being an extended spectral family. However, since E^o preserves all meets, all we need to do is to apply Lemma 5.1.1. In particular, we note that

$$E^o(-\infty) \overset{\text{Theorem 5.1.3}}{=} \bigvee\{\hat{P} \in P(\mathcal{N}) | o(\hat{P}) \leq -\infty\} .$$

However,

$$o(\hat{0}) = o(\bigvee \emptyset) = \sup \emptyset = -\infty$$

which, together with condition (5.1.5), implies that

$$E^o(-\infty) = \hat{0} .$$

Moreover we have that

$$\bigvee_{r \in \mathbb{R}} E^o(r) \overset{\text{Definition 5.1.5,(5.1.6)}}{\geq} \bigvee_{r \in \{o(\hat{P}_i) | i \in I\}} E^o(r) = \bigvee_{i \in I} E^o(o(\hat{P}_i)) .$$

Given the adjunction $o \dashv E^o$ it follows that, for all $\hat{P} \in P(\mathcal{N})$,

$$o(\hat{P}) \leq o(\hat{P}) \Rightarrow \hat{P} \leq E^o(o(\hat{P}))\,,$$

hence

$$\bigvee_{i \in I} E^o(o(\hat{P}_i)) \geq \bigvee_{i \in I} \hat{P}_i = \hat{1}\,.$$

We conclude that since, $\bigvee_{r \in \mathbb{R}} E^o(r) \geq \hat{1}$ and $\hat{1} \geq \bigvee_{r \in \mathbb{R}} E^o(r)$, then $\bigvee_{r \in \mathbb{R}} E^o(r) = \hat{1}$.

Note that since adjoins are unique we have that, for all $o \in QO(P(\mathcal{N}), \overline{\mathbb{R}})$ and $E \in SF(\overline{\mathbb{R}}, P(\mathcal{N}))$, the following holds:

$$E^{o^E} = E \qquad \text{and} \qquad o^{E^o} = o\,.$$

□

Theorem 5.1.5 *Given a von Neumann algebra $\mathcal{N}$, there exists the following bijection:*

$$SA(\mathcal{N}) \simeq QO(P(\mathcal{N}), \overline{\mathbb{R}})$$

$$\hat{A} \mapsto o^{\hat{A}}$$

$$\int_{-\infty}^{+\infty} r\, dE^o(r) \leftarrow\!\shortmid o\,.$$

Proof Given a self-adjoint operator $\hat{A} \in SA(\mathcal{N})$, then we know that the spectral theorem, uniquely, associates to it an extended right-continuous spectral family $E^{\hat{A}}$. Applying Theorem 5.1.4 we obtain the unique q-observable function $o^{E^{\hat{A}}} = o^{\hat{A}}$. On the other hand, given a q-observable function, Theorem 5.1.4 defines the unique extended right-continuous spectral family E^o and, by the spectral theorem, the unique operator $\hat{A}^{E^o} = \int_{-\infty}^{+\infty} r dE^o(r)$. □

q-Observable functions are intimately related to the spectrum of self-adjoint operators. In particular, in [20] it was shown that:

Lemma 5.1.2 *Given a self-adjoint operator $\hat{A}$ affiliated with a von Neumann algebra $\mathcal{N}$, whose corresponding q-observable function is $o^{\hat{A}}$, then*

$$o^{\hat{A}}(P_0(\mathcal{N})) = \mathrm{sp}\hat{A}$$

where $P_0(\mathcal{N})$ is the set of non-zero projection operators in $\mathcal{N}$. If $\hat{A}$ is unbounded from above then $o^{\hat{A}}(\hat{1}) = +\infty$ is in $o^{\hat{A}}(P_0(\mathcal{N}))$.

Proof It is worth at this point recalling the fact that sp$\hat{A}$ is a non-empty set consisting of those elements $s \in \mathbb{R}$ for which $E^{\hat{A}} = (E_r^{\hat{A}})_{r \in \mathbb{R}}$ is non constant on any open neighbourhood of s. This fact, together with the right-continuity property of $E^{\hat{A}}$ implies that, for $r \in$ sp$\hat{A}$ then

$$o^{\hat{A}}(\hat{E}_r^{\hat{A}}) = \inf\{s \in \overline{\mathbb{R}} | \hat{E}_s^{\hat{A}} \geq \hat{E}_r^{\hat{A}}\} = r .$$

On the other hand, if r is in the image of $o^{\hat{A}}$, then $\hat{E}_s^{\hat{A}} < \hat{E}_r^{\hat{A}}$ for all $s < r$, therefore $E^{\hat{A}}$ is non constant on any neighbourhood of r. This implies that $r \in$ sp$\hat{A}$.

When $\hat{A}$ is unbounded, then $\hat{E}_s^{\hat{A}} < \hat{1}$ for all $s \in \mathbb{R}$ hence $\hat{E}_{+\infty}^{\hat{A}} = \hat{1}$ and $o^{\hat{A}}(\hat{1}) = +\infty$ □

For a bounded operator $\hat{A}$ with compact spectrum it follows that the image $o^{\hat{A}}(P_0(\mathcal{N}))$ is compact. From this fact and Theorem 5.1.5 it follows that:

Lemma 5.1.3 *Given a von Neumann algebra $\mathcal{N}$, there exist a bijection between the set $\mathcal{N}_{sa}$ of self-adjoint operators in $\mathcal{N}$ and the set $QO^c(P(\mathcal{N}), \overline{\mathbb{R}})$ of q-observable functions with compact image.*

5.1.1 Lattice Structure

The poset $(QO(P(\mathcal{N}), \overline{\mathbb{R}}) \leq)$ equipped with the pointwise order forms a conditionally complete lattice.[2] Similarly the poset $(SA(\mathcal{N}), \leq_s)$ equipped with the spectral order forms a conditionally complete lattice. In [20] the authors showed that it is possible to relate these lattices through an order-isomorphism.

Theorem 5.1.6 *Given the posets $(SA(\mathcal{N}), \leq_s)$ and $(QO(P(\mathcal{N}), \overline{\mathbb{R}}) \leq)$, the map*

$$\phi : (SA(\mathcal{N}), \leq_s) \to (QO(P(\mathcal{N}), \overline{\mathbb{R}}) \leq)$$
$$\hat{A} \mapsto o^{\hat{A}}$$

is an order-isomorphism of conditionally complete lattice.

Proof We already know that the map ϕ is an isomorphism so, what remains to be shown is that it preserves the order. To this end, consider two self-adjoint operators $\hat{A}, \hat{B} \in SA(\mathcal{N})$ such that

$$\hat{A} \leq_s \hat{B} \iff \forall r \in \overline{\mathbb{R}} : \hat{E}_r^{\hat{A}} \geq \hat{E}_r^{\hat{B}} .$$

[2] A conditionally complete lattice is a lattice in which every non-empty bounded subset has a least upper bound and a greatest lower bound. As an example of a conditionally-complete lattice one may take the set of all real numbers with the usual order.

It follows that for all $\hat{P} \in P(\mathcal{N})$

$$\{r \in \overline{\mathbb{R}} | \hat{E}_r^{\hat{B}} \geq \hat{P}\} \subseteq \{r \in \overline{\mathbb{R}} | \hat{E}_r^{\hat{A}} \geq \hat{P}\} ,$$

therefore

$$o^{\hat{B}}(\hat{P}) = \inf\{r \in \overline{\mathbb{R}} | \hat{E}_r^{\hat{B}} \geq \hat{P}\} \geq o^{\hat{A}}(\hat{P}) = \inf\{r \in \overline{\mathbb{R}} | \hat{E}_r^{\hat{A}} \geq \hat{P}\} .$$

On the other hand, if $o^{\hat{A}} \leq o^{\hat{B}}$ then, for all $r \in \overline{\mathbb{R}}$, we obtain

$$\{\hat{P} \in P(\mathcal{N}) | r \geq o^{\hat{B}}(\hat{P})\} \subseteq \{\hat{P} \in P(\mathcal{N}) | r \geq o^{\hat{A}}(\hat{P})\} ,$$

hence

$$\begin{aligned} \hat{E}_r^{\hat{B}} &= \bigvee \{\hat{P} \in P(\mathcal{N}) | r \geq o^{\hat{B}}(\hat{P})\} \\ &\leq \bigvee \{\hat{P} \in P(\mathcal{N}) | r \geq o^{\hat{A}}(\hat{P})\} \\ &= \hat{E}_r^{\hat{A}} . \end{aligned}$$

This implies that $\hat{A} \leq_s \hat{B}$. □

The upshot of this theorem is that it is now possible to faithfully represent the poset $(SA(\mathcal{N}), \leq_s)$ in terms of the poset $(QO(P(\mathcal{N}), \overline{\mathbb{R}}) \leq)$.

We now consider the poset $SF(\overline{\mathbb{R}}, P(\mathcal{N}))$ and equip it with the inverse pointwise order which is denoted by $\leq_i$ and it is defined as follows:

$$E^{\hat{A}} \leq_i E^{\hat{B}} \iff \forall r \in \overline{\mathbb{R}} : \hat{E}_r^{\hat{A}} \geq \hat{E}_r^{\hat{B}} .$$

Corollary 5.1.1 *The map*

$$\begin{aligned} (SA(\mathcal{N}), \leq_s) &\to SF(\overline{\mathbb{R}}, P(\mathcal{N})) \\ \hat{A} &\mapsto E^{\hat{A}} \end{aligned}$$

is an order-isomorphism of conditionally complete lattices.

Proof Similarly as above, all that remains to be shown is that the order is preserved by the map. However this follows trivially from the definition of the spectral order:

$$E^{\hat{A}} \leq_i E^{\hat{B}} \iff \forall r \in \overline{\mathbb{R}} : \hat{E}_r^{\hat{A}} \geq \hat{E}_r^{\hat{B}} \iff \hat{A} \leq_s \hat{B}$$

□

Theorem 5.1.7 *The map*

$$(QO(P(\mathcal{N}), \overline{\mathbb{R}}), \leq) \to (SF(\overline{\mathbb{R}}, P(\mathcal{N})), \leq_i)$$
$$o \mapsto E^o$$

is an order-isomorphism of conditionally complete lattices.
The proof is very similar to the one of Theorem 5.1.6, however, for the sake of completeness we will nonetheless report it here.

Proof Let us consider $o, o' \in QO(P(\mathcal{N}), \overline{\mathbb{R}})$ such that $o \leq o'$. It then follows that for all $r \in \overline{\mathbb{R}}$

$$\begin{aligned} E^o(r) = \hat{E}^o_r &= \bigvee\{\hat{P} \in P(\mathcal{N}) | o(\hat{P} \leq r\} \\ &\geq E^{o'}(r) = \hat{E}^{o'}_r = \bigvee\{\hat{P} \in P(\mathcal{N}) | o'(\hat{P} \leq r\}. \end{aligned}$$

On the other hand, if $E^o \leq E^{o'}_r$ then, for all $\hat{P} \in P(\mathcal{N})$, we have

$$\begin{aligned} o(\hat{P}) &= \inf\{r \in \overline{\mathbb{R}} | \hat{E}^o_r \geq \hat{P}\} \\ &\leq \inf\{r \in \overline{\mathbb{R}} | \hat{E}^{o'}_r \geq \hat{P}\} \\ &= o'(\hat{P}). \end{aligned}$$

It follows that $o \leq o'$. □

5.2 Relation to Outer Daseinisation

In this section we will analyse the relation between q-observable functions and the process of daseinisation put forward in [20]. To this end we will first need the following Lemma:

Lemma 5.2.1 *Consider two von Neumann algebras $\mathcal{M}$ and $\mathcal{N}$ with common unit element and such that $\mathcal{M} \subseteq \mathcal{N}$. Then the inclusion map $i : P(\mathcal{M}) \to P(\mathcal{N})$ is a morphism of complete orthomodular lattices*[3] *and as such it has both a left adjoint $\delta^o_{\mathcal{M}} : P(\mathcal{N}) \to P(\mathcal{M})$ and a right adjoint $\delta^i_{\mathcal{M}} : P(\mathcal{N}) \to P(\mathcal{M})$.*

Proof To show that the map i is a morphism of orthomodular lattices we need to show that it preserves orthocomplements, all meets and all joins.

[3]A lattice is complemented if every element a has a complement $a^{\perp}$. It is orthocomplemented if it is equipped with an involution that sends each element to a complement. An orthomodular lattice is an orthocomplemented lattice such that $a \leq c$ implies that $a \vee (a^{\perp} \wedge c) = c$.

i *Orthocomplement*: given any $\hat{P} \in P(\mathcal{M})$ its orthocomplement is given by $\hat{1} - \hat{P}$. Since $\mathcal{M} \subseteq \mathcal{N}$ and the unit elements in $\mathcal{M}$ and $\mathcal{N}$ coincide, we have that

$$i(\hat{1} - \hat{P}) = \hat{1} - \hat{P}.$$

ii *Meets*: given a projection operator $\hat{P}_i$, by definition this projects onto a closed subspace S_i of $\mathcal{H}$, i.e. $\hat{P}_j(S_j) = S_j$ and $\hat{P}_j(S_j^{\perp}) = 0$. The subspace S_j is independent of whether P_j is considered to lie in $P(\mathcal{M})$, $P(\mathcal{N})$ or $P(\mathcal{H})$. Therefore, given a family of projections $(\hat{P}_j)_{j\in J}$ in $P(\mathcal{M})$, their intersection $\bigwedge_{j\in J} \hat{P}_j$ will project onto the closed subspace given by $\cap_{j\in J} S_j$. However, since $i(\hat{P}_j) = \hat{P}_j$, it follows that:

$$i\left(\bigwedge_{j\in J} \hat{P}_j\right) = \bigwedge_{j\in J} i(\hat{P}_j).$$

iii *Joins*: given a family of projections $(\hat{P}_j)_{j\in J}$ in $P(\mathcal{M})$, then by de Morgan's law we have that

$$\bigvee_{j\in J} \hat{P}_j = \hat{1} - \bigwedge_{j\in J} \hat{P}_j.$$

Since i preserves meets it then follows that it also preserves joins.

From Theorem 5.1.2 it follows that i has a left adjoint defined as follows:

$$\begin{aligned} \delta^o_{\mathcal{M}} &: P(\mathcal{N}) \to P(\mathcal{M}) \\ &\hat{P} \mapsto \delta^o_{\mathcal{M}}(\hat{P}) := \bigwedge\{\hat{Q} \in P(\mathcal{M}) | \hat{Q} \geq \hat{P}\}. \end{aligned}$$

This represents *outer daseinisation.*

From Theorem 5.1.3 it follows that i has a right adjoint defined as follows:

$$\begin{aligned} \delta^i_{\mathcal{M}} &: P(\mathcal{N}) \to P(\mathcal{M}) \\ &\hat{P} \mapsto \delta^i_{\mathcal{M}}(\hat{P}) := \bigvee\{\hat{Q} \in P(\mathcal{M}) | \hat{Q} \leq \hat{P}\}. \end{aligned}$$

This represents *inner daseinisation.* □

The relation between outer daseinisation and q-observable functions is then given by the following Theorem [20]:

Theorem 5.2.1 *Consider a von Neumann algebra $\mathcal{N}$ and a self-adjoint operator $\hat{A}$ affiliated with $\mathcal{N}$. If we consider a von Neumann algebra $\mathcal{M} \subseteq \mathcal{N}$, such that the* *unit elements of $\mathcal{N}$ and $\mathcal{M}$ coincide, then the weak q-observable of $\delta^o(\hat{A})_{\mathcal{M}}$ is given by*

$$o^{\delta^o(\hat{A})_{\mathcal{N}}} = o^{\hat{A}}_{\mathcal{N}} \circ i = o^{\hat{A}}_{\mathcal{N}}|_{P(\mathcal{N})} : P(\mathcal{N}) \to \overline{\mathbb{R}}.$$

If $\hat{A}$ is bounded from above, then $o^{\delta^o(\hat{A})_{\mathcal{N}}}$ is a proper q-observable function.

Proof Given the right-continuous extended spectral family $E^{\hat{A}} : \overline{\mathbb{R}} \to P(\mathcal{N})$, we define

$$E^{\delta^o(\hat{A})_{\mathcal{M}}} := \delta^i_{\mathcal{M}} \circ E^{\hat{A}} : \overline{\mathbb{R}} \to P(\mathcal{M})$$

such that for all $r \in \overline{\mathbb{R}}$ we obtain

$$E^{\delta^o(\hat{A})_{\mathcal{M}}}(r) = \delta^i_{\mathcal{M}}(E^{\hat{A}}(r)) = \delta^i_{\mathcal{M}}(\hat{E}^{\hat{A}}_r) = \bigvee\{\hat{Q} \in P(\mathcal{M}) | \hat{Q} \leq \hat{E}^{\hat{A}}_r\}.$$

From the definition it follows that $E^{\delta^o(\hat{A})_{\mathcal{M}}}$ is monotone. Since $\delta^i_{\mathcal{M}}$ is a right adjoint, it preserves all meets, hence

$$\bigwedge_{r \in \mathbb{R}} \delta^i_{\mathcal{M}}(E^{\hat{A}}(r)) = \delta^i_{\mathcal{M}} \bigwedge_{r \in \mathbb{R}} E^{\hat{A}}(r) = \delta^i_{\mathcal{M}}(\hat{0}) = \hat{0}.$$

Moreover

$$\delta^i_{\mathcal{M}}(E^{\hat{A}}(+\infty)) = \delta^i_{\mathcal{M}}(\hat{1}) = \hat{1}.$$

However, $E^{\delta^o(\hat{A})_{\mathcal{M}}}$ fails to be a right-continuous extended spectral family since it is not necessarily the case that $\bigvee_{r \in \mathbb{R}} E^{\delta^o(\hat{A})_{\mathcal{M}}}(r) = \hat{1}$. This is a consequence of the fact that $\delta^i(\hat{E}^{\hat{A}}_r)_{\mathcal{M}} \leq \hat{E}^{\hat{A}}_r$ for all $r \in \mathbb{R}$. Hence in this case we call $E^{\delta^o(\hat{A})_{\mathcal{M}}}$ a *weak right-continuous extended spectral family*. Its left adjoint $o^{\delta^o(\hat{A})_{\mathcal{M}}}$ is then a *weak q-observable function*. In this case we can define

$$\delta^o(\hat{A})_{\mathcal{M}} := \int_{-\infty}^{+\infty} r\, d(E^{\delta^o(\hat{A})_{\mathcal{M}}}) = \int_{-\infty}^{+\infty} r\, d(\delta^o(\hat{E}^{\hat{A}}_r)_{\mathcal{M}})$$

which represents the outer daseinisation of $\hat{A}$. If $\hat{A}$ is bounded from above, then there exists an $r \in \mathbb{R}$ such that $\hat{E}^{\hat{A}}_r = \hat{1}$, therefore $\hat{E}^{\delta^o(\hat{A})_{\mathcal{M}}}_r = \hat{1}$ and $\bigvee_{r \in \mathbb{R}} E^{\delta^o(\hat{A})_{\mathcal{M}}} = \hat{1}$. This implies that $E^{\delta^o(\hat{A})_{\mathcal{M}}}$ is a right-continuous extended spectral family in which case we obtain

$$\delta^o(\hat{A})_{\mathcal{M}} = \bigwedge\{\hat{B} \in SA(\mathcal{M}) | \hat{B} \geq_s \hat{A}\}.$$

Note that $\delta^o(\hat{A})_{\mathcal{M}}$ is a self-adjoint operator affiliated with $\mathcal{M}$ iff $E^{\delta^o(\hat{A})_{\mathcal{M}}}$ is a right-continuous extended spectral family.

The map $E^{\delta^o(\hat{A})_{\mathcal{M}}} := \delta^i_{\mathcal{M}} \circ E^{\hat{A}}$ is the composite of two right adjoints hence it is itself a right adjoint whose left adjoint is

$$o^{\delta^o(\hat{A})_{\mathcal{M}}} = o^{\hat{A}}_{\mathcal{N}} \circ i = o^{\hat{A}}_{\mathcal{N}}|_{P(\mathcal{M})} : P(\mathcal{M}) \to \overline{\mathbb{R}}$$

where the second equality follows since i is an inclusion.

Clearly if $E^{\delta^o(\hat{A})_{\mathcal{M}}}$ is a weak right-continuous extended family, then $o^{\delta^o(\hat{A})_{\mathcal{M}}}$ will be a weak q-observable while, if $E^{\delta^o(\hat{A})_{\mathcal{M}}}$ is a right-continuous extended family, then $o^{\delta^o(\hat{A})_{\mathcal{M}}}$ will be a q-observable function. □

5.3 q-Antonymous Functions

The notion of q-antonymous functions arose when trying to multiply a q-observable function by -1 and noticing that the resulting function was not a q-observable function [20]. In particular, given a q-observable function $o^{\hat{A}}$ associated to the self-adjoint operator $\hat{A} \in SA(\mathcal{N})$ then, for all $\hat{P} \in P(\mathcal{N})$, we obtain

$$\begin{aligned} -o^{\hat{A}}(\hat{P}) &= -\inf\{r \in \overline{\mathbb{R}} | \hat{P} \leq \hat{E}^{\hat{A}}_r\} \\ &= \sup\{-r \in \overline{\mathbb{R}} | \hat{P} \leq \hat{E}^{\hat{A}}_r\} \\ &= \sup\{r \in \overline{\mathbb{R}} | \hat{P} \leq \hat{E}^{\hat{A}}_{-r}\} . \end{aligned}$$

Let us now consider the left-continuous extended spectral family of $-\hat{A}$ denoted by $F^{-\hat{A}} = (\hat{F}^{-\hat{A}}_r)_{r\in\overline{\mathbb{R}}}$. Since

$$\hat{F}^{-\hat{A}} = \hat{1} - \hat{E}^{\hat{A}}_{-r}, \quad \forall\, r \in \overline{\mathbb{R}}$$

we can write $-o^{\hat{A}}(\hat{P})$ as follows:

$$-o^{\hat{A}}(\hat{P}) = \sup\{r \in \overline{\mathbb{R}} | \hat{P} \leq \hat{1} - \hat{F}^{-\hat{A}}_r\} .$$

This function is clearly not the q-observable function associated to $-\hat{A}$, it is however a q-antonymous function associated to $-\hat{A}$. These are defined as follows:

Definition 5.3.1 Given a self-adjoin operator $\hat{A} \in SA(\mathcal{N})$, the q-antonymous function associated with $\hat{A}$ is defined by

$$\begin{aligned} a^{\hat{A}} &: P(\mathcal{N}) \to \overline{\mathbb{R}} \\ &\hat{P} \mapsto \sup\{r \in \overline{\mathbb{R}} | \hat{P} \leq \hat{1} - \hat{F}^{\hat{A}}_r\} . \end{aligned}$$

The set of all such functions is denoted by $QA(P(\mathcal{N}), \overline{\mathbb{R}})$. Clearly each $a^{\hat{A}} : P(\mathcal{N}) \to \overline{\mathbb{R}}$ is order-reversing, i.e. an antitone.

We then obtain the following Theorem:

Theorem 5.3.1 *Given the sets $(SA(\mathcal{N}), \leq_s)$ and $(QA(P(\mathcal{N}), \overline{\mathbb{R}}), \leq)$, where the latter is equipped with the pointwise order, there exists an order-isomorphism defined as follows:*

$$\gamma : (SA(\mathcal{N}), \leq_s) \to (QA(P(\mathcal{N}), \overline{\mathbb{R}}), \leq)$$
$$\hat{A} \mapsto a^{\hat{A}} .$$

Proof We will first of all show that γ is a bijection between the respective sets, then we will show that it preserves the order. The fact that γ is a bijection follows from the fact that it can be constructed as follows:

$$\hat{A} \mapsto o^{\hat{A}} \mapsto -o^{\hat{A}} = a^{-\hat{A}}$$

where the first map is the isomorphism of Theorem 5.1.5, while the second map is the obvious bijection between q-observable functions and q-antonymous functions. To show that γ also preserves the order, consider two self-adjoint operators $\hat{A}, \hat{B} \in SA(\mathcal{N})$ then we obtain

$$\begin{aligned} \hat{A} \leq_s \hat{B} &\Longleftrightarrow -\hat{A} \geq_s -\hat{B} \\ &\Longleftrightarrow o^{-\hat{A}} \geq o^{-\hat{B}} \\ &\Longleftrightarrow -o^{-\hat{A}} \leq -o^{-\hat{B}} \\ &\Longleftrightarrow a^{\hat{A}} \leq a^{\hat{B}} . \end{aligned}$$

□

In [20] the author showed that there exists a relation between q-observable functions and q-antonymous functions. In particular, we have the following Lemma:

Lemma 5.3.1 *Given a self-adjoint operator $\hat{A} \in SA(\mathcal{N})$ with associated q-observable function and q-antonymous functions given by $o^{\hat{A}}$ and $a^{\hat{A}}$, respectively, then for all $\hat{P} \in P(\mathcal{N}) \setminus \{0, 1\}$ we obtain*

$$a^{\hat{A}}(\hat{P}) \leq o^{\hat{A}}(\hat{P}) .$$

Proof The self-adjoint operator $\hat{A}$ has associated with it both a right-continuous extended spectral family $E^{\hat{A}}$ and a left-continuous extended spectral family $F^{\hat{A}}$. In terms of these the q-antonymous function $a^{\hat{A}}$ can be written as follows:

$$\begin{aligned} a^{\hat{A}}(\hat{P}) &= \sup\{r \in \overline{\mathbb{R}} | \hat{P} \leq \hat{1} - \hat{F}^{\hat{A}}_r\} \\ &= \sup\{r \in \overline{\mathbb{R}} | \hat{P} \leq \hat{1} - \hat{E}^{\hat{A}}_r\} . \end{aligned}$$

For each $r \in \mathbb{R}$ such that $\hat{P} \leq \hat{E}^{\hat{A}}_r$ then, $\hat{P} \not\leq \hat{1} - \hat{E}^{\hat{A}}_r$. Given that $a^{\hat{A}}(\hat{P})$ is the least upper bound of the set $\{r \in \overline{\mathbb{R}} | \hat{P} \leq \hat{1} - \hat{F}^{\hat{A}}_r\}$ it follows that $r > a^{\hat{A}}(\hat{P})$. Moreover, since $o^{\hat{A}}(P)$ is the greatest lower bound of the set $\{r \in \overline{\mathbb{R}} | \hat{P} \leq \hat{F}^{\hat{A}}_r\}$ it follows that $o^{\hat{A}}(\hat{P}) \geq a^{\hat{A}}(\hat{P})$. □

5.4 Relation to Inner Daseinisation

In this section we will explain how inner daseinisation is related to q-antonymous functions [20]. To this end consider a self-adjoint operator $\hat{A} \in SA(\mathcal{N})$, with left-continuous extended spectral family $F^{\hat{A}} : \overline{\mathbb{R}} \rightarrow P(\mathcal{N})$. We then construct the composite

$$F^{\delta^i(\hat{A})_{\mathcal{M}}} := \delta^o \circ F^{\hat{A}} : \overline{\mathbb{R}} \rightarrow P(\mathcal{M}) .$$

This map preserves joins since both δ^o and $F^{\hat{A}}$ do and, as such, it is left-continuous. Moreover we have that

$$\begin{aligned} F^{\delta^i(\hat{A})_{\mathcal{M}}}(-\infty) &= \hat{F}^{\delta^i(\hat{A})_{\mathcal{M}}}_{-\infty} \\ &= \delta^o(\hat{F}^{\hat{A}}_{-\infty}) \\ &= \delta^o(\hat{0}) = \hat{0} \end{aligned}$$

and

$$\begin{aligned} \bigvee_{r \in \mathbb{R}} \hat{F}^{\delta^i(\hat{A})_{\mathcal{M}}}_r &= \bigvee_{r \in \mathbb{R}} \delta^o(\hat{F}^{\hat{A}}_r) \\ &= \delta^o\left(\bigvee_{r \in \mathbb{R}} \hat{F}^{\hat{A}}_r\right) \\ &= \delta^o(\hat{1}) = \hat{1} . \end{aligned}$$

However, since for all $r \in \mathbb{R}$, $\delta^o_{\mathcal{M}}(\hat{F}^{\hat{A}}_r) \geq \hat{F}^{\hat{A}}$, then in general $\bigwedge \hat{F}^{\delta^i(\hat{A})_{\mathcal{M}}}_r \neq \hat{0}$. Therefore, similarly as it was the case for the q-observable functions, we will call

$\hat{F}_r^{\delta^i(\hat{A})_{\mathcal{M}}}$ a *weak left-continuous extended spectral family*. In the cases for which $\bigwedge \hat{F}_r^{\delta^i(\hat{A})_{\mathcal{M}}} = \hat{0}$ then $\hat{F}_r^{\delta^i(\hat{A})_{\mathcal{M}}}$ will be a *left-continuous extended spectral family*.

Given a self-adjoint operator $\hat{A} \in SA(\mathcal{N})$, its inner daseinisation is given by

$$\delta^i(\hat{A})_{\mathcal{M}} := \int_{-\infty}^{+\infty} r\, d(F^{\delta^i(\hat{A})_{\mathcal{M}}}) = \int_{-\infty}^{+\infty} r\, d(\delta^o(\hat{F}^{\hat{A}})_{\mathcal{M}}).$$

If $\hat{A} \in SA(\mathcal{N})$ is bounded from below then there exists an $r \in \mathbb{R}$ such that $\hat{F}_r^{\hat{A}} = \hat{0}$, hence $\hat{F}^{\delta^i(\hat{A})_{\mathcal{M}}} = \hat{0}$ and $\bigwedge \hat{F}_r^{\delta^i(\hat{A})_{\mathcal{M}}} = \hat{0}$. It follows that $\delta^i(\hat{A})_{\mathcal{M}}$ is a self-adjoint operator affiliated with $\mathcal{M}$ iff $F^{\delta^i(\hat{A})_{\mathcal{M}}}$ is a left-continuous extended spectral family.

Since $F^{\delta^i(\hat{A})_{\mathcal{M}}} = \delta^o \circ F^{\hat{A}}$ is a composite of two left adjoints, it is itself left adjoint and, as such, it has a right adjoint defined as follows:

$$z^{\delta^i(\hat{A})_{\mathcal{M}}} := z^{\hat{A}} \circ i : P(\mathcal{M}) \to \overline{\mathbb{R}}$$

where $z^{\hat{A}} : P(\mathcal{N}) \to \mathbb{R}$ is the right adjoin of $F^{\hat{A}}$ given by Theorem 5.1.3 and $i : P(\mathcal{M}) \to P(\mathcal{N})$ is the inclusion map. We then obtain the following theorem [20]:

Theorem 5.4.1 *Consider two von Neumann algebras $\mathcal{N}$ and $\mathcal{M}$ such that $\mathcal{M} \subseteq \mathcal{N}$ and the unit elements of both coincide. Then, given a self-adjoint operator $\hat{A}$ affiliated with $\mathcal{N}$, the function $z^{\delta^i(\hat{A})_{\mathcal{M}}}$, corresponding to the inner daseinisation $\delta^i(\hat{A})_{\mathcal{M}}$ of $\hat{A}$, is given by*

$$z^{\delta^i(\hat{A})_{\mathcal{M}}} = z^{\hat{A}} \circ i = z^{\hat{A}}|_{P(\mathcal{N})} : P(\mathcal{N}) \to \overline{\mathbb{R}}.$$

If $\hat{A}$ is bounded from below, then $\delta^i(\hat{A})_{\mathcal{M}} = \bigvee\{\hat{B} \in SA(\mathcal{M}) | \hat{B} \leq_s \hat{A}\} \in SA(\mathcal{M})$.

In order to relate this to q-antonymous functions we note that, given any $\hat{A} \in SA(\mathcal{N})$,

$$a^{\hat{A}}(\hat{P}) = z^{\hat{A}}(\hat{1} - \hat{P}) \qquad \forall\, \hat{P} \in P(\mathcal{N}).$$

In fact, since $z^{\hat{A}}$ is the right adjoint of $F^{\hat{A}}$, from Theorem 5.1.3 it follows that $z^{\hat{A}}$ can be explicitly constructed as

$$\begin{aligned} z^{\hat{A}} : P(\mathcal{N}) &\to \overline{\mathbb{R}} \\ \hat{P} &\mapsto \sup\{r \in \mathbb{R} | \hat{F}_r^{\hat{A}} \leq \hat{P}\}. \end{aligned}$$

However, since $\hat{1} - \hat{F}_r^{\hat{A}} \geq \hat{P}$ iff $\hat{F}_r^{\hat{A}} \leq \hat{1} - \hat{P}$, then $a^{\hat{A}}(\hat{P}) = z^{\hat{A}}(\hat{1} - \hat{P})$.

We can now relate inner daseinisation to q-antonymous functions obtaining the analogue of Theorem 5.2.1.

Theorem 5.4.2 *Consider a von Neumann algebra $\mathcal{N}$ and a self-adjoint operator $\hat{A}$ affiliated with $\mathcal{N}$. If we then consider a von Neumann algebra $\mathcal{M} \subseteq \mathcal{N}$ such that the unit elements of $\mathcal{N}$ and $\mathcal{M}$ coincide, the weak q-antonymous function of $\delta^i(\hat{A})_{\mathcal{M}}$ is given by*

$$a^{\delta^i(\hat{A})_{\mathcal{N}}} = a^{\hat{A}}_{\mathcal{N}} \circ i = a^{\hat{A}}_{\mathcal{N}}|_{P(\mathcal{N})} : P(\mathcal{N}) \to \overline{\mathbb{R}} .$$

If $\hat{A}$ is bounded from below, then $a^{\delta^o(\hat{A})_{\mathcal{N}}}$ is a proper q-antonymous function.

5.5 q-Functions and Quantum Probabilities

In this section we are going to expand on the topic of the previous section and explain how q-functions can be interpreted as generalised quantile functions for quantum observables, seen as random variables. The content of this section is a summary of the results presented in [21]. In order to make this section as self sufficient as possible we will need to introduce a few mathematical notations.

5.5.1 *Mathematical Background*

In classical probability theory, a measure space is given by $(\Omega, B(\Omega), \mu)$ where Ω is a non-empty set called the *sample space*, $B(\Omega)$ is a σ-algebra of μ-measurable subsets of Ω, whose elements are called *events* and $\mu : B(\Omega) \to [0,1]$ is a probability measure. Being a probability measure, μ is such that $\mu(\Omega) = 1$ and, for all countable families $(S_i)_{i\in I}$ of pairwise disjoint events, then

$$\mu\left(\bigcup_{i\in\mathbb{N}} S_i\right) = \sum_{i\in\mathbb{N}} \mu(S_i) .$$

A *random variable* is a measurable function $A : \Omega \to \text{im}A \subseteq \mathbb{R}$ and, for every Borel subset $\Delta \subseteq \mathbb{R}$, we have

$$A^{-1}(\Delta) \in B(\Omega) .$$

Definition 5.5.1 A cumulative distribution function (CDF) $\tilde{C}^A$ of a random variable A, given a probability measure μ, is defined by

$$\begin{aligned} \tilde{C}^A : \mathbb{R} &\to [0,1] \\ r &\mapsto \mu(A^{-1}(-\infty, r]) . \end{aligned}$$

An extended cumulative distribution function (ECDF) C^A of a random variable A, given a probability measure μ, is a map $C^A : \overline{\mathbb{R}} \to [0, 1]$ obtained by extending $\tilde{C}^A$ such that

$$C^A(-\infty) = 0 \qquad \text{and} \qquad C^A(+\infty) = 1\,.$$

Conceptually $C^A(r) = \mu(A^{-1}(-\infty, r])$ represents the probability of the variable A having a value which does not exceed r. The maps $C^A : \overline{\mathbb{R}} \to [0, 1]$ for each random variable A are right-continuous and order preserving. In fact, given $r \leq s$ we have that $(-\infty, r] \subseteq (-\infty, s]$ hence $A^{-1}(-\infty, r] \subseteq A^{-1}(-\infty, s]$ and $\mu(A^{-1}(-\infty, r]) \subseteq \mu(A^{-1}(-\infty, s])$. Conceptually it is clear that when $r \leq s$ then the probability of A having values not greater than r is less than the probability of A having value not greater that s, since the former carries more information than the latter. It then trivially follows that C^A is right-continuous. We can now show that C^A is a map between meet-semilattices which preserves all meets. In fact we have

$$C^A(\inf_{i\in I} r_i) \overset{\text{right continuity}}{=} \bigwedge_{s>\inf_{i\in I} r} C^A(s) \overset{\text{monotonicity}}{=} \bigwedge_{i\in I} C^A(r_i)\,.$$

It is now possible to apply Theorem 5.1.2 to construct the left adjoint of C^A as follows:

$$\begin{aligned} q^A &: [0, 1] \to \overline{\mathbb{R}} \\ p &\mapsto \inf\{r \in \overline{\mathbb{R}} | C^A(r) \geq p\}\,. \end{aligned} \tag{5.5.1}$$

This function is called the *quantile function of* A with respect to μ and it assigns to each probability $p \in (0, 1]$, the smallest value r such that the probability of A having a value not greater than r is p.

From what has been said so far, the CDF functions are probability valued functions, however, it is also possible to construct a variation of CDF functions which take values in a lattice. In particular, for the case in which the lattice in question is $B(\Omega)$ we have the following definition:

Definition 5.5.2 Consider the complete Boolean algebra $B(\Omega)$ of equivalent subsets of Ω and the inverse image $A^{-1} : B(\overline{\mathbb{R}}) \to B(\Omega)$ of the random variable A, such that A^{-1} preserves all meets. The $B(\Omega)$-cumulative distribution function ($B(\Omega)$-CDF) of A^{-1} is given by

$$\begin{aligned} \overline{C}^A &: \overline{\mathbb{R}} \to B(\Omega) \\ r &\mapsto A^{-1}([-\infty, r])\,. \end{aligned}$$

If we then replace $B(\Omega)$ with a general meet-semilattice the definition becomes:

Definition 5.5.3 Given the meet-semilattice L and the L-valued measure[4] A^{-1} such that $A^{-1}(\emptyset) = \bot_L$ and $A^{-1}(\overline{\mathbb{R}}) = \top_L$, then the L-CDF of A^{-1} is given by

$$\overline{C}^A : \overline{\mathbb{R}} \to L \tag{5.5.2}$$

$$r \mapsto A^{-1}([-\infty, r]) . \tag{5.5.3}$$

Lemma 5.5.1 *Given a complete meet-semilattice L and an L-valued measure* $A^{-1} : B(\overline{\mathbb{R}}) \to L$ *which preserves all existing meets, then the L-CDF,* $\overline{C}^A : \overline{\mathbb{R}} \to L$ *of* A^{-1} *preserves all meets.*

Proof Consider any family $(r_i)_{i \in I} \subseteq \overline{\mathbb{R}}$, we then obtain:

$$\begin{aligned}
\overline{C}^A(\inf_{i \in I} r_i) &= A^{-1}([-\infty, \inf_{i \in I} r_i] \\
&= A^{-1}\left(\bigcup_{i \in I} [-\infty, r_i]\right) \\
&= \bigwedge_{i \in I} A^{-1}([-\infty, r_i]) \\
&= \bigwedge_{i \in I} \overline{C}^A(r_i) .
\end{aligned}$$

□

Since $\overline{C}^A$ preserves all meets it will have a left adjoint which, from Theorem 5.1.2, is defined by:

$$\overline{q}^A : L \to \overline{\mathbb{R}} \tag{5.5.4}$$

$$T \mapsto \inf\{r \in \overline{\mathbb{R}} | T \leq \overline{C}^A(r)\} \tag{5.5.5}$$

and it preserves all joins. $\overline{q}^A$ is called the *L-quantile function* of A^{-1}. It is possible to express the usual CDF in terms of the L-CDF as follows:

$$C^A : \mu \circ \overline{C}^A : \overline{\mathbb{R}} \to L \to [0, 1] \tag{5.5.6}$$

$$r \mapsto \mu(A^{-1}([-\infty, r]) . \tag{5.5.7}$$

[4] Here the notation A^{-1} is only symbolic since there may not exist any function A whose inverse is A^{-1}. We used this notation to resemble the Definition in 5.5.2.

Since by assumption A^{-1} is meet preserving, it follows that the L-CDF $\overline{C}^A$ is also meet preserving and, as such, it has a left adjoint

$$\begin{aligned} k : [0,1] &\to L \\ s &\mapsto \bigwedge\{T \in L | s \leq \mu(T)\} . \end{aligned}$$

Hence $C^A = \mu \circ \overline{C}^A$ also preserves meets and has a left adjoint $q^A : [0,1] \to \overline{\mathbb{R}}$ which represents the quantile function of the random variable A. This can be decomposed as

$$q^A = \overline{q}^A \circ k .$$

5.5.2 *Quantum Theory*

In this section we would like to apply what we have learned to the case of quantum theory. To this end we recall that, given a noncommutative von Neumann algebra $\mathcal{N}$ and a self-adjoint operator $\hat{A}$ representing a random variable, then the spectral measure of $\hat{A}$ is given by

$$e^{\hat{A}} : B(\overline{\mathbb{R}}) \to P(V_{\hat{A}}) \hookrightarrow P(\mathcal{N}) .$$

Here $V_{\hat{A}} = \{\hat{A}, \hat{1}\}''$ and $P(V_{\hat{A}})$ is the complete Boolean algebra of projection operators in $V_{\hat{A}}$. This is a subalgebra of the complete orthomodular lattice $P(\mathcal{N})$ of projection operators in $\mathcal{N}$. Clearly $e^{\hat{A}}$ can be seen as a $P(\mathcal{N})$-valued measure.

In previous sections we encountered the extended right-continuous spectral family $E^{\hat{A}}$ of the self-adjoint operator $\hat{A}$. This can be written as

$$\begin{aligned} E^{\hat{A}} : \overline{\mathbb{R}} &\to P(\mathcal{N}) \\ r &\mapsto e^{\hat{A}}([-\infty, r]) . \end{aligned}$$

If we compare this definition with (5.5.2), it follows that $E^{\hat{A}}$ is the $P(\mathcal{N})$-CDF of the projection valued measure $e^{\hat{A}}$. The analogue of (5.5.4) is then given by the q-observable function

$$o^{\hat{A}} : P(\mathcal{N}) \to \overline{\mathbb{R}} .$$

It is in this sense that q-observable functions are interpreted as $P(\mathcal{N})$-quantile functions of quantum variables described by $P(\mathcal{N})$-valued measures.

Now we would like to construct the quantum version of the CDF function defined in (5.5.6). To this end we need the analogue of the probability measure $\mu : B(\Omega) \to [0, 1]$. This is given by the finitely additive measure

$$\mu_\rho : P(\mathcal{N}) \to [0, 1]$$

equivalent to the state $\rho : \mathcal{N} \to \mathbb{C}$ via Gleason's Theorem. In particular we have that

$$\mu_\rho = \rho|_{P(\mathcal{N})} .$$

From now on we will restrict our attention to normal states. These are state such that, for all $\hat{A} \in P(\mathcal{N})$ then $\rho(\hat{A}) = \text{tr}(\tilde{\rho}\hat{A})$, where $\tilde{\rho}$ is some positive trace-class operator of trace 1.

In this setting, a CDF of a quantum random variable $\hat{A}$, given a state ρ, is defined as

$$C^{\hat{A}} = \mu_\rho \circ E^{\hat{A}} : \overline{\mathbb{R}} \to [0, 1] \tag{5.5.8}$$
$$r \mapsto \mu_\rho(\hat{E}^{\hat{A}}_r) .$$

Lemma 5.5.2 *The CDF $C^{\hat{A}}$ defined in* (5.5.8) *preserves meets.*

Proof We know that the extended spectral family $E^{\hat{A}}$ preserves meets since it is monotonic and right-continuous, hence we only need to show that μ_ρ preserves meets. To this end let us consider an arbitrary family $(r_i)_{i\in I}$ of real numbers. Then

$$\begin{aligned}
\mu_\rho\left(\bigwedge_{i\in I} \hat{E}^{\hat{A}}_{r_i}\right) &= \mu_\rho\left(\hat{1} - \bigvee_{i\in I}(\hat{1} - \hat{E}^{\hat{A}}_{r_i})\right) \\
&= 1 - \mu_\rho\left(\hat{1} - \hat{E}^{\hat{A}}_{r_i}\right) \\
&= 1 - \sup_{i\in I} \mu_\rho(\hat{1} - \hat{E}^{\hat{A}}_{r_i}) \\
&= \inf_{i\in I} \mu_\rho(\hat{E}^{\hat{A}}_{r_i})
\end{aligned}$$

where the third equality follows from the fact that normal states preserve suprema of increasing nets. □

From the above proof we have discovered that μ_ρ preserves meets, hence it has a left adjoint given by

$$k_\rho : [0, 1] \to P(\mathcal{N})$$
$$s \mapsto \bigwedge\{\hat{P} \in P(\mathcal{N}) | s \leq \mu_\rho(\hat{P})\} .$$

This, together with the left adjoint $o^{\hat{A}} : P(\mathcal{N}) \to \overline{\mathbb{R}}$ of $E^{\hat{A}}$, allows us to define the left adjoint of $C^{\hat{A}} = \mu_\rho \circ E^{\hat{A}}$ as

$$q^{\hat{A}} = o^{\hat{A}} \circ k_\rho : [0, 1] \to \overline{\mathbb{R}}$$
$$s \mapsto o^{\hat{A}}(k_\rho(s)) .$$

$q^{\hat{A}}$ represents the quantile function of the quantum random variable $\hat{A}$ with respect to the normal state ρ.

It is also possible to define $q^{\hat{A}}$ by applying Theorem 5.1.2. In this case we would obtain

$$q^{\hat{A}} : [0, 1] \to \overline{\mathbb{R}}$$
$$s \mapsto \inf\{r \in \overline{\mathbb{R}} | s \leq C^{\hat{A}}(r)\} .$$

Lemma 5.5.3 ([21]) *The two expressions for $q^{\hat{A}}$ are equivalent.*

Proof To show that the two expressions for the quantile function coincide we need to show that, for all $a \in [0, 1]$ then $(o^{\hat{A}} \circ k_\rho)(s) = \inf\{r \in \overline{\mathbb{R}} | s \leq C^{\hat{A}}(r)\}$. From the definition of $C^{\hat{A}}$ we have that

$$\inf\{r \in \overline{\mathbb{R}} | s \leq C^{\hat{A}}(r)\} = \inf\{r \in \overline{\mathbb{R}} | s \leq \mu_\rho(\hat{E}^{\hat{A}}_r)\} .$$

On the other hand, applying the definition given in (5.1.4) we obtain

$$o^{\hat{A}}(k_\rho(s)) = \inf\{r \in \overline{\mathbb{R}} | \bigwedge\{\hat{P} \in P(\mathcal{N}) | s \leq \mu_\rho(\hat{P}) \leq \hat{E}^{\hat{A}}_r\} .$$

Clearly if $\hat{P} \leq \hat{E}^{\hat{A}}_r$ then $\mu_\rho(\hat{P}) \leq \mu_\rho(\hat{E}^{\hat{A}}_r)$. Conversely given $\hat{P} = k_\rho(s) = \bigwedge\{\hat{Q} \in P(\mathcal{N}) | s \leq \mu_\rho(\hat{P})\}$ which is the smallest projection for which $\mu_\rho(\hat{P}) \geq s$, then $\mu_\rho(\hat{E}^{\hat{A}}_r) \geq \mu_\rho(\hat{P}) \geq s$ implies that $\hat{E}^{\hat{A}}_r \geq \hat{P} = k_\rho(s)$. It follows that the two expressions for $q^{\hat{A}}$ coincide. □

5.5.3 The Case for Topos Quantum Theory

In [21] it was shown how to apply the ideas of the previous section to the case of topos quantum theory. As a first step we recall that in this setting a random variable is defined as a function from the spectra presehaf $\underline{\Sigma}$ to the presehaf representing the quantity value object $\underline{\mathbb{R}}^{\leftrightarrow}$:

$$\breve{\delta}(\hat{A}) : \underline{\Sigma} \to \underline{\mathbb{R}}^{\leftrightarrow} .$$

The inverse image of a random variable is then given by

$$\breve{A}^{-1} : B(\overline{\mathbb{R}}) \to \text{Sub}_{cl}\underline{\Sigma}$$
$$\Delta \mapsto \underline{\delta^o}(e^{\hat{A}}(\Delta)) =: \underline{S}(\hat{A}, \Delta)\,.$$

Here

$$e^{\hat{A}} : B(\overline{\mathbb{R}}) \to P(\mathcal{N})$$
$$\Delta \mapsto \hat{E}[A \in \Delta] =: e^{\hat{A}}(\Delta)$$

is the extended spectral measure of $\hat{A}$ and the projection $e^{\hat{A}}(\Delta)$ represents the proposition "if a measurement on A is performed, the result will lie in the Borel set Δ". The clopen sub-object $\underline{\delta^o}(e^{\hat{A}}(\Delta)) =: \underline{S}(\hat{A}, \Delta)$ is obtained via the daseinisation map defined in Definition 2.2.2.

Definition 5.5.4 Given a random variable $\breve{\delta}(\hat{A}) : \underline{\Sigma} \to \underline{\mathbb{R}}^{\leftrightarrow}$ with associated inverse image $\breve{A}^{-1} : B(\overline{\mathbb{R}} \to \text{Sub}_{cl}\underline{\Sigma}$ we define the $\text{Sub}_{cl}\underline{\Sigma}$-valued CDF function by:

$$E^{\breve{A}} : \overline{\mathbb{R}} \to \text{Sub}_{\text{cl}}\underline{\Sigma}$$
$$r \mapsto \breve{A}^{-1}([-\infty, r]) = \underline{\delta^o}(\hat{E}^{\hat{A}}_r) = \underline{S}(\hat{A}, [-\infty, r])\,,$$

where $E^{\breve{A}}(-\infty) = \underline{\emptyset}$ since $\hat{E}^{\hat{A}}_{-\infty} = \hat{0}$ and $E^{\breve{A}}(+\infty) = \underline{\Sigma}$ since $\hat{E}^{\hat{A}}_{+\infty} = \hat{1}$. Clearly the map $E^{\breve{A}}$ can be decomposed as follows:

$$E^{\breve{A}} = \underline{\delta} \circ E^{\hat{A}} : \overline{\mathbb{R}} \to \text{Sub}_{\text{cl}}\underline{\Sigma}$$

Theorem 5.5.1 *Given a von Neumann algebra $\mathcal{N}$ and a self-adjoint operator $\hat{A} \in \mathcal{N}_{sa}$, then the map*

$$E^{\breve{A}} : \overline{\mathbb{R}} \to \text{Sub}_{cl}\underline{\Sigma}$$

is right-continuous.

Proof To show that $E^{\breve{A}}$ is right-continuous we need to show that

$$\bigwedge_{s>r} E^{\breve{A}}(s) = E^{\breve{A}}(r)\,.$$

Since $E^{\breve{A}}$ is a map between presheaves, we need to show that the above equality holds for each context $V \in \mathcal{N}$. In particular, for all $V \in \mathcal{N}$ we have that

$$\begin{aligned}
\Big(\bigwedge_{s>r} E^{\breve{A}}(s)\Big)_V &= \Big(\bigwedge_{s>r} \underline{\delta(\hat{E}^{\hat{A}}_s)}\Big)_V \\
&= \bigwedge_{s>r} \delta^o(\hat{E}^{\hat{A}}_s)_V \\
&= \bigwedge_{s>r} \bigwedge \{\hat{Q} \in P(V) | \hat{E}^{\hat{A}}_s \le \hat{Q}\} \\
&= \bigwedge \{\hat{Q} \in P(V) | \bigwedge_{s>r} \hat{E}^{\hat{A}}_s \le \hat{Q}\} \\
&= \bigwedge \{\hat{Q} \in P(V) | \hat{E}^{\hat{A}}_r \le \hat{Q}\} \\
&= \delta^o(\hat{E}^{\hat{A}}_r)_V \\
&= \underline{\delta(\hat{E}^{\hat{A}}_r)}_V .
\end{aligned}$$

□

Clearly $E^{\breve{A}}$ preserves meets therefore it has a left given by

$$\begin{aligned}
o^{\hat{A}} : \mathrm{Sub}_{cl}\underline{\Sigma} &\to \overline{\mathbb{R}} \\
\underline{S} &\mapsto \inf\{r \in \overline{\mathbb{R}} | \underline{S} \le E^{\breve{A}}(r)\}
\end{aligned}$$

which represents the $\mathrm{Sub}_{cl}\underline{\Sigma}$-quantile function associated to $E^{\breve{A}}$. Applying such a function to the special case for which $\underline{S} = \underline{\delta(\hat{P})}$ we obtain

$$\begin{aligned}
o^{\breve{A}}(\underline{\delta(\hat{P})}) &= \inf\{r \in \overline{\mathbb{R}} | \underline{\delta(\hat{P})} \le E^{\breve{A}}\} \\
&= \inf\{r \in \overline{\mathbb{R}} | \underline{\delta(\hat{P})} \le \underline{\delta(\hat{E}^{\hat{A}}_r)}\} \\
&= \inf\{r \in \overline{\mathbb{R}} | \hat{P} \le \hat{E}^{\hat{A}}_r\} \\
&= o^{\hat{A}}(\hat{P}) .
\end{aligned}$$

Now we would like to define the topos quantum analogue of both the CDF function $C^{\hat{A}}$ and the quantile function $q^{\hat{A}}$. This requires the notion of a *probability measure* on $\underline{\Sigma}$ given in Definition 3.4.4.

Definition 5.5.5 Given a state ρ with associated probability measure μ_ρ and the inverse image $\breve{A}^{-1}$ of a random variable $\breve{A}$, then the corresponding CDF is defined by

$$\begin{aligned} C^{\breve{A}} &: \overline{\mathbb{R}} \to [0,1] \\ r &\mapsto \min_{V \in V(\mathcal{N})} \mu_\rho(E^{\breve{A}}(r)), \end{aligned}$$

where the minimum is obtained at those contexts $V \in V(\mathcal{N})$ such that $\hat{A} \in V$. From Definition 5.5.4, we know that $E^{\breve{A}}(r) = \breve{A}^{-1}([-\infty, r]) = \underline{\delta(\hat{E}_r^{\hat{A}})}$.

The corresponding quantile function is then given by the left adjoint of $C^{\breve{A}}$ which is defined as follows:

$$\begin{aligned} q^{\breve{A}} &: [0,1] \to \overline{\mathbb{R}} \\ s &\mapsto \inf\{r \in \overline{\mathbb{R}} | s \le C^{\breve{A}}(r)\}. \end{aligned}$$

Chapter 6
What Information Can Be Recovered from the Abelian Subalgebras of a von Neumann Algebra

As explained in [26], the motivation for constructing topos quantum theory is to render quantum theory more "realist". This is done by expressing quantum theory in terms of similar mathematical constructs used to formalise classical physics, in the hope that the ensuing interpretation would be more "realist" (as is the case for classical theory). This is achieved by describing quantum objects in terms of presheaves in the topos $\mathbf{Sets}^{\mathcal{V}(\mathcal{H})}$, where $\mathcal{V}(\mathcal{H})$ is the poset of abelian von Neumann subalgebras of the algebra of bounded operators $\mathcal{B}(\mathcal{H})$. Each such abelian subalgebra $V \in \mathcal{V}(\mathcal{H})$ represents a classical snapshot since it contain only simultaneously measurable observables. Hence a quantum object can be seen as a collection of classical approximation "glued" together by the categorical structure of the base category $\mathcal{V}(\mathcal{H})$. The quantum information is then contained in the categorical structure of $\mathcal{V}(\mathcal{H})$ which is also reflected at the level of presheaves. The question that then comes to mind is the following: given a general von Neumann algebra $\mathcal{N}$ and its collection of abelian subalgebras $\mathcal{V}(\mathcal{N})$, how much of $\mathcal{N}$, if any, can be reconstructed from $\mathcal{V}(\mathcal{N})$? This question was asked in [37]. There it was shown that, if the initial von Neumann algebra $\mathcal{N}$ is abelian, then it can be completely reconstructed from the poset of its abelian von Neumann subalgebras. However, if the algebra $\mathcal{N}$ is not abelian, then it can only be reconstructed up to its Jordan structure. This is because both $\mathcal{N}$ and its opposite op($\mathcal{N}$) have the same collection of subalgebras but they are not necessarily isomorphic to each other [9].

6.1 Mathematical Preliminaries

In the following we will explain the results obtained in [37]. In order to make the exposition more compatible with the original article we will utilise, whenever possible, the notation put forward in [37].

C. Flori, *A Second Course in Topos Quantum Theory*,
Lecture Notes in Physics 944, https://doi.org/10.1007/978-3-319-71108-9_6

For clarity purpose we will recall a few definition regarding von Neumann algebras which will be useful in understanding the content of this chapter. We will start with the notion of the *commutant* of a given set. In particular consider the C^* algebra of all bounded operators on a Hilbert space $\mathcal{H}$. This is denoted by $\mathcal{B}(\mathcal{H})$. Given a subset S of such an algebra, the commutant of S, which is denoted by S', is the set of elements in $\mathcal{B}(\mathcal{H})$ which commute with each element of S. It follows that the double commutant of S, which is denoted by $S^{''}$, is the set of element in $\mathcal{B}(\mathcal{H})$ which commute with each element in S'. It is possible to define a von Neumann algebra in terms of the notion of double commutant. In particular, a subset $\mathcal{N} \subseteq \mathcal{B}(\mathcal{H})$ is a von Neumann algebra if $\mathcal{N} = \mathcal{N}^{''}$, i.e. a von Neumann algebra is equal to its own double commutant. The set of projections in a von Neumann algebra $\mathcal{N}$ is denoted by $Proj(\mathcal{N})$ and it forms an orthomodular lattice. It turns out that a von Neumann algebra is uniquely determined by its set of projection. In particular we have that

$$\mathcal{N} = (Proj(\mathcal{N}))^{''}.$$

Given a von Nuemann algebra $\mathcal{N}$, a von Neumann subalgebra $\mathcal{V}$ of $\mathcal{N}$ is defined to be a subset $\mathcal{V} \subseteq \mathcal{N}$ which is itself a von Neumann algebra. In the following sections we will consider particular types of von Nuemann subalgebras of a given von Neumann algebra $\mathcal{N}$.

Definition 6.1.1 Given a von Neumann algebra $\mathcal{N}$, $Sub(\mathcal{N})$ will denote the set of all von Neumann subalgebras of $\mathcal{N}$, $AbSub(\mathcal{N})$ will denote the set of all abelian subalgebras of $\mathcal{N}$, and $FAbSub(\mathcal{N})$ will denote the set of all abelian subalgebras of $\mathcal{N}$ containing only finitely many projections. Each of these sets are ordered by set inclusion.

Each of the above sets can be shown to be a lattice. In particular, $Sub(\mathcal{N})$ is a complete lattice where meets are given by intersection and, given a family $(V_i)_{i \in I}$ of subalgebras, the join is the weak closure of the algebra generated by S_i, $i \in I$.

$AbSub(\mathcal{N})$ is a complete meet semilattice where joins are defined only for those subsets which are closed under finite joins. If $\mathcal{N}$ is not abelian, then $AbSub(\mathcal{N})$ has no top element.

$FAbSub(\mathcal{N})$ is a complete meet semilattice for which every meet is finite. If $\mathcal{N}$ is not abelian, then $FAbSub(\mathcal{N})$ does not have a top element.

Similarly, for an orthomodular lattice $Proj(\mathcal{N})$ of projections in $\mathcal{N}$ we have the following:

Definition 6.1.2 *Sub*($Proj(\mathcal{N})$) denotes the poset of subalgebras of $Proj(\mathcal{N})$ ordered by subset inclusion; $BSub(Proj(\mathcal{N}))$ denotes the poset of Boolean subalgebras of $Proj(\mathcal{N})$ ordered by subset inclusion and $FBSub(Proj(\mathcal{N}))$ denotes the poset of finite Boolean subalgebras of $Proj(\mathcal{N})$ ordered by subset inclusion.

Given two von Neumann algebras $\mathcal{M}$ and $\mathcal{N}$, a *-isomorphism between them is a map $f : \mathcal{M} \to \mathcal{N}$ which is linear, bijective, preserves the involution $*$ and is such that $f(ab) = f(a)f(b)$. On the other hand a *-antiisomorphism is a map $f : \mathcal{M} \to \mathcal{N}$ which is linear, bijective, preserves the involution $*$ and is such that $f(ab) = f(b)f(a)$. In previous sections we have seen that to each von Neumann algebra $\mathcal{N}$ one can associate its Jordan algebra where the Jordan product is defined as $a \circ b = \frac{1}{2}(ab + ba)$. Such a product is commutative but not necessarily associative, as opposed to the von Nuemann product which is associative but not necessarily commutative. If we now consider two von Neumann algebras $\mathcal{M}$ and $\mathcal{N}$, a Jordan isomorphism between them is a linear bijective function which preserves the involution and it is such that $f(a \circ b) = f(a) \circ f(b)$. The following is a well known result.

Theorem 6.1.1 *Every Jordan isomorphism $f : \mathcal{M} \to \mathcal{N}$ between von Neumann algebras can be decomposed as a sum of a *-isomorphism and a *-antiisomorphism.*

Of particular importance for the next section is a result by Dye.

Theorem 6.1.2 ([25]) *Consider two von Neumann algebras $\mathcal{M}$ and $\mathcal{N}$ without type I_2 summands. To each orthomodular lattice isomorphism $\psi : Proj(\mathcal{M}) \to Proj(\mathcal{N})$ there corresponds a unique Jordan *-isomorphism $\Psi : \mathcal{M} \to \mathcal{N}$, such that $\Psi(\hat{P}) = \psi(\hat{P})$ for all $\hat{P} \in Proj(\mathcal{M})$.*

This theorem essentially tells us that Ψ extends the action of ψ on projections. Given the spectral theorem by defining how Ψ acts on projections, one is able to define its action on any other element in the algebra. Hence Ψ will be the unique Jordan *-isomorphism which extends ψ.

Another important result which we will use in the next section is the following theorem:

Theorem 6.1.3 ([38]) *Given two orthomodular lattices (OML) L and M which have no blocks with four elements, and an isomorphism $\phi : BSub(L) \to BSub(M)$ of posets, then there is a unique isomorphism $\Phi : L \to M$ with $\phi(A) = \Phi[A]$ for each Boolean subalgebra A of L.*

We are interested in the particular case in which the orthomodular lattices in question are $Proj(\mathcal{M})$ and $Proj(\mathcal{N})$ for von Neumann algebras $\mathcal{M}$ and $\mathcal{N}$, respectively. For the theorem to apply we require both $Proj(\mathcal{M})$ and $Proj(\mathcal{N})$ not to have any blocks with four elements. However, this is equivalent to the condition that neither $\mathcal{M}$ nor $\mathcal{N}$ are isomorphic to $\mathbb{C} \oplus \mathbb{C}$ or $\mathcal{B}(\mathbb{C} \oplus \mathbb{C})$. Therefore, rephrasing the above theorem for the case at hand we obtain:

Corollary 6.1.1 *Consider two von Neumann algebras $\mathcal{M}$ and $\mathcal{N}$ which are not isomorphic to $\mathbb{C} \oplus \mathbb{C}$ nor $\mathcal{B}(\mathbb{C} \oplus \mathbb{C})$, then given an isomorphism $\psi : BSub(Proj(\mathcal{M})) \to BSub(Proj(\mathcal{N}))$ there exists a unique isomorphism $\Phi : Proj(\mathcal{M}) \to Proj(\mathcal{N})$ with $\phi(A) = \Phi[A]$ for each Boolean subalgebra A of $Proj(\mathcal{M})$.*

6.2 Reconstructing the Jordan Structure

In this section we will show how it is possible to retrieve the Jordan information of a von Nuemann algebra $\mathcal{N}$ given the poset of abelian subalgebras of $\mathcal{N}$.

Theorem 6.2.1 ([37]) *Consider two von Neumann algebras $\mathcal{M}$ and $\mathcal{N}$, neither isomorphic to $\mathbb{C} \oplus \mathbb{C}$ and without type I_2 summand. Given an order-isomorphism $f : AbSub(\mathcal{M}) \to AbSub(\mathcal{N})$ there exists a unique Jordan *-isomorphism $F : \mathcal{M} \to \mathcal{N}$ such that $f(S) = F[S]$ for all $S \in \mathcal{N}$. Here $F[S]$ denotes the image $\{ F(s) | s \in S\}$ of the set S under F.*

Before proving this theorem we will need a few more results.

Lemma 6.2.1 *Given a von Neumann algebra $\mathcal{N}$, then there exists an order isomorphism*

$$\Psi : FAbSub(\mathcal{N}) \to FBSub(Proj(\mathcal{N}))$$
$$S \mapsto \Psi(S) := S \cap Proj(\mathcal{N}).$$

Proof First of all we need to show that the map Ψ is well defined. It is a known result that the projections of any abelian subalgebra of $\mathcal{N}$ form a Boolean subalgebra of $Proj(\mathcal{N})$ [1], hence $S \cap Proj(\mathcal{N}) \in FBSub(Proj(\mathcal{N}))$. To show that Ψ is order preserving we need to show that if $\Psi(S) \subseteq \Psi(T)$ then $S \subseteq T$. Assuming that indeed $\Psi(S) \subseteq \Psi(T)$ then, since $\Psi(S) = Proj(S)$ and $S = (Proj(S))''$, it follows that $S = (\Psi(S))'' \subseteq (\Psi(T))'' = T$. Clearly Ψ is injective by construction, hence all that remains to be shown is that it is also onto. That is, for each $T \in FBSub(Proj(N))$ we need to show that $T = \Psi(S)$ for some $S \in FAbSub(\mathcal{N})$. Assume we are given $T \in FBSub(Proj(\mathcal{N}))$ whose minimal elements are $\hat{P}_1, \hat{P}_2, \ldots, \hat{P}_n$, then we define the map $\gamma : \mathbb{C}^n \to \mathcal{N}$ such that $\gamma(\lambda_1, \lambda_2, \ldots, \lambda_n) = \sum_{i=1}^n \lambda_i \hat{P}_i$. Clearly such a map is a unital *-isomorphism which maps into $\mathcal{N}$. In [1, Lemma 2.100] it was shown that, given a normal unital *-isomorphism γ from a von Neumann algebra $\mathcal{M}$ into a von Neumann algebra $\mathcal{N}$, then $\gamma(\mathcal{M})$ is a σ-weakly closed subalgebra of $\mathcal{N}$, hence a von Neumann algebra. This implies that $\gamma(\mathbb{C}^n) = S$ is a von Neumann subalgebra of $\mathcal{N}$. By construction S has finitely many projections and it is abelian. Moreover $\Psi(S) = S \cap Proj(\mathcal{M}) = B$ hence Ψ is onto. □

Lemma 6.2.2 *Given two orthomodular lattices M and N, then each order-isomorphism $\alpha : FBSub(N) \to FBsub(M)$ extends uniquely to an isomorphism $\beta : BSub(N) \to BSub(M)$.*

Proof To prove the above Lemma we need to recall the definition of an ideal of a partially ordered set. This is a non-empty subset I such that $(\forall x \in I) y \leq x \Rightarrow y \in I$ and $\forall x, y \in I \;\; \exists z \in I$ s.t. $(x \leq z) \wedge (y \leq z)$, i.e. I is a downset. In the case at hand we define I to be that downset of $FBSub(N)$, such that any two elements in I have a join and that join belongs to I. In particular, for any element $x \in BSub(N)$, we define an ideal I of $FBSub(N)$ to be the set $A_x = \downarrow x \cap FBSub(N) = \{z \in FBSub(N) | z \subseteq x\}$. Clearly if $w, y \in A_x$ then $w \vee y \in A_x$ since $w \subseteq x$ and $y \subseteq x$. Moreover $y \leq w \vee y$ and $w \leq w \vee y$ hence the second condition of an ideal is satisfied. Regarding the first condition, for any $y \in A_x$ if $w \subseteq y$ then clearly $w \subseteq x$ and $w \in A_x$. It is easy to see that the join of A_x in $BSUb(N)$ is x. Since each finitely generated subalgebra in a subalgebra lattice is compact, all ideals in $FBSub(N)$ are of the form defined above.

Next, given an order-isomorphism $\alpha : FBSub(N) \to FBsub(M)$, we construct $\beta : BSub(N) \to BSub(M)$ as follows:

$$\beta(x) = \bigvee \alpha[\downarrow x \cap FBSub(N)].$$

Since α is an isomorphism it preserves ideals hence the join is well defined since $\alpha[\downarrow x \cap FBSub(N)]$ is an ideal in $FBSub(\mathcal{M})$. To show that β is order preserving we need to show that if $\beta(x) \leq \beta(y)$ then $x \leq y$. If $\beta(x) \leq \beta(y)$, then from the definition $\bigvee \alpha[\downarrow x \cap FBSub(N)] \leq \bigvee \alpha[\downarrow y \cap FBSub(N)]$, therefore for each $z \in \downarrow x \cap FBSub(N)$ it follows that $\alpha(z) \leq \bigvee \alpha[\downarrow x \cap FBSub(N)] \leq \bigvee \alpha[\downarrow y \cap FBSub(N)]$. Since α is an order isomorphism and each finite Boolean subalgebra is compact[1] it follows that $z \leq \bigvee[\downarrow y \cap FBSub(N)] = y$. This is true for all $z \in \downarrow x \cap FBSub(N)$, hence $x \leq y$.

The fact that β is one-two-one is obvious, hence all that remains to be shown is that it is onto. Clearly given any $x \in BSub(M)$ with associated ideal $I_x = \downarrow x \cap FBSub(M) = \{z \in FBSub(M) | z \subseteq x\}$ in $FBSub(M)$, then $x = \bigvee I_x$. Since α is an order-isomorphism $\alpha^{-1}(I_x)$ is an ideal in $FBSub(N)$ and, as such, it has a join $y \in BSub(N)$, i.e. $y = \bigvee \alpha^{-1}(I_x) = \bigvee[\downarrow y \cap FBSub(N)]$. Then $\beta(y) = x$.

We have then showed that β, as defined above, is an order-isomorphism and clearly it extends α. For uniqueness, consider any other order-isomorphism $\beta' : BSub(N) \to BSub(M)$ which extends α, then by definition β' preserves joins, hence $\beta'(x) = \beta'(\bigvee[\downarrow x \cap FBSub(N)]) = \bigvee \beta'[\downarrow x \cap FBSub(N)] = \bigvee \alpha[\downarrow x \cap FBSub(N)] = \beta(x)$. □

We are now ready to prove Theorem 6.2.1

Proof As a first step we will show that there exists a functorial assignment for each order-isomorphism $f : AbSub(\mathcal{M}) \to AbSub(\mathcal{N})$ of a Jordan *-isomorphism $F : \mathcal{M} \to \mathcal{N}$. In particular, assume we have an order-isomorphism $f : AbSub(\mathcal{M}) \to AbSub(\mathcal{N})$, this clearly restrict to an order-isomorphism $g : FAbSub(\mathcal{M}) \to FAbSub(\mathcal{N})$ since the elements in $FAbSub(\mathcal{M})$ are those elements in $AbSUb(\mathcal{M})$ which have only finitely many elements beneath them. From Lemma 6.2.1 it follows that there exist two order-isomorphisms $\Psi_{\mathcal{M}} : FAbSub(\mathcal{M}) \to FBSub(Proj(\mathcal{M}))$; $S \mapsto S \cap Proj(\mathcal{M})$ and $\Psi_{\mathcal{N}} : FAbSub(\mathcal{N}) \mapsto FBSub(Proj(\mathcal{N}))$; $T \mapsto T \cap Proj(\mathcal{N})$. We then obtain the following diagram:

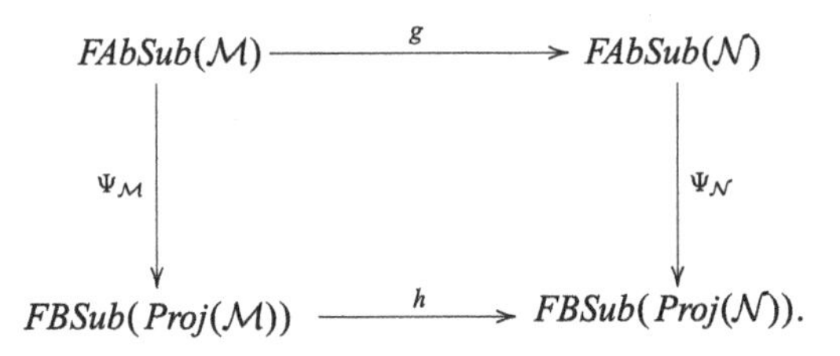

[1]Given a partially ordered set L, then an element $a \in L$ is compact if for any directed subset $P \subseteq L$, if P has a supremum and $a \leq Sup(P)$, then $a \leq d$ for some $d \in Sup(P)$.

Clearly the unique $h : FBSub(Proj(\mathcal{M})) \to FBSub(Proj(\mathcal{N}))$, which makes the diagram commute, will be defined as $h(S \cap Proj(\mathcal{M})) = g(S) \cap Proj(\mathcal{N})$ for each $S \in FBSub(Proj(\mathcal{M}))$. From Lemma 6.2.2 this extends uniquely to an order-isomorphism $j : BSub(Proj(\mathcal{M})) \to BSub(Proj(\mathcal{N}))$. As shown in [38] and in Theorem 6.1.3: given two orthomodular lattices (OML) M, N without any 4-element blocks, then to any order-isomorphism $\alpha : BSub(N) \to BSub(M)$ there corresponds a unique OML-isomorphism $\beta : N \to M$, such that $\alpha(A) = \beta[A]$. We can use this result for our map j. In fact, since neither $\mathcal{M}$ nor $\mathcal{N}$ are isomorphism to $\mathbb{C} \oplus \mathbb{C}$ nor $\mathcal{B}(\mathbb{C} \oplus \mathbb{C})$, then $Proj(\mathcal{M})$ and $Proj(\mathcal{N})$ have no 4-elements blocks. It follows that the map j determines a unique map $k : Proj(\mathcal{M}) \to Proj(\mathcal{N})$ such that $j(A) = k[A]$ for all $A \in Proj(\mathcal{M})$.

In [25] it was shown that, given two von Neumann algebras $\mathcal{M}$ and $\mathcal{N}$, without type I_2 summand, then for any OML-isomorphism $\gamma : Proj(\mathcal{M}) \to Proj(\mathcal{N})$ there is a unique Jordan *-isomorphism $\Gamma : \mathcal{M} \to \mathcal{N}$ with $\gamma(\hat{P}) = \Gamma(\hat{P})$ for all $\hat{P} \in \mathcal{M}$ (See Theorem 6.1.2). This result shows that k uniquely extends to a Jordan *-isomorphism $F : \mathcal{M} \to \mathcal{N}$.

Next we consider two categories vNa_1 and vNa_2 the objects of which are both von Neumann algebras but whose morphisms are defined differently. In particular, consider two algebras $\mathcal{M}$ and $\mathcal{N}$ in vNa_1, then a morphism between $\mathcal{M}$ and $\mathcal{N}$ is defined to be an order-isomorphism $f : AbSub(\mathcal{M}) \to AbSub(\mathcal{N})$. On the other hand, given the same two objects $\mathcal{M}, \mathcal{N} \in vNa_2$, then a morphisms between them is defined to be a Jordan *-isomorphism $F : \mathcal{M} \to \mathcal{N}$. Let us then construct a functor between these to categories which acts trivially on object and which assigns to each order-isomorphism $f : AbSub(\mathcal{M}) \to AbSub(\mathcal{N})$ the associated Jordan *-isomorphism $F : \mathcal{M} \to \mathcal{N}$. Clearly F extends the action of f on projections, since we defined it that way. Moreover, since a Jordan homomorphism is uniquely determined by how it acts on projections it follows that the assignment $f \mapsto F$ is unique. Functoriality then follows from this uniqueness. In particular if F and G extend the action of f and g on projections, it follows that $G \circ F$ extends the action of $g \circ f$ and Id does the same as id.

Since f extends to F, it follows that f^{-1} extends to F^{-1}. We will now show that for every projection $\hat{P} \in \mathcal{N}$, $\hat{P} \in F[S]$ iff $p \in f(S)$ for any $S \in AbSub(\mathcal{M})$. To this end let $< \hat{P} >$ be the abelian subalgebra generated by $\hat{P}$. If $\hat{P} \in f(S)$ then $< \hat{P} > \subseteq f(S)$. Since f is an order-isomorphism then $f^{-1} < \hat{P} > \subseteq S$. However, since F^{-1} extends f^{-1}, then one has that $f^{-1}(< \hat{P} >) = F^{-1}(< \hat{P} >)$, therefore $\hat{P} \in f(S)$ is equivalent to $F^{-1}(< \hat{P} >) \subseteq S$ which is equivalent to $F^{-1}(\hat{P}) \in S$. Since F is a unital order-isomorphism then $F^{-1}(\hat{P}) \in S$ is equivalent to $\hat{P} \in F(S)$.

Next we need to show that both $F[S]$ and $f(s)$ are von Neumann subalgebras of $\mathcal{N}$. We do this by claiming that given any $S \in AbSub(\mathcal{M})$ then $F[S] \in AbSub(\mathcal{N})$. Clearly, since S is abelian and F is a Jordan *-isomorphism, it follows that $F|_S$ preserves the associative product. Moreover from [1, Prop. 4.19] it follows that F is a unital order-isomorphism and its restriction $F|_S : S \to \mathcal{N}$ is a normal unital *-isomorphism into $\mathcal{N}$. From [1, Lemma 2.100] it follows that $F|_S[S]$ is a von Neumann subalgebra of $\mathcal{N}$ which is clearly abelian hence $F[S] = F|_S[S] \in AbSub(\mathcal{N})$.

Since both $F[S]$ and $f(s)$ are von Neumann subalgebras of $\mathcal{N}$, and since they contain the same projection they have to be the same, hence $F[S] = f(S)$.

Finally we need to show that F is unique, i.e. given any other Jordan *-isomorphism $G : \mathcal{M} \rightarrow \mathcal{N}$ such that $f(S) = G[S]$ for each $S \in AbSub(\mathcal{M})$, then $F = G$. Because of the spectral theorem, if $F[\hat{P}] = G[\hat{P}]$ for all $\hat{P} \in Proj(\mathcal{M})$, then $F = G$ since any other operator can be written in terms of the projections operators in the algebras. Therefore we only need to show that F and G agree on $Proj(\mathcal{M})$. However from the result in [38] and Theorem 6.1.3 it suffices to show that F and G agree on each Boolean subalgebra of $Proj(\mathcal{M})$. From the uniqueness in Lemma 6.2.1 it is enough to show that $F[T \cap Proj(\mathcal{M})] = G[T \cap Proj(\mathcal{M})]$ for each $T \in FAbsub(\mathcal{M})$. But since we have assumed that $G[S] = F[S] = f(S)$ for all $S \in AbSub(\mathcal{M})$ then clearly for $T \in FAbsub(\mathcal{M})$, $G[T] = F[T]$, therefore $F = G$. □

Interestingly enough the converse of the above theorem is also true, namely:

Theorem 6.2.2 *Given two von Neumann algebras $\mathcal{M}$ and $\mathcal{N}$ and a Jordan *-isomorphism $F : \mathcal{M} \rightarrow \mathcal{N}$ between them, then F induces a unique order-isomorphism $f : AbSub(\mathcal{M}) \rightarrow AbSub(\mathcal{N})$ such that for all $S \in AbSub(\mathcal{M})$, $f(S) = F[S]$.*

Proof The proof is a straightforward consequence of the fact that a Jordan *-isomorphism preserves commutativity, hence F maps abelian subalgebras of $\mathcal{M}$ to abelian subalgebras of $\mathcal{N}$. Moreover since it is an isomorphism it does so in a bijective order preserving way hence $f : AbSub(\mathcal{M}) \rightarrow AbSub(\mathcal{N})$ is an order-isomorphism. □

The result of Theorem 6.2.1 is very important since it allows us to recover the Jordan structure of a von Nuemann algebra from its poset of abelian sub algebras. The question that still remains to be answered is if and how the full von Neumann structure can be recovered. This could be done by adding additional information on the poset of abelian subalgebras which would allow us to retrieve the full von Neumann algebra. A step in this direction is given by *orientation theory* [1].

Chapter 7
Grothendieck Topoi

In this chapter we will describe the topos of sheaves over a category $\mathcal{C}$ equipped with a Grothendieck topology. In [26] we came across the definition of sheaf over a topological space X. This definition relied solely on the lattice of open sets of the topological space X, i.e. on the topology. In this chapter we would like to extend the notion of topology so as to be able to define sheaves on this more general 'topology'.

7.1 Grothendieck Topology

In this section we will try to generalise the notion of a topology. As a first step we will consider the concept of a *covering* which is central to topologies. The rigorous definition of a covering is as follows:

Definition 7.1.1 Given a topological space X, a cover C of X is a collection of subsets $U_i \subseteq X$ such that $X = \bigcup_i U_i$. If each U_i is open then C is an open cover. The notion of a cover also extends to subsets of the entire space X.

Definition 7.1.2 Given a subset $Y \subseteq X$, then a cover of Y is a collection of subsets $U_i \subseteq X$ such that $Y \subseteq \bigcup_i U_i$.
If we now consider the above definition in categorical language, then the notion of a subset would be replaced with that of a monic arrow whose codomain is X or Y. The particular category we will consider is the category $\mathcal{O}(X)$, which has as objects open subsets of the topological space X, while a morphism $V \to U$ in $\mathcal{O}(X)$ is defined iff $V \subseteq U$. Given such a category, the definition of a covering is as follows:

Definition 7.1.3 Given the category $\mathcal{O}(X)$ of open subsets of the topological space X, a covering of an open $U \in \mathcal{O}(X)$ is a family $\{U_i | i \in I\}$ of opens in X, such that $U_i \to U$ for all $i \in I$ and $U \subseteq \bigvee_i U_i$.
In other words $\{U_i \to U | i \in I\}$ covers U iff $U \subseteq \bigcup_i U_i$.

C. Flori, *A Second Course in Topos Quantum Theory*,
Lecture Notes in Physics 944, https://doi.org/10.1007/978-3-319-71108-9_7

From the definition of a cover it is easy to see that if we are given a cover $\{U_i \to U|i \in I\}$ of U and $V \subseteq U$, then $\{U_i \cap V \to V|i \in I\}$ is a cover of V. In fact, $V = V \cap U \subseteq V \cap \bigcup_i U_i = \bigcup_i V \cap U_i$. This implies that any map $V \to U$ in $\mathcal{O}(X)$ can be used to 'pullback' covers.

We now recall the definition of a sieve

Definition 7.1.4 A sieve on an object $A \in \mathcal{C}$ is a collection S of morphisms in $\mathcal{C}$ whose codomain is A and such that, if $f : B \to A \in S$ then, given any morphisms $g : C \to B$ we have $f \circ g \in S$, i.e. S is closed under left composition:

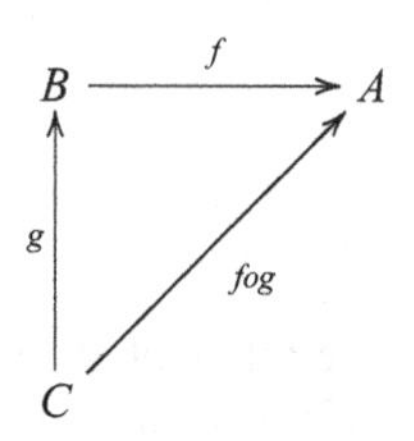

Given a sieve S on A, if $f : B \to A$ belongs to S then the pullback of S by f determines a sieve on B, i.e.

$$f^*(S) := \{h : C \to B|f \circ h \in S\} = \{h : C \to B\}$$

is a sieve on B. Such a sieve is called a principal sieve (i.e. the sieve which contains the identity arrow) and it is denoted by $\downarrow B$.

If we apply the definition of a sieve in the context of the category $\mathcal{O}(X)$, then sieve a S on U is a set $S = \{V \in \mathcal{O}(X)|V \subseteq U\}$ such that, if $V \in S$ and $V' \subseteq V$, then $V' \in S$. Given this definition of a sieve it is clear that S is a cover of U iff $U \subseteq \bigcup_{V \in S} V$. We will now show several properties of a covering sieve.

1. `Maximal sieve`. The sieve S which contains U itself (the principal sieve) is a covering of U.
   Proof: Clearly if $U \in S$ then $U \subseteq \bigcup_{V \in S} V$.
2. `Stability axiom`. If S is a covering of U, then for any $U' \subseteq U$, $U' \cap S$ is a cover of U'.
   Proof: $U' = U' \cap U \subseteq U' \cap \bigcup_{V \in S} V = \bigcup_{V \in S} V \cap U'$.
3. `Transitivity axiom`. If S is a covering of U and R is any sieve on U such that for all $U' \in S$, $U' \cap R$ is a covering on U', then R is a covering of U.
   Proof: Since S covers U then $U \subseteq \bigcup_{U' \in S} U'$. From the fact that for each $U' \in S$, $U' \cap R$ is a covering on U', it follows that $U' \subseteq \bigcup_{V \in (U' \cap R)} V = \bigcup_{V' \in R} V' \cap U'$. Putting the two results together we have that $U \subseteq \bigcup_{U' \in S} \bigcup_{V' \in R} V' \cap U' = \bigcup_{U' \in S} U' \cap \bigcup_{V' \in R} V'$. It then follows that $U \subseteq \bigcup_{V' \in R} V'$, i.e. R is a covering of U.

The above discussion reveals how, as far as the category $\mathcal{O}(X)$ is concerned, the (canonical) notion of a covering can be given in terms of sieves. Generally, however, given a topological space X, the covering defined in Definition 7.1.1 is not necessarily a sieve but it does generate one, namely the sieve consisting of all those opens $V \subseteq X$, such that $V \subseteq U_i$ for some U_i in the cover. In this case the coverings $\{U_i \to U | i \in I\}$ generate covering sieves.

We would now like to generalise the definition of coverings in terms of sieves for a general category, not only for $\mathcal{O}(X)$. It is precisely this generalisation that represents an extension of the notion of a topology on a category $\mathcal{C}$ and goes by the name of *Grothendieck topology*.

Definition 7.1.5 ([55]) Given a category $\mathcal{C}$, a Grothendieck topology is a function J which assigns for each object $C \in \mathcal{C}$ a collection $J(C)$ of sieves on C, such that the following conditions hold:

C.1 `Maximal sieve`. $J(C)$ contains the maximal sieve on C.
C.2 `Stability axiom`. If $S \in J(C)$, given any arrow $h : D \to C$, then $h^*(S) \in J(D)$.
C.3 `Transitivity axiom`. Given $S \in J(C)$ and R is any sieve on C such that, for all $h : D \to C$ in S $h^*(R) \in J(D)$, then $R \in J(C)$.

When $S \in J(C)$ then we say that S is a covering sieve of C or is a J-cover. A straightforward consequence of the above definition is the following:

Corollary 7.1.1 *Given a sieve $S \in J(C)$ then if $S \subseteq R$ it follows that $R \in J(C)$.*

Proof Given any $f : D \to C$ in S, then f^*S is the principal sieve on C and thus $f^*S \in J(C)$. Moreover, since $S \subseteq R$ then $f^*S \subseteq f^*R$, thus f^*R is also the principal sieve on C and $f^*R \in J(C)$. Since this is true for any $f \in S$, from the transitivity axiom it follows that $R \in J(C)$. □

The above Lemma uncovers the fact that if a topology J on a category $\mathcal{C}$ contains a sieve S on C, then it also contains all sieves on C, which are coarser (bigger) then S. Therefore, given two Grothendieck topologies J and J' on the same category $\mathcal{C}$, such that $J(C) \subseteq J'(C)$, then J' contains all sieves on C which are finer than those of J. The fact that we can compare sieves allows us to compare Grothendieck topologies.

Definition 7.1.6 Given two topologies J and J' on $\mathcal{C}$, we say that J' is finer than J if for all objects $C \in \mathcal{C}$, then $J(C) \subseteq J'(C)$. In this case the topology J is called coarser than J' and it is denoted by $J \subseteq J'$.

The two extreme examples of topologies on a category $\mathcal{C}$ are the *discrete topology* and the *trivial topology*. The *discrete topology* is the topology for which, given any $C \in \mathcal{C}$, then $J(C)$ contains all sieves on C. On the other hand, the *trivial topology* is the topology for which, given any $C \in \mathcal{C}$, then $J(C)$ contains only the principal sieve on C.

We have now the necessary tools to define a *site*.

Definition 7.1.7 A site is a pair $(\mathcal{C}, J)$ consisting of a category $\mathcal{C}$ and a Grothendieck topology J.

It is also possible to define covering sieves of arrows rather than objects. In particular, we say that a sieve S on C covers an arrow $f : D \to C$ if f^*S covers D. It follows that $S \in J(C)$ iff S covers the identity arrow on C. It is then possible to re-write the axioms of a Grothendieck topology in terms of sieves covering arrows rather than objects.

Definition 7.1.8 ([55]) Given a category $\mathcal{C}$, a Grothendieck topology is a function J which assigns, for each object $C \in \mathcal{C}$, a collection $J(C)$ of sieves on C such that the following conditions hold:

C'.1 `Maximal sieve`. If S is a sieve on C and $f \in S$ then S covers f.
C'.2 `Stability axiom`. If S covers an arrow $f : D \to C$, given any other arrow $h : E \to D$, then S covers $f \circ h : E \to C$.
C'.3 `Transitivity axiom`. If S covers an arrow $f : D \to C$ and R is any sieve on C such that it covers all arrows in S, then R covers f.

We will now prove that Definitions 7.1.5 and 7.1.8 are equivalent.

We will start buy assuming that conditions C'.1, C'.2 and C'.3 hold and show that this entails that conditions C.1, C.2 and C.3 also hold. In particular, if C'.1 holds, then $f^*S \in J(dom(f)$ and it is the principal sieve on $dom(f)$. By choosing $f = id_C$, then condition C.1 follows. Similarly if C'.2 holds then if $f^*S \in J(D)$, given any other arrow $h : E \to D$, it follows that $(f \circ h) * S = h * (f * S) \in J(E)$. By choosing $f = id_C$ then condition C.2 holds. Finally we assume condition C'.3 holds. If $f * (S) \in J(D)$ and R is any sieve on C such that for any g in S, then $g * (R) \in J(dom(g))$, it follows that $f * (R) \in J(D)$. By taking f to be id_C then condition C.3 follows.

Conversely, let us assume that condition C.1–C.3 hold. To prove C'.1 we need to show that given a sieve S on C and $f \in S$ then S covers f. However, from C.2 we know that $f^*(S) \in J(dom(f))$, therefore S covers f. To prove C'.2 we need to show that if S covers an arrow $f : D \to C$ (if $f^*S \in J(D)$), given any other arrow $h : E \to D$, then $(f \circ h)^*S = h^*(f^*(S)) \in J(D)$. By applying C.2 with S replaced by f^*S the result follows. Finally to prove condition C'.3 let us assume that S covers an arrow $f : D \to C$ and R is any sieve on C such that it covers all arrows in S. This means that $f^*S \in J(D)$ and for all $g \in S$, $g^*R \in J(dom(g))$. We now need to show that R covers f, i.e. $f^*R \in J(D)$. Now given any $h \in f^*S$, then from the definition of a pullback of a sieve it follows that $f \circ h \in S$. Therefore our condition tells us that $(f \circ h) * R = h * (f * (R)) \in J(dom(h))$. Since this is valid for any h it follows from condition C.3 that $f^*R \in J(D)$.

We now give some useful results concerning sieves.

Lemma 7.1.1 (Common Refinement) *If $R, S \in J(C)$ then $S \cap R \in J(C)$.*

Proof Given any map $f \in S$ then, from the stability axiom, $f^*(S \cap R) = f^*(S) \cap f^*(R) = f^*(R) \in J(C)$. Since this is valid for any f, from the transitivity axiom it follows that $S \cap R \in J(C)$. □

As with the definition of a Grothendieck topology the common refinement Lemma can also be stated in terms of arrows.

Lemma 7.1.2 *If both R and S cover $f : D \to C$, then $R \cap S$ covers f.*

Proof The condition that both R and S cover f implies that $f^*R, f^*S \in J(D)$. Now consider $g \in f^*(S)$, then by definition $f \circ g \in S$ thus, by C'.2 we obtain $(f \circ g)^*(S \cap R) = g^*(f^*(R \cap S)) \in J(dom(g))$. Now since the set of sieves on an object is a complete lattice, then $R \cap S$ is a sieve on C and $f^*(S \cap R)$ is a sieve on D. We then have that $f^*(S) \in J(D)$ and $f^*(S \cap R)$ is a sieve on D such that for any $g \in f^*S$, $g^*(f^*(S \cap R)) \in J(dom(g))$. It then follows from C.3 that $f^*(S \cap R) \in J(D)$, i.e. $S \cap R$ covers f. □

Similarly as in general topology, also in the context of Grothendieck topologies we have the notion of a *basis*. In order to introduce such a notion we first need to explain what a *generated sieve* and *covering family* are. To this end we note that a sieve S on C can be seen as a sub-object of $\mathbf{y}(C)$ in $\mathbf{Sets}^{\mathcal{C}^{op}}$ where $\mathbf{y} : \mathcal{C} \to \mathbf{Sets}^{\mathcal{C}^{op}}$; $C \mapsto Hom_{\mathcal{C}}(-C)$ is the Yoneda embedding (Lemma A.7.2) [55]. Having said that we can define the generated sieve as a particular presheaf.

Definition 7.1.9 (Generated Sieve) Consider a family of morphisms $D = \{f_i : D_i \to C\}_{i \in I}$ in $\mathcal{C}$. This family generates a sieve S_D on C defined on:

1. Objects: given any object $A \in \mathcal{C}$ we define a subset $S_D(A) \subseteq \mathbf{y}(C)(A)$ as

$$S_D(A) := \bigcup_{i \in I} (f_i)_* \mathbf{y}(D_i)(A).$$

2. Morphisms: Given a morphisms $g : B \to A$ in $\mathcal{C}$, then

$$\begin{aligned} S_D(g) : S_D(A) &\to S_D(B) \\ h &\mapsto h \circ g. \end{aligned}$$

Clearly the above is a sub-object of $\mathbf{y}(C)$ in $\mathbf{Sets}^{\mathcal{C}^{op}}$, i.e. it is a sieve. In particular, S_D is that sieve on C whose elements all factor via some element of the family D, i.e. as a set we have $S_D := \{f \circ g | f \in D,\ dom(f) = cod(g)\}$.

Definition 7.1.10 (Covering Family) Given a site $(\mathcal{C}, J)$. A family of morphisms $D = \{f_i : D_i \to C\}$ in $\mathcal{C}$ is called a J-covering family of C if the sieve S_D generated by D is in $J(C)$, i.e. if S_D is a J-covering.

From this definition it is straightforward to see that if we are given two Grothendieck topologies J and J' on $\mathcal{C}$, such that $J \subseteq J'$, any J-covering family is also a J'-covering family. In fact, if $D = \{f_i : D_i \to C\}$ is a J-covering family, then $S_D \in J(C) \subseteq J'(C)$.

As we will explain shortly, Grothendieck topologies can be generated by covering families. As an example consider an object $C \in \mathcal{C}$, and the set $\mathcal{D}(C)$ of families of morphisms $D = \{f_i : D_i \to C\}$ in $\mathcal{C}$. Then there exists a coarsest Grothendieck topology J on $\mathcal{C}$ for which all the families in $\mathcal{D}(C)$ are J-coverings. Such a topology is called the topology generated by the families $\mathcal{D}(C)$.

An interesting exercise to better understand how Grothendieck topologies work is to construct the relative topology on an slice category In particular, given a site $(\mathcal{C}, J)$, then for any element $B \in \mathcal{C}$, J induces a topology J' on $\mathcal{C}/B$, defined as follows:
Consider an object $x : X \to B \in \mathcal{C}/B$ and let S be a sieve on X, then we define the sieve S_x on x as[1] the functor

$$\begin{aligned}
&\mathbf{S}_x : (\mathcal{C}/B)^{\mathrm{op}} \to \mathbf{Sets}\\
&y : Y \to B \mapsto \mathbf{S}_x(y) := \{g : Y \to X \in \mathbf{S}(Y) | \, y = x \circ g\}\\
&h : y' \to y \mapsto h^* : \mathbf{S}_x(y) \to \mathbf{S}_x(y') \text{ s.t. } h^*(g) := g \circ h.
\end{aligned}$$

We now need to show that the above map is well defined and is a functor. For being well defined we require that the object $h^*(g) := g \circ h$ is indeed in $\mathbf{S}_x(y')$. However, since $g \in \mathbf{S}_x(y)$ then $y = x \circ g$. Moreover, the map $h : y' \to y \in \mathcal{C}/B$ induces the commutative diagram

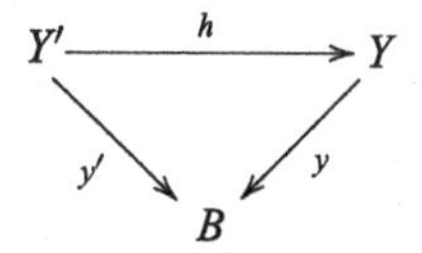

therefore $y' = y \circ h = x \circ g \circ h$, thus $g \circ h \in \mathbf{S}_x(y')$.

Next we need to show that the functor $\mathbf{S}_x$ is well behaved under composition and with respect to the identity. Let us consider $f : x' \to y'$ and $h : y' \to y$. Then $\mathbf{S}_x(h \circ f)(g) = (h \circ f)^*(g) = f^*(h^*(g)) = g \circ h \circ f$ and $\mathbf{S}_x(f) \circ \mathbf{S}_x(h)(g) = f^* \circ h^*(g) = g \circ h \circ f$. Similarly, one can show that $\mathbf{S}_x(id_y) = id_{\mathbf{S}_x(y)}$. Now that we know that the above functor is well defined we need to show that it is a sieve. This follows trivially by noting that $\mathbf{S}_x$ is a subfunctor of $\mathbf{y}(x) : (\mathcal{C}/B)^{\mathrm{op}} \to \mathbf{Sets}$.

Exercise 7.1.1 Show that the associated set $S_x = \bigcup_{y \in \mathcal{C}/B} \mathbf{S}_x(y)$ is

$$S_x := \{g \in S | \text{ given any } y : dom(g) \to B \in \mathcal{C}/B, \; y = x \circ g\}.$$

We then define the induced topology, for each $x \in \mathcal{C}/B$ as $J'_x = \{\mathbf{S}_x | S \in J(X)\}$.

As a next step we now need to show that conditions C.1–C.3 of Definition 7.1.5 for a Grothendieck topology hold for J'. We will do this in terms of the set S_x rather than the functor $\mathbf{S}_x$. First of all we need to show that indeed S_x, as defined in Exercise 7.1.1, is a sieve. Therefore we need to show that given any $g : y \to x \in S_x$ and $f : x' \to g \in \mathcal{C}/B$, where $x' : X' \to B$, then $g \circ f \in S_x$. However, if $g \in S_x$ then $g \in S$, therefore, given $f : X' \to Y$, then $g \circ f \in S$. To show that $g \circ f \in S_x$, we need to show that, given any $x' : X' \to B$, then $x' = x \circ g \circ f$. But we know that $g \in S_x$

[1]Here we will frequently use that fact that, given two objects $x : X \to B$ and $y : Y \to B$ in $\mathcal{C}/B$, then a map $g : y \to x$ in $\mathcal{C}/B$ is defined in terms of the map $g : Y \to X$ in $\mathcal{C}$.

implies that for any $y : Y \to B$, $y = x \circ g$. The result then follows by simply defining $x' := y \circ f$ and chasing the commutative diagram

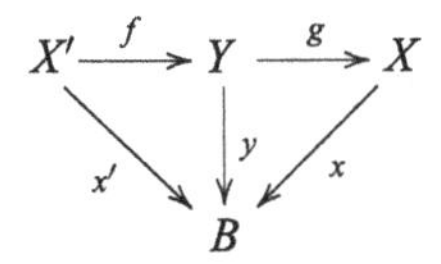

Let us now define the induced topology J' for each $x \in \mathcal{C}/B$ as

$$J'(x) := \{S_x | S \in J(X)\}.$$

It remains to show that J' is indeed a Grothendieck topology, i.e. that it satisfies conditions 1–3 in Definition 7.1.5. For 1. we note that, given any $X \in \mathcal{C}$ the principal sieve R belongs to $J(X)$, then the associated sieve R_x will be the principal sieve on x by construction.

For condition C.2, let us assume that $S_x \in J'(x)$ and let us consider a map $f : x' \to x$ in $\mathcal{C}/B$ given by $f : X' \to X$, such that $x' = x \circ f$. In terms of the topology J this translates in $S \in J(X)$, but since J is a Grothendieck topology it follows that $f^*(S) \in J(X')$ and thus $(f^*(S))_x \in J'(x')$. What we wanted to show was that $f^*(S_x) \in J'(x')$. For this to be the case we need to prove that $(f^*(S))_{x'} = f^*(S_x)$. On the one hand

$$\begin{aligned}
f^*(S_x) &= \{g | cod(g) = X', \ f \circ g \in S_x\} \\
&\overset{\text{Definition of } S_x}{=} \{g | cod(g) = X', \ f \circ g \in S, \text{ for any } y : dom(g) \to B \in \mathcal{C}/B, \ y = x \circ f \circ g\} \\
&\overset{\text{Definition of } f^* S}{=} \{g \in f^*(S) | \text{ for any } y : dom(g) \to B \in \mathcal{C}/B, \ y = x \circ f \circ g\} \\
&\overset{\text{Definition of } x'}{=} \{g \in f^*(S) | \text{ for any } y : dom(g) \to B \in \mathcal{C}/B, \ y = x' \circ g\} \\
&= f^*(S)_{x'}.
\end{aligned}$$

For condition C.3 let us assume that $S_x \in J'(x)$ and let S'_x be a sieve on x such that for all $f : x' \to x \in S_x$, then $f^*(S'_x) \in J'(x')$. We want to show that $S'_x \in J'(x)$. In terms of the topology J the assumptions we have imply that for all $f : X' \to X$ in S with $x' = x \circ f$, then $f^*(S) \in J(X')$. From the definition of maps in the comma category we can define $x' = x \circ f$, then, applying condition C.3 to the Grothendieck topology J we obtain $S' \in J(X)$. It then follows that $S'_x \in J'(x)$.

Exercise 7.1.2 Prove that the functor $\mathbf{S}_x$ satisfies conditions C.1–C.3 of Definition 7.1.5.

Exercise 7.1.3 Show that $S = S_x$ for all $x \in \mathcal{C}/B$.

Now that we have introduced the notion of a generating sieve, we will define the notion of a basis of a Grothendieck topology. For this we will assume that the category we are working with has pullbacks.

Definition 7.1.11 Given a category $\mathcal{C}$ will pullbacks, a basis for a Grothendieck topology on $\mathcal{C}$ is a function K which assigns to each object $C \in \mathcal{C}$ a set $K(C)$, whose elements are families $D = \{f_i : D_i \to C\}_{i \in I}$ of morphisms with codomain C subject to the following conditions:

B.1 For each $C \in \mathcal{C}$, the set $\{f : C' \to C | f \text{ is an isomorphisms}\}$ is in $K(C)$;
B.2 Given $\{f_i : D_i \to C\}_{i \in I} \in K(C)$, for any morphisms $g : B \to C$, then the family of pullbacks $\{D_i \times_C B \to B\}_{i \in I}$ is in $K(B)$;
B.3 Given a family $\{f_i : D_i \to C\}_{i \in I} \in K(C)$ and, for each $i \in I$, a family $\{g_{ij} : B_{ij} \to D_i\}_{j \in I_i} \in K(D_i)$, then the family of composites $\{f_i \circ g_{ij} : B_{ij} \to C | i \in I, j \in I_i\}$ is in $K(C)$.

The pair $(\mathcal{C}, K)$ is called a *site* and the elements of each set $K(C)$ for $C \in \mathcal{C}$ are the *covering families*.

Exercise 7.1.4 Show that a Grothendieck topology J satisfies B.2, B.3 in Definition 7.1.11 above but not B.1.

Definition 7.1.12 Given a basis K on a category $\mathcal{C}$, this generates a Grothendieck topology J as follows:

$$S \in J(C) \text{ iff } \exists\, D \in K(C) \text{ s.t. } D \subseteq S.$$

We now show that indeed J, as defined above, is a Grothendieck topology.

Proof We need to show that J as defined in Definition 7.1.12 satisfies condition 1–3 in Definition 7.1.5.

C.1 Assume that S is the principal sieve on C, we want to show that there exists an element $R \in K(C)$ such that $R \subseteq S$. From condition B.1 we know that $R = \{ id_C : C \to C\} \in K(C)$. Clearly $R \subseteq S$, therefore $S \in J(C)$.
C.2 Assume that $S \in J(C)$, therefore there exists an $R \in K(C)$ such that $R \subseteq S$. Now given any $h : D \to C$ we want to show that $h^*(S) \in J(D)$. This is equivalent to showing that there exists an $R' \in K(D)$ such that $R' \subseteq h^*(S)$. We choose R' to be constructed as in B.2, i.e. $R' = \{g : D \times_C C_i \to D\}_{i \in I}$. That is, R' is the set of all those arrows which are constructed via the following pullback diagram for some $f_i \in R$

$$\begin{array}{ccc} D \times_C C_i & \longrightarrow & C_i \\ \downarrow g & & \downarrow f_i \\ D & \xrightarrow{h} & C \end{array}$$

Clearly $R' \subseteq h^*(S) := \{k | cod(h) = D,\ g \circ k \in S\}$, therefore $h^*(S) \in J(D)$.

C.3 Assume that $S \in J(C)$, therefore there exists a $T \in K(C)$ such that $T \subseteq S$. Next we have a sieve R on C such that for all $h : D \to C \in S$, then $h^*R \in J(D)$. This implies that there exists a $T' \in K(D)$ such that $T' \subseteq h^*(R)$. Since $T \subseteq S$, then for each $h_i : C_i \to C \in T \subseteq S$ we construct T' as in B.3, i.e. $T' = \{g_{ij} : D_{ij} \to C_i | j \in I_j\} \in K(C_i)$. Clearly from the definition of $h^*(R)$ and the property of sieves being closed under left composition, it follows that $T' \subseteq h^*(R)$. Now we can apply B.3 to obtain the composite $T'' = \{h_i \circ g_{ij} : D_{ij} \to C | i \in I,\ j \in I_j\} \in K(C)$. It remains to show that $T'' \subseteq R$. But since $T' \subseteq h_i^*(R) = \{g | cod(g) = D,\ h_i \circ g \in R\}$, then for any g_{ij} it follows that $h_i \circ g_{ij} \in R$, therefore $T'' \subseteq R$ and $R \in J(C)$.

□

Exercise 7.1.5 Show that Definition 7.1.12 is equivalent to the following definition:

Definition 7.1.13 Given a basis K, the Grothendieck topology generated by K is the coarsest topology on $\mathcal{C}$ such that all $R \in K(C)$ (for all $C \in \mathcal{C}$) are J-covering families as defined in Definition 7.1.10.

Definition 7.1.14 Given two families $R = \{f_i : D_i \to C\}$ and $R' = \{g_j : D_j \to C\}$, then we say that R refines R' if every element $f_i \in R$ factors through an element $g_i \in R'$.
Note that, for a given topology J, there exists a maximum basis K which generates J. This is given by:

$$R \in K(C) \Leftrightarrow S_R \in J(C) \quad \text{where } S_R = \{f \circ g | f \in R, dom(f) = cod(g)\}.$$

Lemma 7.1.3 *Given any two families of covers $R, R' \in K(C)$ then there exists a common refinement.*

Proof Given $R, R' \in K(C)$, then $S_R, S_{R'} \in J(C)$. From Lemma 7.1.1 it follows that $S_R \cap S_{R'} \in J(C)$, therefore there exists a $T \in K(C)$ such that $T \subseteq S_R \cap S_{R'}$. Therefore $T \subseteq S_R$ and $T \subseteq S_{R'}$. This, in turn, means that T refines both R and R'. □

As an example consider again the category $\mathcal{O}(X)$ for some topological space X. We defined the basis K on $\mathcal{O}(X)$ by $\{f_i : U_i \to X | i \in I\} \in K(X)$ iff each U_i is open in X and $\bigcup_{i \in I} U_i = X$. This is the so-called *open cover* topology (see Definition 7.1.1). We now need to show that, indeed, this definition of K satisfies conditions B.1–B.3 of being a basis. Clearly $\{ id_U : U \to U\}$ satisfies the requirement of belonging to $K(U)$. For condition B.2, consider a family $\{f_i : U_i \to U\}_{i \in I} \in K(U)$, then given a morphisms $g : V \to U$ we want to show that $\{V \times_U U_i \to V\} \in K(V)$. Now in $\mathcal{O}(X)$ the arrows are given by subset inclusion and the pullback by intersections, hence the pullback $V \cap U_i \to V$ is monic. Since finite intersections of opens are open, $V \cap U_i$ is open. Finally, since $V \subseteq U = \bigcup_i U_i$, it follows that indeed $\bigcup_i V \cap U_i = V$, hence $\{V \times_U U_i \to V\} \in K(V)$. To show that B.3 holds, consider the family $\{f_i : U_i \to U\}_{i \in I} \in K(U)$ such that for each $i \in I$ the family $\{g_{ij} : V_{ij} \to U_i\}_{j \in I_i} \in K(U_i)$. We want to show that $\{f_i \circ g_{ij} : V_{ij} \to U | i \in I,\ j \in I_i\} \in K(U)$. Since compositions of monics are monics, $V_{ij} \subseteq U$. Moreover, given $U = \bigcup_{i \in I} U_i$ and $U_i = \bigcup_{j \in I_j} V_{ij}$, it follows that $U = \bigcup_{i \in I} \bigcup_{j \in I_j} V_{ij}$.

7.2 Grothendieck Sheaves

We now will introduce the definition of a sheaf on a site. In order to do so, we will first recall the definition of a sheaf on a general topological space which relies solely upon the lattice of open sets of that space.

Roughly speaking, a sheaf can be thought of as a fibre bundle in which the fibres may vary from point to point. Formally a sheaf is a presheaf with values in the category of sets that satisfies the following two axioms:

1. Given an open set U with open covering U_i, if $s, t \in F(U)$ are such that $s|_{U_i} = t|_{U_i}$ for all i, then $s = t$.
2. Given an open set U with open covering U_i, and $s_i \in F(U_i)$ for all i, such that for each pair U_i and U_j, $s_i|_{U_i \cap U_j} = s_j|_{U_i \cap U_j}$, then there exists $s \in F(U)$ such that $s|_{U_i} = s_i$ for each i. "s" is called the *gluing*, while the s_i (for each i) are called *compatible*.

Axioms (1) and (2) state that compatible sections can be uniquely glued together. The definition of a sheaf given in [26] was:

Definition 7.2.1 A sheaf of sets F on a topological space X is a functor $F : \mathcal{O}(X)^{op} \to Sets$, such that each open covering $U = \bigcup_i U_i$, $i \in I$ of an open set U of X determines an equaliser

$$F(U) \overset{e}{\rightarrowtail} \prod_i F(U_i) \underset{q}{\overset{p}{\rightrightarrows}} \prod_{i,j} F(U_i \cap U_j)$$

where for $t \in F(U)$ we have $e(t) = \{t|_{U_i} | i \in I\}$ and for a family $t_i \in F(U_i)$ we obtain

$$p\{t_i\} = \{t_i|_{U_i \cap U_j}\}, \quad q\{t_i\} = \{t_j|_{U_i \cap U_j}\}. \tag{7.2.1}$$

Given the definition of product in a category [26, 55] it follows that the maps e, p, and q above are determined through the diagram

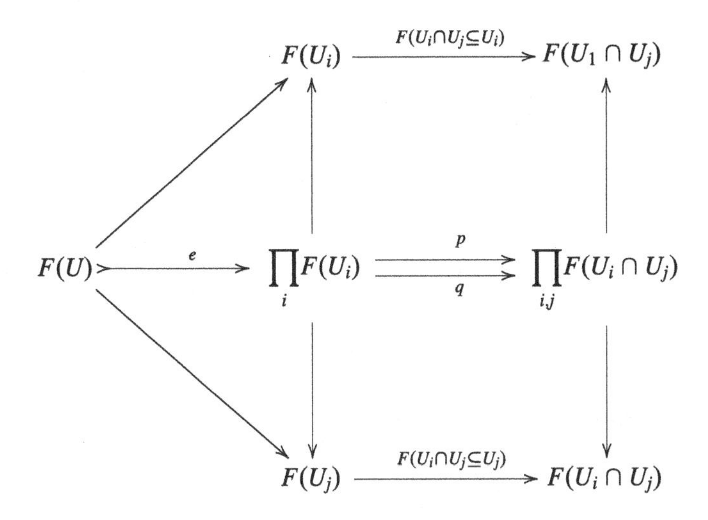

An equivalent way of defining a sheaf is as follows:

Lemma 7.2.1 ([55]) *Given a presheaf* $\mathbf{P}$ *on a space* X*, then* $\mathbf{P}$ *is a sheaf iff for every open* $U \subseteq X$ *and every covering sieve* S *on* U *the inclusion* $i_{\mathbf{S}} : \mathbf{S} \to \mathbf{y}(U)$ *induces an isomorphism*

$$Hom(\mathbf{y}(U), \mathbf{P}) \simeq Home(\mathbf{S}, \mathbf{P})$$

In the above, $\mathbf{y}$ represents the Yoneda embedding (Lemma A.7.2).

Proof Given a covering $\bigcup_i U_i = U$ of U we can construct an equaliser

$$E \rightarrowtail^{e} \prod_i \mathbf{P}(U_i) \overset{p}{\underset{q}{\rightrightarrows}} \prod_{i,j} \mathbf{P}(U_i \cap U_j)$$

where E consists of all those elements $t_i \in \mathbf{P}(U_i)$ such that $t_i|_{U_i \cap U_j} = t_j|_{U_i \cap U_j}$. We then consider the associated sieve of the cover U_i, i.e. the sieve S consisting of all subsets V of U, such that $V \subseteq U_i$ for some U_i in the cover. This allows us to define, for any $V \in S$, such that $V \subseteq U_i$, $t_V := t_i|_V$. Since $t_i|_{U_i \cap U_j} = t_j|_{U_i \cap U_j}$, it follows that the definition of t_V is independent of the i, therefore E becomes the set of all those family of elements $t_V \in \mathbf{P}(V)$ with $V \in S$, such that $t_V|_{V'} = t_{V'}$ whenever $V' \subseteq V$.

From previous discussion we have seen that a sieve S on an element U can be seen as a subfunctor $\mathbf{S}$ of $\mathbf{y}(U) = Hom(-, U) \in \mathbf{Sets}^{\mathcal{O}(X)^{\mathrm{op}}}$. In particular, for each $V \in \mathcal{O}(X)$ we have $\mathbf{S}(V) = 1$ if $V \in S$ and $\mathbf{S}(V) = \emptyset$ otherwise. It follows that for each $V \in S$, then the component $\mathbf{S}(V) \to \mathbf{P}(V)$ of the natural transformation $\mathbf{S} \to \mathbf{P}$ represents the element t_V. Therefore each $t_V \in E$ gets replaced by a natural transformation and E becomes $Hom(\mathbf{S}, \mathbf{P})$.

The inclusion $i_{\mathbf{S}} : \mathbf{S} \to \mathbf{y}(U)$ induces a map

$$(i_{\mathbf{S}})^* : Hom(\mathbf{y}(U), \mathbf{P}) \to Hom(\mathbf{S}, \mathbf{P})$$
$$(f : \mathbf{y}(U) \to \mathbf{P}) \mapsto (f|_{\mathbf{S}} : \mathbf{S} \to \mathbf{P}).$$

We can now construct the following diagram

$$\begin{array}{ccccc} Hom(\mathbf{S},\mathbf{P}) & \xrightarrow{d} & \prod_i \mathbf{P}(U_i) & \overset{p}{\underset{q}{\rightrightarrows}} & \prod_{i,j} \mathbf{P}(U_i \cap U_j) \\ \uparrow{\scriptstyle (i_{\mathbf{S}})^*} & & \uparrow{\scriptstyle e} & & \\ Hom(\mathbf{y}U,\mathbf{P}) & \xrightarrow{\cong} & \mathbf{P}(U) & & \end{array}$$

where the arrows e, p, q are defined as in Definition 7.2.1 and $Hom(\mathbf{y}U, \mathbf{P}) \cong \mathbf{P}(U)$ is a consequence of the Yoneda embedding (see Lem. 8.4.3 [26], Lemma A.7.2). The map d is the equaliser such that for all $g : \mathbf{S} \to \mathbf{P}$, then $d(g) = g|_{U_i}(1) \in \mathbf{P}(U_i)$. The property of d being an equaliser implies that the square commutes. Since e is also an equaliser, then its universal property requires $(i_{\mathbf{S}})^*$ to be an isomorphism. It then follows that $\mathbf{P}$ is a sheaf iff for each covering U_i, $(i_{\mathbf{S}})^*$ is an isomorphism, where S is the corresponding covering sieve. □

Having discussed the definition of a sieve, we will now define sheaves on a site, first with respect to a Grothendieck topology J and then with respect to a basis K. In order to prove the equivalence of these two definition we will utilise Lemma 7.2.1.

First of all we need to define a presheaf $\mathbf{P}$ on a cite $(\mathcal{C}, J)$. However, since the definition of a presheaf does not depend on the topology of the base category, $\mathbf{P}$ is simply a functor $\mathbf{P} : \mathcal{C}^{\text{op}} \to \mathbf{Sets}$. Next, in order to define a sheaf on $(\mathcal{C}, J)$, we need to introduce the notions of *matching families* and *amalgamation point*.

Definition 7.2.2 (Matching Family) Given a presheaf $\mathbf{P} : \mathcal{C}^{\text{op}} \to \mathbf{Sets}$ on $(\mathcal{C}, J)$ and a sieve $S \in J(C)$, a matching family for S of elements of $\mathbf{P}$ is a function which assigns to each $f \in S$ an element $x_f \in \mathbf{P}(dom(f))$ such that given any $g \in \mathcal{C}$ with $cod(g) = dom(f)$ then,

$$\mathbf{P}(g)(x_f) = x_{f \circ g}.$$

We will often denote a matching family for a sieve S by $\{x_f\}_{f \in S}$.

Definition 7.2.3 (Amalgamation) Given a matching family $\{x_f\}_{f \in S}$, where S is a sieve on C, an amalgamation is an element $x \in \mathbf{P}(C)$ such that:

$$\mathbf{P}(f)(x) = x_f, \quad \forall f \in S.$$

Given these two notions we can now define a sheaf on a site.

Definition 7.2.4 Given a site $(\mathcal{C}, J)$, a presheaf $\mathbf{P} : \mathcal{C}^{op} \rightarrow \mathbf{Sets}$ is a sheaf iff each matching family for any cover of any object in $\mathcal{C}$ has a unique amalgamation point.

Diagrammatically, what the above definition states is that for each object $C \in \mathcal{C}$ and each cover $S \in J(C)$, if the diagram

$$\mathbf{P}(C) \xrightarrow{e} \prod_{f \in S} \mathbf{P}(dom(f)) \overset{p}{\underset{q}{\rightrightarrows}} \prod_{\substack{f,g\, f \in S \\ cod(g)=dom(f)}} \mathbf{P}(dom(g))$$

is an equalizer for all covers S of all objects $C \in \mathcal{C}$, then $\mathbf{P}$ is a presheaf. Here $e(x) := \{\mathbf{P}(f)(x)\}_{f \in S}$, $p(\{x_f\}_{f \in S})_{f,g} = x_{fg}$ and $q(\{x_f\}_{f \in S})_{f,g} = \mathbf{P}(g)(x_f)$.

We would now like to compare Definitions 7.2.1 and 7.2.4. To this end we note that, for $\{x_f\}_{f \in S} \in \prod_{f \in S} \mathbf{P}(dom(f))$, the requirement of each family having a unique amalgamation point is equivalent to stating that $Hom(\mathbf{S}, \mathbf{P}) \cong Hom(\mathbf{y}(C), \mathbf{P})$. In fact, since a site S on C can be seen as a subfunctor of $\mathbf{y}(C) = Hom(-, C)$, then, for $f \in S \in J(C)$, the assignment $f \mapsto x_f$ is actually a component of the natural transformation $\mathbf{S} \rightarrow \mathbf{P}$ at $dom(f) \in \mathcal{C}$. Therefore, the fact that the family $\{x_f\}_{f \in S}$ has a unique amalgamation point is equivalent to the fact that $\mathbf{S} \rightarrow \mathbf{P}$ can be uniquely extended as follows:

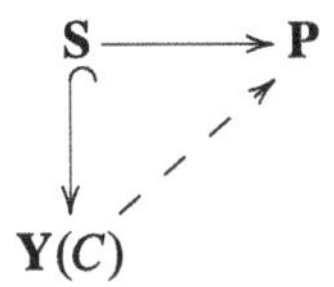

Hence $\mathbf{P}$ is a sheaf iff for every covering S of C, $S \hookrightarrow \mathbf{y}(C)$ induces the isomorphism $Hom(\mathbf{S}, \mathbf{P}) \cong Hom(\mathbf{Y}(C), \mathbf{P})$.

We would now like to give the definition of a sieve in terms of a basis K for a topology J. In order to do this we need to define the notion of a matching family and an amalgamation point with respect to K.

Definition 7.2.5 Given a cover K and a family of morphisms $R = \{f_i : D_i \rightarrow C\}_{i \in I} \in K(C)$ a family of elements $\{x_i\}_{i \in I}$, where $x_i \in \mathbf{P}(D_i)$, is said to be matching for R iff

$$\mathbf{P}(pr_1)(x_i) = \mathbf{P}(pr_2)(x_i), \quad \forall\, i, j \in I.$$

Here $pr_1 : D_i \times_C D_j \rightarrow D_i$ and $pr_2 : D_i \times_C D_j \rightarrow D_j$.

Definition 7.2.6 Given a matching family $\{x_i\}_{i \in I}$ for $R = \{f_i : D_i \rightarrow C\}_{i \in I} \in K(C)$ an amalgamation point is an element $x \in \mathbf{P}(C)$, such that for all $i \in I$, $\mathbf{P}(f_i)(x) = x_i$.

We now have the tools to define the sheaf condition with respect to a basis K.

Lemma 7.2.2 *Given a site $(\mathcal{C}, J)$, a presheaf $\mathbf{P} : \mathcal{C}^{op} \rightarrow \mathbf{Sets}$ is a sheaf for J iff for any family of morphisms $R = \{f_i : D_i \rightarrow C\}_{i \in I} \in K(C)$ then, any matching family $\{x_i\}_{i \in I}$ has a unique amalgamation point.*

We will now report a proof of this Lemma which can be found in [55].

Proof We will first prove the "if then" direction then the "only if".

⇒ We will assume that **P** is a sheaf for J and show that this implies that for any cover in K, any matching family has a unique amalgamation point. The assumption that **P** is a sheaf implies that given any sieve S in J and any matching family, this will have a unique amalgamation point. However we know how to construct sieve given covers in K. In particular, given a cover $R = \{f_i : D_i \to C\}_{i\in I} \in K(C)$, this generates a sieve $S_R = \{g : D \to C | g = f_i \circ h\, f_i \in R\}$. If we now consider the matching family $\{x_i\}_{i\in I}$ for R, we can construct a matching family $\{y_g\}_{g\in S}$ by

$$y_g = \mathbf{P}(h)(x_i), \quad h : D \to D_i \text{ s.t. } g = f_i \circ h.$$

We need to check that this definition does not depend on the choice of i and h. To this end consider the pullback diagram

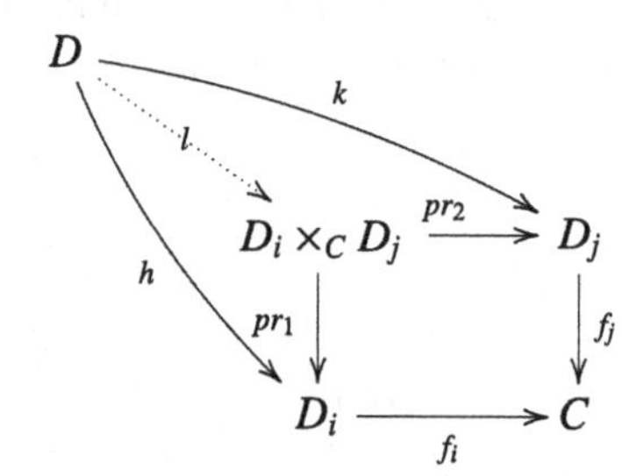

For $g : D \to C$, if $g = f_i \circ h = f_j \circ k$ then, the pullback property together with the fact that $\{x_i\}$ are matching for R, imply that

$$\mathbf{P}(h)(x_i) = \mathbf{P}(l)(\mathbf{P}(pr_1)(x_i)) = \mathbf{P}(l)(\mathbf{P}(pr_2)(x_j)) = \mathbf{P}(k)(x_j).$$

Given this matching family, since **P** is a sheaf with respect to J, it follows that there exists a unique amalgamation point $y \in \mathbf{P}(C)$ such that $\mathbf{P}(C)(g)(y) = y_g$ for all $g \in S_R$. Since $R \subseteq S_R$, then for all $f_i \in R$ we have that $\mathbf{P}(C)(f_i)(y) = y_{f_i} = x_i$, thus y is an amalgamation point for the family $\{x_i\}_{i\in I}$. It remains to show that such point is unique. This can be done by showing that an amalgamation point for $\{x_i\}_{i\in I}$ is also an amalgamation point for the cover S_R and we know this to be unique. In particular assume that $\{x_i\}_{i\in I}$ has another amalgamation point y', then $\mathbf{P}(C)(f_i)(y') = x_i$. Given $g = f_i \circ h \in S$ we then have $\mathbf{P}(g)(y') = \mathbf{P}(h)(\mathbf{P}(f_i)(y')) = \mathbf{P}(h)(x_i) = y_g$, hence y' is also an amalgamation point for S_R and we know this to be unique.

⇐ We now assume that each matching family for any cover R in K has a unique amalgamation point and then show that this implies that **P** is a sheaf with respect to J. To this end consider a cover $S \in J(C)$. By definition there exists an $R \in K(C)$ such that $R \subseteq S$. Next consider a matching family $\{y_g | g \in S\}$, then the sub-family $\{y_f | f \in R\}$ is matching and, by assumption, it has a unique amalgamation point y. We next want to show that such a point is also the unique amalgamation point for $\{y_g | g \in S\}$. Take any $g \in S$ and $f \in R$ and construct the pullback diagram

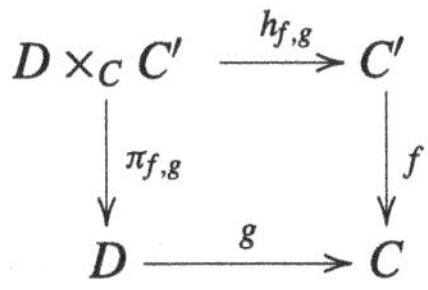

$$\begin{array}{ccc} D \times_C C' & \xrightarrow{h_{f,g}} & C' \\ \downarrow{\scriptstyle \pi_{f,g}} & & \downarrow{\scriptstyle f} \\ D & \xrightarrow{g} & C \end{array}$$

then by B.2 in Definition 7.1.11, $\{\pi_{f,g} | f \in R\} \in K(D)$. Given any $f \in R$ we also have that

$$\begin{aligned}
\mathbf{P}(g \circ \pi_{f,g})(y) &\overset{\text{commutativity}}{=} \mathbf{P}(f \circ h_{f,g})y \\
&\overset{f \in R}{=} \mathbf{P}(h_{f,g})(y_f) \\
&\overset{\text{matching family}}{=} y_{f \circ h_{f,g}} \\
&\overset{\text{commutativity}}{=} y_{g \circ \pi_{f,g}} \\
&\overset{\text{matching family}}{=} \mathbf{P}(\pi_{f,g})y_g
\end{aligned}$$

Next, construct the following pullback for any $f'' : C'' \to C \in R(C)$

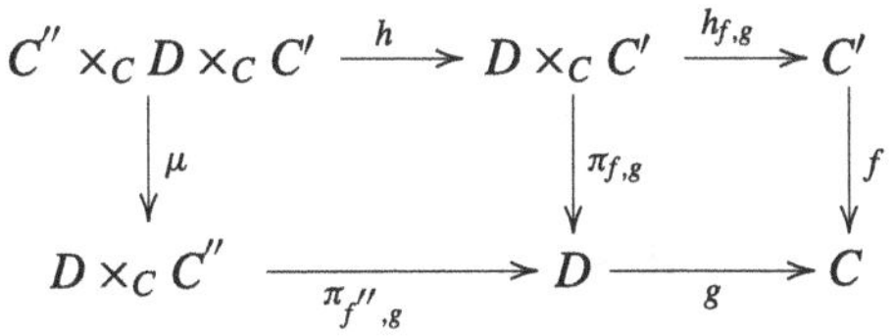

$$\begin{array}{ccccc} C'' \times_C D \times_C C' & \xrightarrow{h} & D \times_C C' & \xrightarrow{h_{f,g}} & C' \\ \downarrow{\scriptstyle \mu} & & \downarrow{\scriptstyle \pi_{f,g}} & & \downarrow{\scriptstyle f} \\ D \times_C C'' & \xrightarrow[\pi_{f'',g}]{} & D & \xrightarrow[g]{} & C \end{array}$$

Chasing the diagram around shows that $\mathbf{P}(g \circ \pi_{f,g})$ matches $\mathbf{P}(g \circ \pi_{f'',g})$, therefore the family $\{y_{g \circ \pi_{f,g}} | f \in R\}$ is matching for the cover $R' = \{\pi_{f,g} | f \in R\} \in K(D)$. By assumption, this family has a unique amalgamation point $y_g \in \mathbf{P}(D)$ such that $\mathbf{P}(\pi_{f,g})y_g = y_{g \circ \pi_{f,g}}$ for each $\pi_{f,g} \in R'$. But we know that $y_{g \circ \pi_{f,g}} = \mathbf{P}(g \circ \pi_{f,g})(y)$ and since the amalgamation point is unique it follows that $y_g = \mathbf{P}(g)(y)$. Hence y is also the unique amalgamation point for $\{y_g | g \in S\}$.

□

The following Corollary allows us to express sheaves in terms of the more familiar notion of equaliser.

Corollary 7.2.1 *Given a site $(\mathcal{C}, J)$, a presheaf $\mathbf{P} : \mathcal{C}^{op} \to$ **Sets** is a sheaf for J iff for any cover $\{f_i : D_i \to C\}_{i\in I} \in K(C)$, then the following diagram is an equaliser:*

$$\mathbf{P}(C) \xrightarrow{e} \prod_{i\in I} \mathbf{P}(D_i) \underset{q}{\overset{p}{\rightrightarrows}} \prod_{i,j\in I} \mathbf{P}(D_i \cap D_j) \tag{7.2.2}$$

where $e(x) = \{\mathbf{P}(f_i)(x)\}_{i\in I}$, $p(\{x_i\}_{i\in I})_{i,j} = \mathbf{P}(pr_1)(x_i)$ *and* $q(\{x_i\}_{i\in I})_{i,j} = \mathbf{P}(pr_2)(x_j)$.

The proof is just a matter of spelling out the definitions but for the sake of completeness we will nonetheless report it below.

Proof

- $\Rightarrow$ Let us assume that $\mathbf{P}$ is a sheaf for J. We know from Lemma 7.2.2 that any matching family for any basis K has a unique amalgamation point. Therefore consider any cover $R = \{f_i : D_i \to C\}_{i\in I} \in K(C)$, a matching family $\{x_i\}_{i\in I}$ for R will be such that $\mathbf{P}(pr_1)(x_i) = \mathbf{P}(pr_2)(x_j)$ for all $i, j \in I$. The unique amalgamation point is then an element $x \in P(C)$, such that $\mathbf{P}(f_i)(x) = x_i$ for all $i \in I$. But this is precisely the statement that the diagram (7.2.2) is an equaliser.
- $\Leftarrow$ On the other hand, if the diagram (7.2.2) is an equaliser, then for any element $\{\mathbf{P}(f_i)(x)\}_{i\in I} \in \prod_{i\in I} \mathbf{P}(D_i)$ such that $\mathbf{P}(pr_1)(x_i) = \mathbf{P}(pr_2)(x_j)$ for all $i, j \in I$, there exists a unique element $x \in P(C)$ such that $\mathbf{P}(f_i)(x) = x_i$ for all $i \in I$.

$\square$

The collection of all sheaves defined on a site $(\mathcal{C}, J)$ forms a category and it is denoted by $Sh(\mathcal{C}, J)$. It turns out that $Sh(\mathcal{C}, J)$ is actually a topos. To show that this is the case, we will first analyse how the sub-object classifier is defined in $Sh(\mathcal{C}, J)$. This will be the topic of the next section.

7.3 Sub-object Classifier

We would now like to construct a sheaf on a site $(\mathcal{C}, J)$ which plays the role of a sub-object classifier $\mathbf{\Omega}$. To this end we first recall that given a topological space X, then for each $U \in \mathcal{O}(X)$ the sheaf $\mathbf{\Omega}$ has been defined as

$$\mathbf{\Omega}(U) = \{\downarrow V | V \subseteq U\} \tag{7.3.1}$$

where $\downarrow V$ indicates the principal sieve on V. We will now utilise a variation of the definition of a principal sieve given by the following Lemma:

Lemma 7.3.1 *Given a sieve S on $U \in \mathcal{O}(X)$ if S satisfies the following condition*

$$\forall\ W \subseteq U, \quad S \text{ covers } W \Rightarrow W \in S$$

then S is a principal sieve on U.

Proof To prove this lemma simply take $W = U$. □

Essentially what the above Lemma says is that S is closed under arbitrary union of its elements. We now extend this property of a sieve being *closed* to an arbitrary site $(\mathcal{C}, J)$.

Definition 7.3.1 Given a site $(\mathcal{C}, J)$, a sieve S on $C \in \mathcal{C}$ is closed for J iff for all $f : D \rightarrow C$ in $\mathcal{C}$, then

$$S \text{ covers } f \Rightarrow f \in S. \tag{7.3.2}$$

Recall that "S covers f" means that $f^*S \in J(D)$ and $f \in S$ means that f^*S is a principal sieve on D, hence Eq. (7.3.2) is equivalent to

$$f^*S \text{ covers } D \Rightarrow f^*S \text{ is maximal on } D.$$

Corollary 7.3.1 *Given a sieve S on C and any morphism $g : B \rightarrow C$ then if S is closed so is g^*S, i.e. the property of being closed is preserved under pullback.*

Proof Assume that S is closed, we want to show that g^*S is also closed, i.e. given any $f : A \rightarrow B$ then $g^*(S)$ covers $f \Rightarrow f \in g^*(S)$. Let us assume that indeed $g^*(S)$ covers f, this means that $f^*(g^*(S)) \in J(A)$. But since S is closed it follows that $g \circ h \in S$, therefore by definition $f \in g^*(S)$. □

Given any sieve S one can construct the closure of S which is denoted as $\overline{S}$ and it is defined as follows:

Definition 7.3.2 (Closure) Given any sieve S on $C \in \mathcal{C}$, we define its closure to be the sieve

$$\overline{S} = \{h | cod(h) = C, \ S \text{ covers } h\}. \tag{7.3.3}$$

We now need to check that indeed $\overline{S}$ is a sieve and it is closed. To show that it is a sieve we need to show that it is closed under left composition. To this end assume that $h \in \overline{S}$ and take g such that $cod(g) = dom(h)$, we then want to show that $h \circ g \in \overline{S}$. Since $h \in \overline{S}$, then S covers h. From condition C'.2 it then follows that S also covers $f \circ h$, therefore $f \circ h \in \overline{S}$.

Next we need to show that $\overline{S}$ is closed, i.e. if $\overline{S}$ covers f then $f \in \overline{S}$. Assume that $\overline{S}$ covers f. By definition the sieve S covers every arrow in $\overline{S}$. Applying condition C'.3 it then follows that S also covers f, hence $f \in \overline{S}$.

From the above definition it transpires that $\overline{S}$ is the smallest closed sieve which contains S. We will now show that, for any $g : D \rightarrow C$,

$$\overline{g^*(S)} = g^*(\overline{S}). \tag{7.3.4}$$

In fact, since $\overline{S}$ is closed, from Lemma 7.3.1, it follows that $g^*(\overline{S})$ is also closed, therefore $\overline{g^*(S)} \subseteq g^*(\overline{S})$. On the other hand, if $f \in g^*(\overline{S})$, then $g \circ f \in \overline{S}$ which, form Eq. (7.3.3), is equivalent to S covering $g \circ f$ and $g^*(S)$ covering f. Therefore $f \in \overline{g^*(S)}$.

If we replace in Eq. (7.3.1) the notion of a principal sieve by that of a closed sieve we can define, for each object $C \in \mathcal{C}$

$$\mathbf{\Omega}(C) := \{S | S \text{ is a closed sieve on } C\}.$$

This definition actually turns $\mathbf{\Omega}$ into a presheaf $\mathbf{\Omega} : \mathcal{C}^{op} \to \mathbf{Sets}$ which is defined:

1. on objects: $C \mapsto \mathbf{\Omega}(C) := \{S|S \text{ is a closed sieve on } C\}$.
2. On morphisms: Given $f : D \to C$ then $\mathbf{\Omega}(f) : \mathbf{\Omega}(C) \to \mathbf{\Omega}(D)$; $S \mapsto f^*S$.

We would now like to show that $\mathbf{\Omega}$ is actually a sheaf for $(\mathcal{C}, J)$. Recall from Definition 7.2.4 that the condition of $\mathbf{\Omega}$ being a sheaf for $(\mathcal{C}, J)$ is equivalent to requiring the following diagram to be an equaliser.

$$\mathbf{\Omega}(C) \xrightarrow{e} \prod_{f \in S} \mathbf{\Omega}(dom(f)) \underset{q}{\overset{p}{\rightrightarrows}} \prod_{\substack{f, g\, f \in S \\ cod(g) = dom(f)}} \mathbf{\Omega}(U_i \cap U_j).$$

To show that indeed this is an equaliser we need to show two things: (1) that e is injective; (2) every matching family has a unique amalgamation point.

Proof ([55]) We start by showing that e is injective. To this end consider two closed sieves $M, N \in \mathbf{\Omega}(C)$ and assume that for each $f \in S \in J(C)$, then $f^*M = f^*N$, i.e. the images of M, N via e coincide in $\prod_{f \in S} \mathbf{\Omega}(dom(f))$. Now if $f^*M = f^*N$ for all $f \in S$ it follows, from the definition of the pullback, that $S \cap N = S \cap M$. Next consider any $g \in M$. By condition $C'.2$ it follows that M covers g. Since S covers C it also covers g, hence by Lemma 7.1.1 it follows that $M \cap S$ covers g. We know that $M \cap S = N \cap S \subseteq N$, hence N covers g and, since N is closed, it follows that $g \in N$. Similarly one can apply the same reasoning for any $h \in N$ obtaining that $M = N$. Hence e is injective.

Next we need to show that any matching family has a unique amalgamation point. Consider a cover $S \in J(C)$, then for any $f : D \to C$ in S the $M_f \in \mathbf{\Omega}(D)$ form a matching family $\{M_f\}_{f \in S}$ of closed sieves such that $g^*M_f = M_{f \circ g}$ for any g with $cod(g) = dom(f)$. To construct the amalgamation point we first construct the sieve

$$M = \{f \circ g | g \in M_f,\ f \in S\}.$$

Let us now define its closure $\overline{M}$ as in Definition 7.3.2. We claim such a closure to be the unique amalgamation point of $\{M_f\}_{f \in S}$, i.e. $f^*(\overline{M}) = M_f$ for all $f \in S$.

For any $f \in S$, then $f^*(M) = \{h | f \circ h \in M\}$, which clearly implies that $f^*(M) \supseteq M_f$. On the other hand if $h \in f^*M$, then by definition $f \circ h \in M$, which means that there exists a $f' \in S$ and $g' \in M_{f'}$, such that $f \circ h = f' \circ g'$. But since M_f and $M_{f'}$ belong to the matching family, then $M_{f \circ h} = h^*(M_f) = g^*(M_{f'}) = M_{f' \circ g'}$. However, since $g' \in M_{f'}$, then $g^*(M_{f'})$ is a maximal sieve and so is $h^*(M_f)$ therefore $f \in M_f$ and $f^*(M) \subseteq M_f$.

From Eq. (7.3.4) we know that $f^*(\overline{M}) = \overline{f^*(M)}$, but we have just shown that $f^*(M) = M_f$ and since M_f is closed we obtain that $f^*(\overline{M}) = M_f$. Hence $\overline{M}$ is an amalgamation point. □

We would now like to show that, as defined above, $\boldsymbol{\Omega}$ is indeed a sub-object classifier. To this end we recall the definition of a sub-object classifier [26] to be:

Definition 7.3.3 Given a category with a terminal object 1, a **sub-object classifier** is an object Ω, together with a monic arrow $\mathcal{T} : 1 \rightarrow \Omega$ (topos analogue of the set theoretic arrow *true*) such that, given a monic $\mathcal{C}$-arrow $f : a \rightarrow b$, there exists one and only one χ_f arrow, which makes the following diagram:

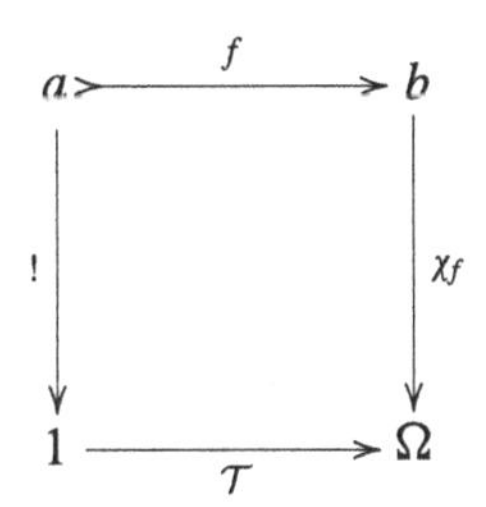

a pullback.
For the case at hand we first of all need to define the analogue of the arrow $\mathcal{T} : 1 \rightarrow \Omega$. We define it to be $C \mapsto t_C = \{f|cod(f) = C\}$, i.e. for each object $C \in \mathcal{C}$ the arrow $\mathcal{T}$ picks out the principal sieve on C. From the definition it is easy to see that t_C is closed. Given a map $g : D \rightarrow C$, then the following diagram clearly commutes

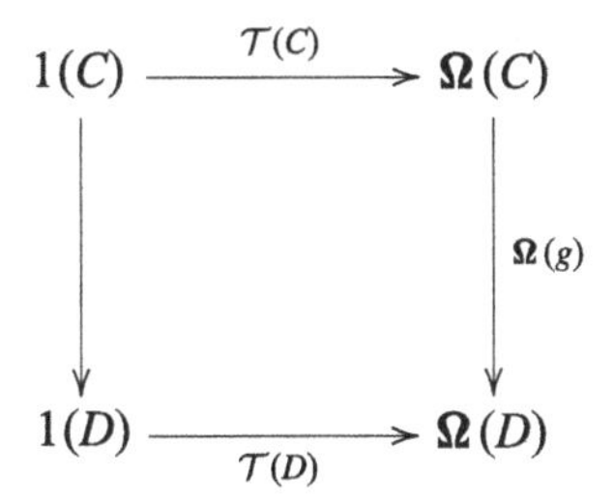

In fact $\boldsymbol{\Omega}(g)(t_C) = g^*(t_C) = t_D$. It follows that $\mathcal{T}$ is a natural transformation.

Given the definition of the arrow $\mathcal{T}$, we claim that the sheaf $\boldsymbol{\Omega}$ together with $\mathcal{T}$ is a sub-object classifier for the category $Sh(\mathcal{C}, J)$.

From Definition 7.3.3, in order to prove that the pair $\boldsymbol{\Omega}$ and $\mathcal{T}$ define a sub-object classifier we first of all need to define the analogue of the map χ_f, then show that such a definition would make the analogue of the diagram in Definition 7.3.3 a pullback. However in our setting, instead of considering a monic arrow f as in Definition 7.3.3, we will consider a subsheaf $\mathbf{Q}$ of a sheaf $\mathbf{P}$, which essentially is the same thing.

Definition 7.3.4 Consider a sheaf $\mathbf{P}$ on $(\mathcal{C}, J)$, and a subpresheaf $\mathbf{Q} \subseteq \mathbf{P}$. $\mathbf{Q}$ is a subsheaf iff for all $C \in \mathcal{C}$, for all covers S of C and for any $x \in \mathbf{P}(C)$, then $x \in \mathbf{Q}(C)$ whenever $\mathbf{P}(f)(x) \in \mathbf{Q}(dom(f))$ for all $f \in S$.

Essentially what this definition states is that any matching family of elements in $\mathbf{Q}$ has a unique amalgamation point since $\mathbf{P}$ is a sheaf and this amalgamation point lies in $\mathbf{Q}$. Clearly this means that $\mathbf{Q}$ is a sheaf.

Now that we know what a subsheaf is we can prove that the pair $\mathbf{\Omega}$ and $\mathcal{T}$ as defined above is a sub-object classifier.

Proof Given a subpresheaf $\mathbf{Q} \subseteq \mathbf{P}$ we define the arrow $\chi_{\mathbf{Q}} : \mathbf{P} \to \mathbf{\Omega}$, for any $C \in \mathcal{C}$ and $x \in \mathbf{P}(C)$ to be

$$(\chi_{\mathbf{Q}})_C(x) = \{f : D \to C | \mathbf{P}(f)(x) \in \mathbf{Q}(D)\}. \tag{7.3.5}$$

This is clearly a sieve. In fact given $f : D \to C$ in $(\chi_{\mathbf{Q}})_C(x)$ then $\mathbf{P}(f)(x) \in \mathbf{Q}(D)$. Now consider $g : B \to D$ then $f \circ g : B \to C$ will give rise to the presheaf map $\mathbf{P}(f \circ g)$, such that $\mathbf{P}(f \circ g)(x) \in \mathbf{Q}(B)$ since $\mathbf{P}(f)(x) \in \mathbf{Q}(D)$, therefore $f \circ g \in (\chi_{\mathbf{Q}})_C(x)$.

Consider a map $h : D \to C$ and assume that $(\chi_{\mathbf{Q}})_C(x)$ covers h, i.e. $h^*((\chi_{\mathbf{Q}})_C(x)) \in J(D)$. We want to show that $h \in (\chi_{\mathbf{Q}})_C(x)$, i.e. that $\chi_{\mathbf{Q}}(x)$ is closed and hence an element of $\mathbf{\Omega}(C)$. To this end consider any $k \in h^*((\chi_{\mathbf{Q}})_C(x)) \in J(D)$, then by definition $k \circ h \in (\chi_{\mathbf{Q}})_C(x)$, which implies that $\mathbf{P}(k \circ h)(x) \in \mathbf{Q}(dom(h))$. However since $\mathbf{Q} \subseteq \mathbf{P}$, then we can apply the definition of a subpresheaf obtaining $\mathbf{P}(h)(x) \in \mathbf{Q}(D)$ as desired.

Next we want to show that $\chi_{\mathbf{Q}}$ satisfies the naturality condition. To this end consider a map $g : B \to C$ and construct the diagram

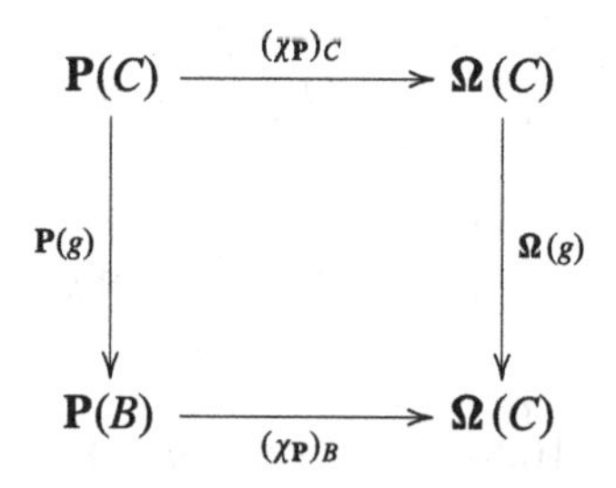

We want to show that it commutes, i.e. given $x \in \mathbf{Q}(C) \subseteq \mathbf{P}(C)$, then $(\chi_{\mathbf{Q}})_B(\mathbf{P}(g)(x)) = \mathbf{\Omega}(g) \circ (\chi_{\mathbf{Q}})_C(x)$. Consider an $f \in (\chi_{\mathbf{Q}})_B(\mathbf{P}(g)(x))$ then, since $\mathbf{P}(f)(\mathbf{P}(g)(x)) = (\mathbf{P}(g \circ f)(x)$, this implies that $g \circ f \in (\chi_{\mathbf{Q}})_C$ therefore $f \in g^*(\chi_{\mathbf{Q}})_C$. But from the definition of $\mathbf{\Omega}$ we know that $\mathbf{\Omega}(g) \circ (\chi_{\mathbf{Q}})_C(x) := g^*(\chi_{\mathbf{Q}})_C$ hence $\chi_{\mathbf{Q}}$ is a natural transformation.

As a final step we need to show that $\chi_{\mathbf{Q}}$ is the unique arrow such that the square

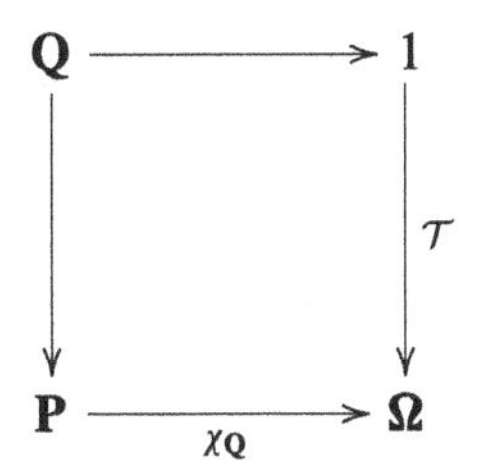

is a pullback. Since in $Sh(\mathcal{C}, J)$ pullbacks are computed component-wise, we need to show that for all $C \in \mathcal{C}$ then

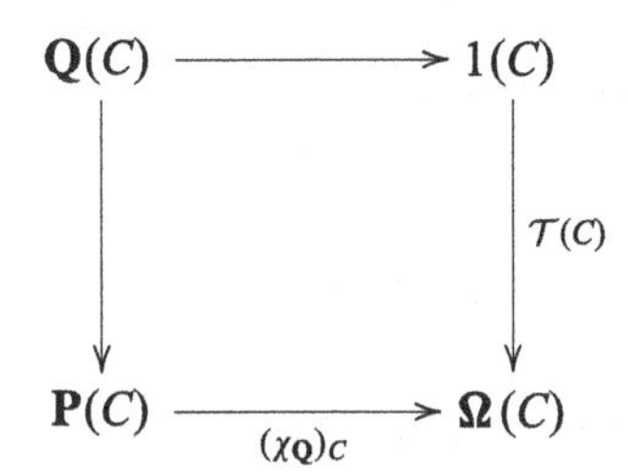

is a pullback. The pullback condition is equivalent to stating that, given any $x \in \mathbf{P}(C)$ then $x \in \mathbf{Q}(C)$ iff $(\chi_{\mathbf{Q}})_C = t_C$. However, the latter condition follows from the definition of $(\chi_{\mathbf{Q}})_C$, therefore $\mathbf{Q}(C)$ is the pullback of $\chi_{\mathbf{Q}}$ along $\mathcal{T}$. Next we need to show that $(\chi_{\mathbf{Q}})_C$ is the unique arrow for which the pullback condition is satisfied. This will be shown by proving that the condition "$id_C \in (\chi_{\mathbf{Q}})_C(x)$ iff $x \in \mathbf{Q}$", implies that $(\chi_{\mathbf{Q}})_C(x)$ is defined as in (7.3.5). In fact given any $f : D \to C \in \mathcal{C}$, then $f \in (\chi_{\mathbf{Q}})_C(x)$ iff $id_D \in f^*(\chi_{\mathbf{Q}})_C(x))$. But by naturality of $\chi_{\mathbf{Q}})_C$ and the condition that $id_C \in (\chi_{\mathbf{Q}})_C(x)$ iff $x \in \mathbf{Q}$, it follows that $f^*(\chi_{\mathbf{Q}})_C(x)) = (\chi_{\mathbf{Q}})_C(\mathbf{P}(f)(x))$ (see diagram below). But this in turn implies that $\mathbf{P}(f)(x) \in \mathbf{Q}(D)$ hence $(\chi_{\mathbf{Q}})_C(x)$ is defined as in (7.3.5).

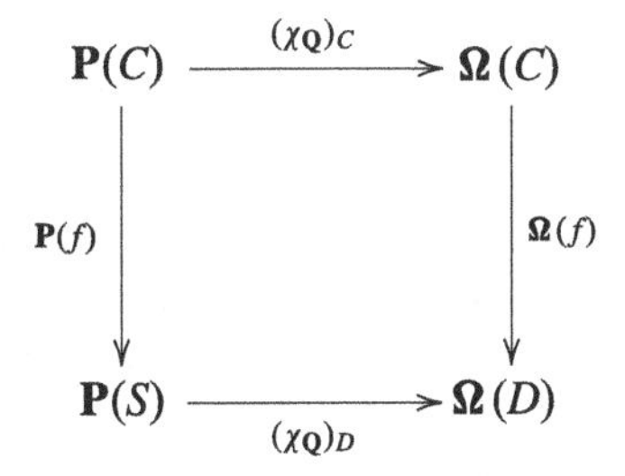

□

7.4 $Sh(\mathcal{C}, J)$ Is an Elementary Topos

In this section we would like to prove that the category of sheaves over a site is actually an elementary topos. We recall from [26] that an elementary topos τ is a category such that:

1. τ has all finite limits and colimits;
2. τ has exponentials;
3. τ has a sub-object classifier.

So far we have seen that indeed $Sh(\mathcal{C}, J)$ has a sub-object classifier, so it remains to show that also (1) and (2) are satisfied.

We will start by showing that $Sh(\mathcal{C}, J)$ is closed under taking finite limits. In order to do this we will recall briefly how limits are defined in a category. For a more in depth discussion the reader is referred to [26, 55].

Before defining limits we need to introduce the notion of a diagram of a category. Given category $\mathcal{C}$ and a small category I called the index category, a diagram D in $\mathcal{C}$ of type I is a functor $(I \to \mathcal{C}) \in \mathcal{C}^I$. Alternatively one can define it as in Definition 4.5.10 [26].

Definition 7.4.1 Given a category $\mathcal{C}$, a diagram D in $\mathcal{C}$ is defined to be a collection of $\mathcal{C}$-objects $a_i \in C$ $(i \in I)$ and a collection of $\mathcal{C}$-arrows $a_i \to a_j$ between some of the $\mathcal{C}$-objects above.

A special type of diagram is the D-cone, i.e. a cone for a diagram D. This consists of a $\mathcal{C}$-object c and $\mathcal{C}$-arrows $f_i : c \to a_i$, one for each $a_i \in D$, such that

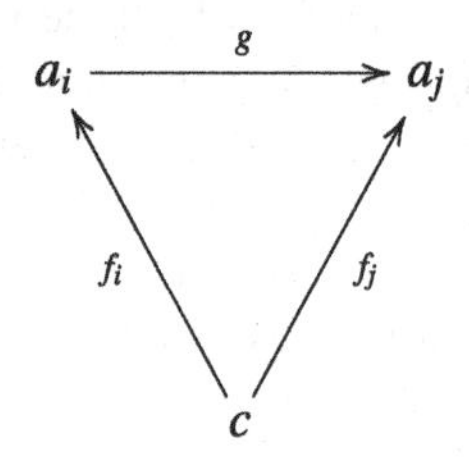

commutes whenever g is an arrow in the diagram D.

A cone is denoted by $\{f_i : c \to a_i\}$ and c is called the vertex of the cone. It is also possible to define a cone in terms of functors. To this end consider the functor $\mathbf{\Delta_I} : \mathcal{C} \to \mathcal{C}^I$ which takes each object $C \in \mathcal{C}$ to the constant diagram $\mathbf{\Delta_I}(C) \in \mathcal{C}^I$, which has value C for all $i \in I$. Next, given any other diagram $A \in \mathcal{C}^I$, then a natural transformation $f : \mathbf{\Delta_I}(C) \to A$ has components $f_i : C \to A_i$ for each $i \in I$ such that, given $(g : i \to j) \in I$, the following diagram commutes

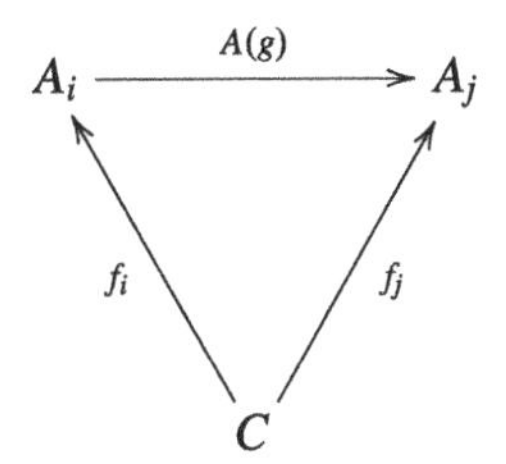

Here $f : \mathbf{\Delta}_I(C) \to A$ is called a cone of the diagram A with vertex C. Often such cones are simply denoted by $f : C \to A$.

We can now introduce the notion of a limit [26, 55].

Definition 7.4.2 A limit for a diagram D is a D-cone $\{f_i : c \to a_i\}$ such that, given any other D-cone $\{f_i^{'} : c^{'} \to a_i\}$, there is only one $\mathcal{C}$-arrow $g : c^{'} \to c$ such that, for each $a_i \in D$ the following diagram commutes:

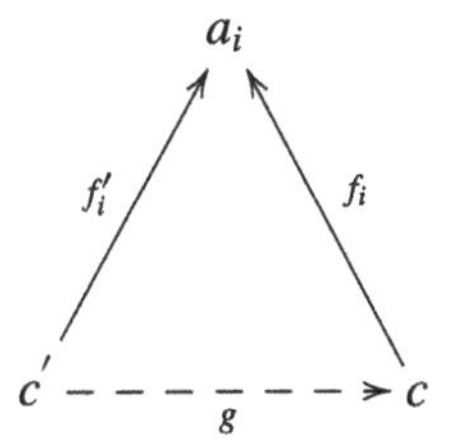

The limiting cone of a diagram D has the *universal property* with respect to all other D-cones, in the sense that any other D-cone factors out through the limiting cone.

It is also possible to define limits in terms of functors. Consider the diagram A, a limit of such a diagram is a cone $f : C \to A$ with vertex C, such that given any other $f' : C' \to A$ there exists a unique map $g : C' \to C$ which make the following diagram commutes for any $(h : i \to j) \in I$

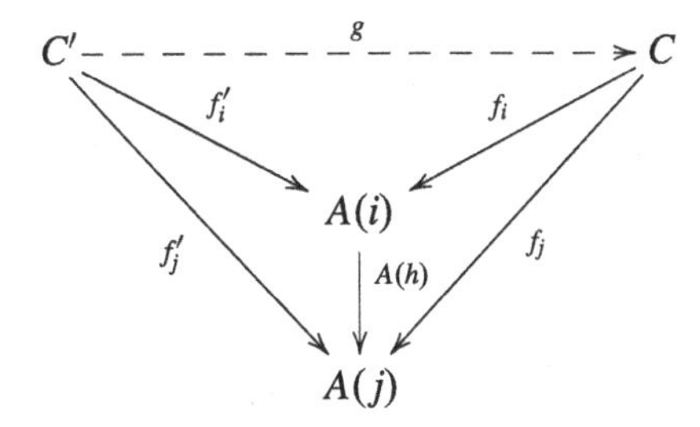

The limit of a diagram A is denoted by $\lim_{\leftarrow_I} A$. This implies that "taking the limit" is a functor of the form

$$\lim_{\leftarrow_I} : \mathcal{C}^I \to \mathcal{C}.$$

It turns out that $\lim_{\leftarrow I}$ is right adjoint to $\mathbf{\Delta_I}$.

An important result for presheaves is that limits are computed pointwise. In particular, consider the functor category $\mathcal{C}^{\mathcal{D}}$ for any two categories $\mathcal{D}$ and $\mathcal{C}$. A diagram of type I in $\mathcal{C}^{\mathcal{D}}$ is a map $A : I \to \mathcal{C}^{\mathcal{D}}$. Given any such diagram define $A_D : I \to \mathcal{C}; i \mapsto A_D(i)$ by

$$A_D(i) := A(i)(D).$$

This map then defines a diagram $A_D : I \to \mathcal{C}$ in $\mathcal{C}$ for each object $D \in \mathcal{D}$. Next assume that each such diagram has a limit $L_D = \lim_{\leftarrow J} A_D$, then the presheaf structure combines all these limits together to get a limit for the original diagram A. Such a limit is represented by a functor $\lim_{\leftarrow I} A : \mathcal{D} \to \mathcal{C}$ in $\mathcal{C}^{\mathcal{D}}$, whose components are given by the individual limits in $\mathcal{C}$, i.e.

$$(\lim_{\leftarrow I} A)(D) = \lim_{\leftarrow I} A_D. \tag{7.4.1}$$

Now that we have revised the notion of a limit and some of its properties let us go back and analyse limits in $Sh(\mathcal{C}, J)$. From the discussions of the previous section it is easy to see that the category of sheaves on a site $(\mathcal{C}, J)$ is a full sub-category of the functor category $\mathbf{Sets}^{\mathcal{C}^{op}}$. In $\mathbf{Sets}^{\mathcal{C}^{op}}$ a diagram is simply a functor $I \to \mathbf{Sets}^{\mathcal{C}^{op}}$; $i \to \mathbf{P}_i$, therefore we have the following theorem:

Theorem 7.4.1 ([55]) *Consider a site $(\mathcal{C}, J)$ and a diagram $I \to \mathbf{Sets}^{\mathcal{C}^{op}}$ of presheaves $\mathbf{P}_i$. If all $\mathbf{P}_i$ $(i \in I)$ are sheaves then so is $\lim_{\leftarrow I} \mathbf{P}_i$.*

Proof Consider the limit $\mathbf{P} = \lim_{\leftarrow I} \mathbf{P}_i$ of presheaves in $\mathbf{Sets}^{\mathcal{C}^{op}}$. Given any $C \in \mathcal{C}$, from Eq. (7.4.1) it follows that $\mathbf{P}(C) = \lim_{\leftarrow I} \mathbf{P}_i(C)$. We want to show that $\mathbf{P}$ satisfies the sheaf condition when all $\mathbf{P}_i$, $i \in I$ satisfy it. Let us spell out the sheaf condition for $\mathbf{P}_i$. This states that, given a cover S of $C \in \mathcal{C}$, the following diagram is an equaliser

$$\mathbf{P}_i(C) \xrightarrow{e} \prod_{f \in S} \mathbf{P}_i(dom(f)) \underset{q}{\overset{p}{\rightrightarrows}} \prod_{\substack{f,g\, f \in S \\ cod(g)=dom(f)}} \mathbf{P}_i(dom(g)). \tag{7.4.2}$$

However, we know that the functor $\lim_{\leftarrow I}$ is a right adjoint and, as such, it preserves limits. Since the equaliser is a limit, applying $\lim_{\leftarrow I}$ to the above equaliser gives us the equaliser

$$\mathbf{P}(C) \xrightarrow{e} \prod_{f \in S} \mathbf{P}(dom(f)) \underset{q}{\overset{p}{\rightrightarrows}} \prod_{\substack{f,g\, f \in S \\ cod(g)=dom(f)}} \mathbf{P}(dom(g)),$$

hence $\mathbf{P}$ is a sheaf. □

This proves that $Sh(\mathcal{C}, J)$ has finite limits.

Next we need to show that all small colimits exist in $Sh(\mathcal{C}, J)$. Colimits are dual to limits and, as such, are defined in terms of cocones. Given a diagram D a cocone consists of an object c and arrows $\{f_i : a_i \to c\}$, one for each element $a_i \in D$, such that

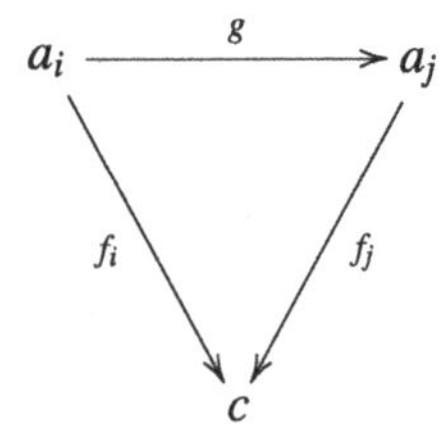

commutes whenever g is an arrow in D. A cocone is often denoted by $\{f_i : a_i \to c\}$. Alternatively, given a diagram $A : I \to \mathcal{C}$, a cocone with vertex C is a map $A \to \mathbf{\Delta}_I(C)$ in $\mathcal{C}^I$. The universal cocone of A is the colimit of A and it is denoted by $\lim_{\to_I} A$. Similarly, as for the case of the limit, "taking the colimit" gives rise to a functor

$$\lim_{\to_I} : \mathcal{C}^I \to \mathcal{C}$$

whose left adjoint is the diagonal functor $\mathbf{\Delta}_I : \mathcal{C} \to \mathcal{C}^I$.

In order to show that all small colimits exist in $Sh(\mathcal{C}, J)$ we need to introduce the adjunction

$$Sh(\mathcal{C}, J) \underset{\mathbf{a}}{\overset{i}{\rightleftarrows}} \mathbf{Sets}^{\mathcal{C}^{\mathrm{op}}}$$

Clearly from the definition of $Sh(\mathcal{C}, J)$ the map $i : Sh(\mathcal{C}, J) \to \mathbf{Sets}^{\mathcal{C}^{\mathrm{op}}}$ is simply an inclusion. Its left adjoint $\mathbf{a} : \mathbf{Sets}^{\mathcal{C}^{\mathrm{op}}} \to Sh(\mathcal{C}, J)$ is called the associated sheaf functor.

Before proving the adjunction $\mathbf{a} \dashv i$ we need to understand how the functor $\mathbf{a}$ acts. Consider any presheaf $\mathbf{P} \in \mathbf{Sets}^{\mathcal{C}^{\mathrm{op}}}$, we then define the associated presheaf by

$$\mathbf{P}^+(C) := \lim_{\to R \in J(C)} \mathrm{Match}(R, \mathbf{P})$$

where $\mathrm{Match}(R, P)$ are the matching families for the cover R, and the colimit is computed with respect to all the covers of $C \in \mathcal{C}$, which are ordered by reverse inclusion. As an example of how $\mathbf{P}^+(C)$ is constructed let us consider two covers R, S, which have a common refinement $T \subseteq R \cap S$ with $T \in J(C)$. We then obtain the maps $S \to T$ and $R \to T$. These can be combined into the following pullback diagram:

$$\begin{array}{ccc} S \cap R & \longrightarrow & R \\ \downarrow & & \downarrow \\ S & \longrightarrow & T \end{array}$$

At the presheaf level this translates into.

$$\begin{array}{ccc} & & \text{Match}(R, P) \\ & & \uparrow \\ \text{Match}(S, P) & \longleftarrow & \text{Match}(T, P) \end{array}$$

The colimit of this diagram is then given by the pushforward

$$\begin{array}{ccc} \text{Match}(S, P) \coprod_{\text{Match}(T,P)} \text{Match}(R, P) & \longleftarrow & \text{Match}(R, P) \\ \uparrow & & \uparrow \\ \text{Match}(S, P) & \longleftarrow & \text{Match}(T, P) \end{array}$$

Essentially the object $\text{Match}(S, P) \coprod_{\text{Match}(T,P)} \text{Match}(R, P) \simeq \text{Match}(S \cap R, P)$ consists of equivalence classes of matching families where two matching families belong to the same equivalence relation, if their restriction to a common refinement T coincides. In particular, an element of $P^+(C)$ would be an equivalence class $[\{x_f\}_{f \in R}]$ where

$$\{x_f\}_{f \in R} \quad \text{s.t}\, x_f \in \mathbf{P}(D) \quad \text{and}\, \forall g : E \to D, \;\; \mathbf{P}(g)(x_f) = x_{f \circ g}.$$

Given two families $\{x_f\}_{f \in R}$ and $\{y_g\}_{g \in S}$, these belong to the same equivalence class if, given a common refinement $T \subseteq S \cap R$ in $J(C)$, then $x_k = y_k$ for all $k \in T$.

Next we need to define the presheaf maps. Given a map $h : D \to C$ in $\mathcal{C}$, at the level of presheaves we obtain

$$\begin{aligned} &\mathbf{P}^+(C) \to \mathbf{P}^+(D) \\ &[\{x_f\}_{f \in R}] \mapsto \mathbf{P}^+(h)([\{x_f\}_{f \in R}]) = [\{x_{h \circ f'}\}_{f' \in h^*(R)}]. \end{aligned} \tag{7.4.3}$$

Note that $\mathbf{P}^+(h)([\{x_f\}_{f \in R}]) = ([\{\mathbf{P}(h) x_{f'}\}_{f' \in h^*(R)}])$.

Clearly such a map preserves equivalence classes that is, if $\{x_f\}_{f \in R} \sim \{y_g\}_{g \in S}$, then $= \{x_{h \circ f'}\}_{f' \in h^*(R)} = \{x_{h \circ g'}\}_{g' \in h^*(S)}$. Since $T \subseteq S \cap R$, then $h^*(T) \subseteq h^*(S \cap R)$ and by property C.2 of a Grothendieck topology $h^*(T) \in J(D)$. Let us consider next the two families $\{x_{h \circ f'}\}_{f' \in h^*(R)}$ and $\{x_{h \circ g'}\}_{g' \in h^*(S)}$. For them to be equivalent we need to find a common refinement on which they coincide. Choose the common refinement to be $h^*(T) = \{l | h \circ l \in T\}$. We then want that for all $l \in h^*(T)$, $x_{h \circ l} = y_{h \circ l}$, which is indeed the case since $h \circ l \in T$ and by assumption $\{x_f\}_{f \in R} \sim \{y_g\}_{g \in S}$.

Having defined $\mathbf{P}^+$ we need to show that it is indeed a presheaf. In particular we need to show that

1. $\mathbf{P}^+(h \circ g) = \mathbf{P}^+(g) \circ \mathbf{P}^+(h)$ whenever $cod(g) = dom(h)$;
2. $\mathbf{P}^+(id_C) = id_{\mathbf{P}^+(C)}$.

We will start with condition (1).

$$\begin{aligned}\mathbf{P}^+(h \circ g)([\{x_f\}_{f \in R}]) &= ([\{\mathbf{P}(h \circ g)x'_f\}_{f' \in g^*h^*(R)}])\\ &= ([\{\mathbf{P}(g) \circ \mathbf{P}(h)x'_f\}_{f' \in g^*h^*(R)}])\\ &= \mathbf{P}^+(g) \circ \mathbf{P}^+(h)([\{x_f\}_{f \in R}]).\end{aligned}$$

Condition (2) is a straightforward consequence of the definition of $\mathbf{P}^+$.

Now that we have constructed the presheaf $\mathbf{P}^+$ we would like to analyse whether it is also a sheaf. It turns out that this is not always the case. However, what is always true, is that $\mathbf{P}^+$ is "almost" a sheaf in the sense that it is *separated*. Essentially, what it means for a presheaf to be separated is that it satisfies the requirement of uniqueness for the amalgamation point but not necessarily the requirement of existence. In other words, for $\mathbf{P}$ separated, the diagram (7.4.2) might fail to be an equaliser but the map e is still injective.

Lemma 7.4.1 ([55]) *Given any presheaf* $\mathbf{P}$, *then* $\mathbf{P}^+$ *is separated.*

Proof To show that $\mathbf{P}^+$ is separated we need to show that the map e in the diagram the diagram (7.4.2) is injective, i.e. we need to show that the following map is injective:

$$\mathbf{P}^+(C) \xrightarrow{e} \prod_{h \in Q} \mathbf{P}^+(dom(h))$$

To this end let us consider two elements $\{x_f\}_{f \in R}$ and $\{y_g\}_{g \in S}$ in $\mathbf{P}^+(C)$ such that $\mathbf{P}^+(h)\{x_f\}_{f \in S} = \mathbf{P}^+(h)\{y_g\}_{g \in R}$ for all $(h : D \to C) \in Q$, where Q is a cover of C. Our task is to show that the matching families $\{x_f\}_{f \in S}$ and $\{y_g\}_{g \in R}$ are the same. From the definition of the presheaf maps (7.4.3), we know that the equality $\mathbf{P}^+(h)\{x_f\}_{f \in S} = \mathbf{P}^+(h)\{y_g\}_{g \in R}$, means that $\{x_{hf'} | f' \in h^*(S)\} = \{x_{hg'} | g' \in h^*(R)\}$. This, in turn, implies that for each $h \in Q$ there exists some cover $T_h \subseteq h^*(R) \cap h^*(S)$ of D, such that $x_{ht} = y_{ht}$ for all $t \in T_h$. Let us consider now the set

$$T = \{ht | h \in Q, t \in T_h\}$$

This is clearly a sieve on C since Q is and all sieves are closed under composition, then the pullback $h^*(T)$ for all $h \in Q$ defines a sieve on $dom(h)$. From the transitivity axiom of Grothendieck topology C'.3 in Definition 7.1.8 it follows that $T \in J(C)$. Next consider $h^*(T) = \{f|h \circ f \in T\}$. Clearly if $f \in h^*T$, then $h \circ f \in T$, therefore $f \in T_h$. This means that $T_h \subseteq h^*(T)$. On the other hand if $f \in T_h$ then $h \circ f \in T$ and $f \in h^*(T)$. This implies that $T_h \supseteq h^*(T)$. Putting these two results together we obtain that $T_h = h^*(T)$, then $T \subseteq R \cap S$. Now to show that $\{x_f\}_{f \in S} = \{y_g\}_{g \in R}$ we need to show that there exists a subcover $W \subseteq S \cap R$ such that $x_j = y_j$ for all $j \in W$. But this is precisely what T is. In fact any element of T has the form ht for $h \in Q$ and $t \in T_h$. Then for each such element we have indeed that $x_{ht} = y_{ht}$, therefore $\{x_f\}_{f \in S} = \{y_g\}_{g \in R}$ and the map e is indeed injective. □

The idea is now to define $\mathbf{a}$ by $\mathbf{a}(P) := (P^+)^+$ and show that this turns $\mathbf{a}$ into a left adjoint for i. In order to prove the adjunction we first need to show that $(P^+)^+$ is actually a sheaf. That this is the case is elucidated by the following Lemma:

Lemma 7.4.2 *If* $\mathbf{P}$ *is a separated presheaf then* $\mathbf{P}^+$ *is a sheaf.*

Proof In order to show that $\mathbf{P}^+$ is a sheaf we need to show that for any matching family there exists an amalgamation point and this point is unique. To start with, let us define a matching family in $\mathbf{P}^+$. From now on we will denote an equivalence class of matching families $\{x_g\}_{g \in S}$ as $\mathbf{x}$. Given a cover $R \in J(C)$, the family $\{\mathbf{x}_f|f \in R\}$, for $\mathbf{x}_f \in \mathbf{P}^+(D)$, is a matching family in $\mathbf{P}^+$ if for any morphisms $h : E \rightarrow D$, $\mathbf{P}^+(h)\mathbf{x}_f = \mathbf{x}_{fh}$, where $f : D \rightarrow C$, $\mathbf{x}_f = \{x_{f,g}|g : E \rightarrow D \in S_f\}$ and $x_{f,g} \in P(E)$. The condition $\mathbf{P}^+(h)\mathbf{x}_f = \mathbf{x}_{fh}$ means that there exists an equivalence of families

$$\{x_{f,hg'}|g' \in h^*(S_f)\} \sim \{x_{fh,g}|g \in S_{fh}\}.$$

This equivalence implies that there exists a cover $T_{f,h} \subseteq h^*(S_f) \cap S_{fh}$, such that $x_{f,hk} = x_{f \circ h,k}$ for all $k \in T_{f,h}$.

Our first task is to find an amalgamation point for the family $\{\mathbf{x}_f|f \in R\}$. This will be a point $\mathbf{y} \in \mathbf{P}^+(C)$, such that $\mathbf{P}(f)\mathbf{y} = \mathbf{y}_f$ for all $f \in R$. Since our amalgamation point is an element of $\mathbf{P}^+(C)$ it has to be a matching family, so we need to construct it in terms of the matching families we have at our disposal. Clearly a possibility is to define $\mathbf{y}$ in terms of the matching family $\mathbf{x}_f$. First of all we need to choose a cover of C in terms of which to define $\mathbf{y}$. To this end we construct the sieve $Q = \{f \circ g|f \in R, g \in S_f\}$, where $R \in J(C)$ and $S_f \in J(D)$. Then for all $h \in R$, clearly $h^*(Q) \in J(D)$, thus by the transitivity axiom of Grothendieck topology we have that $Q \in J(C)$. Given such a cover, then for all $f \circ g \in Q$ we define

$$y_{f \circ g} := x_{f,g} \tag{7.4.4}$$

and

$$\mathbf{y} = \{y_h|h \in Q\}.$$

Clearly if $\mathbf{y}$ is well defined then it automatically is a matching family since $\mathbf{x}_f$ is one. So what we need to check is that $\mathbf{y}$ is indeed well defined. What this means is that the definition we gave does not have to depend on the choice of factorisation of

$f \circ g$, i.e. if $f \circ g = f' \circ g'$ then we need to have that $y_{f \circ g} = y_{f' \circ g'}$. The strategy we will use to show this it to show that the points $x_{f,g}, x_{f',g'} \in \mathbf{P}(E)$ are amalgamation points and, since $\mathbf{P}$ is separated, they have to be the same. To this end we consider the cover $T_{f,g} \cap T_{f',g'}$ of E. Then for all $k \in T_{f,g} \cap T_{f',g'}$ we have

$$\begin{aligned}
\mathbf{P}(k)(x_{f,g}) &\overset{\mathbf{x}_f \text{matching family}}{=} x_{f,gk} \\
&= x_{f \circ g,k} \\
&= x_{f' \circ g',k} \\
&= x_{f',g'k} \\
&\overset{\mathbf{x}_{f'} \text{matching family}}{=} \mathbf{P}(k)(x_{f',g'}).
\end{aligned}$$

Because of the uniqueness of amalgamation point it follows that $x_{f,g} = x_{f',g'}$, thus $y_{f \circ g}$ is well defined and $\mathbf{y} = \{y_h | h \in Q\}$ is a matching family and hence an element of $\mathbf{P}^+(C)$. As the final step in our proof we need to show that $\mathbf{y}$ is the amalgamation point of the family $\{\mathbf{x}_f | f \in R\}$. What is means is that we need to show that for each $f : D \to C$ in R, then $\mathbf{P}^+(C)(f)\mathbf{y} = \mathbf{x}_f$. To see that this is indeed the case we need to spell out the action of the presheaf map

$$\mathbf{P}^+(C)(f)\mathbf{y} = \{y_{f \circ h} | h \in f^*Q\}.$$

Let us assume that $y_{f \circ h} \in \mathbf{P}^+(C)(f)\mathbf{y}$, then from Definition (7.4.4) and that of $f^*(Q) = \{h | f \circ h \in Q\}$ it follows that indeed $y_{f \circ h} = x_{f,g}$. On the other hand if $x_{f,g} \in \mathbf{x}_f$, then $g \in S_f$. However, from the definition of Q, it follows that $S_f \subseteq f^*(Q)$, therefore for all $g \in S_f$, $x_{f,g} = y_{f \circ g}$. This shows that $\mathbf{y}$ is an amalgamation point. The fact that it is unique follows from the fact that $\mathbf{P}^+$ is separated. □

Given the above Lemmas we now have the right tools to prove the adjunction

$$Sh(\mathcal{C}, J) \underset{\mathbf{a}}{\overset{i}{\rightleftarrows}} \mathbf{Sets}^{\mathcal{C}^{op}}.$$

Proof We want to show that $\mathbf{a}$ is left adjoint to the inclusion functor $i : \mathbf{Sets}^{\mathcal{C}^{op}} \to Sh(\mathcal{C}, J)$. Recall that the adjunction tells us that for a given map $\mathbf{P} \to i(\mathbf{F})$ there corresponds a unique map $a(\mathbf{P}) \to \mathbf{F}$. Consider the natural transformation $\eta : \mathbf{P} \to \mathbf{P}^+$, which, for each $C \in \mathcal{C}$ is defined as:

$$\begin{aligned}
\eta_C : \mathbf{P}_C &\to \mathbf{P}^+(C) \\
x &\mapsto \eta_C(x) := \{\mathbf{P}(f)x | f \in t_C\}
\end{aligned} \tag{7.4.5}$$

where t_C is the maximal sieve on C. Clearly two applications of η would result in a map from $\mathbf{P}$ to $\mathbf{P}^{++} = a(\mathbf{P})$. We want to show that any map from $\mathbf{P}$ to a sheaf $\mathbf{F}$ factors uniquely through the map $\mathbf{P} \xrightarrow{\eta \circ \eta} a(\mathbf{P})$ which, pictorially translates to the following diagram:

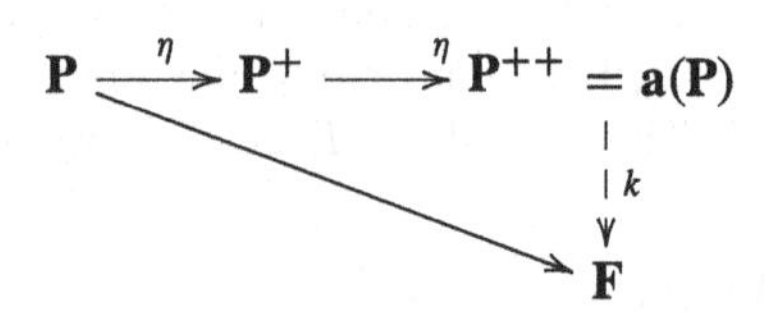

Since we are applying the map η two times it suffices to show that the factorisation

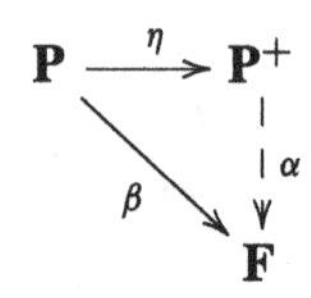

is unique, i.e. we want to construct a unique map α which makes the above diagram commute. Let us consider an element $\{x_f|f \in R\} \in \mathbf{P}(C)$ for some cover $R \in J(C)$. Then, given any map $g : D \to C$ in R we have that $\eta_D(x_g) = \{\mathbf{P}(h)x_g|h \in t_D\}$. However, since $\{x_f|f \in R\}$ is a matching family of $\mathbf{P}$, then $\mathbf{P}(g)\left(\{x_f|f \in R\}\right) = \{x_{gf'}|f' \in g^*R\}$. Since $g \in R$, then $g^*R = t_D$, therefore we obtain $\eta_D(x_g) = \mathbf{P}(g)\left(\{x_f|f \in R\}\right)$. This is true for all $g \in R$. If the map α were to exist, it should preserve such equality, that is there should exist a unique $\alpha(\{x_f|f \in R\}) \in F(C)$ such that

$$\mathbf{F}(g)\alpha(\{x_f|f \in R\}) = \alpha([\mathbf{P}(g)\{x_f|f \in R\}]) = \alpha(\eta_D(x_g)) = \beta(x_g) \tag{7.4.6}$$

for all $g \in R$. However, since F is a sheaf and $\{\beta(x_g)|g \in R\}$ is a matching family, then indeed there exists a unique element $\alpha(\{x_f|f \in R\}) \in F(C)$ which satisfies condition (7.4.6). Hence α exists and it is unique. If we then consider the map $l : \mathbf{P} \to i(\mathbf{F})$, the above discussion tells us that l uniquely determines a $h : \mathbf{a}(\mathbf{P}) \to \mathbf{F}$ such that the following diagram commutes:

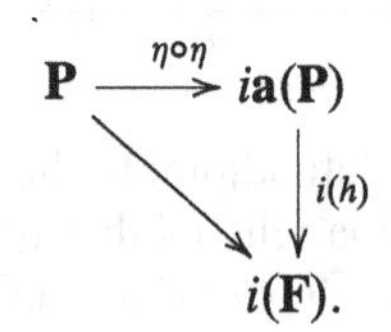

This means that the map $\eta \circ \eta : \mathbf{P} \to \mathbf{P}^+ \to \mathbf{P}^{++}$ is the unit of the adjunction. □
Since $\mathbf{a}$ is a left adjoint, it preserves colimits, hence all small colimits exist in $Sh(\mathcal{C}, J)$ since they exist in $\mathbf{Sets}^{\mathcal{C}^{op}}$.

As a last step in proving that $Sh(\mathcal{C}, J)$ is a topos, we need to show that it has an exponential object. Recall that in standard set theory an exponential Z^X is the function set consisting of all functions $f : X \to Z$. This set is completely determined by the bijection

$$Hom(Y \times X, Z) \to Hom(Y, Z^X)$$
$$f : Y \times X \mapsto f' : Y \to Z^X$$

such that for $y \in Y$ then $f'y(x) := f(y, x)$.
In the context of presheaves, given $\mathbf{P}, \mathbf{F} \in \mathbf{Sets}^{\mathcal{C}^{op}}$, then we know from the Yoneda embedding that $\mathbf{F}^{\mathbf{P}}(C) \cong Hom_{\mathbf{Sets}^{\mathcal{C}^{op}}}(\mathbf{y}(C), \mathbf{F}^{\mathbf{P}})$, where $\mathbf{y}(C) = Hom_{\mathcal{C}}(-, C)$. Then the above bijection is given by

$$Hom_{\mathbf{Sets}^{\mathcal{C}^{op}}}(\mathbf{y}(C), \mathbf{F}^{\mathbf{P}}) \simeq Hom_{\mathbf{Sets}^{\mathcal{C}^{op}}}(\mathbf{y}(C) \times P, \mathbf{F}).$$

Given such a bijection the exponential object is defined by

$$\mathbf{F}^{\mathbf{P}}(C) := Hom_{\mathbf{Sets}^{\mathcal{C}^{op}}}(\mathbf{y}(C) \times P, \mathbf{F})$$

Therefore the elements of $\mathbf{F}^{\mathbf{P}}(C)$ are natural transformations $\tau : \mathbf{y}(C) \times \mathbf{P} \to \mathbf{F}$. Here naturality implies that, given any $h : E \to D$, then the following diagram commutes:

$$\begin{array}{ccc} \mathbf{y}(C)(D) \times \mathbf{P}(D) & \xrightarrow{\tau} & \mathbf{F}(D) \\ \Big\downarrow{\scriptstyle \mathbf{y}(C)(h) \times \mathbf{P}(h)} & & \Big\downarrow{\scriptstyle \mathbf{F}(h)} \\ \mathbf{y}(C)(E) \times \mathbf{P}(E) & \xrightarrow{\tau} & \mathbf{F}(E) \end{array}$$

This implies that for any $(g, x) \in \mathbf{y}(C)(D) \times \mathbf{P}(D)$, then $\tau(\mathbf{y}(C)(h)g \times \mathbf{P}(h)x) = \mathbf{F}(h)\tau(g, x)$.

Given a map $f : C' \to C$, then at the level of the presheaf $\mathbf{F}^{\mathbf{P}}$ we have the corresponding presheaf map

$$\mathbf{F}^{\mathbf{P}}(C) \to \mathbf{F}^{\mathbf{P}}(C')$$
$$\tau \mapsto \mathbf{F}^{\mathbf{P}}(f)(\tau).$$

Here $\tau : \mathbf{y}(C) \times P \to \mathbf{F}$ while $\mathbf{F}^{\mathbf{P}}(f)(\tau) : \mathbf{y}(C') \times P \to \mathbf{F}$. Therefore, given a map $(g' : D \to C') \in Hom_{\mathcal{C}'}(D, C') = \mathbf{y}(C')(D)$ and an element $x \in \mathbf{P}(D)$, the action of $\mathbf{F}^{\mathbf{P}}(f)(\tau)$ is defines as

$$\mathbf{F}^{\mathbf{P}}(f)(\tau)\left(g', x\right) := \tau(f \circ g', x). \tag{7.4.7}$$

Applications of the adjunction $\mathbf{a} \vdash i$ and Yoneda lemma (Lemma A.7.2) imply that $i(\mathbf{F}^{\mathbf{P}}) \simeq i(\mathbf{F})^{i(\mathbf{P})}$. In fact, given that $\mathbf{a} \circ i \cong id$ we have:

$$\begin{aligned}
\left(\mathbf{G} \to i(\mathbf{F}^{\mathbf{P}})\right) &\overset{\text{adjunction}}{\cong} \left(\mathbf{a}(\mathbf{G}) \to \mathbf{F}^{\mathbf{P}}\right) \\
\left(\mathbf{a}(\mathbf{G}) \to \mathbf{F}^{\mathbf{P}}\right) &\overset{\text{Yoneda's lemma}}{\cong} \mathbf{a}(\mathbf{G}) \times \mathbf{P} \to \mathbf{F} \\
\mathbf{a}(\mathbf{G}) \times \mathbf{P} &\overset{\mathbf{a}\text{ preserves products}}{\cong} \mathbf{a}\,(\mathbf{G} \times i(\mathbf{P})) \\
\mathbf{a}\,(\mathbf{G} \times i(\mathbf{P})) \to \mathbf{F} &\overset{\text{adjunction}}{\cong} \mathbf{G} \times i(\mathbf{P}) \to i(\mathbf{F}) \\
\mathbf{G} \times i(\mathbf{P}) \to i(\mathbf{F}) &\overset{\text{Yoneda's lemma}}{\cong} \mathbf{G} \to i(\mathbf{F})^{i(\mathbf{P})}.
\end{aligned}$$

This implies that the exponential object in $Sh(\mathcal{C}, J)$, if it exists, will be constructed in the same way as it is constructed in $\mathbf{Sets}^{\mathcal{C}^{op}}$.

Now that we know how exponentials are constructed in $Sh(\mathcal{C}, J)$ we need to show that, given $\mathbf{P}, \mathbf{F} \in \mathbf{Sets}^{\mathcal{C}^{op}}$, if $\mathbf{F} \in Sh(\mathcal{C}, J)$ then $\mathbf{F}^{\mathbf{P}} \in Sh(\mathcal{C}, J)$. The proof will consist in first showing that the presheaf $\mathbf{F}^{\mathbf{P}}$ is separated and then that it has an amalgamation point.

To show that the presheaf $\mathbf{F}^{\mathbf{P}}$ is separated we need to show that, given a cover $S \in J(C)$, the following map is injective:

$$\mathbf{F}^{\mathbf{P}}(C) \to \prod_{f \in S} \mathbf{F}^{\mathbf{P}}(dom(f)).$$

Choose any $\tau, \sigma \in \mathbf{F}^{\mathbf{P}}(C)$ such that $\mathbf{F}^{\mathbf{P}}(f)\tau = \mathbf{F}^{\mathbf{P}}(f)\sigma$ for all $f \in S$. Applying the definition of the presheaf maps given in (7.4.7), this means that $\tau(f \circ g', x) = \sigma(f \circ g', x)$ for all $g' : D \to C'$ and $x \in \mathbf{P}(D)$. Choosing $g' = id_D$ then we have

$$\tau(f, x) = \sigma(f, x). \tag{7.4.8}$$

We then need to show that $\tau = \sigma$. In order to prove this we will utilise the fact that $\mathbf{F}$ is separated. To this end consider any map $k : C' \to C$ with codomain C, then $k^*(S)$ is a cover of C'. Since $\mathbf{F}$ is separated, for any $x \in \mathbf{F}(C')$ we have that

$$\mathbf{F}(C') \to \prod_{g' \in k^*(S)} \mathbf{F}(dom(k^*(S))) \tag{7.4.9}$$

is injective, therefore given two elements $\tau(k, x), \sigma(k, x) \in \mathbf{F}(C')$, then

$$\begin{aligned}
\mathbf{F}(g')\tau(k, x) &\overset{\text{Naturality of }\tau}{=} \tau(kg', xg') \\
&\overset{(7.4.8)}{=} \sigma(kg', xg') \\
&\overset{\text{Naturality of }\sigma}{=} \mathbf{F}(g')\sigma(k, g).
\end{aligned}$$

Since we have assumed that $\mathbf{F}$ is separated (i.e. the map in (7.4.9) is injective), it follows that $\tau(k, x) = \sigma(k, x)$. Since this result holds for any k and x, it follows that $\tau = \sigma$ and $\mathbf{F}^{\mathbf{P}}$ is separated.

The final step in proving that $\mathbf{F}^{\mathbf{P}}$ is a sheaf is to show that any matching family has an amalgamation point. Its uniqueness is a consequence of the fact that $\mathbf{F}^{\mathbf{P}}$ is separated. Let us first construct a matching family. To this end consider a cover $S \in J(C)$ such that for all $f : D \to C$ in S the natural transformations $\tau_f : \mathbf{y}(D) \times \mathbf{P} \to \mathbf{F}$ form a matching family. By this we mean that for any $g : E \to D$ then

$$\mathbf{F}^{\mathbf{P}}(g)\tau_f = \tau_{fg}.$$

Hence, given an element $h : E' \to E \in \mathbf{y}(E)$ and $x \in \mathbf{P}(E')$ we obtain that

$$\tau_{fg}(h, x) = (\mathbf{F}^{\mathbf{P}}(g)\tau_f)(h, x) = \tau_f(gh, x). \tag{7.4.10}$$

The fact that an amalgamation point for the family $\{\tau_f\}_{f \in S}$ exists can be seen from the following commuting diagram:

Diagram 7.4.1

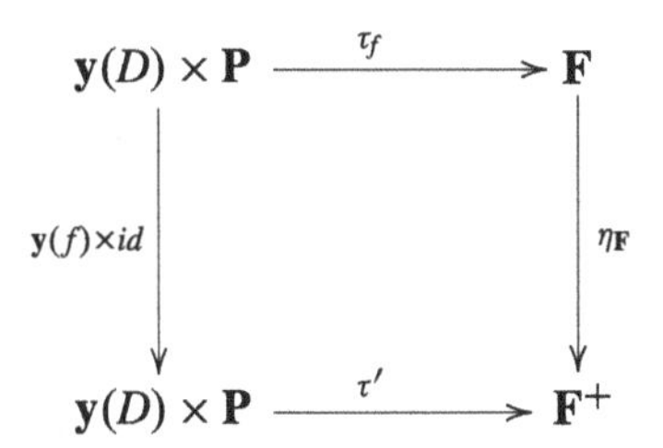

Since $\mathbf{F}$ is a sheaf, it is straightforward to show that the map $\eta_{\mathbf{F}}$, defined in (7.4.5), is an isomorphism. Therefore the desired amalgamation point is $\eta_{\mathbf{F}}^{-1} \circ \tau'$.

But how does one actually define τ'? To this end let us consider an object $B \in \mathcal{C}$, then we have $\tau'_B : \mathbf{Y}(C)(B) \times \mathbf{P}(B) \to \mathbf{F}^+(B)$ which we define, for any $k : B \to C$ and $x \in \mathbf{P}(B)$ as

$$\tau'_B(k, x) = \{\tau_{kh}(id, \mathbf{P}(h)x) | h \in k^* S\}.$$

To show that this is well defined we need to show that it represents a matching family of elements of $\mathbf{F}$ for the cover $k^*(S) \in J(B)$, i.e. an element of $\mathbf{F}^+(B)$. Therefore we need to show that for any $h \in k^*(S)$, then $\tau_{kh}(id, \mathbf{P}(h)x)$ is such that, for all m for which the composite $h \circ m$ is defined, then $\mathbf{F}^+(m)(\tau_{kh}(id, \mathbf{P}(h)(x)) = \tau_{khm}(1, \mathbf{P}(hm)(x))$. This is indeed the case, in fact we have:

$$\mathbf{F}^+(m)(\tau_{kh}(id, \mathbf{P}(h)(x)) \overset{\text{naturality}}{=} \tau_{kh}(m, \mathbf{P}(hm)(x))$$
$$\overset{(7.4.10)}{=} \tau_{khm}(id, \mathbf{P}(hm)(x)).$$

We have thus shown that $\tau'_B(k, x)$, as defined above, is indeed an element of $\mathbf{F}^+$. However we still need to show that with such a definition for τ' diagram 7.4.1 commutes. To this end consider a map $f : D \to C \in S$, then $f^*S = t_D$ is the principal sieve on D. Given an element $(k, x) \in \mathbf{Y}(D)(B) \times \mathbf{P}(E)$ where $k : B \to D$, going around the left hand side of diagram 7.4.1 we obtain

$$(\tau' \circ (\mathbf{Y}(f) \times id))(k, x) = \tau'(fk, x) = \{\tau_{fkh}(id, \mathbf{P}(h)(x)|h \in (fk)^*(S)\}$$

where $(fk)^*(S) = k^*f^*(S) = k^*(t_D) = t_B$. On the other hand, going round the right side of diagram 7.4.1, we obtain

$$(\eta_\mathbf{F} \circ \tau_f)(k, x) = \eta_\mathbf{F}(\tau_{fk}(id, x)) = \{\tau_{fkh}(id, \mathbf{P}(h)(x)|h \in t_B\}.$$

Therefore the diagram commutes. This ends our proof.

From the discussion above we can conclude that the category $Sh(J, C)$ of sheaves over a site is a topos.

Chapter 8
Locales

8.1 Locales and Their Construction

As a first step, we introduce the notion of a frame [73] which is a lattice L with all finite meets and all joins which satisfies the following distributive law:

$$U \wedge \bigvee_i V_i = \bigvee_i (U \wedge V_i), \quad \forall\, U, V_i \in L. \tag{8.1.1}$$

An example of a frame is given by the collection of all open subsets $\mathcal{O}(X)$ of a topological space X. Given the definition of a frame it follows that each frame has a largest element $\bigwedge \emptyset$ and a smallest element $\bigvee \emptyset$. In fact the definition of meet ($\bigwedge$) states that $\bigwedge \emptyset$ is the biggest element such that for all $x \in \emptyset$ then $\bigwedge \emptyset \leq x$. However the last condition is vacuously true, hence all that remains is that $\bigwedge \emptyset$ is the biggest element. Similarly for the definition of join, $\bigvee \emptyset$ is the smallest element such that for all $x \in \emptyset$, $x \leq \bigvee \emptyset$. Again the last condition is vacuous and what remains is that $\bigwedge \emptyset$ is the smallest element.

A morphism between frames $\phi : L_1 \to L_2$ is a map of partially ordered sets which preserves both finite meets and infinite joins, hence we also call it a frame homomorphism. Such homomorphism has to satisfy the following conditions:

$$\phi(0) = 0, \;\; \phi(1) = 1, \;\; \phi(U \wedge V) = \phi(U) \wedge \phi(V), \;\; \phi(\bigwedge_i V_i) = \bigwedge_i (\phi V_i). \tag{8.1.2}$$

For example, given a continuous map between two topological spaces $f : X_1 \to X_2$, the inverse image map $f^{-1} : \mathcal{O}(X_2) \to \mathcal{O}(X_1)$ is a morphism of frames.

A morphism of frames $\phi : L_1 \to L_2$ has a right adjoint $\psi : L_2 \to L_1$ defined by:

$$\forall\, U \in L_2, \;\; \psi(U) := \bigvee \{V \in L_1 | \phi(V) \leq U\}.$$

C. Flori, *A Second Course in Topos Quantum Theory*,
Lecture Notes in Physics 944, https://doi.org/10.1007/978-3-319-71108-9_8

To show that $\phi \dashv \psi$ is an adjoint pair we need to show that for each map $\phi(V) \leq U$ there corresponds a map $V \leq \psi(U)$. We start by assuming that we have a map $\phi(V) \leq U$, this implies that $V \in \{V \in L_1 | \phi(V) \leq U\}$. Since $\psi(U)$ is the join of the set $\{V \in L_1 | \phi(V) \leq U\}$, it follows that $V \leq \psi(U)$. On the other hand, given a map $V \leq \psi(U)$, then since ϕ is a frame homomorphism it follows that $\phi(V) \leq \phi(\bigvee\{V \in L_1 | \phi(V) \leq U\}) = \bigvee\{\phi(V) | \phi(V) \leq U\}$, therefore $\phi(V) \leq U$. Being an adjoint pair, ψ preserves all meets, i.e. $\psi(\bigwedge_i U_i) = \bigwedge_i \psi(U_i)$.

As an example of the above mentioned adjoint pair, consider two topological spaces X and Y, with a continuous map between them $f : X \to Y$. Then the frame map is $f^{-1} : \mathcal{O}(Y) \to \mathcal{O}(X)$ with right adjoint $f_* : \mathcal{O}(X) \to \mathcal{O}(Y)$, such that

$$U \mapsto f_*(U) := \bigcup\{V | f^{-1}(V) \subseteq U\}. \tag{8.1.3}$$

It turns out that f_* needs not to be a frame homomorphism, since it needs not to preserve suprema.

Proposition 8.1.1 *A frame is equivalent to a complete Heyting algebra (**cHa**),*

Proof Recall that a **cHa** is a complete lattice L equipped with an implication operation $\Rightarrow: L^{op} \times L \to L$ such that for any $U, V, W \in L$ we have $W \leq (U \Rightarrow V)$ iff $W \wedge U \leq V$. However this condition states precisely that each $U \Rightarrow (-)$ is the right adjoint of $U \wedge (-)$. Since each $U \wedge (-)$ is a left adjoint it preserves all joins. This is indeed the condition (8.1.1) in the definition of a frame. Therefore, given a frame L, the implication relation is defined as $U \Rightarrow V := \bigvee\{W \in L | W \wedge U \leq V\}$. □

Definition 8.1.1 Given a frame L, a nucleus is a function $f : L \to L$ such that the following conditions are satisfied:

$$f(a \wedge b) = f(a) \wedge f(b) \tag{8.1.4}$$

$$a \leq f(a) \tag{8.1.5}$$

$$f(f(a)) \leq f(a). \tag{8.1.6}$$

Alternatively it is possible to define a nucleus in terms of a subset S of L, which satisfies the following conditions:

1. $\bigwedge A \in S$ whenever $A \subseteq S$.
2. $a \Rightarrow b \in S$ whenever $b \in S$.

The map f is then defined by $f(a) := \bigwedge\{b \in L \mid b \in S,\ a \leq b\}$ and $S = \{a \in L \mid f(a) = a\}$.

Clearly frames have a topological flavour, but sometimes it is the algebraic aspect of a frame which one is more interested in. This algebraic aspect is encoded in the notion of a locale. In particular, a locale is the same thing as a frame, but such that to each morphism between frames there corresponds a morphism between locales going the other way. What this means in mathematical terms is that the category of locales is the opposite of the category of frames: **Loc** $=$ $(\textbf{Frames})^{op}$. It is a common convention to denote the frame by $\mathcal{O}(X)$ and the corresponding locale by X. Moreover, given two locales X, Y with corresponding frames denoted as $\mathcal{O}(X)$ and $\mathcal{O}(Y)$ respectively, a continuous map of locales $l : X \rightarrow Y$ is defined to be a frame homomorphism $l^{-1} : \mathcal{O}(Y) \rightarrow \mathcal{O}(X)$.

From the above, it is clear that there exists a covariant functor which maps topological space to locales [50, 55]:

$$\text{Loc} : \textbf{Spaces} \rightarrow \textbf{Loc} \tag{8.1.7}$$
$$X \mapsto \text{Loc}(X).$$

The corresponding frame of Loc(X) is $\mathcal{O}(\text{Loc}(X)) := \mathcal{O}(X)$. Therefore, given any topological space X, we can construct the locale Loc(X), whose frame consists of all the open subsets of X and, to each map $f : X \rightarrow Y$ between spaces, there corresponds the locale map $\text{Loc}(f) : \text{Loc}(X) \rightarrow \text{Loc}(Y)$ with associated frame map $f^{-1} : \mathcal{O}(Y) \rightarrow \mathcal{O}(X)$.

From the above description one can infer that the elements of a locale are "extended regions" rather than points. It is however possible to define the notion of a point in a locale [50, 55].

Definition 8.1.2 A point in a locale X is defined as a morphism $p : 1 \rightarrow X$ from the locale corresponding to a one-point space (the terminal object in **Loc**) to the locale in question.

In terms of frames, a point is defined as a frame homomorphism $p^{-1} : \mathcal{O}(X) \rightarrow \mathcal{O}(1)$ where $\mathcal{O}(1)$ is the frame consisting only of the bottom element 0 and the top element 1, therefore $\mathcal{O}(1) \simeq \{0, 1\} = \Omega$.

Alternatively, a point P in a frame $\mathcal{O}(X)$ is a proper prime element of $\mathcal{O}(X)$, i.e. $1 \neq P$ and $U \wedge V \leq P$ iff $U \leq P$ or $V \leq P$. To understand how these definitions are equivalent, let us analyse the kernel of the homomorphism $p^{-1} : \mathcal{O}(X) \rightarrow \mathcal{O}(1)$. This is given by the subset $K := \{U | p^{-1}(U) = 0\}$ such that:

1. $1 \notin K$;
2. $U \wedge V \in K$ iff $U \in K$ or $V \in K$;
3. $\bigvee U_i \in K$ iff $U_i \in K$ for all i.

Given any such subset $K \in \mathcal{O}(X)$, this defines an element $P = \left(\bigvee_{U \in K} U\right) \in \mathcal{O}(X)$. Translating conditions (1); (2); (3) above, to conditions on P, we obtain:

(a) $1 \neq P$;
(b) $U \wedge V \leq P$ iff $U \leq P$ or $V \leq P$;
(c) $U \leq P$ iff $U \in K$, therefore $K = \downarrow P$.

From the discussion above it is clear that a point in a locale can be defined in three equivalent ways: (1) as a frame morphism; (2) as a subset (K) of the locale in question; (3) as a proper prime element of the corresponding frame.

Let us see how these equivalent definitions apply in the case of a locale defined in term of the map Loc, defined in (8.1.7). To this end let us consider a topological space X, then a point $x \in X$ determines a point in the locale Loc(X).

Frame Morphism x can be defined in terms of a frame morphisms $p_x^{-1} : \mathcal{O}(X) \to \mathcal{O}(1)$ with associated locale map $p_x : 1 \to \mathrm{Loc}(X)$, such that $p_x^{-1}(U) = 0$ iff $x \notin U$.

Subset Alternatively x can be defined in terms of the subset $K_x = \{U\mathcal{O}(X)|x \notin U\}$. Clearly K_x is such that (1) $X \notin K_x$; (2) if $U \wedge V \in K_x$, then $x \notin U \wedge V$. Therefore $x \notin U$ or $x \notin V$, hence $U \notin K_x$ or $V \notin K_x$; (3) if $\bigvee_i U_i \in k_x$ then $x \notin \bigvee_i U_i$, hence $x \notin U_i$ for all i. This implies that $U_i \in K_x$ for all i.

Prime Element Finally x can be defined in terms of a proper prime element of $\mathcal{O}(X)$. In this case this element is identified with the open $X - \overline{\{x\}}$. Clearly $X \neq X - \overline{\{x\}}$ and $U \wedge V \subseteq X - \overline{\{x\}}$ iff $U \subseteq X - \overline{\{x\}}$ or $V \subseteq X - \overline{\{x\}}$. This last condition follows from the fact that $U \wedge V \subseteq X - \overline{\{x\}}$ implies that either $x \notin V$ or $x \notin U$. Assume the latter, then $x \in X - U$, therefore $\overline{\{x\}} \in X - U$ (since $\overline{\{x\}}$ is the smallest closed subset containing x), hence $U \in X - \overline{\{x\}}$.

We have seen so far that there is a close connection between topological spaces and locals, however the question still remains on how much of the topological space can be reconstructed from its lattice of open subsets. It turns out that a particular class of topological spaces called *sober* spaces are determined up to homeomorphism by their lattice of open subsets. A sober space is defined as follows

Definition 8.1.3 A topological space X is said to be sober if for any open subset $P \subseteq X$ satisfying the following conditions:

i) $P \neq X$.
ii) If $U \cap V \subseteq P$ then either $U \subseteq P$ or $V \subseteq P$,

then there is a unique point $x \in X$ with $P = X - \overline{\{x\}}$.

Clearly the above definition is equivalent to the condition that there is a bijection between the points $x \in X$ and points of the locale $\mathrm{Loc}(X)$, given by the maps $p_x^{-1} : \mathcal{O}(X) \to \mathcal{O}(1)$. An alternative definition of a sober space is as follows:

Definition 8.1.4 A topological space X is said to be sober if every non-empty irreducible closed subset of X is the closure of exactly one point of X.

To see that the two definitions are equivalent we first assume Definition 8.1.3, then for each $x \in X$, $P = X - \overline{\{x\}} = \overline{\{x\}}^c$. We now show that this implies that $\overline{\{x\}} = X - \overline{\{x\}}^c$ is an irreducible closed set. Since $P \neq X$, then $\overline{\{x\}}$ is non-empty. Moreover, since $U \cap V \subseteq P$, then $P^c \subseteq (U \cap V)^c = U^c \cup V^c$, but $U \subseteq P$ implies that $P^c \subseteq U^c$, therefore $P^c = \overline{\{x\}}$ is irreducible.

On the other hand let us assume that Definition 8.1.4 holds, i.e. for all $x \in X$, $\overline{\{x\}}$ are the only non-empty irreducible closed subsets of X. It then follows that, for each $x \in X$, the sets $P = \overline{\{x\}}^c = X - \overline{\{x\}}$ are open. Moreover, since $\overline{\{x\}}$ is non-empty, it follows that $P \neq X$. Now assume that $U \cap V \subseteq P$, then $P^c = \overline{\{x\}} \subseteq U^c \cup V^c$. However, since $\overline{\{x\}}$ is irreducible it follows that either $P^c \subseteq U^c$ or $P^c \subseteq V^c$, hence $U \subseteq P$ or $V \subseteq P$.

There exist many examples of familiar spaces which are sober as shown by the following theorem:

Theorem 8.1.1 *All sober spaces are T_0*[1] *and all Hausdorff spaces are sober.*

Proof Assume that the space X is sober, we then want to show that for each pair of points $x, y \in X$ there exists an open U, such that $x \in U$ and $y \notin U$. Since X is sober the map $x \rightarrow \overline{\{x\}}$, from a point to a non-empty closed irreducible subsets, is a bijection. Therefore, given any to points $x, y \in X$ two distinct closed sets there are associated to them, namely $\overline{\{x\}}$ and $\overline{\{y\}}$, respectively. Therefore there exists an open $\overline{\{y\}}^c$ such that $y \notin \overline{\{y\}}^c$ and $x \in \overline{\{y\}}^c$.

We now assume that X is Hausdorff, that is for any two points $x, y \in X$ there exists opens $U_x \ni x$ and $U_y \ni y$ such that $U_x \cap U_y = \emptyset$. We want to show that X is sober, i.e. each non-empty closed irreducible subset is the closure of a singleton. Let us assume that P is a non-empty closed irreducible subset such that $x, y \in P$, then $P = (P - U_x) \cup (P - U_y)$. But this contradicts the assumption, hence P needs to be irreducible. □

So far we have seen that given a topological space X we can define the associated locale using the map Loc in (8.1.7). It is only natural to ask if one can also do the reverse, namely: given a locale is it possible to define a topological space? It turns out that this is indeed the case and to achieve this one utilises the points of a local, i.e. the maps $p : 1 \rightarrow (X)$. In fact, the set of all points $\text{pt}(X) = \{p : 1 \rightarrow X\}$ of a locale X is equipped with a topological structure where the opens are defined by

$$\text{pt}(U) = \{p \in \text{pt}(X) | p^{-1}(U) = 1\} \tag{8.1.8}$$

where $U \in \mathcal{O}(X)$.

To show that the collection of subsets of this form constitute, indeed, a topology we note that for $X \in \mathcal{O}(X)$, then $\{p \in \text{pt}(X) | p^{-1}(X) = 1\} = \text{pt}(X)$ is the whole space, while for $\emptyset \in \mathcal{O}(X)$ then $\{p \in \text{pt}(X) | p^{-1}(\emptyset) = 1\} = \text{pt}(\emptyset) = \emptyset$ is empty. Now, given two opens $U, V \in \mathcal{O}(X)$, then $\text{pt}(U \wedge V) = \{p \in \text{pt}(X) | p^{-1}(U \wedge V) = 1\}$. However, since p^{-1} is a frame morphisms then $p^{-1}(U \wedge V) = p^{-1}(U) \wedge p^{-1}(V) = 1$ which implies that $p^{-1}(U) = 1$ and $p^{-1}(V) = 1$. Therefore $\text{pt}(U \wedge V) = \text{pt}(U) \wedge \text{pt}(V)$. Similarly one can show that $\text{pt}(\bigvee_i U_i) = \bigvee_i \text{pt}(U_i)$.

This shows that given a locale X it is possible to define a topological space $\text{pt}(X)$, whose opens are given by the sets of the form of (8.1.8). Topological spaces, defined in such a way, share the property of being sober.

Lemma 8.1.1 *Given a locale A, then* $\text{pt}(A)$ *is sober.*

Proof In order to show that $\text{pt}(A)$ is sober we will show that the map $x \rightarrow \overline{\{x\}}$ is a bijection. To this end consider a non-empty closed irreducible subset F of $\text{pt}(A)$, then F^c is a prime open, i.e. if $U \cap V \subseteq F^c$ for U, V open and not disjoint, then either

[1] Recall that a T_0 space is a space X such that for any two points $x, y \in X$ there exists an open set U such that, either $x \in U$ and $y \notin U$ or $y \in U$ and $x \notin U$.

$U \subseteq F^c$ or $V \subseteq F^c$. Clearly F^c, then, is a proper prime element of $\mathcal{O}(\text{pt}(A))$. From the definition of points of a local, F^c corresponds to a point $p_{F^c} : 1 \to A$, therefore the map $x \to \overline{\{x\}}$ is surjective. It remains to show that this map is also injective. To this end we will first show that the space pt(A) is T_0. In fact if we consider two distinct points $p \neq q : 1 \to A$, then $q^{-1}(U) \neq p^{-1}(U)$ for some $U \in \mathcal{O}(\text{pt}(A))$. This implies that either $p \in \text{pt}(U)$ or $q \in \text{pt}(U)$, hence pt(A) is T_0. Hence the map $x \mapsto \overline{\{x\}}$ is injective. □

Given two locales X and Y and a map $f : X \to Y$, we would like to define a map between the topological spaces pt(X) and pt(Y). This can be done by a simple composition:

$$\begin{aligned} \text{pt}(f) &: \text{pt}(X) \to \text{pt}(Y) \\ (p : 1 \to X) &\mapsto (f \circ p : 1 \to X \to Y). \end{aligned} \tag{8.1.9}$$

However, we need to check that this map is continuous. To this end consider an open pt$(V) \subseteq$ pt(Y), then the inverse image is $\text{pt}(f)^{-1}\text{pt}(V) = \text{pt}(f)^{-1}\{q \in \text{pt}(Y)|q^{-1}(V) = 1\} = \{p \in \text{pt}(X)|p^{-1}(f^{-1}(V)) = 1\} = \text{pt}(f^{-1}(V))$, which is open.

This discussion uncovers the fact that the operation of defining points in a locale is a covariant functor

$$\begin{aligned} \text{pt} : \mathbf{Loc} &\to \mathbf{Spaces} \\ X &\mapsto \text{pt}(X). \end{aligned} \tag{8.1.10}$$

Theorem 8.1.2 ([55]) pt : ***Loc*** $\to$ ***Spaces*** *is the right adjoint of* Loc : ***Spaces*** $\to$ ***Loc***.

Proof To prove the adjunction Loc $\dashv$ pt we need to show that, given a topological space X and locale A there exists an isomorphism $Hom_{\mathbf{Loc}}(\text{Loc}(X), A) \simeq Hom_{\mathbf{spaces}}(X, \text{pt}(A))$. To show this we will construct a map $g : X \to \text{pt}(A)$ from a map $f : \text{Loc}(X) \to A$ and vice versa. Let us first assume that we have a map $f : \text{Loc}(X) \to A$, with associated frame map $f^{-1} : \mathcal{O}(A) \to \mathcal{O}(X)$, we then construct the map $g : X \to \text{pt}(A); x \mapsto g(x)$ such that, for all $x \in X$, $g(x)^{-1}(U) = 1$ iff $x \in f^{-1}(U)$. We now need to show that, as defined, g is continuous. To this end consider an open pt$(U) \subseteq$ pt(A), then $g^{-1}(\text{pt}(U)) = g^{-1}\{p \in \text{pt}(X)|p^{-1}(U) = 1\} = \{x \in X|g(x)^{-1}(U) = 1\} = f^{-1}(U)$, therefore g is continuous. Now we do the reverse. Let us assume we are given a $g : X \to \text{pt}(A)$ and we try constructing an $f : \text{Loc}(X) \to A$ with associated frame map $f^{-1} : \mathcal{O}(A) \to \mathcal{O}(X)$. We define $f^{-1}(U) := \{x \in X|g(x)^{-1}(U) = 1\}$. Clearly $f^{-1}(U) = g^{-1}(\text{pt}(U))$. We now need to show that f^{-1}, so defined, is a frame morphism, i.e. it preserves finite meets and arbitrary joins.

$$\begin{aligned} f^{-1}(U \wedge V) &\overset{definition}{=} g^{-1}(\text{pt}(U \wedge V))) \overset{pt \text{ preserves meets}}{=} g^{-1}(\text{pt}(U) \cap \text{pt}(V)) \\ &\overset{\text{continuity of } g}{=} g^{-1}(\text{pt}(U)) \cap g^{-1}(\text{pt}(V)) = f^{-1}(U) \cap f^{-1}(V). \end{aligned}$$

On the other hand

$$f^{-1}(\bigvee_i U_i) \overset{definition}{=} g^{-1}(\text{pt}(\bigvee_i U_i)) \overset{pt \text{ preserves joins}}{=} g^{-1}(\bigcup_i \text{pt}(U_i))$$
$$\overset{\text{continuity of } g}{=} \bigcup_i (g^{-1}(\text{pt}(U_i)) = \bigcup_i (f^{-1}(U_i)).$$

We have now constructed the maps $\phi : Hom_{\mathbf{Loc}}(\text{Loc}(X), A) \to Hom_{\mathbf{spaces}}(X, \text{pt}(A))$; $f \mapsto g$ and $\psi : Hom_{\mathbf{spaces}}(X, \text{pt}(A)) \to Hom_{\mathbf{Loc}}(\text{Loc}(X), A)$; $g \mapsto f$. What remains to show is that these are inverse of each other. Let us start with $\psi(\phi(f))$, we want to show that this is equivalent to f. Consider an open $U \subseteq \mathcal{O}(A)$, then

$$(\psi(\phi(f)))^{-1}(U) \overset{definition}{=} \{x \in X | \phi(f)(x)^{-1}(U) = 1\}$$
$$\overset{\text{definition of } \phi(f)}{=} \{x \in X | x \in f^{-1}(U)\} = f^{-1}(U),$$

therefore $\psi(\phi(f)) = f$.

On the other hand, we want to show that $\phi(\psi(g)) \cong g$. Here $\phi(\psi(g)) : X \to \text{pt}(A)$ is such that for all $x \in X$, $\phi(\psi(g))(x) : 1 \to A$ is the locale map such that:

$$\phi(\psi(g))(x)^{-1}(U) = 1 \Leftrightarrow x \in \psi(g)^{-1}(U)$$
$$x \in \psi(g)^{-1}(U) \Leftrightarrow x \in \{x \in X | g(x)^{-1}(U) = 1\},$$

therefore

$$\phi(\psi(g))(x)^{-1}(U) = 1 \Leftrightarrow g(x)^{-1}(U) = 1.$$

□

An alternative way of proving the above theorem is by constructing the unit and counit of the adjunction.

Proof Consider the map $\eta : X \to \text{pt}(Loc(X))$ which takes each $x \in X$ to the corresponding point $p_x : 1 \to \text{Loc}(X)$. We will now show that this map satisfies the triangular identities for the unit of the adjunction. In particular consider a map $g : X \to pt(A)$, we need to show that there exists a unique $f : \text{Loc}(X) \to A$ in **Loc**, such that the following diagram commutes

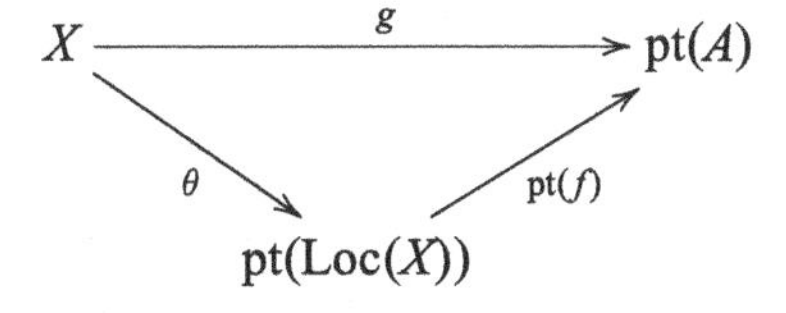

Applying the functor pt to f we obtain the map $\text{pt}(f) : \text{pt}(\text{Loc}(X)) \to \text{pt}(A)$ which, as explained before, is defined though composition $\text{pt}(f) : (p : 1 \to \text{Loc}(X)) \to (f \circ p : 1 \to A)$. If we now compose $pt(f)$ with ϕ we obtain the map

$$\begin{aligned}\text{pt}(f) \circ \theta : X &\to \text{pt}(A)\\ x &\mapsto f \circ p_x : 1 \to A.\end{aligned}$$

$f \circ p_x : 1 \to A$ is s point such that, given any $U \in \mathcal{O}(A)$, $(f \circ p_x)^{-1}(U) = 1$ if $x \in f^{-1}(U)$ and $(f \circ p_x)^{-1}(U) = 0$ if $x \notin f^{-1}(U)$. Therefore, for the diagram to commute we need $g(x) = f \circ p_x$, i.e. $x \in f^{-1}(U)$ iff $x \in g(x)^{-1}(U)$. This means that the map f should be defined, for any $U \in \mathcal{O}(A)$, as $f^{-1}(U) = \{x \in X | g(x)^{-1}(V) = 1\}$. Clearly such an f is unique.

On the other hand let us consider the map $\epsilon : \text{Loc}(\text{pt}(A)) \to A$ such that $\epsilon^{-1} : \mathcal{O}(A) \to \mathcal{O}(\text{pt}(A))$; $U \mapsto \text{pt}(U)$. We want to show that given an $f : \text{Loc}(X) \to A$, then there exists a unique $g : X \to \text{pt}(A)$ such that the following diagram commutes:

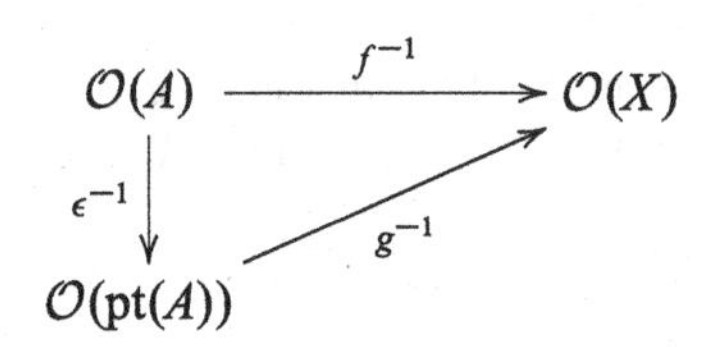

That is, we require that for any $U \in \mathcal{O}(A), f^{-1}(U) = g^{-1}(\epsilon^{-1}(U)) = g^{-1}(\text{pt}(U)$. This is equivalent to the condition that $x \in f^{-1}(U) \Leftrightarrow x \in g^{-1}(\text{pt}(U) \Leftrightarrow g(x)^{-1}(U) = 1$. To account for this we define the map $g : X \to \text{pt}(A)$ which assigns to each $x \in X$ a point $g(x) : 1 \to A$ such that, for all $U \in \mathcal{O}(A)$, $g(x)^{-1}(U) = 1$ iff $x \in f^{-1}(U)$ and $g(x)^{-1}(U) = 0$ otherwise. Clearly such a defined g is unique. □

Summarising, given a topological space X, the unit the adjunction is defined by

$$\begin{aligned}\eta : X &\to \text{pt Loc}(X)\\ x &\mapsto (p_x : 1 \to \text{Loc}(X)).\end{aligned} \tag{8.1.11}$$

On the other hand, the counit ϵ : Loc $\text{pt}(X) \to X$ is defined in terms of the corresponding frame map

$$\begin{aligned}\epsilon^{-1} : \mathcal{O}(X) &\to \mathcal{O}(\text{pt}(X))\\ U &\mapsto \text{pt}(U).\end{aligned} \tag{8.1.12}$$

Both the unit and the counit of the adjunction are important in understanding certain properties of locales. In particular the property of a locale being sober is equivalent to the condition that the unit is a homeomorphism. On the other hand the property of a locale having enough points (notion to be defined later) is equivalent to the condition that the counit is an isomorphisms of locales. These properties are expressed by two theorems.

Theorem 8.1.3 *Given a topological space X, the following statements are equivalent:*

1. *X is sober.*
2. *The unit $\eta : X \to$ pt Loc (X) is a homeomorphism.*
3. *There exists a locale A such that* $\text{pt}(A) \simeq X$.

Proof

$1 \Rightarrow 2$ Assume that S is sober, from Definition 8.1.3 we know that a space is sober if there exists a bijection between the points $x \in X$ and the points of the corresponding locale Loc(X). Thus the condition of being sober is equivalent to the condition of the unit being a bijection. What remains to be shown is that η is both open and continuous. To this end consider an open $U \subseteq X$, then $U \in \mathcal{O}\text{Loc}(X)$. To such an open there corresponds the open $\text{pt}(U) \subseteq \text{pt}(\text{Loc}(X))$. Given a $x \in X$, then $\eta(x) \in \text{pt}(U)$ iff $\eta^{-1}(U) = 1$ iff $x \in U$. This shows that η is continuous, but it also shows that it is open since $\eta(U) = pt(U)$.

$2 \Rightarrow 3$ Follows by choosing $X = \text{Loc}(X)$.

$3 \Rightarrow 1$ Follows from Lemma 8.1.1.

□

As stated in previous sections, the main idea behind locales is to consider open regions as primary entities rather than points. Given this, it is natural to ask whether locales have points and 'how many' do they have. It turns out that some locales have points while others don't. Of those that do, some are said to have 'enough points' while others do not. Locales that have 'enough points' are called *spacial*. Having enough points means that the points present are enough to distinguish elements of the corresponding lattice. In particular, consider a locale X, this is spacial iff for any $U, V \in \mathcal{O}(X)$ then $\text{pt}(U) = \text{pt}(V)$ implies that $U = V$. Alternatively we say that X is spacial if for any two $U, V \in \mathcal{O}(X)$ there exists a point $p : 1 \to X$ such that $p^{-1}(U) \neq p^{-1}(V)$. The property of being spacial is given in terms of the counit though the following theorem.

Theorem 8.1.4 *Given a locale A, the following statements are equivalent:*

1. *A is spacial.*
2. *The counit ϵ :* Loc pt(A) $\to$ *A is an isomorphism of locales.*
3. *There exists a topological space X such that $A \simeq$* Loc(X).

Proof

$1 \Rightarrow 2$ Assume that the locale A is sober, then for any $U, V \in \mathcal{O}(A)$ if $\text{pt}(U) = \text{pt}(V)$ it follows that $U = V$. Next consider the frame map $\epsilon^{-1} : \mathcal{O}(A) \to \mathcal{O}(\text{pt}(A))$; $U \mapsto \text{pt}(U)$. By construction such a map is onto, since each open in $\mathcal{O}(\mathrm{pt}(A))$ is of the form $\mathrm{pt}(U)$ for some $U \in \mathcal{O}(A)$. Moreover, from the condition of A being sober it follows that ϵ^{-1} is injective, hence ϵ^{-1} is a frame isomorphism and ϵ is an isomorphism of locales.

$2 \Rightarrow 3$ Follows from choosing $X = \mathrm{pt}(A)$.

$3 \Rightarrow 1$ Assume that $A \simeq \mathrm{Loc}(X)$ for some topological space X. Then given two distinct opens $U, V \subseteq X$, these are distinguishable in terms of the concrete points of X, but this implies that they are also distinguishable using points of $\mathrm{Loc}X$.

□

A direct consequence of the above theorems is that the adjunction $\mathrm{Loc} \dashv \mathrm{pt}$ of Theorem 8.1.2 becomes an equivalence of categories when restricted to the sub-categories of spacial locales and sober spaces.

Lemma 8.1.2 *The adjunction*

$$\mathbf{Spaces} \underset{\mathrm{pt}}{\overset{\mathrm{Loc}}{\underset{\longleftarrow}{\overset{\longrightarrow}{\bot}}}} \mathbf{Loc} \ ,$$

when restricted to the full sub-category of spacial locales and the full sub-category of sober spaces, is an equivalence of categories.

Before proving this Lemma we will briefly recall the definition of an equivalence of categories.

Definition 8.1.5 Given two categories $\mathcal{C}$ and $\mathcal{D}$, they are said to be equivalent if there exists an equivalence between them. An equivalence consists of a pair of functors

$$\mathcal{C} \underset{G}{\overset{F}{\rightleftarrows}} \mathcal{D} \ ,$$

and two natural transformations

$$F \circ G \cong Id_{\mathcal{D}} \qquad \text{and} \qquad G \circ F \cong Id_{\mathcal{C}} \ .$$

We will now prove the above Lemma.

Proof From Lemma 8.1.1 (and also Theorem 8.1.3), we know that, given a locale A, then $\mathrm{pt}(A)$ is sober, therefore the image of $\mathrm{pt} : \mathrm{Loc} \to \mathrm{Space}$ is contained in the sub-category of sober spaces. On the other hand, from Theorem 8.1.4, given a space X, then $\mathrm{Loc}(X)$ is spacial, hence the image of $\mathrm{Loc} : \mathbf{Spaces} \to \mathbf{Loc}$ is contained in the sub-category of spacial locales. This implies that $\mathrm{Loc}_{|\mathrm{sober}} \dashv \mathrm{pt}_{|\mathrm{spacial}}$. But from Theorems 8.1.3 and 8.1.4, the unit and counit to this restricted adjunction are isomorphisms. This proves the desired equivalence. □

8.2 Maps Between Locales

In this section we would like to characterise various maps between locales. We will pay particular attention to embedding maps, since these will pave the road for the definition of sublocales in the next section.

As a first step let us consider a map $f : A \to B$ between two locales A and B. Let us also assume that the map $f^{-1} : \mathcal{O}(B) \to \mathcal{O}(A)$ is surjective, then it turns out that the map $\mathrm{pt}(f) : \mathrm{pt}(A) \to \mathrm{pt}(B)$ is injective. To see this consider two points $p_x, p_y \in \mathrm{pt}(A)$, such that $\mathrm{pt}(f)(p_x) = \mathrm{pt}(f)(p_y)$. Given $U \in \mathcal{O}(A)$, then from the surjectivity of f^{-1} it follows that there exists a $V \in \mathcal{O}(B)$ such that $U = f^{-1}(V)$. It then follows that

$$\begin{aligned} p_x^{-1}(U) &= p_x^{-1}(f^{-1}(V)) = (f \circ p_x)^{-1}(V) \overset{\text{definition of pt}(f)}{=} \mathrm{pt}(f)(p_x)(V) \\ &= \mathrm{pt}(f)(p_y)(V) = (f \circ p_y)^{-1}(V) = p_y^{-1}(f^{-1}(V)) \\ &= p_y^{-1}(U). \end{aligned}$$

Thus it seems that there is a close connection between surjective frame maps and injective topological maps. This connection is encoded in the following Lemma:

Lemma 8.2.1 *Given two topological spaces X, Y, a continuous map $i : X \to Y$ is an embedding iff the frame map $i^{-1} : \mathcal{O}(Y) \to \mathcal{O}(X)$ is surjective and* pt(i) *is injective.*

Proof

- $\Rightarrow$ Assume that i is an embedding, then the open sets in X are of the form $X \cap U$ for U open in Y. This implies that $i^{-1} : \mathcal{O}(Y) \to \mathcal{O}(X)$ is surjective. Moreover, since i is injective it follows trivially, that $\mathrm{pt}(i) : \mathrm{pt}(X) \to \mathrm{pt}(Y)$ is injective.
- $\Leftarrow$ We now assume that $i^{-1} : \mathcal{O}(Y) \to \mathcal{O}(X)$ is surjective and $\mathrm{pt}(i)$ is injective. We then want to show that i is an embedding. Injectivity of i is a direct consequence of the injectivity of $\mathrm{pt}(i)$. To show that the map i is also open we note that surjectivity of i^{-1} implies that any open $U \in \mathcal{O}(X)$ is of the form $i^{-1}(U')$ for some open $U' \in \mathcal{O}(Y)$. Therefore, consider two opens $U, V \subseteq X$, then from the surjectivity of i^{-1} these are of the form $U = i^{-1}(U')$ and $V = i^{-1}(V')$ for some opens $U', V' \subseteq Y$. Hence both $ii^{-1}(U') = U'$ and $ii^{-1}(V') = V'$ are open in Y.

□

It is possible to relax the condition of injectivity of the map $\mathrm{pt}(i) : \mathrm{pt}(X) \to \mathrm{pt}(Y)$ in the above Lemma, but then certain topological requirements of X are necessary. In particular we would have:

Lemma 8.2.2 *Given two topological spaces X, Y such that X is T_0, then a continuous map $i : X \to Y$ is an embedding iff the frame map $i^{-1} : \mathcal{O}(Y) \to \mathcal{O}(X)$ is surjective*

Proof

- $\Rightarrow$ Same as the proof of Lemma 8.2.5.
- $\Leftarrow$ We now assume that $i^{-1} : \mathcal{O}(Y) \to \mathcal{O}(X)$ is surjective. We want to show that the map i is injective. To this end consider two points $x, y \in X$ such that $i(x) = i(y)$. Next, consider the collection of opens $A = \{U \subseteq Y | i(x) = i(y) \in U\}$. Because of surjectivity, all opens in X are of the form $i^{-1}(U)$ for some open $U \subseteq Y$, therefore $i^{-1}i(x) = x \in U'$ iff $U' = i(U)$ for some open $U \in A$. Similarly $i^{-1}i(y) = y \in U'$ iff $U' = i(U)$ for some open $U \in A$. This implies that the collection of opens containing x is the same as the collection of opens containing y. Since the space is T_0 this implies that $x = y$.

□

Similarly there is a Lemma relating surjective maps between topological spaces and injective frame morphisms.

Lemma 8.2.3 *Given two topological spaces X, Y such that Y is T_1, then a map $f : X \to Y$ is surjective iff the corresponding frame map $f^{-1} : \mathcal{O}(Y) \to \mathcal{O}(X)$ is injective.*

Proof

- $\Rightarrow$ Assume that f is surjective. Then consider two opens $U, V \subseteq Y$ such that $f^{-1}(U) = f^{-1}(V)$. Since f is surjective, $f(f^{-1}(U)) = U = f(f^{-1}(V)) = V$, hence f^{-1} is injective.
- $\Leftarrow$ Assume that f^{-1} is injective, then since Y is T_1, each singleton $\{y\}$ is closed. Hence $\{y\}^c$ is open and $f^{-1}(\{y\}^c) \subset f^{-1}(Y) = Y$. This implies that there exists a point $x \in X$ such that $f(y) = x$. Since this proof holds for all $y \in Y$ it follows that f is surjective.

□

Given the above two Lemmas, it is now easy to define embeddings and surjections of locales.

Definition 8.2.1 Given two locales A, B, a map $f : A \to B$ is an embedding (respectively a surjection) iff the corresponding frame map $f^{-1} : \mathcal{O}(B) \to \mathcal{O}(A)$ is surjective (respectively injective).

Clearly one could also define embeddings and surjections of topological spaces in terms of locale maps.

Definition 8.2.2 Given two topological spaces X, Y then a map $f : X \to Y$ is an embedding (respectively a surjection) iff the map $\text{Loc}(f) : \text{Loc}X \to \text{Loc}(Y)$ is an embedding (respectively, a surjection) and Y is T_1 (respectively, X is T_0).

As a corollary of the above definition we have that:

Corollary 8.2.1 *Given locales A, B, C, with maps $f : A \to B$ and $g : C \to A$, then:*

- i. *If the map $f : A \to B$ is both an embedding and a surjection then f is an isomorphism.*
- ii. *If $f \circ g$ is a surjection then so is f.*
- iii. *If $f \circ g$ is an embedding, then so is g.*

Proof

i. Assume the map f is both an embedding and a surjection. This implies that the frame homomorphism f^{-1} is both surjective and injective, which means that it is an isomorphism. Therefore f is an isomorphism.

ii. If $f \circ g$ is a surjection then $(f \circ g)^{-1}$ is injective, i.e. if $g^{-1}(f^{-1}(U)) = g^{-1}(f^{-1}(V))$, then $U = V$. Now assume that f^{-1} is not injective, then there exists $U \neq V$ such that $f^{-1}(U) = f^{-1}(V)$. But this contradicts the assumption that $(f \circ g)^{-1}$ is injective.

iii. If $f \circ g$ is an embedding, then $(f \circ g)^{-1}$ is a surjection, therefore for all $U \in \mathcal{O}(C)$, $U = g^{-1}(f^{-1}(U'))$ for $U' \in \mathcal{O}(B)$. Now assume that g^{-1} is not surjective, then there exists a $V \in \mathcal{O}(C)$ such that $V \neq g^{-1}(W)$ for $W \in \mathcal{O}(A)$. For such a V it then follows that $V \neq g^{-1}(f^{-1}(V'))$ for $V' \in \mathcal{O}(B)$ contradicting the assumption.

□

It is interesting to note that it is possible to define both surjections and embeddings of locales in terms of the adjunction defined in Eq. (8.1.3).

Lemma 8.2.4 *Given two locales A, B and a map $f : A \to B$ between them, then the following statements are equivalent:*

1. *f is surjective.*
2. *$f_* f^{-1} = id : \mathcal{O}(B) \to \mathcal{O}(B)$.*
3. *The right adjoint $f_* : \mathcal{O}(A) \to \mathcal{O}(B)$ is a surjection of posets.*

Proof

1. $1 \Rightarrow 2$. If we assume that f is surjective then from Definition 8.2.1 it follows that $f^{-1} : \mathcal{O}(B) \to \mathcal{O}(A)$ is injective. If we now consider the triangular inequalities of the adjunction $f^{-1} \dashv f_*$ we get $f^{-1} f_* f^{-1} = f^{-1}$ and $f_* f^{-1} f_* = f_*$. Since f^{-1} is an injection and hence left cancellable, we get from the first inequality that $f_* f^{-1} = id$.
2. $2 \Rightarrow 3$. If we assume that $f_* f^{-1} = id$, then for all $U \in \mathcal{O}(B)$, $U = f_* f^{-1}(U)$ where $f^{-1}(U) \in \mathcal{O}(A)$. Therefore each $U \in \mathcal{O}(B)$ is of the form $U = f_* U'$ for some $U' \in \mathcal{O}(A)$.
3. $3 \Rightarrow 1$. Let us assume that f_* is surjective, then take $U, V \in \mathcal{O}(B)$ such that $f^{-1}(U) = f^{-1}(V)$. Since f^* is surjective, then $U = f_*(U')$ and $V = f_*(V')$ for some $U', V' \in \mathcal{O}(A)$, obtaining $f^{-1} f_*(U') = f^{-1} f_*(V')$. Then clearly $f_* f^{-1} f_*(U') = f_* f^{-1} f_*(V')$. Form the triangular inequality $f_* f^{-1} f_* = f_*$ we obtain that $U = f_*(U') = f_*(V') = V$, hence f_* is surjective.

□

A similar lemma holds for embedding of locales

Lemma 8.2.5 *Given two locales A, B and a map $f : A \to B$ between them, then the following statements are equivalent:*

1. *f is an embedding.*
2. *$f^{-1} f_* = id : \mathcal{O}(A) \to \mathcal{O}(A)$*
3. *f_* is injective.*

The proof of this Lemma is very similar to the prove of the previous Lemma but for completeness reasons we will nevertheless report it.

Proof

$1 \Rightarrow 2$ If f is an embedding, then $f^{-1} : \mathcal{O}(B) \to \mathcal{O}(A)$ is surjective (right cancellable), then, from the triangular equality $f^{-1}f_*f^{-1} = f^{-1}$ we get that $f^{-1}f_* = id$.

$2 \Rightarrow 3$ Assume that $f^{-1}f_* = id$ and that $f_*(U) = f_*(V)$, then clearly $f^{-1}f_*(U) = U = f^{-1}f_*(V) = V$, hence f_* is injective.

$3 \Rightarrow 1$ Assume that f_* is injective (left cancellable), then from the triangular equality $f_*f^{-1}f_* = f_*$ it follows that $f^{-1}f_* = id$. Then, given any $U \in \mathcal{O}(A)$, we have that $U = f^{-1}(f_*(U))$, hence f_* is surjective.

□

8.3 Sublocales

We now have the necessary tools to define the notion of a sublocale. In fact, given an embedding $f : B \to A$ between locales, then the sublocale B will be defined in terms of the adjunction $f^{-1} \dashv f_*$. More precisely it will be defined in terms of the monad $j : f_*f^{-1} : \mathcal{O}(A) \to \mathcal{O}(B)$ of the adjunction. To understand why this is the case we should go back to the definition of a subspace of a topological space. In particular, consider a topological space X with a subspace Y, then the topology on Y is given by all those opens U, such that $U = Y \cap V$ for some open $V \subseteq X$. However, it can be the case that for $V \neq W$ we still have that $U = V \cap Y = W \cap Y$, therefore each open set U in Y is associated with a collection of open sets V_i, such that $U = Y \cap V_i$ for all i. Alternatively we can define U to be $U = Y \cap (\bigcup_i V_i) = \bigcup_i (Y \cap V_i)$. In this case we can say that each open in Y is associated bijectively to a union of open sets in X. Moreover, for each V_i, we have that $f^{-1}(V_i) = Y \cap V_i \subseteq \bigcup_i (Y \cap V_i) = U$, therefore each open U in Y is associated with the union of opens $V \subseteq X$, which satisfy $f^{-1}(V) \subseteq U$. We now recall that the map $f : Y \to X$ gives rise to the adjunction $f^{-1} \dashv f_*$ where the right adjoint f_* is defined on each open $U \subseteq Y$ as $f_*(U) = \bigcup\{V \subseteq X | f^{-1}(V) \subseteq U\}$. These unions are precisely the ones which are in bijective correspondence to the open sets in Y and can be characterised through the triangular identity $f_*f^{-1}f_* = f_*$. In fact $f_*f^{-1}f_* = f_*$ implies that $f_*f^{-1}f_*(U) = f_*(U)$. This means that the opens in Y are in bijective correspondence to opens in X which are invariant under the operator $f_*f^{-1} : \mathcal{O}(X) \to \mathcal{O}(X)$, i.e. opens sets of Y correspond to fixed points of j.

Keeping this in mind we now go back to the case of interest and consider two locales A and B with an embedding map $f : B \to A$. We know from Lemma 8.2.5, that $f_* : \mathcal{O}(Y) \to \mathcal{O}(X)$ is injective and, from the triangular identity $f_*f^{-1}f_* = f_*$, it follows that the image of f_* consists of all those V which are invariant under the operator $f_*f^{-1} : \mathcal{O}(A) \to \mathcal{O}(A)$. Therefore, for f an embedding, it follows that

$\mathcal{O}(B)$ is isomorphic to the set of fixed points of j, i.e. $\mathcal{O}(B) \simeq \{U \in \mathcal{O}(A) | j(U) = U\}$. It turns out the converse is also true. However, before showing this, we will characterise the operator j in more details. As mentioned above, j is the monad of the adjunction $f^{-1} \dashv f_*$ hence it comes equipped with two natural transformations: unit and multiplication. These are defined, for any open $U \in \mathcal{O}(A)$, by

$$U \leq j(U), \quad \text{and} \tag{8.3.1}$$

$$jj(U) \leq U, \tag{8.3.2}$$

respectively. Because of functoriality of j, we have that $j(U) \leq jj(U)$, hence Eq. (8.3.2) becomes

$$jj(U) = U. \tag{8.3.3}$$

Moreover, since f_* is a right adjoint and f^{-1} is a frame morphisms, they both preserve meets therefore

$$j(U \wedge V) = j(U) \wedge j(V).$$

We are now in a position to fully characterise the operator j.

Definition 8.3.1 Given a locale A, an operator $j : \mathcal{O}(A) \to \mathcal{O}(A)$ is called a nucleus on A if for each $U \in \mathcal{O}(A)$, then the following conditions are satisfied:

$$U \leq j(U) \tag{8.3.4}$$

$$jj(U) = j(U) \tag{8.3.5}$$

$$j(U \wedge V) = j(U) \wedge j(V). \tag{8.3.6}$$

We have seen so far that, given an embedding $f : B \to A$, the frame $\mathcal{O}(B)$ is isomorphic to the set $\{U \in \mathcal{O}(A) | j(U) = U\}$ of fixed points of j. We now would like to prove that the reverse also holds. Once we do that we can completely characterise sublocales of a locale in terms of its nucleuses.

Lemma 8.3.1 *Given a nucleus $j : \mathcal{O}(A) \to \mathcal{O}(A)$ of a locale A, then the set $\mathcal{O}(A_j) = \{U \in \mathcal{O}(A) | j(U) = U\}$ of fixed points of j is a frame and the map $i^{-1} : \mathcal{O}(A) \to \mathcal{O}(A_j)$; $U \mapsto jU$ is a surjection of frames.*

Proof As a first step we need to show that $\mathcal{O}(A_j)$ is indeed a frame, i.e. it has to be closed under finite meets and arbitrary joins and condition (8.1.1) has to be satisfied. From the definition of j we know that it preserves finite meets, hence $\mathcal{O}(A_j)$ is closed under finite meets. To show that it is closed under arbitrary joins, just consider a family of element $\{U_i\}$ in $\mathcal{O}(A_j)$. One can then define $sup\{U_i\} = j(\bigvee U_i)$ where $\bigvee U_i$ is the supremum in $\mathcal{O}(A)$. We then need to show that $j \bigvee (V \wedge U_i) = V \wedge j \bigvee U_i$:

$$\begin{aligned} V \wedge j \bigvee U_i &\overset{V \in \mathcal{O}(A_j)}{=} jV \wedge j \bigvee U_i \\ &\overset{\text{property of } j}{=} j(V \wedge \bigvee U_i) \\ &\overset{\mathcal{O}(A) \text{ is a frame}}{=} j \bigvee (V \wedge U_i). \end{aligned}$$

As a last step we need to show that i^{-1} is surjective, i.e. for all $U \in \mathcal{O}(A_j)$ then $U = j(U')$ with $U' \in \mathcal{O}(A)$. However we know that $U = j(U)$ and, because of Eq. (8.3.5), $jj(U) = j(U)$. Therefore the map is surjective. The preservation of finite meets and arbitrary joints follows from the fact that j preserves such objects. □

We know from Lemma 8.2.5 that, if the frame map $i^{-1} : \mathcal{O}(A_j) \to \mathcal{O}(A)$ is surjective, then the corresponding locale map $i : A_j \to A$ is an embedding. Therefore it follows that nuclei of locales give rise to embedding of locales. Moreover, it is easy to see that $i^{-1} : \mathcal{O}(A) \to \mathcal{O}(A_j)$ is the left adjoint of the inclusion map $i_* : \mathcal{O}(A_j) \to \mathcal{O}(A); U \mapsto U$.

Now that we have proved Lemma 8.3.1 we can easily define sublocales as follows:

Definition 8.3.2 Given a locale A with nucleous $j : \mathcal{O}(A) \to \mathcal{O}(A)$, the locale A_j with associated frame $\mathcal{O}(A_j) = \{U \in \mathcal{O}(A) | j(U) = U\}$ of fixed points of j is a sublocale of A.

Alternatively, one can define a sublocale B of a locale A as a subset satisfying the following conditions:

1. B is closed under all meets.
2. For all $b \in B$ and all $a \in A$, the pseudo compliment $a \to b$ belongs to B.

From this alternative but equivalent definition it is straightforward to see that sublocales are always non-empty since $1 = \bigwedge \emptyset$ for every B satisfying 1.

Exercise 8.3.1 Show that the two definitions of sublocale are equivalent.

An alternative way of proving that sublocales are in bijective correspondence with nuclei is by the following Theorem [49]:

Theorem 8.3.1 *Given a locale A, there exists a bijective correspondence between nuclei of A and embeddings $f : B \to A$.*

Proof Given an embedding f we know that this gives rise to the nucleus $f_* f^{-1} : \mathcal{O}(A) \to \mathcal{O}(A)$. On the other hand let us assume that one has a nucleus $j = f_* f^{-1}$. This gives rise to an inclusion map $i_* : \mathcal{O}(A_j) \to \mathcal{O}(A)$. To show that f is an embedding we need to show that B is isomorphic to A_j. However, since f is an embedding we know from Lemma 8.2.5 that f^{-1} is a surjection. Therefore $f^{-1} f_* = id : \mathcal{O}(B) \to \mathcal{O}(B)$ and $f_* : \mathcal{O}(B) \to \mathcal{O}(A)$ is injective. Next we consider, with some abuse of notation, the map $f_{*|\mathcal{O}(A_j)} : \mathcal{O}(B) \to \mathcal{O}(A_j); U \mapsto f_*(U)$ and show that this is an isomorphism. First of all, because of the triangular equality $f_*f^{-1}f_* = f_*$, it follows that the above map is well defined. Secondly, since f_* is injective, and each $f_*(U) = f_*f^{-1}f_*(U)$, it follows that $f_{*|\mathcal{O}(A_j)}$ is injective. To show that $f_{*|\mathcal{O}(A_j)}$ is also surjective, we need to show that for each $U \in \mathcal{O}(A_j)$, then $U = f_*(U')$ for $U' \in \mathcal{O}(B)$, but $U = f_*f^{-1}(U)$ and f^{-1} is surjective, hence $f^{-1}(U) \in \mathcal{O}(B)$. □

Definition 8.3.3 Given a locale A, a sublocale B is said to be dense if $0 \in B$, i.e. $j(0) = 0$.

Lemma 8.3.2 *Every locale has a smallest dense sublocale.*

Proof As a first step we will consider the map $\neg\neg : A \to A$ which takes any element a of the locale A and maps it to its double negation[2] $\neg\neg a$. We want to show that such a map is a nucleus, therefore we need to show that it satisfies Eqs. (8.3.4)–(8.3.6). From the definition of $\neg a := a \Rightarrow 0$, it follows that $b \leq \neg a$ iff $b \wedge a = 0$. Clearly $a \wedge \neg a = 0$, therefore $a \leq \neg\neg a$. This proves Eq. (8.3.4).

Next, since $\neg\neg a \wedge \neg\neg\neg a = 0$, it follows that $\neg\neg a \leq \neg\neg\neg\neg a$. On the other hand, since $a \leq \neg\neg a$ holds for all elements then $\neg a \leq \neg\neg\neg a$. Hence $\neg a \wedge \neg\neg\neg\neg a \leq \neg\neg\neg a \wedge \neg\neg\neg\neg a = 0$ and $\neg\neg\neg\neg a \leq \neg\neg a$. These two results put together show that $\neg\neg\neg\neg a = \neg\neg a$ and also Eq. (8.3.5) is proved. Finally consider the inequality $a \wedge b \leq a \leq \neg\neg a$, it follows that $a \wedge b \wedge \neg a = 0$, therefore $\neg a \leq \neg(a \wedge b)$. Applying the same reasoning again we obtain that $\neg\neg(a \wedge b) \leq \neg\neg a$. Similarly we also obtain that $\neg\neg(a \wedge b) \leq \neg\neg b$, therefore $\neg\neg(a \wedge b) \leq \neg\neg a \wedge \neg\neg b$. To prove the converse inequality, we start by noting that

$$\neg(a \wedge b) \wedge a \wedge b = 0 \Leftrightarrow \neg(a \wedge b) \wedge a \leq \neg b.$$

Since $a \leq \neg\neg a$, then $a \wedge \neg\neg\neg a \leq \neg\neg a \wedge \neg\neg\neg a = 0$, which implies that $\neg\neg\neg a \leq \neg a$. On the other hand $\neg\neg a \wedge \neg a = 0 \Leftrightarrow \neg a \leq \neg\neg\neg a$, thus $\neg a = \neg\neg\neg a$. We then obtain that

$$\begin{aligned}
\neg(a \wedge b) \wedge a \leq \neg\neg\neg b &\Leftrightarrow \neg\neg b \wedge \neg(a \wedge b) \wedge a = 0 \\
&\Leftrightarrow \neg\neg b \wedge \neg(a \wedge b) \leq \neg a = \neg\neg\neg a \\
&\Leftrightarrow \neg\neg b \wedge \neg(a \wedge b) \wedge \neg\neg a = 0 \\
&\Leftrightarrow \neg\neg a \wedge \neg\neg b \leq \neg\neg(a \wedge b).
\end{aligned}$$

It follows that $\neg\neg a \wedge \neg\neg b = \neg\neg(a \wedge b)$. We have now shown that the map $\neg\neg : A \to A$ is a nucleus. Clearly it is dense since $\neg\neg 0 = \neg 1 = 0$. Now take any nucleus j such that $0 \in A_j$, we know from condition 2 of the definition of a sublocale that, for any $a \in A$, then $a \to 0 \in A_j$ and $((a \to 0) \to 0) = \neg\neg a \in A_j$, therefore $\neg\neg A \subseteq A_j$. □

[2]Here we define the negation $\neg$ in terms of the Heyting algebra negation, i.e. $\neg a := a \Rightarrow 0$. Recall that $a \Rightarrow 0$ is defined as the least upper bound of all those elements b with $b \wedge a \leq 0$.

8.4 Defining Sheaves on a Locale

From Definition 7.2.1 it is clear that the notion of a sheaf on a topological space solely depends on the lattice of open sets of that space. Moreover in Sect. 8.1 we have seen that a locale is just a lattice which mimics the properties of the lattice of open subsets of a topological space, therefore it can be considered as a generalised notion of a topological space. It then follows that it is also possible to define sheaves on a locale. We will start with the notion of a presheaf on a locale then extend it to that of a sheaf.

Definition 8.4.1 Given a locale X, a presheaf on X is a functor:

$$\mathcal{O}(X) \to \mathbf{Sets}$$
$$U \mapsto F(U)$$

Such that for any subset inclusion $V \subseteq U$, the corresponding presheaf map $F(U) \to F(V)$ is given by restriction.

Definition 8.4.2 Given a locale X, a presheaf on X is a sheaf if, given a family $\{x_i \in F(U_i)|i \in I\}$, such that for any pair (i, j), then $x_i|_{U_i \cap U_j} = x_j|_{U_i \cap U_j}$ as elements in $F(U_i \cap U_j)$, then there exists a unique element $x \in F(\bigcup_{i \in I} U_i)$ such that $x|_{U_i} = x_i$ for all $i \in I$.

Let us consider a locale X, with associated frame $\mathcal{O}(X)$. We want to define the notion of a covering family for an open $U \in \mathcal{O}(X)$.

Definition 8.4.3 A family $\{U_i|i \in I\}$ of opens in X such that $U_i \leq U$ is a covering family for U iff $U = \bigvee_i U_i$.

Given such a covering we can define a basis K, for each $U \in \mathcal{O}(X)$ as

$$K(U) = \{\{U_i \to U\}_{i \in I}|U_i \subseteq U \text{ is open, } U = \bigcup_i U_i\}. \tag{8.4.1}$$

As discussed at the end of Sect. 7.1 this indeed satisfies the requirements for being a Grothendieck basis. We will indicate the corresponding Grothendieck topology by J_k. Since $\mathcal{O}(X)$ is ultimately a local, the sheaf defined on the site $(\mathcal{O}(X), J_k)$ is essentially the sheaf defined on the locale $\mathcal{O}(X)$.

Next we would like to show that, given any map $X \to Y$ between locals, there corresponds a geometric morphism $Sh(X) \to Sh(Y)$. As a first step we need to show that the locale X can be recovered from the topos $Sh(X)$. In particular we have that:

Lemma 8.4.1 *Given a locale X, and the terminal object $1 \in Sh(X)$, then*

$$Sub_{Sh(X)}(1) \cong \mathcal{O}(X).$$

Before proving this Lemma we need a few definitions and results. The first definition we need is that of a subsheaf.

Definition 8.4.4 A sub-object A of an object $P \in Sh(X)$ is a functor $A : \mathcal{O}(X) \rightarrow$ **Sets**, such that

i) $A(U) \subseteq P(U)$.
ii) A morphism $A(C) \rightarrow A(D)$ is simply the restriction of the morphism $P(C) \rightarrow P(D)$.
iii) For each object $U \in \mathcal{O}(X)$, each cover S of U and each $x \in P(C)$ then $A(f)(e) \in A(D)$ for all $f : D \rightarrow C$ in S implies that $x \in A(C)$. The last condition simply states that A is actually a sheaf.

Next we will show that given any sheaf, the collection of its subsheaves forms a complete Heyting algebra [55].

Lemma 8.4.2 *Given a site* $(\mathcal{C}, J)$ *and a sheaf* $P \in Sh(\mathcal{C}, J)$, *then* $Sub(P)$ *for a complete Heyting algebra.*

Proof To show that $Sub(P)$ is a complete Heyting algebra we need to show that it has the structure of a lattice and that the operations of join and meet distribute. First of all we note that $Sub(P)$ is endowed with a partial ordering as follows:

$$A \leq B \text{ iff } A(C) \subseteq B(C), \quad \forall C \in \mathcal{C}.$$

Given $A, B \in Sub(P)$ we want to show that $A \wedge B \in Sub(P)$, where for any $C \in \mathcal{C}$ we have $(A \wedge B)(C) = A(C) \cap B(C)$. To see that indeed $A \wedge B \in Sub(P)$ we need to prove that conditions i)–iii) of Definition 8.4.4 hold.

i) If $A(C) \subseteq P(C)$ and $B(C) \subseteq P(C)$ then clearly $A(C) \cap B(C) \subseteq P(C)$.
ii) Given a map $f : C' \rightarrow C$ then $P(f)|_A : A(C) \rightarrow A(C')$ and $P(f)|_B : B(C) \rightarrow B(C')$. It follows that $P(f)|_{A\wedge B} : (A \wedge B)(C') \rightarrow (A \wedge B)(C)$ is such that, given $x \in (A \wedge B)(C')$, then $P(f)|_{A\wedge B}(x) = P(f)|_A(x) \cap P(f)|_B(x) \in A(C) \cap B(C)$.
iii) Assume that for each object $C \in \mathcal{C}$ and each cover $S \in J(C)$ and $x \in P(C)$ both implications

$$\forall f \in S, \quad P|_A(x) \in A(D) \Rightarrow x \in A(C) \tag{8.4.2}$$

$$\forall f \in S, \quad P|_B(x) \in B(D) \Rightarrow x \in B(C) \tag{8.4.3}$$

hold. It then follows that $\forall f \in S$, $P|_{A\wedge B}(f)(x) = P|_A(f)(x) \cap P|_B(x) \in A(D) \cap B(D) = (A \wedge B)(D)$. Therefore, $x \in A(D) \cap B(D) = (A \wedge B)(D)$.

This discussion shows that indeed $A \wedge B \in Sub(P)$. In fact, given any family $\{A_i\}$ of sub-objects in P, the infimum $\bigwedge_i A_i$ exists in $Sub(P)$.

$$\bigwedge_i A_i(C) = \bigcap_i A_i(C).$$

The supremum can now be defined in terms of the infimum as follows:

$$\bigvee_i A_i = \bigwedge\{B|A_i \subseteq B \;\forall\; i\}.$$

In $Sh(C, J)$ it is also possible to define the supremum as follows: for any $C \in \mathcal{C}$ and $x \in P(C)$ then

$$x \in (\bigwedge_i A_i)(C) \text{ iff } \{f : D \to C|\, P(f)x \in A_i(D) \text{ for some } i\} \in J(C). \tag{8.4.4}$$

As a final step we need to show that the following equality holds for any $A, B \in Sub(P)$:

$$B \wedge \bigvee_i A_i = \bigvee_i B \wedge A_i.$$

Given a $C \in \mathcal{C}$ we fist assume that $x \in \bigvee_i B \wedge A_i(C)$. From (8.4.4) it follows that the sieve $S = \{f : D \to C|\, P(f)x \in (B \wedge A_i)(D) \text{ for some } i\}$ covers C. From the definition of infimum it follows that $P(f)x \in B(D)$ and $P(f)x \in A_i(D)$ for some i. Since $B \subseteq P$ then $P(f)x \in B(D) \Rightarrow x \in B(C)$. Moreover the above sieve S is such that for any $f \in S$ then $P(f)x \in A_i(D)$ for some i, therefore $x \in \bigvee_i A_i(C)$ and $x \in B(C) \wedge \bigvee_i A_i(C)$. This proves that for all $C \in \mathcal{C}$, $\left(\bigvee_i B \wedge A_i\right)(C) \subseteq \left(B \wedge \bigvee_i A_i\right)(C)$.

On the other hand let us assume that, given a $C \in \mathcal{C}$ then $x \in \left(B \wedge \bigvee_i A_i\right)(C)$. This implies that $x \in B(C)$ and $x \in \bigvee_i A_i(C)$. Therefore the sieve $S = \{f : D \to C|\, P(f) \in A_i(D) \text{ for some } i\}$ covers C. Given such a sieve and given the fact that $B \subseteq P$ then, for all $f \in S$, we have that $P(f)x \in (B \wedge A_i)(D)$ for some i. Therefore S is the sieve such that $x \in \left(\bigvee_i B \wedge A_i\right)(C)$. This proves that $\left(B \wedge \bigvee_i A_i\right)(C) \subseteq \left(\bigvee_i B \wedge A_i\right)(C)$. □

From the above Lemma it follows that given a locale X and $1 \in Sh(X)$, then $Sub_{Sh(X)}(1)$ is a Heyting algebra, hence a frame. Let us analyse in more details how a sub-object $S \subseteq 1$ behaves.

$$S : \mathcal{O}(X)^{\text{op}} \to \mathbf{Sets}$$
$$U \mapsto A(U) \subseteq \{0\}.$$

Next consider a cover $\{U_i\}_{i\in I}$ of U, i.e. $U = \bigvee_i U_i$, then if $0 \in S(U_i)$ for all i then $0 \in S(U)$, since S is a sheaf. It follows that any sub-object $S \subseteq 1$ is completely determined by the element $W = \bigvee\{U|0 \in S(U)\}$ of $\mathcal{O}(X)$, therefore $Sub_{Sh(X)}(1) \cong \mathcal{O}(X)$. We have thus proved Lemma (8.4.1).

It was shown in [55] that given any topos τ with all small colimits, a geometric morphism $\tau \to Sh(Y)$ for a locale Y, corresponds to a left-exact[3] functor

[3] A functor F is said to be left-exact if it preserves finite limits.

$F : \mathcal{O}(Y) \to \tau$. We are interested in the case when $\tau = Sh(X)$ for some locale X. The left-exact functor in this case is $F : \mathcal{O}(Y) \to Sh(X)$. We know from Lemma 8.4.1 that $\mathcal{O}(Y) \cong Sub_{Sh(Y)}(1)$. Since F if left exact then $Im(F) \subseteq Sub_{Sh(X)}(1) = \mathcal{O}(X)$. Therefore F becomes $F : \mathcal{O}(Y) \to \mathcal{O}(X)$ which corresponds to the locale map $f : X \to Y$. This implies that, for each geometric morphism $F : Sh(X) \to Sh(Y)$ there corresponds a unique locale map $f : X \to Y$. The above discussion formalises in the following Lemma:

Theorem 8.4.1 *Given any two locales X, Y, there is an equivalence of categories*

$$\mathbf{Maps}(X, Y) \xrightarrow{\sim} Hom(Sh(X), Sh(Y))$$

induced by the functor $X \mapsto Sh(X)$ from locales to topoi.

Here $\mathbf{Maps}(X, Y)$ is the category whose objects are locale maps $f : X \to Y$ and morphisms are the natural transformations $f \to g$. In particular $\mathbf{Maps}(X, Y)$ is a poset such that $f \leq g$ if $f^{-1}(U) \leq g^{-1}(U)$ for $U \in \mathcal{O}(Y)$.

Chapter 9
Internalizing Objects in Topos Theory

In this chapter we will explain how to define categorical notions internally within a topos. This internal description of objects is needed to understand the covariant approach to topos quantum theory explained in the next chapter.

9.1 Internal Category

Given a topos τ, we would like to define the notion of a category internal to τ.

Definition 9.1.1 An internal category C in τ consists of the following elements:

1. three objects C_0, C_1, C_2 representing the objects of objects, the object of morphisms and the object of composable pairs, respectively.
2. The codomain morphism $d_0^1 : C_1 \to C_0$ and the domain morphism $d_1^1 : C_1 \to C_0$ which assign to each morphism its domain and codomain, respectively.
3. A morphism $s_0^0 : C_0 \to C_1$ called inclusion of identities which assigns to each object the identity morphism.
4. Three morphisms: $d_0^2, d_1^2, d_2^2 : C_2 \to C_1$ which represent the first member, the composite and the second member of the composable pair, respectively.

The above elements are subject to the following conditions:

1. The law of composite morphisms which is represented by the following pullback diagram

C. Flori, *A Second Course in Topos Quantum Theory*,
Lecture Notes in Physics 944, https://doi.org/10.1007/978-3-319-71108-9_9

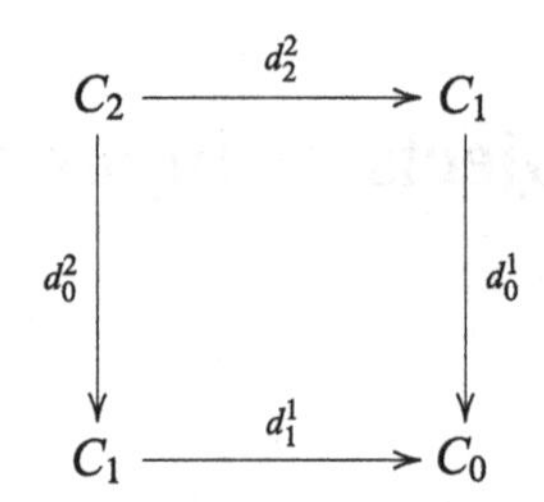

2. Laws specifying the source and target of identity morphisms which are represented by the following commuting diagrams

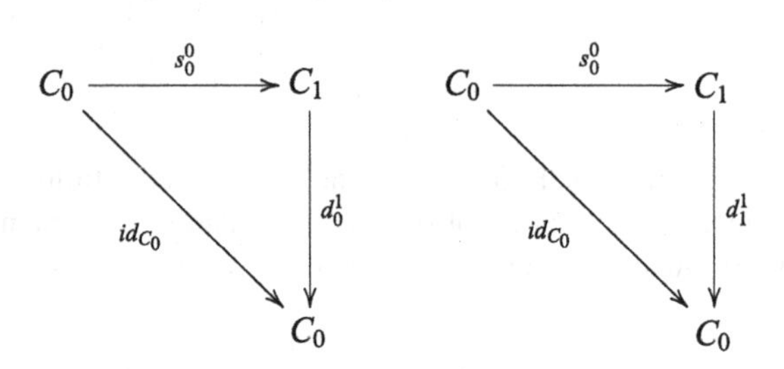

3. Laws specifying the source and target of composite morphisms which are represented by the following commuting diagrams

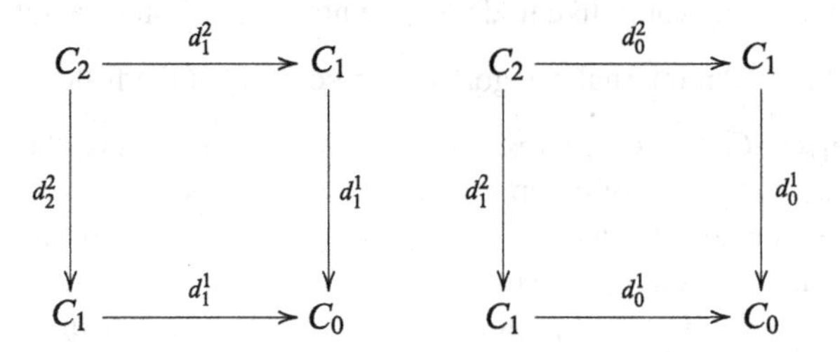

4. Left and right unit laws for composition of morphisms are given by $d_1^2 \circ s_0^1 = id_{C_1} = d_1^2 \circ s_1^1$ where $s_0^1 : C_1 \to C_2$ is defined via

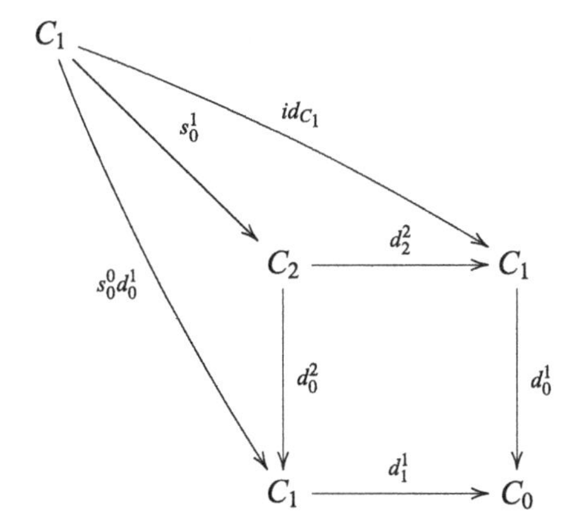

so that, for a particular $f \in C_1$ we obtain

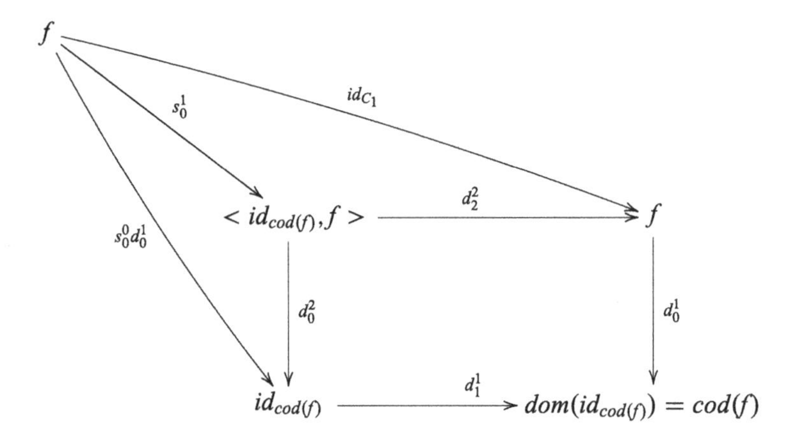

Therefore $s_0^1 : C_1 \to C_2; f \to < id_{cod(f)}, f >$. This implies that $d_1^2 \circ s_0^1 = id_{C_1}$ is equivalent to $id_{cod(f)} \circ f = f$.
Similarly $s_1^1 : C_1 \to C_2$ is defined via

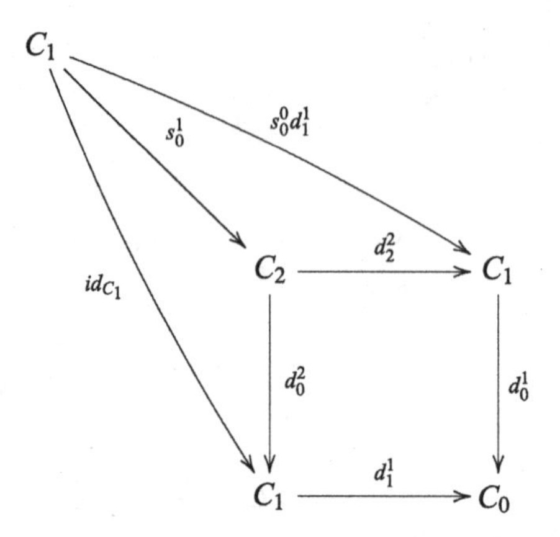

therefore $s_1^1 : C_1 \to C_2; f \mapsto < f, id_{dom(f)} >$. This implies that $id_{C_1} = d_1^2 \circ s_1^1$ is equivalent to $f \circ id_{dom(f)} = f$.

5. Associativity law for composition of morphisms is given by the following commuting diagram

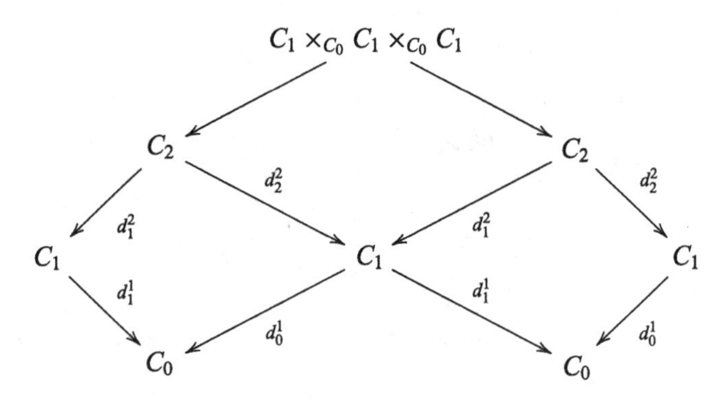

Given the definition of internal category, we also have a definition of internal functor and internal natural transformation.

Definition 9.1.2 Given two internal categories $\mathcal{C}$ and $\mathcal{D}$, an internal functor $F : \mathcal{C} \to \mathcal{D}$ consists of morphisms $F_0 : C_0 \to D_0$, $F_1 : C_1 \to D_1$ and $F_2 : C_2 \to D_2$, which commute with the appropriate structure relations defined above.

Definition 9.1.3 Given two internal functors $F, G : \mathcal{C} \to \mathcal{D}$, an *internal natural transformation* $\alpha : F \to G$ consists of a morphisms $\alpha : C_0 \to D_1$ such that

$$d_0^1 \alpha = g_0, \quad d_1^1 \alpha = f_0, \quad d_1^2(\alpha d_0^1, f_1) = d_1^2(g_1, \alpha d_1^1) : C_1 \to D_1.$$

9.2 Internal C^*-Algebra

In order to define an internal C^*-algebra for a topos τ we will have to explain the notion of rational numbers and integers in a topos [55]. These can be defined in any topos τ which has a natural number object $\underline{\mathbb{N}}_\tau$. In particular, given $\underline{\mathbb{N}}_\tau$, then the object $\underline{\mathbb{Z}}_\tau$ of integers is defined as the co-equaliser

$$\underline{E} \rightrightarrows \underline{\mathbb{N}}_\tau \times \underline{\mathbb{N}}_\tau \longrightarrow \underline{\mathbb{Z}}_\tau$$

where E is the pullback

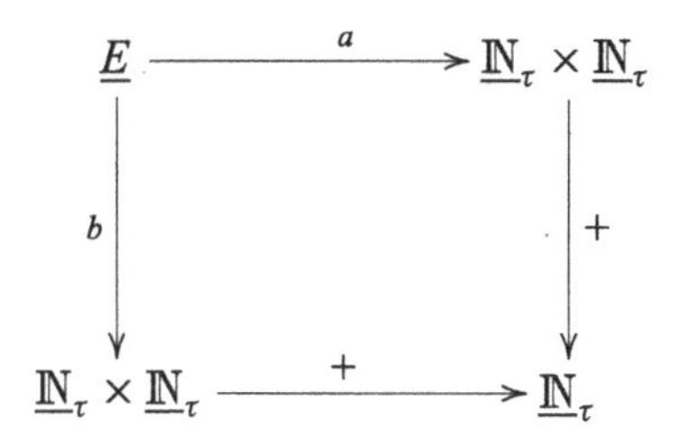

$\underline{E}$ represents the set of 4-tuples (n, m, n', m') such that $n + m' = n' + m$, and $a(n, m, n', m') = (n, m')$ while $b(n, m, n', m') = (n', m)$. This construction reflects the idea that in set theory integers are defined in terms of an equivalence relations on the natural numbers as follows:

$$\mathbb{Z} = \{(n, m)|n, m \in \mathbb{N}\}/ \sim$$

where $(n, m) \sim (n', m')$ iff $n + m' = n' + m$. Similarly, the construction of rational numbers in a topos reflects the idea that in **Sets**, $\mathbb{Q}$ is defined as the quotient $\{(n, m)|n \in \mathbb{Z}, m \in \mathbb{N}\}/ \sim$ where $(n, m) \sim (n'm')$ if $n(m' + 1) = n'(m + 1)$. Therefore the pair (n, m) represent the rational $n/(m + 1)$. This construction translated within a topos τ amounts to defining the rational number object $\underline{\mathbb{Q}}_\tau$ as the following co-equaliser

$$\underline{F} \rightrightarrows \underline{\mathbb{Z}}_\tau \times \underline{\mathbb{N}}_\tau \longrightarrow \underline{\mathbb{Q}}_\tau$$

where $\underline{F}$ is defined via the pullback

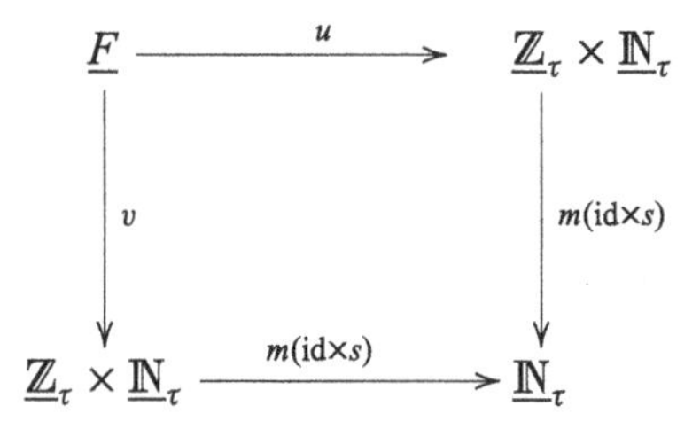

and $s : \underline{\mathbb{N}}_\tau \to \underline{\mathbb{N}}_\tau$ is the successor while $m : \underline{\mathbb{Z}}_\tau \times \underline{\mathbb{Z}}_\tau \to \underline{\mathbb{Z}}_\tau$ represents multiplication. Given the rational $\underline{\mathbb{Q}}_\tau$ we denote by $\underline{\mathbb{Q}}_\tau[i] = \{p + iq | p.q \in \underline{\mathbb{Q}}_\tau\}$ the complexified rational numbers in τ. A vector space over $\underline{\mathbb{Q}}_\tau[i]$ is defined as an object $\underline{A}$ in τ together with the following morphisms:

$$+ : \underline{A} \times \underline{A} \to \underline{A}$$
$$\cdot : Q_\tau[i] \times \underline{A} \to \underline{A}$$
$$0 : \underline{1} \to \underline{A}$$

which represent addition, scalar multiplication and the constant 0, respectively. These maps satisfy the usual axioms of a vector space. Given the notion of an internal vector space over $\underline{\mathbb{Q}}_\tau[i]$ we can now define the notion of an internal C^*-algebra.

Definition 9.2.1 A $*$-algebra $\underline{A}$ in a topos τ is a vector space over $\underline{\mathbb{Q}}_\tau[i]$ together with an associative bilinear map $\cdot : \underline{A} \times \underline{A} \to \underline{A}$ and a map $(-)^* : \underline{A} \to \underline{A}$ such that for all $a, b \in \underline{A}$ and $z \in \underline{\mathbb{Q}}_\tau[i]$

$$(a + b)^* = a^* + b^*$$
$$(z \cdot a)^* = \bar{z} \cdot a$$
$$(a \cdot b)^* = b^* \cdot a^*$$
$$a^{**} = a.$$

If for all $a, b \in \underline{A}$, $a \cdot b = b \cdot a$ then $\underline{A}$ is commutative, while it is unital if there is a neutral element $1 : \underline{1} \to \underline{A}$ for the multiplication such that, for all $a \in \underline{A}$,

$$a \cdot 1 = a = 1 \cdot a.$$

In order to turn $\underline{A}$ into a C^*-algebra we need to introduce the notion of a norm in τ. In **Sets**, a norm N on an algebra A is defined by a subset $N \subseteq A \times \mathbb{Q}^+$ where $(a, p) \in N$ iff $||a|| < p$, and $|| \cdot || : A \to [0, \infty)$. Similarly one defines a *norm* on $\underline{A}$ as a sub-object $\underline{N} \subseteq \underline{A} \times \underline{\mathbb{Q}}^+$ which satisfies the following axioms [74]:

1. $\forall p \in \underline{\mathbb{Q}}^+,\ (0, p) \in \underline{N}$ which expresses the fact that $||0|| = 0$
2. $\forall a \in \underline{A},\ \exists p \in \underline{\mathbb{Q}}^+$ s.t. $(a, p) \in \underline{N}$ which expresses the fact that $||a||$ can not be equal to ∞.
3. $\forall a \in \underline{A},\ ((\forall p \in \mathbb{Q}^+\ (a, p) \in \underline{N}) \Rightarrow (a = 0))$ which expresses the fact that $||a|| = 0$ implies that $a = 0$. This states that $\underline{N}$ is a norm, not just a semi-norm.
4. $\forall a \in \underline{A}$ and $\forall p \in \underline{Q}^+,\ (a, p) \in \underline{N} \iff (\exists q \in \underline{Q}^+\ (p > q) \wedge ((a.q) \in \underline{N}))$. This axiom together with the previous one indicates that the norm $\underline{N}$ can be seen as a map $|| \cdot || : \underline{A} \to \underline{[0, \infty]}_u$.

5. $\forall a \in \underline{A}$ and $\forall p \in \underline{Q}^+$, $((a,p) \in \underline{N} \Rightarrow (a^*,p) \in \underline{N})$ which, together with the involution property of $*$ implies that $||a|| = ||a^*||$.
6. $\forall a,b \in \underline{A}$ and $\forall p,q \in \underline{Q}^+$, $((a,p) \in \underline{N} \wedge (b,q) \in \underline{N}) \Rightarrow (*a+b, p+q) \in \underline{N}$ which expresses the triangle inequality $||a+b|| \leq ||a|| + ||b||$.
7. $\forall a.b \in \underline{A}$ and $\forall p,q \in \underline{\mathbb{Q}}^+$, $((a,p) \in \underline{N} \wedge (b,q) \in \underline{N}) \Rightarrow (a \cdot b, p \cdot q) \in \underline{N}$ which expresses the fact that $||a \cdot b|| \leq ||a|| \cdot ||b||$.
8. $\forall a \in \underline{A}$, $\forall x \in \underline{\mathbb{Q}[i]}$ and $\forall p,q \in \underline{\mathbb{Q}}$, $(((a,p) \in \underline{N} \wedge (|x| < q)) \Rightarrow (x \cdot a, p \cdot q) \in \underline{N})$ which expresses the fact that $||x \cdot a|| = |x| \cdot ||a||$. Here $|\cdot| : \underline{\mathbb{Q}[i]} \to \underline{\mathbb{Q}}$ takes a complex number and assigns its module, i.e. $x + iy \mapsto x^2 + y^2$.
9. $\forall a \in \underline{A}$ and $\forall p \in \underline{Q}^+$, $((a,p) \in \underline{N} \iff (a \cdot a^*, p^2) \in \underline{N}$ which expresses the fact that $||a||^2 = ||a \cdot a^*||$.

Note that the third axiom, because of the presence of the existential quantifier $\forall$, can not be expressed in terms of the geometric logic explained in Chap. 10.
One needs $\underline{A}$ to be complete with respect to the norm $\underline{N}$. This can be done in terms of conversions of Cauchy approximations. For an in depth explanation the reader should refer to [74].

The notion of an internal *unital *-homomorphism* between internal C^*-algebras is quite straightforward. In particular, given two such algebras $\underline{A}, \underline{B}$, an internal *unital *-homomorphism* between them is a linear map $f : \underline{A} \to \underline{B}$, such that $f(a,b) = f(a)f(b), f(a^*) = f(a)^*$ and $f(1_{\underline{A}}) = 1_{\underline{B}}$.

The collection of internal unital C^*-algebras, together with internal unital *-homomorphism, form an internal category which we denote **CStar**. If we consider commutative C^*-algebras these form a full sub-category of **CStar** which we denote by **cCStar**.

Moreover, there exists a Gelfand duality theorem internal to any topos τ. We will report the theorem below since it is an essential ingredient for the covariant formulation of topos quantum theory. However, for a detailed discussion and proof the reader should refer to [3–5]

Theorem 9.2.1 *Given any topos τ, there exists the following categorical duality*[1]

(9.2.1)

where $\Sigma \vdash C(-, \mathbb{C}_\tau)$.

The above adjunction associates, to each $\underline{A} \in$ **cCStar**, the locale $\Sigma(\underline{A})$ which represents the Gelfand spectrum of $\underline{A}$, while for each completely regular compact

[1] **KRegLoc** represents the category of compact regular locales. A locale L is compact if every subset $S \subseteq L$ with $\bigvee S = \top$ has a finite subset F with $\bigvee F = \top$. It is regular if every element of L is the join of the elements well inside itself. Given two elements, a, b then a is well inside b (denoted $a \ll b$) if there exists c with $c \wedge a = \bot$ and $c \vee b = \top$.

locale X, $C(X, \mathbb{C}_\tau)$ represents the collection of locale maps from X to the locale of Dedekind complex numbers in the topos τ (See Appendix A.1).

9.3 Internal Locales

In this section we investigate the notions of internal frame and internal locale.

Definition 9.3.1 Given a topos τ, an internal lattice L is an object in τ together with two arrows $\bigwedge : L \times L \to L$ and $\bigvee : L \times L \to L$, such that the following diagrams commute

- Associativity

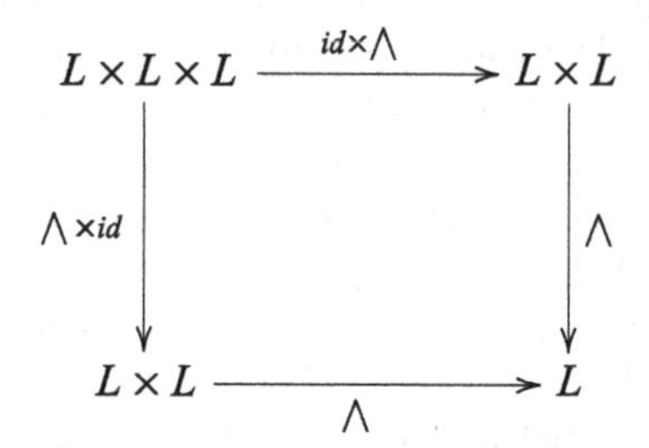

  and a similar diagram for $\bigvee$.
- Commutativity

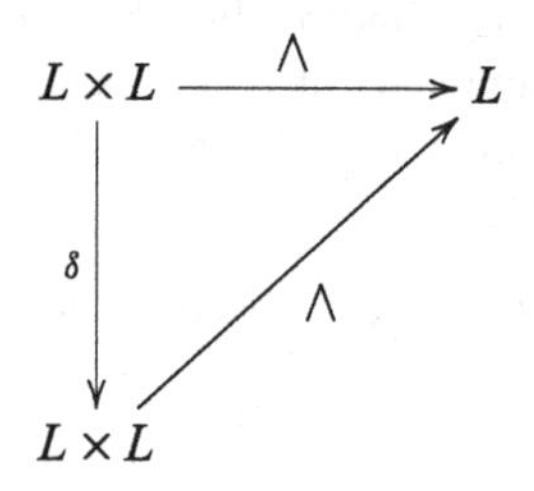

  and a similar diagram for $\bigvee$. Here $\delta : L \times L \to L \times L$ is defined by $\delta(a, b) = (b, a)$.
- Idempotent

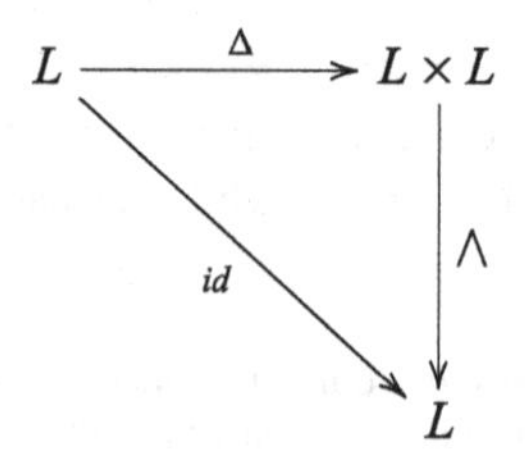

and a similar diagram for $\bigvee$. Here $\Delta : L \to L \times L$ is defined by $\Delta(x) = (x, x)$

- Absorption

$$\begin{array}{ccccc} L & & \xleftarrow{\wedge} & & L \times L \\ \uparrow{\scriptstyle pr_1} & & & & \uparrow{\scriptstyle id \times \vee} \\ L \times L & \xrightarrow{\Delta \times id} & L \times L \times L & \xrightarrow{id} & L \times L \times L \\ \downarrow{\scriptstyle pr_1} & & & & \downarrow{\scriptstyle \wedge \times id} \\ L & & \xleftarrow[\vee]{} & & L \times L \end{array}$$

Given the terminal object $1 \in \tau$, the top and bottom elements of L are defined by the maps $\top : 1 \to L$ and $\bot : 1 \to L$ such that both the compositions

$$L \simeq L \times 1 \xrightarrow{id \times \top} L \times L \xrightarrow{\bigwedge} L$$

$$L \simeq L \times 1 \xrightarrow{id \times \bot} L \times L \xrightarrow{\bigvee} L$$

give the identity.

Definition 9.3.2 An internal frame F is an internal lattice with all finite meets and all joins and for which the diagrammatic equivalent of the following equation is satisfied

$$U \wedge \bigvee_i V_i = \bigvee_i (U \wedge V_i).$$

Given a topos τ, an internal frame is an internal complete Heyting algebra which is define as follows:

Definition 9.3.3 Given a topos τ, an internal Heyting algebra is an internal lattice $L \in \tau$ together with a binary operation $\Rightarrow : L \times L \to L$ such that the following conditions are satisfied:

- Identifying an element of the lattice x with the arrow $x : 1 \to L$, then $\top : 1 \to L$ is equivalent to the composition

$$1 \xrightarrow{x} L \xrightarrow{\Delta} L \times L \xrightarrow{\Rightarrow} L.$$

- The following diagrams commute

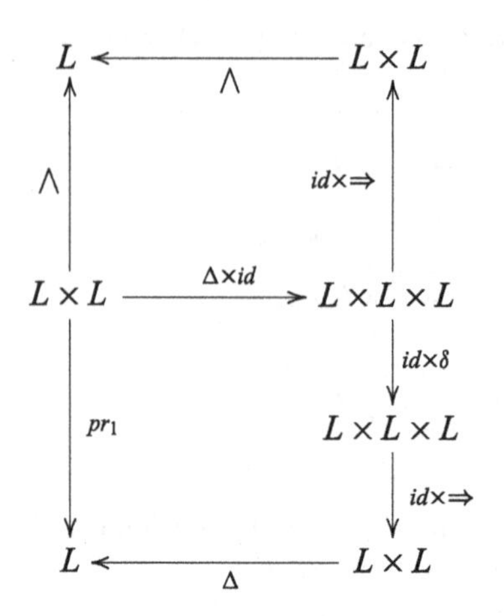

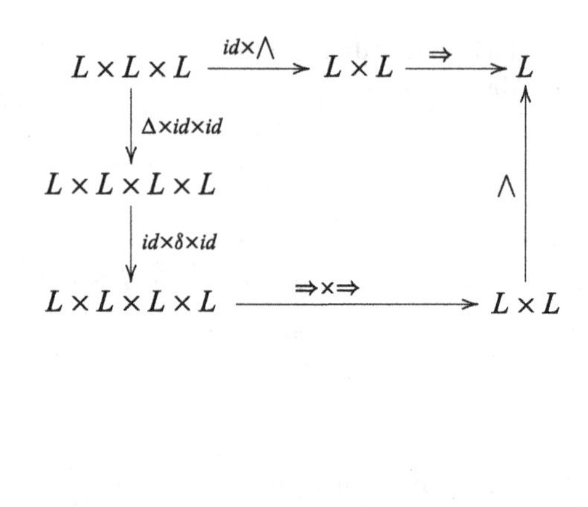

An *internal complete Heyting algebra* is an internal Heyting algebra which is a complete internal lattice. As such it is equivalent to an internal frame.

Similarly as for external locales, also internal locales are identified with internal frames. Moreover, given a frame map $f : \mathcal{O}(X) \to \mathcal{O}(Y)$ the corresponding locale map is $Y \to X$.

A point of an internal locale X is given by the frame map $\mathcal{O}(X) \to \Omega$, where now Ω is the sub-object classifier of the topos τ. Opens are defined by maps $1 \to X$ where 1 is the terminal object in the topos τ. Similarly as for sets, the collection of all points pt(X) is given the topology expressed by the analogue of Eq. (8.1.8),

$$\mathrm{pt}(U) = \{p \in \mathrm{pt}(X) | p^{-1} \circ (U) = \top\}, \tag{9.3.1}$$

where $U \in \mathcal{O}(X)$ and $\top$ is the 'maximal' element of Ω.

We are particularly interested in locales internal to the topos $Sh(X)$ for some topological space X. In what follows we will analyse some results pertaining the category **Loc**$(Sh(X))$ of internal locales in $Sh(X)$. In particular, we will prove that, for any locale X, the category **Loc**$(Sh(X))$ of internal locales in $Sh(X)$ is equivalent to the slice category **Loc**$/X$ [50]. The slice category **Loc**$/X$ has as objects locale maps $f : Y \to X$ where Y is a locale in **Sets**. Given two objects $W \to X$ and $Y \to X$ in **Loc**$(Sh(X))$, a morphism between them consists of a commuting triangle of the form

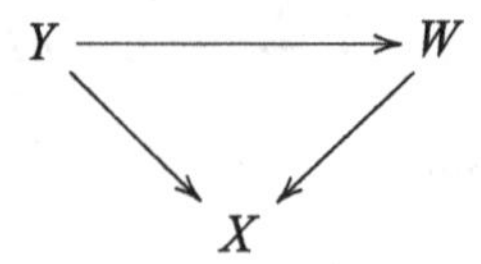

where the map $Y \to W$ is a morphisms of locales.
Before proving the above equivalence we briefly recall what an equivalence actually is.

Definition 9.3.4 Given two categories $\mathcal{C}$ and $\mathcal{D}$, an equivalence between them is a pair of functors

$$\mathcal{C} \underset{G}{\overset{F}{\rightleftarrows}} \mathcal{D} \ ; \qquad F \dashv G$$

such that there are natural isomorphisms

$$F \circ G \cong Id_{\mathcal{D}} \quad \text{and} \quad G \circ F \cong Id_{\mathcal{C}}.$$

Applying the above definition to the case at hand, we need to construct two functors $\mathcal{I} : \mathbf{Loc}/X \to \mathbf{Loc}(Sh(X))$ and $\mathcal{E} : \mathbf{Loc}(Sh(X)) \to \mathbf{Loc}/X$ and show that their compositions are isomorphic to the identity functor on the appropriate category. We note that the notation $\mathcal{I}$ indicate the fact that we are *internalising*, while the notation $\mathcal{E}$ indicates that we are *externalising*.

To construct the functor $\mathcal{I} : \mathbf{Loc}/X \to \mathbf{Loc}(Sh(X))$ consider an object $f : Y \to X \in \mathbf{Loc}/X$. This induces a geometric morphism $f : Sh(Y) \to Sh(X)$ such that the direct image functor f_* preserves Heyting algebras and complete internal posets. Consider now the sub-object classifier[2] $\Omega^Y \in Sh(Y)$. This is defined for each open set $U \in \mathcal{O}(Y)$ as $\Omega^Y(U) := \{V \in \mathcal{O}(Y) | V \leq U\}$ such that given an inclusion $U' \subseteq U$, the corresponding restriction map $\Omega^Y(U) \to \Omega^Y(U')$ is given by $- \cap U'$. If we apply the functor f_* to Ω^Y then, because of the above mentioned properties of f_*, the sheaf $f_*(\Omega^Y)$ is a complete Heyting algebra (internal frame) in $Sh(X)$. Given any open set $W \in \mathcal{O}(X)$, then $f_*(\Omega^Y)(W) = \Omega^Y(f^{-1}(W)) = \{V \in \mathcal{O}(Y) | V \leq f^{-1}(W)\}$. Therefore, starting from the object $f : Y \to X$, we have defined the internal locale $f_*(\Omega^Y) \in Sh(X)$ whose associated frame we denote by $\mathcal{O}(Y)$.

We now need to show that the assignment

$$\mathcal{I} : \mathbf{Loc}/X \to \mathbf{Loc}(Sh(X))$$
$$f \mapsto f_*(\Omega^Y)$$

is indeed a functor [50, Prop. C1.6.1]. As a first step we need to define its action on morphisms. To this end, let us consider a map $h \in \mathbf{Loc}/X$ given by the commutative triangle

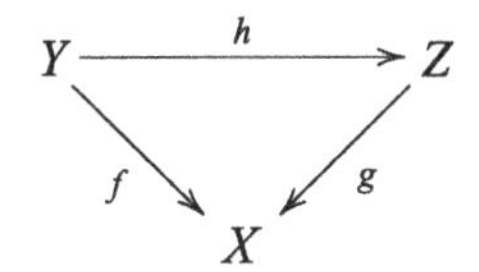

[2]It is a standard result that in every topos the sub-object classifier is a Heyting algebra.

This induces the commutative diagrams

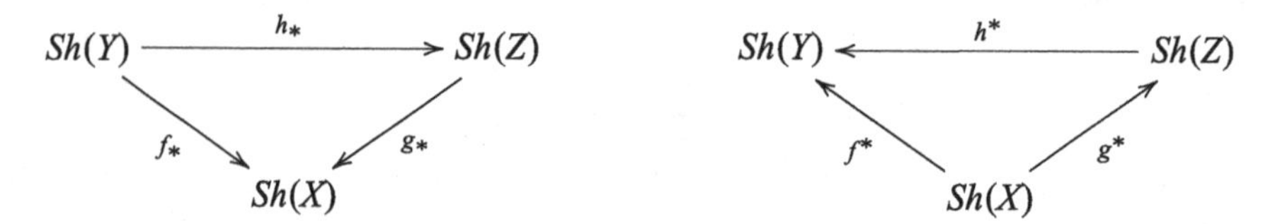

Therefore, given $g_*(\Omega_Z)(U) = \{V \in \mathcal{O}(Z) | V \leq g^{-1}(U)\}$ for any $U \in \mathcal{O}(X)$, then $h^*(g_*(\Omega_Z)(U)) = \{h^{-1}V \in \mathcal{O}(Z) | h^{-1}V \leq h^{-1}g^{-1}(U)\} = \{W \in \mathcal{O}(Y) | W \leq f^{-1}(U)\} = f_*(\Omega_Z)(U)$. This implies that for each $U \in \mathcal{O}(X)$, h^* restricts to frame homomorphism

$$h^*_U : g_*(\Omega_Z)(U) \to f_*(\Omega_Y)(U).$$

Since h^* is the inverse image of the geometric morphism $h : Sh(Z) \to Sh(Y)$ it is the left adjoint of h_* and it is left exact, which implies that h^* preserves both small colimits and finite limits.[3] Therefore, given $U' \subseteq U$ the following diagram commutes

$$\begin{array}{ccc} g_*(\Omega_Z)(U) & \xrightarrow{h^*_U} & f_*(\Omega_Y)(U) \\ \Big\downarrow {\scriptstyle -\cap U'} & & \Big\downarrow {\scriptstyle -\cap U'} \\ g_*(\Omega_Z)(U') & \xrightarrow{h^*_{U'}} & f_*(\Omega_Y)(U') \end{array}$$

This shows that $g_*(\Omega_Z)(U) \to f_*(\Omega_Z)(U)$ is an internal lattice homomorphism in $Sh(X)$. However we would like $g_*(\Omega_Z)(U) \to f_*(\Omega_Z)(U)$ to be a map in $\mathbf{Loc}(Sh(X))$, hence we also need to show that it is an internal frame homomorphism. To this end it suffice to show that the right adjoint of the h^*_U also forms a natural transformation so that $g_*(\Omega_Z)(U) \to f_*(\Omega_Z)(U)$ would be both a left adjoint and left exact. To this end we note that h^*_U sends $V \leq f^{-1}(U)$ to $h(V) \cap h(f^{-1}(U)) = h(V) \cap (h \circ h^{-1} \circ g^{-1})(U) = h(V) \cap g^{-1}(U)$. This action is clearly natural in U since given $U' \subseteq U$ then

$$\begin{aligned} h(V \cap f^{-1}(U')) \cap g^{-1}(U') &= (h(V) \cap hh^{-1}g^{-1}(U')) \cap g^{-1}(U') \\ &= h(V) \cap g^{-1}(U') \cap g^{-1}(U') \\ &= (h(V) \cap g^{-1}(U)) \cap g^{-1}(U') \end{aligned}$$

[3]In the case at hand it means that h^* commutes with arbitrary unions and finite intersections, respectively.

Therefore we have shown that, given the map $h \in \mathbf{Loc}/X$ as described above, then $\mathcal{I}(h) : \mathcal{I}(f) \to \mathcal{I}(g)$ is an internal locale morphisms which is functorial in h by construction.

As a next step we need to construct a functor $\mathcal{E} : \mathbf{Loc}(Sh(X)) \to \mathbf{Loc}/X$ which assigns to each internal locale $L \in \mathbf{Loc}(Sh(X))$ a morphism $f : Y \to X$ of locales. Clearly we would like such an assignment to be such that $L \simeq \mathcal{I}(f)$. The strategy that we will adopt is as follows: we start with an internal frame $L \in Sh(X)$, then the locale Y will be defined by setting $\mathcal{O}(Y) = L(X)$. The fact that $L(X)$ is a frame will be explained later on. Given this setting, the map f will then be identified with the frame homomorphism $f^{-1} : \mathcal{O}(X) \to L(X)$ such that for each open $V \in \mathcal{O}(X)$, $f^{-1}(V) = \sigma^V_X(\top_V) \in L(X)$. Here, for each $V \leq U \in \mathcal{O}(X)$, $\sigma^V_U : L(V) \to L(U)$ is the left adjoint to the restriction map $\rho^U_V : L(U) \to L(V)$.

To understand the above construction let us consider an internal frame $L \in Sh(X)$. The global section functor $\Gamma_X(-) : Sh(X) \to \mathbf{Sets}; L \mapsto L(X)$ is the direct image of the geometric morphisms $Sh(X) \to \mathbf{Sets}$, hence it preserves Heyting algebras. This means that $L(X)$ is a frame in **Sets**.
Next consider subterminal objects (sub-presheaves of the terminal object $\underline{1}$) in $Sh(X)$. These identified with elements of $\mathcal{O}(X)$. In fact the terminal object $\underline{1}$ in $Sh(X)$ is nothing but the constant functor with value 1, hence a sub-presheaf $\underline{S}$ of $\underline{1}$ is determined by the set $\{U \in \mathcal{O}(X) | \underline{S}(U) = 1\}$. Since the maps in $\mathcal{O}(X)$ are inclusion maps this set must be a downward-closed subset in $\mathcal{O}(X)$. Moreover, since $\underline{S}$ is a sheaf, it has to be closed under coverings meaning that the set $\{U \in \mathcal{O}(X) | \underline{S}(U) = 1\}$ is actually a principle ideal $\downarrow U$ for some $U \in \mathcal{O}(X)$. Therefore $Sub(\underline{1}) \simeq \mathcal{O}(X)$. This implies that elements in $L(U)$ are actually morphisms $U \to L$ in $Sh(X)$. Moreover, since L is an internal poset, it has the property of being complete. This means that the map $\downarrow : L \to \underline{\Omega}^L$ which for each $U \in \mathcal{O}(X)$ is defined by

$$\downarrow_U : L(U) \to \underline{\Omega}^L(U)$$
$$p \mapsto \{p' \in L(U) | p' \leq p\}$$

has a left adjoint $\bigvee : \underline{\Omega}^L \to L$. Given two objects $V, U \in \mathcal{O}(X)$ such that $V \leq U$, we then obtain

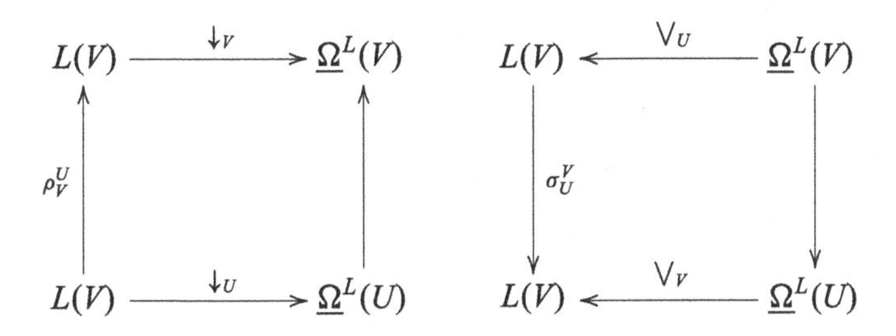

Since subterminal objects are the same as maps $\underline{1} \to \underline{\Omega}$, i.e. points in $\underline{\Omega}$, this adjointness relation reduced to the fact that the restriction map $\rho^U_V : L(U) \to L(V)$ has left adjoint $\sigma^V_U : L(U) \to L(V)$. It then follows that, for any $V_1, V_2 \subseteq U$, given

the pullback square

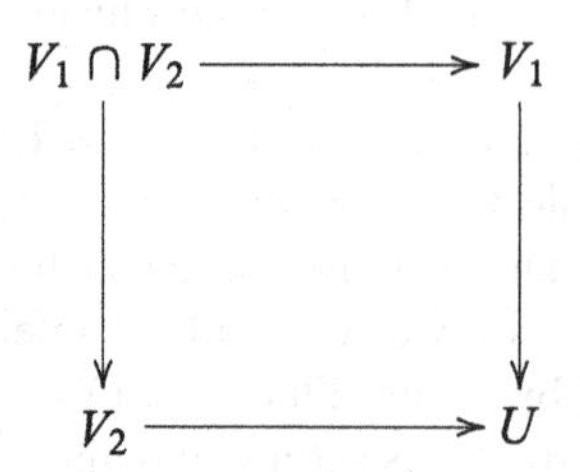

then the square

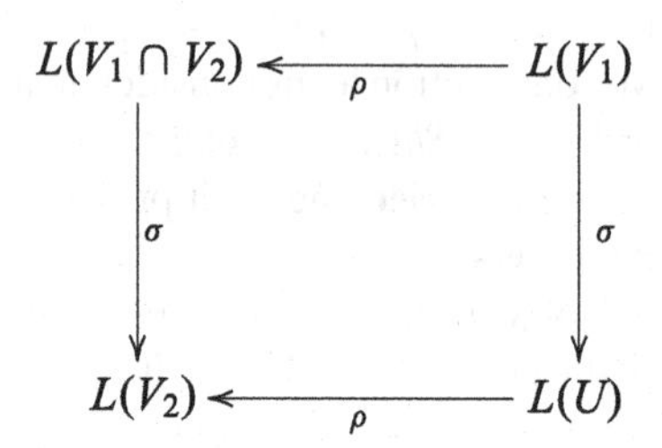

commutes. This is known in the literature as the Beck-Chevalley condition [50]. For $V_1 = V_2 = V$ we then obtain the commuting square

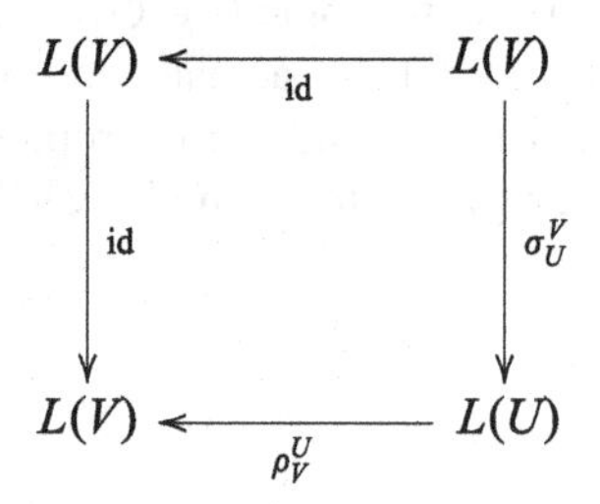

which implies that $\rho^U_V \circ \sigma^V_U = \mathrm{id}_{L(V)}$. Since L is an internal Heyting algebra, the maps ρ^U_V are algebra homomorphisms. However, since a Heyting algebra is a bicartesian closed poset,[4] the maps ρ^U_V are cartesian closed which implies that the Frobenius reciprocity law holds [50, A.1.5.8]:

$$\sigma^V_U(y \wedge \rho^U_V(x)) = \sigma^V_U(y) \wedge x.$$

[4] A bicartesian closed poset is a poset which (when thought of as a thin category) is (a) finitely complete, (b) finitely cocomplete and cartesian closed.

If we now assume $U = X$ and $y = \top_V$ is the top element of the lattice $L(V)$, then Frobenius reciprocity law tells us that

$$\sigma_X^V \rho_V^X(x) = \sigma_X^V(\top_V) \wedge x.$$

Since the maps ρ_V^X are restriction maps it follows that any element $\rho_V^X(x) = y \in L(V) \subseteq L(X)$ will have to be such that $y \leq \sigma_X^V(\top_V)$. This is because, when we mapped y to $L(X)$ via σ_X^V we obtain $\sigma_X^V \rho_V^X(x) = \sigma_X^V(\top_V) \wedge x$. This allows us to identify $L(V)$ as the principal ideal $\{y \in L(X) | y \leq \sigma_X^V(\top_V\}$ of $L(X)$.

Next we claim that the map $s : \mathcal{O}(X) \to L(X)$ defined for all $V \in \mathcal{O}(X)$ as $s(V) = \sigma_X^V(\top_V)$ is a frame homomorphism. To prove this claim we need to show that s preserves binary meets and arbitrary joins. To this end consider the following commuting diagram

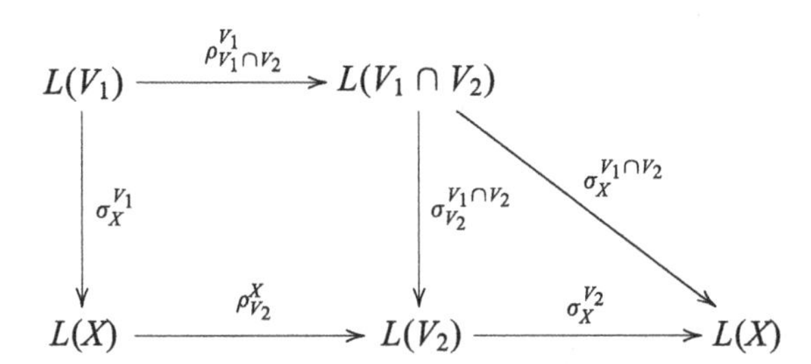

Given the element $\top_{V_1}$ we then have that

$$\begin{aligned}
\sigma_X^{V_2} \circ \rho_{V_2}^X \circ \sigma_X^{V_1}(\top_{V_1}) &= \sigma_X^{V_2} \circ \sigma_{V_2}^{V_1 \cap V_2} \circ \rho^{V_1}_{V_1 \cap V_2}(\top_{V_1}) \\
&= \sigma_X^{V_1 \cap V_2} \circ \rho^{V_1}_{V_1 \cap V_2}(\top_{V_1}) \\
&= \sigma_X^{V_1 \cap V_2}(\top_{V_1 \cap V_2}) \\
&= s(V_1 \cap V_2)
\end{aligned}$$

where the third equality follows since ρ is a Heyting algebra homomorphism. However, Frobenius reciprocity law tells us that

$$\sigma_X^{V_2} \circ \rho_{V_2}^X \circ \sigma_X^{V_1}(\top_{V_1}) = \sigma_X^{V_1}(\top_{V_1}) \wedge \sigma_X^{V_2}(\top_{V_2}) = s(V_1) \wedge s(V_2)$$

Hence, for any $V_1, V_2 \in \mathcal{O}(X)$ we have that

$$s(V_1 \cap V_2) = s(V_1) \wedge s(V_2).$$

This shows that s preserves binary meets. To show that it preserves arbitrary joins we consider a covering $U = \bigcup_{i \in I} U_i \in \mathcal{O}(X)$. It is easy to see that $\top_U = \bigvee_{i \in I} \sigma_U^{U_i}(\top_{U_i})$. This follows from the fact that L is a sheaf over X and compatible sections can be uniquely glued together. In particular, restricting $\top_U$ and

$\bigvee_{i\in I}\sigma_U^{U_i}(\top_{U_i})$ to $L(U_i)$, for any $i \in I$, we obtain on, the one hand $\rho^U_{U_i}(\top_U) = \top_{U_i}$ and, on the other, $\rho^U_{U_i}(\bigvee_{i\in I}\sigma_U^{U_i}(\top_{U_i})) = \rho^U_{U_i}(\sigma_U^{U_i}(\top_{U_i})) = \top_{U_i}$ were the last equality follows since $\rho^U_{U_i} \circ \sigma_U^{U_i} = \mathrm{id}_{L(U_i)}$ as showed above.
Since σ_X^U is left adjoint we obtain

$$\begin{aligned} s(U) &= \sigma_X^U(\top_U) \\ &= \sigma_X^U(\bigvee_{i\in I}\sigma_U^{U_i}(\top_{U_i})) \\ &= \bigvee_{i\in I}(\sigma_X^U\sigma_U^{U_i}(\top_{U_i})) \\ &= \bigvee_{i\in I}(\sigma_X^{U_i}(\top_{U_i})) \\ &= \bigvee_{i\in I} s(U_i) \end{aligned}$$

were the fourth equality is given by simple composition. Thus s preserves arbitrary joins.

We are now able to define the desired map $\mathcal{E} : \mathbf{Loc}(Sh(X)) \to \mathbf{Loc}/X$. Namely, for each internal locale $L \in \mathbf{Loc}(Sh(X))$, the locale map $\mathcal{E}(L) = f : Y \to X$ is defined by setting $Y = L(X)$ and $f^{-1} = s$. Then, for each $U \in \mathcal{O}(X)$, we have $L(U) \simeq \{x \in L(X) | x \leq f^{-1}(U)\}$. Clearly these isomorphisms are natural in U. In fact, given $V \leq U$ in $\mathcal{O}(X)$, then the diagram

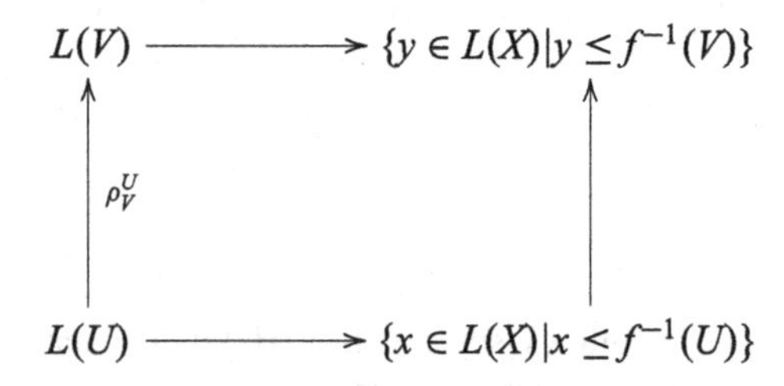

commutes since $V \leq U$ implies that $f^{-1}V \leq f^{-1}U$, therefore if $y \leq f^{-1}(V)$ then $y \leq f^{-1}(U)$ and the map $\{x \in L(X)|x \leq f^{-1}(U)\} \to \{y \in L(X)| y \leq f^{-1}(V)\}$ is given by restriction. We have thus proved the following Lemma:

Lemma 9.3.1 *Given an internal frame $L \in Sh(X)$ there exists a locale morphisms $f : Y \to X$ such that $L \simeq \mathcal{O}(\mathcal{I}(f))$, i.e. $\mathcal{I} \circ \mathcal{E} \simeq Id$*

Next we need to show that the map $\mathcal{E} : \mathbf{Loc}(Sh(X)) \to \mathbf{Loc}/X$ as defined above is indeed a functor. We already know how it acts on objects but we still need to understand how it acts on morphisms. In particular, given a map of locales $k : L' \to L$ we would like to define a map $\mathcal{E}(k)$ such that the following diagram commutes

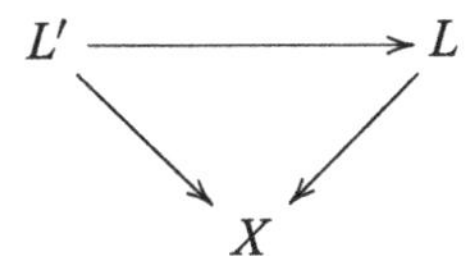

We will do this buy considering the respective frame maps, i.e. we would like to show that the following diagram of frames commutes

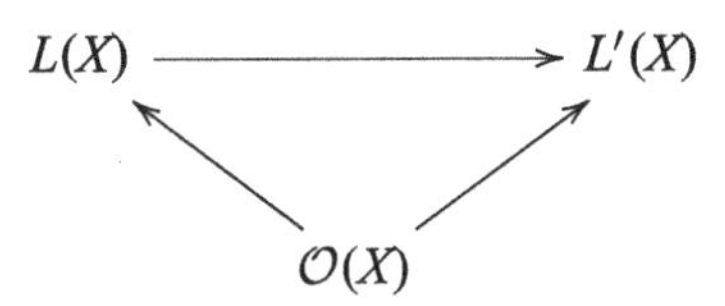

Each internal frame homomorphism $k^{-1} : L \to L'$ in $Sh(X)$ induces an external frame homomorphism $k^{-1}(X) : L(X) \to L'(X)$ and, since k^{-1} preserve joins, the diagram

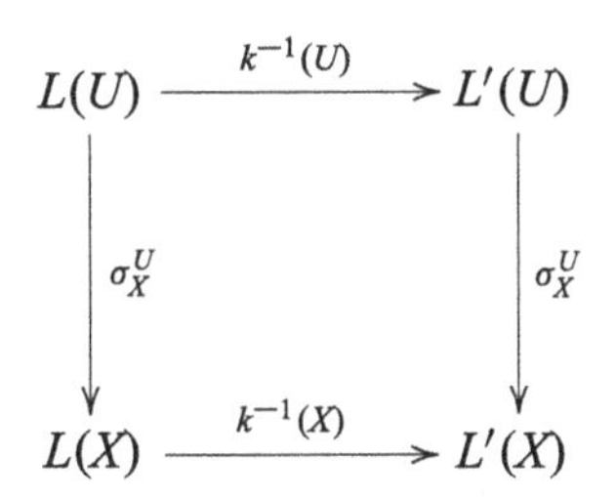

commutes. Moreover, since $k^{-1}(U)$ preserves the top element we have that $k^{-1}(X)(s(U)) = k^{-1}(X)(\sigma_X^U(\top_U)) = \sigma_X^U(k^{-1}(U)(\top_U)) = \sigma_X^U(\top'_U) = s'(U)$. It then follows that $k^{-1}(X)$ is a frame homomorphism under $\mathcal{O}(X)$ and indeed the diagram

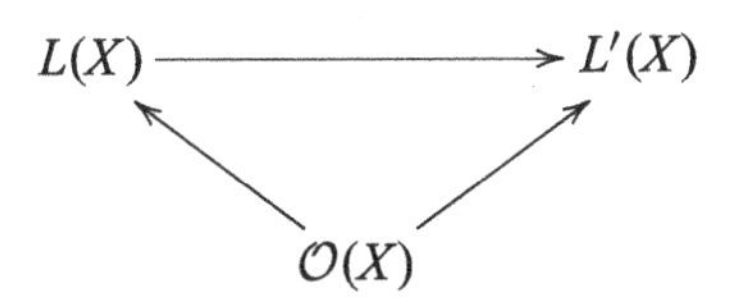

commutes. Functoriality follows by construction.

So far we have managed to define both functors $\mathcal{I} : \mathbf{Loc}/X \to \mathbf{Loc}(Sh(X))$ and $\mathcal{E} : \mathbf{Loc}(Sh(X)) \to \mathbf{Loc}/X$ and show that $\mathcal{I} \circ \mathcal{E} \simeq Id$. What remains to be done is to show that $\mathcal{E} \circ \mathcal{I}$ is isomorphic to the identity on $\mathbf{Loc}/X$. Given any map of locales $f : Y \to X$, then by construction $\mathcal{I}(f) = f_*(\Omega_Y)$ and $\mathcal{O}(dom(\mathcal{E}(\mathcal{I}(f))) = (f_*(\Omega_Y))(X) \simeq \mathcal{O}(Y)$. To finish the proof we need to show that the frame homomorphism $s : \mathcal{O}(X) \to (f_*(\Omega_Y))(X)$ coincides with f^{-1}. However, by constructions, $f_*(\Omega_Y)(U) = \{V \in (f_*(\Omega_Y))(X) | V \leq f^{-1}(U)\}$ which implies that its top element is $\top_U = f^{-1}(U)$ and σ_X^U is the inclusion map.

We have thus proved the following theorem

Theorem 9.3.1 *Given any locale L, the categories* **Loc**$(Sh(X))$ *and* **Loc**$/X$ *are equivalent.*

Chapter 10
Geometric Logic

In this chapter we will introduce the notion of a *geometric logic*. This is the logic of implication between geometric formulas (see Definition 10.2.3). Before introducing the notion of a geometric logic we need to recall the notion of a *higher order type language* $\mathcal{L}$. This was already introduced in [26, A.3], however we will briefly summarise the main points in the following section.

10.1 The Higher Order Type Language $\mathcal{L}$

In this section we will define, in general terms, the notion of a first order type language which we denote by $\mathcal{L}$. Such a language consists of a set of symbols and terms.

Symbols

1. A collection of "sorts" or "types". If $T_1, T_2, \cdots, T_n$, $n \geq 1$, are type symbols, then so is $T_1 \times T_2 \times \cdots \times T_n$. If $n = 0$ then $T_1 \times T_2 \times \cdots \times T_n = 1$.
2. If T is a type symbol, then[1] so is PT.
3. Given any type T there are a countable set of variables of type T.
4. There is a special symbol $*$.
5. A set of function symbols for each pair of type symbols, together with a map which assigns to each function its type. This assignment consists of a finite, non-empty list of types. For example, if we have the pair of type symbols (T_1, T_2), the associated set of function symbols will be $F_{\mathcal{L}}(T_1, T_2)$. An element $f \in F_{\mathcal{L}}(T_1, T_2)$ has type T_1, T_2. This is indicated by writing $f : T_1 \to T_2$.
6. A set of relation symbols R_i together with a map which assigns the type of the arguments of the relation. This consists of a list of types. For example, a relation

[1] PT indicates the collection of all subsets of T.

C. Flori, *A Second Course in Topos Quantum Theory*,
Lecture Notes in Physics 944, https://doi.org/10.1007/978-3-319-71108-9_10

taking an argument $x_1 \in T_1$ of type T_1 to an argument $x_2 \in T_2$ of type T_2 is denoted as $R = R(x_1, x_2) \subseteq T_1 \times T_2$.

Terms

1. The variables of type T are terms of type T, $\forall T$.
2. The symbol $*$ is a term of type 1.
3. A term of type Ω is called a formula. If the formula has no free variables then we call it a sentence.
4. Given a function symbol $f : T_1 \to T_2$ and a term t of type T_1, then $f(t)$ is a term of type T_2.
5. Given $t_1, t_2, \cdots, t_n$ which are terms of type $T_1, T_2, \cdots, T_n$, respectively, then $\langle t_1, t_2, \cdots, t_n \rangle$ is a term of type $T_1 \times T_2 \times \cdots \times T_n$.
6. If x is a term of type $T_1 \times T_2 \times \cdots \times T_n$, then for $1 \leq i \leq n$, x_i is a term of type T_i.
7. If ω is a term of type Ω and α is a variable of type T, then $\{\alpha | \omega\}$ is a term of type PT.
8. If x_1, x_2 are terms of the same type, then $x_1 = x_2$ is a term of type Ω.
9. If x_1, x_2 are terms of type T and PT respectively, then $x_1 \in x_2$ is a term of type Ω.
10. If x_1, x_2 are terms of type PT and PPT respectively, then $x_1 \in x_2$ is a term of type Ω.
11. If x_1, x_2 are both terms of type PT, then $x_1 \subseteq x_2$ is a term of type Ω.

The entire set of formulas in the language $\mathcal{L}$ are defined, recursively, through repeated applications of formation rules, which are the analogues of the standard logical connectives. In particular, we have *atomic formulas* and *composite formulas* The former are:

1. The terms of relation.
2. Equality terms defined above.
3. Truth $\top$ is an atomic formula with empty set of free variables.
4. False $\bot$ is an atomic formula with empty set of free variables.

We can now build more complicated formulas through the use of the logical connectives $\vee$, $\wedge$, $\Rightarrow$ and $\neg$. These are the *composite formulas*:

1. Given two formulas α and β then $\alpha \vee \beta$ is a formula such that, the set of free variables is defined to be the union of the free variables in α and β.
2. Given two formulas α and β then $\alpha \wedge \beta$ is a formula such that, the set of free variables is defined to be the union of the free variables in α and β.
3. Given a formula α its negation $\neg\alpha$ is still a formula with the same amount of free variables.
4. Given two formulas α and β, then $\alpha \Rightarrow \beta$ is a formula with free variables given by the union of the free variables in α and β.

It is interesting to note that the logical operations just defined can actually be expressed in terms of the primitive symbols as follows:

1. *true*:= $* = *$.
2. $\alpha \wedge \beta := \langle \alpha, \beta \rangle = \langle true, true \rangle = \langle * = *, * = * \rangle$.
3. $\alpha \Leftrightarrow \beta := \alpha = \beta$.
4. $\alpha \Rightarrow \beta := \Big((\alpha \wedge \beta) \Leftrightarrow \alpha \Big) := \langle \alpha, \beta \rangle = \langle true, true \rangle = \alpha$.
5. $\forall x \alpha := \{x : \alpha\} = \{x : \text{true}\}$.
6. *false*:= $\forall w w := \{w : w\} = \{w : \text{true}\}$.
7. $\neg \alpha := \alpha \Rightarrow \text{false}$.
8. $\alpha \wedge \beta := \forall w[(\alpha \Rightarrow w \wedge \beta \Rightarrow w) \Rightarrow]$.
9. $\exists x \alpha := \forall w[\forall x(\alpha \Rightarrow w) \Rightarrow w]$.

In the above the notation $\{x : y\}$ indicates the set of all x, such that y.

10.2 Geometric Theories

As was already explained in [26], topos theory is intimately related to first order predicate logic. As seen in the previous section, such languages are comprised of the following objects:

1. *terms* which denote the "atomic variables'.
2. *formulae* which are logical expressions denoting predicates pertaining the terms.

Logical connectives are then utilised to construct *compound terms* and *compound formulae*. It is possible to represent a first-order predicate language within a category. In this context, terms are given by morphisms while formulae are given by sub-objects. Clearly the category in question needs to be rich enough in categorical structures so as to give meaning to the logical connectives. A topos represents precisely such a rich category which allows the construction of logical connectives (true, false), standard connectives ($\wedge$, $\vee$, $\neg$, $\implies$) and quantifiers ($\forall$, $\exists$). In particular, the language $\mathcal{L}$ we will consider, will be an *infinitary, first-order, many-sorted predicate logic with equality*. Here *infinitary* means that infinite conjunctions and disjunctions are allowed, while *many-sorted* means that the *terms* are grouped into different *sorts*.

We will now explain, in more technical details, how a language $\mathcal{L}$ is constructed. A central ingredient in defining $\mathcal{L}$ is its *first order signature* Σ[2] which essentially consists of the non-logical symbols in the language.

Definition 10.2.1 A first order signature Σ comprises the following objects:

1. A set S of sorts.

[2]Note, the signature Σ should not be confused with the Gelfand spectrum. The two objects are completely unrelated, although they have similar symbols.

2. A set F of function symbols and a map which assigns to each such function symbol $f \in F$ its type which is a non-empty finite list of sorts $A_1, A_2, A_3, \ldots A_n, B$. The assignment of a type to a function is given by

$$f : A_1 \times A_2 \times A_3, \times \cdots \times A_n \to B.$$

This tells us that f is of type $A_1, A_2, A_3, \ldots A_n, B$. Here n is called the arity of f. When $n = 0$ f is a constant of sort B which is denoted by $f : 1 \to B$, where 1 is the singleton set.
3. A set Q of relations and a map which assigns to each relation $R \in Q$ its type, which is a finite list of sorts.

$$R \hookrightarrow A_1 \times A_2 \times \cdots \times A_n$$

indicates that R has type $A_1, A_2, \ldots A_n$. Similarly as for functions, n is the arity of R and when $n = 0$, R is an atomic proposition which we denote by $P \subseteq 1$.

Each sort A of signature Σ has associated to itself a fixed countably infinite set of variable V_A.

Next we introduce the *terms* over Σ.

Definition 10.2.2 The collection of terms over Σ is defined recursively by the following steps:

1. x is a term of sort A (dented by $x : A$) if x is a variable of sort A.
2. $f(t_1, t_2 \ldots t_n)$ is a term of sort B $(f(t_1, t_2 \ldots t_n) : B)$ if $f : A_1 \times A_2 \times \cdots \times A_n \to B$ is a function symbol and $t_1 : A_1, t_2 : A_2, \ldots t_n : A_n$. Every constant is a term. The free variables of $f(t_1, t_2 \ldots t_n)$ are the free variables of $t_1, t_2 \ldots t_n$.

As far as the *formulae* over Σ are concerned we will only consider *geometrical formulae*.

Definition 10.2.3 The class of geometrical formulae ϕ over Σ, together with the finite set of free variables for each ϕ, is the smallest class closed under the following rules:

1. If $R \subseteq A_1 \times A_2 \times \cdots \times A_n$ is a relation and $t_1 : A_1, t_2 : A_2, \ldots t_n : A_n$ are terms, then $R(t_1, t_2, \ldots t_n)$ is a formula. The free variables of $R(t_1, t_2, \ldots t_n)$ are the free variables of $t_1, t_2 \ldots t_n$. Every propositional symbol is a formula without free variables.
2. If s and t are two terms of the same sort, then $s = t$ is a formula. The free variables of $s = t$ are the free variables of s and t.
3. Truth T is a formula without free variables.
4. If ϕ and ψ are formulae, then $\phi \vee \psi$ is a formula. The free variables of $\phi \vee \psi$ are those of ϕ together with those of ψ.
5. False $\perp$ is a formula without free variables.
6. If ϕ and ψ are formulae, then $\phi \wedge \psi$ is a formula. The free variables of $\phi \wedge \psi$ are those of ϕ together with those of ψ.

7. If ϕ is a formula with free variables, $x, x_1, x_2, \ldots, x_n$, then $(\exists x)\phi$ is a formulae with free variable $x_1, x_2, \ldots, x_n$
8. If each ϕ_i, for i in some index set I, is a formula and the collection of all the free variables is finite, then $\bigvee_{i \in I} \phi$ is a formula. The free variables of $\bigvee_{i \in I} \phi$ are the free variables of all the ϕ_i.

Given a formula ϕ, the variables x in ϕ which are in the scope of some quantifier[3] for example $(\exists x)$, are called *bounded variables*. Two formulas ϕ and ψ are said to be α-equivalent if they differ only in the names of bounded variables. Two α-equivalent formulae are considered indistinguishable for all practical purposes. If a formula has no free variable it is called a *closed formula* or *sentence*. A *closed term* instead is a term that contains no variables at all.

Formulae are never considered on their own but always with respect to a given *context* which, in this setting, is defined as follows:

Definition 10.2.4 A context is a finite list $\vec{x} = x_1, x_2, \ldots, x_n$ of distinct variables of some sort. When $n = 0$ then we have the empty context denoted by $[]$. Each context can be enlarged with the acquisition of a new variable not already present. For example, if y is a variable not in $\vec{x}$, then we can form the new context $\vec{x}, y$ by appending the variable y to the list. Similarly, if two contexts $\vec{y}$ and $\vec{x}$ are disjoint, one can form the composite context $\vec{x}, \vec{y}$ by concatenating the two.

The type of a context is given by the string of (not necessarily distinct) sorts of the variables appearing in it.

Given a context $\vec{x}$, we say that $\vec{x}$ is *suitable* for a formula ψ if all the free variables of ϕ occur in $\vec{x}$. A *formula-in-context* is then denoted by $\vec{x}.\phi$ where $\vec{x}$ is a suitable context for ϕ. The *canonical context* for a formula ϕ is the context consisting of only the free variables in ϕ, listed in order of appearance. Similarly a *term-in-context* is denoted by $\vec{x}.t$ where $\vec{x}$ is a suitable context for t, i.e. all the free variables of t occur in $\vec{x}$.

A formal expression of the form $\psi \vdash_{\vec{x}} \phi$ is called a *sequent* over a signature Σ where both ϕ and ψ are formula over Σ and $\vec{x}$ is a context suitable for both. The meaning of $\psi \vdash_{\vec{x}} \phi$ is that ϕ is a logical consequence of ψ in the context $\vec{x}$. A set of such sequents $\psi \vdash_{\vec{x}} \phi$ comprises a theory $\mathbb{T}$ over Σ.

Definition 10.2.5 A theory over a signature Σ is a set of sequents $\mathbb{T}$ over Σ whose elements are called the axioms of $\mathbb{T}$. If all the sequents in $\mathbb{T}$ are geometric then we say that $\mathbb{T}$ is a geometric theory.

10.3 Interpreting a Geometric Theory in a Category

Now that we have briefly described the expressions in a first order language, we would like to understand how these are interpreted in a given category. Since, ultimately, we are interested in the representation of this language in the topos

[3]For geometric theories we only have the existential quantifier $\exists$.

$\mathbf{Sets}^{C(\mathcal{A})}$, we will consider out category to be a topos τ.

Definition 10.3.1 Given a topos τ, a Σ-structure M in τ is given by the following assignments:

1. To each sort A in σ an object $MA \in \tau$.
2. To each function symbol $f : A_1 \times A_2 \times \cdots \times A_n \to B$ a morphism $Mf : MA_1 \times MA_2 \times \cdots \times MA_n \to MB$ in τ. A constant $c : 1 \to B$ is given by an arrow $Mc : M1 \to MB$ where $M1$ is the terminal object in τ.
3. To each relation $R \hookrightarrow A_1 \times A_2 \times \cdots \times A_n$ a sub-object $MR \subseteq MA_1 \times MA_2 \times \cdots \times MA_n$.

The collection of Σ-structures in τ forms a category denoted by Σ-$\mathbf{Str}(\tau)$ whose morphisms are called the Σ-structure homomorphisms. Given two Σ-structures M and N, a morphism $h : M \to N$ is given by a collection of morphisms $h_A : MA \to NA$ one for each sort of Σ, such that it correctly maps function symbols and relations symbols. In particular, we require that for each function symbol $f : A_1 \times A_2 \times \cdots \times A_n \to B$, the following diagram commutes

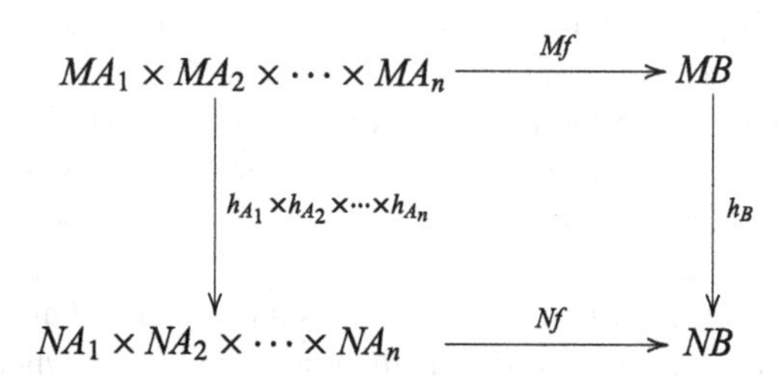

This can be seen as a naturality condition with respect to function symbols.

Moreover, we also require that for each relation symbol $R \hookrightarrow A_1 \times A_2 \times \cdots \times A_n$ the following diagram is a pullback

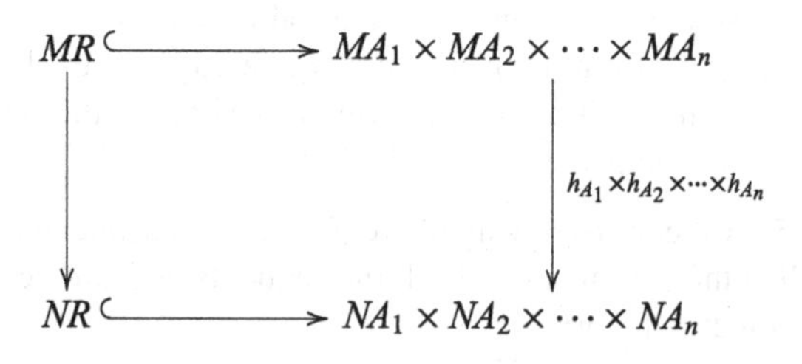

Given two topoi τ_1 and τ_2 with a left exact functor $F : \tau_1 \to \tau_2$ between them, this induces a functor Σ-$\mathbf{Str}(F) : \Sigma$-$\mathbf{Str}(\tau_1) \to \Sigma$-$\mathbf{Str}(\tau_2)$. Moreover, any natural transformation $\alpha : F \to G$ between functor $F, G : \tau_1 \to \tau_2$ induces a natural transformation Σ-$\mathbf{Str}(\tau_1) \to \Sigma$-$\mathbf{Str}(\tau_2)$. This implies that the construction Σ-$\mathbf{Str}(-)$ is 2-functorial.

Now we need to represent both terms and formulas in the topos τ.

Definition 10.3.2 Given a Σ-structure M in τ, a term-in-context $\vec{x}.t$ over Σ (where $x_i : A_i$, $i = 1, \ldots, n$) is represented in τ by the morphisms:

$$[[\vec{x}.t]]_M : MA_1 \times MA_2 \times \cdots \times MA_n \to MB.$$

This is defined recursively as follows:

1. If t is a variable it will necessarily be of type x_i for some unique $i \leq n$, hence $[[\vec{x}.t]]_M = pr_i$.
2. If t is a function symbol $f(t_1, t_2, \ldots, t_m)$ for $t_i : C_i$ then $[[\vec{x}.t]]_M$ becomes

$$MA_1 \times MA_2 \times \cdots \times MA_n \xrightarrow{[[\vec{x}.t_1]]_M \times [[\vec{x}.t_2]]_M \times \cdots \times [[\vec{x}.t_n]]_M} MB_1 \times MB_2 \times \cdots \times MB_n \xrightarrow{Mf} MB.$$

The naturality condition for morphisms $h : M \to N$ of Σ-structure we encountered for function symbols also extends to terms. In particular, given a term-in-context $\vec{x}.t$ over Σ, such that $x_i : A_i$ and $t : B$, then the following diagram commutes:

$$\begin{array}{ccc} MA_1 \times MA_2 \times \cdots \times MA_n & \xrightarrow{[[\vec{x}.t]]_M} & MB \\ \downarrow{\scriptstyle h_{A_1} \times h_{A_2} \times \cdots \times h_{A_n}} & & \downarrow{\scriptstyle h_B} \\ NA_1 \times NA_2 \times \cdots \times NA_n & \xrightarrow{[[\vec{x}.t]]_N} & NB \end{array}$$

On the other hand, formulas are represented recursively through the following definition:

Definition 10.3.3 Given a Σ-structure M in τ, a formula in context $\vec{x}.\phi$ over Σ (where $x_i : A_i$, $i = 1, \ldots, n$) is represented in τ by a sub-object

$$[[\vec{x}.\phi]] \hookrightarrow MA_1 \times MA_2 \times \cdots \times MA_n$$

such that:

1. If ϕ is a relation $R(t_1, \ldots, t_n)$ of type $B_1, \ldots, B_n$, then $[[\vec{x}.\phi]]$ is represented by the pullback

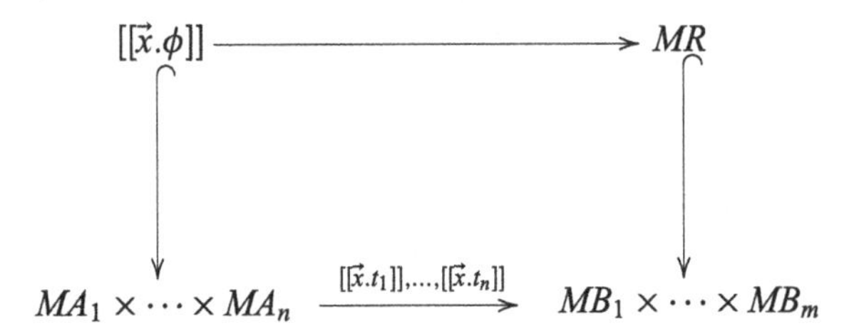

2. If ϕ is of the form $s = t$ where $s, t : B$, then $[[\vec{x}.\phi]]$ is represented by the equaliser

$$MA_1 \times \ldots MA_n \underset{[[\vec{x}.t]]}{\overset{[[\vec{x}.s]]}{\rightrightarrows}} MB.$$

3. If ϕ is $\top$, then $[[\vec{x}.\phi]]$ is the top element of $Sub(MA_1 \times \ldots MA_n)$.
4. If ϕ is of the form $\alpha \wedge \beta$, then $[[\vec{x}.\phi]]$ is represented by the pullback

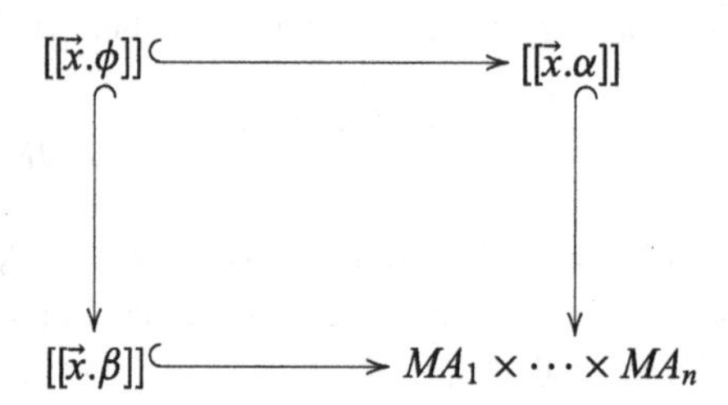

5. If ϕ is $\perp$, then $[[\vec{x}.\phi]]$ is the bottom element of $Sub(MA_1 \times \cdots \times MA_n)$.
6. If ϕ is of the form $\alpha \wedge \beta$, then $[[\vec{x}.\phi]]$ is represented by the pushout

$$\begin{array}{ccc} [[\vec{x}.\beta]] \cap [[\vec{x}.\alpha]] & \hookrightarrow & [[\vec{x}.\alpha]] \\ \downarrow & & \downarrow \\ [[\vec{x}.\beta]] & \longrightarrow & [[\vec{x}.\phi]] \end{array}$$

7. If ϕ is of the form $(\exists y)\psi$ where $y : B$, then $[[\vec{x}.\phi]]$ is represented by the image of the composite map

$$[[\vec{x}, y.\phi]] \hookrightarrow MA_1 \times \cdots \times MA_n \times MB \xrightarrow{\pi} MA_1 \times \cdots \times MA_n.$$

8. If ϕ is of the form $\bigwedge_{i \in I} \psi_i$, then $[[\vec{x}.\phi]]$ is represented by the union of all the $[[\vec{x}.\psi]]$ in $Sub(MA_1 \times \cdots \times MA_n)$.

Now that we have understood how to represent formulas and terms in a given topos τ, the next step is to understand how axioms are represented in τ and what it means for an axiom to be valid in a given structure.

Definition 10.3.4 Given a Σ-structures M in τ, a sequent $\sigma = \phi \vdash_{\vec{x}} \psi$ over Σ is satisfied in M if $[[\vec{x}.\phi]] \subseteq [[\vec{x}.\psi]]$ as elements of $Sub(MA_1 \times \cdots \times MA_n)$.

We have seen above that a (geometric) theory $\mathbb{T}$ over a structure Σ is comprised by a set of sequents over Σ. We then say that a structure M is a model for a given theory $\mathbb{T}$, if every axiom in $\mathbb{T}$ is satisfied in M. Clearly each theory $\mathbb{T}$ can have more than one model. The collection of all such models forms a full sub-category $\mathbb{T}$-**Mod**(τ) of Σ-**Str**(τ).

We have seen above, that any left exact functor $F : \tau_1 \to \tau_2$ induces a functor between the respective Σ structure as follows: $\Sigma\text{-}\mathbf{Str}(F) : \Sigma\text{-}\mathbf{Str}(\tau_1) \to \Sigma\text{-}\mathbf{Str}(\tau_2)$. However, such a functor does not restrict to a functor $\mathbb{T}\text{-}\mathbf{Mod}(\tau_1) \to \mathbb{T}\text{-}\mathbf{Mod}(\tau_2)$, but when considering geometric morphisms $f : \tau_1 \to \tau_2$ we have the following result:

Theorem 10.3.1 *Given a geometric morphism $f : \tau_1 \to \tau_2$ and a Σ-structure M in τ_1, such that f^*M is the induced structure in τ_2, then for any geometric formula $\vec{x}.\phi$ over Σ (where $x_i : A_i$) we have that*

$$f^*([[\vec{x}.\phi]]_M) = [[\vec{x}.\phi]]_{f^*M}$$

as sub-objects of $f^*(MA_1 \times \cdots \times MA_n) \simeq f^*MA_1 \times \cdots \times f^*MA_n$.

Proof The proof is by induction on the construction of each formulas as defined in Definition 10.3.3. In particular, since f^* is the inverse image of a geometric morphism it preserves small colimits and finite limits, hence we obtain:

1. when ϕ is a relation $R(t_1, \ldots, t_n)$ the pullback representing $[[\vec{x}.\phi]]_M$ in τ_1 gets mapped via f^* to a pullback in τ_2 representing the formula $[[\vec{x}.\phi]]_{f^*M}$.
2. When $\phi := (s = t)$, the equaliser representing $[[\vec{x}.\phi]]_M$ in τ_1 is preserved via the action of f^*.
3. Both the top and bottom elements of $Sub(MA_1 \times \cdots \times MA_n)$ are preserved under the action of f^*.
4. Since f^* preserves both meets and joins we have that $f^*([[\vec{x}.\alpha]]_M \wedge [[\vec{x}.\beta]]_M) = [[\vec{x}.\alpha]]_{f^*M} \wedge [[\vec{x}.\beta]]_{f^*M}$ and $f^*([[\vec{x}.\alpha]]_M \vee [[\vec{x}.\beta]]_M) = [[\vec{x}.\alpha]]_{f^*M} \vee [[\vec{x}.\beta]]_{f^*M}$. Similar reasoning holds when ϕ is of the form $\bigwedge_{i\in I} \psi_i$.
5. Since f^* preserves images, then $f^*[[\vec{x}.(\exists y)\psi]]_M = [[\vec{x}.(\exists y)\psi]]_{f^*M}$.

□

A consequence of the above theorem is the following:

Theorem 10.3.2 *Given a geometric theory $\mathbb{T}$, any geometric morphism $f : \tau_1 \to \tau_2$ induces a functor $f^* : \mathbb{T}\text{-}\mathbf{Mod}(\tau_2) \to \mathbb{T}\text{-}\mathbf{Mod}(\tau_1)$.*

Proof Assume that M is a model for $\mathbb{T}$ in τ_2, hence all axioms $(\phi \vdash_{\vec{x}} \psi)$ of $\mathbb{T}$ are valid in M. The validity of such axioms is represented by the relation $[[\vec{x}.\phi]]_M \subseteq [[\vec{x}.\psi]]_M$ in $Sub(MA_1 \times \cdots \times MA_2)$ (where $x_i : A_i$). Since f^* preserves inclusion of sub-objects we obtain that $f^*([[\vec{x}.\phi]]_M) \subseteq f^*([[\vec{x}.\psi]]_M)$, which, by applying the Theorem 10.3.1, is equivalent to $[[\vec{x}.\phi]]_{f^*M} \subseteq [[\vec{x}.\psi]]_{f^*M}$. □

The above theorem allows one to determine which Σ-structures are $\mathbb{T}$-models for a topos τ_1 by determining which Σ-structure are $\mathbb{T}$-models for some other topos τ_2, provided we have a geometric morphism $f : \tau_1 \to \tau_2$. This fact is of considerable importance in situations in which the topos τ_1 might be particularly intractable whereas τ_2 might not. In the case for which $\tau_1 = [\mathcal{C}, \mathbf{Sets}]$ and $\tau_2 = \mathbf{Sets}$ we have the following Lemma:

Lemma 10.3.1 ([50], D1.2.14) *Given a geometric theory* $\mathbb{T}$ *over a signature* Σ *and a small category* $\mathcal{C}$, *then a* Σ*-structure* M *in* $[\mathcal{C}, \mathbf{Sets}]$ *is a* $\mathbb{T}$*-model iff for each* $C \in \mathcal{C}$, $ev_C(M)$ *is a* $\mathbb{T}$*-model in* **Sets**. *Here* $ev_C : [\mathcal{C}, \mathbf{Sets}] \to \mathbf{Sets}$ *represents the evaluation functor for the object C, i.e. it assigns the set associated to C. Furthermore we have the following isomorphism of categories:*

$$\mathbb{T}\text{-}\mathbf{Mod}([\mathcal{C}, \mathbf{Sets}]) \simeq [\mathcal{C}, \mathbb{T}\text{-}\mathbf{Mod}(\mathbf{Sets})].$$

Proof We recall that a point of a topos τ is identified with a geometric morphism $\mathbf{Sets} \to \tau$. Hence, given the topos $\tau = [\mathcal{C}, \mathbf{Sets}]$ and an object $C \in \mathcal{C}$, a point is defined by a geometric morphism $ev_C : \mathbf{Sets} \to [\mathcal{C}, \mathbf{Sets}]$. The inverse image part of such geometric morphism is precisely the evaluation functor. By Theorem 10.3.2 it follows that any $\mathbb{T}$-model on $[\mathcal{C}, \mathbf{Sets}]$ induces a $\mathbb{T}$-model on **Sets**. The "if" part follows trivially from the definition of $\mathbb{T}$-models in $[\mathcal{C}, \mathbf{Sets}]$.

The isomorphism of categories $\mathbb{T}\text{-}\mathbf{Mod}([\mathcal{C}, \mathbf{Sets}]) \simeq [\mathcal{C}, \mathbb{T}\text{-}\mathbf{Mod}(\mathbf{Sets})]$ is given by the fact that homomorphisms of Σ-structure on $[\mathcal{C}, \mathbf{Sets}]$ are equivalent to natural transformations between functor $F : \mathcal{C} \to \Sigma\text{-}\mathbf{Str}(\mathbf{Sets})$. □

Each geometric theory $\mathbb{T}$ has associated to it a *Lindenbaum* algebra $\mathcal{O}([\mathbb{T}])$ which is defined as the poset of formulae of $\mathbb{T}$-modulo provable equivalence ordered by provable entailment. Provable equivalence means that two formulae ϕ and ψ are equivalent ($\psi \sim \phi$) if the theory $\mathbb{T}$ proves that each proposition implies the other. For geometric theories, $\mathcal{O}([\mathbb{T}])$ turns out to be a frame.

As stated above, a structure M is a model for a given theory $\mathbb{T}$, if every axiom in $\mathbb{T}$ is satisfied in M, therefore a model for a theory $\mathbb{T}$ is equivalent to a consistent assignment of either true or false to each of the axioms/formulae in $\mathbb{T}$. Given that the collection of formulae forms a frame $\mathcal{O}([\mathbb{T}])$, a model for $\mathbb{T}$ can be seen as a frame map $\mathcal{O}([\mathbb{T}]) \to \{0, 1\}$ and hence a locale map $* \to [\mathbb{T}]$, i.e. a point of the locale $\mathbb{T}$. This can be generalised to any locale Y, and one can consider a model of $\mathbb{T}$ in a frame $\mathcal{O}(Y)$ to be a locale map $Y \to [\mathbb{T}]$.

We will now give an example of how to construct objects in a topos τ using a geometric theory $\mathbb{T}$. The object we choose to describe is the complex number object. In particular, the geometric theory of complex numbers in τ is obtained by assigning to each pair (r, s) of rational complex numbers in τ an atomic formulae

$$z \in (r, s),$$

which intuitively represents the statement that the complex number z lies in the complex rational open rectangle spanned by s and r. All the other formulae of the theory are obtained by taking arbitrary disjunctions of finite conjunctions, etc. of the above formulae. The sequence (or axioms) of this theory are given by:

1. $z \in (r, s) \vdash \bot$ if $(r, z) < 0$.
2. $\top \vdash \bigvee_{(r,s)} z \in (r, s)$.

3. $z \in (r, s) \vdash z \in (p, q) \vee z \in (p', q')$ if $(r, s) \lhd (p, q) \vee (p'q')$.[4]
4. $z \in (p, q) \wedge z \in (p'q') \vdash z \in (r, s)$ if $(p, q) \wedge (p', q') \lhd (r, s)$.
5. $z \in (r, s) \vdash \bigvee_{(r', s') \lhd (r, s)} z \in (r', s')$.
6. $\bigvee_{(r', s') \lhd (r, s)} z \in (r', s') \vdash z \in (r, s)$.

Given the above geometric theory $\mathbb{T}$, the complex number object $\mathbb{C}$ in the topos τ is identified with the locale given by the Lindenbaum algebra associated to $\mathbb{T}$. Therefore, the objects in $\mathbb{C}$ are given by the formulae of the theory modulo provable equivalence in the theory, partially ordered by provable entailment in the theory.

Clearly $\mathbb{C}$ can also be defined in terms of frames and generators, in particular we consider the atomic formulae to be the generators while the relations are given by the above axioms upon replacing $\vdash$ by $\leq$.

[4]The symbol $\lhd$ indicates the *rather below* relation in a locale Y. in particular, we say that $v \lhd u$ if there exists an element $w \in Y$ such that $v \wedge w = 0$ and $u \vee w = 1$.

Chapter 11
Brief Introduction to Covariant Topos Quantum Theory

In this chapter we will describe a different way in which topos theory was utilised to describe quantum theory. This approach is called *covariant topos quantum theory* and it was first put forward in [42]. The aim of this approach is to combine, on the one hand, algebraic quantum theory by describing a system via a C^*-algebra $\mathcal{A}$ and, on the other, Bohr's idea of classical snapshots which enables one to talk about physical quantities, only with respect to a suitable context of compatible physical quantities. Such contexts are defined in terms of commuting subalgebras of $\mathcal{A}$ which form a poset $C(\mathcal{A})$ ordered by inclusion. The topos utilised in this approach is the topos of covariant functors $C(\mathcal{A}) \rightarrow \mathbf{Sets}$ which we will denote by $[C(\mathcal{A}), \mathbf{Sets}]$. The main mathematical object in this topos is the covariant functor[1]

$$\overline{\mathcal{A}} : C(\mathcal{A}) \rightarrow \mathbf{Sets}; \quad \overline{\mathcal{A}}(C) = C,$$

such that, for each map $D \hookrightarrow C$ the corresponding covariant presheaf map is given by inclusion. This covariant functor $\overline{\mathcal{A}}$ is an internal commutative unital C^*-algebra which is called the *Bohrification* of A. Given the internal version of Gelfand duality, which is valid in any (Grothendieck) topos, there exists a covariant functor $\overline{\Sigma}_{\overline{\mathcal{A}}}$ in $[\mathcal{C}_{\mathcal{A}}, \mathbf{Sets}]$, which represents the Gelfand spectrum of $\overline{\mathcal{A}}$ and $\overline{\mathcal{A}}$ is isomorphic to the algebra of continuous complex values functions on $\overline{\Sigma}_{\overline{\mathcal{A}}}$. In this setting $\underline{\Sigma}_{\underline{A}}$ is a compact completely regular local, rather than a compact Hausdorff space and it represents the state-space of the theory as seen internally in the topos $[C(\mathcal{A}), \mathbf{Sets}]$. Given such a state-space, a state is represented, internally, as a probability valuation $\underline{\mu} : \mathcal{O}(\overline{\Sigma}_{\overline{\mathcal{A}}}) \rightarrow \overline{[0,1]}_l$ where $\overline{[0,1]}_l$ represents the covariant functor of lower reals in $[C(\mathcal{A}), \mathbf{Sets}]$. Self-adjoints operators, instead, are represented internally by locale maps $\overline{\Sigma}_{\overline{\mathcal{A}}} \rightarrow \overline{\mathbb{IR}}$, where $\overline{\mathbb{IR}}$ is the interval domain in $[C(\mathcal{A}), \mathbf{Sets}]$.

[1] Note that we have denoted covariant functors by an overline $\overline{X}$ to distinguish them from contravariant functors which we denote by an underline $\underline{X}$.

C. Flori, *A Second Course in Topos Quantum Theory*,
Lecture Notes in Physics 944, https://doi.org/10.1007/978-3-319-71108-9_11

There are many similarities between this approach and the topos approach delineated in [26]. The reader is referred to [75] for an in depth analysis. The main difference, however, is the fact that in the covariant approach one defines all physical quantities internally to the topos $[\mathcal{C}(\mathcal{A}), \mathbf{Sets}]$ and any reasoning is done utilising the internal language of $[\mathcal{C}(\mathcal{A}), \mathbf{Sets}]$, which is the Mitchell-Benabou language with the Kripke-Joyal semantics [50]. In what follows we will try and describe the main building blocks of the covariant topos quantum theory for a more detailed exposition the reader should refer to [42, 43].

11.1 Internal Topos Quantum Theory

In the covariant approach to topos quantum theory the authors start with a non-commutative C^*-algebra $\mathcal{A}$ and, then, construct a topos associated to each such algebra $\mathcal{A}$ as follows:

Theorem 11.1.1 *The map* $C : \mathbf{cStar} \rightarrow \mathbf{Poset}$ *is defined*

1. *on objects:* $\mathcal{A} \mapsto \{\mathcal{C} \subseteq \mathcal{A} | \mathcal{C} \in \mathbf{cCStar}\} =: \mathcal{C}(\mathcal{A})$ *which are ordered by inclusion.*
2. *On morphisms:* $f : \mathcal{A} \rightarrow \mathcal{B}$ *gets mapped to* $\mathcal{C}(f) : \mathcal{C}(\mathcal{A}) \rightarrow \mathcal{C}(\mathcal{B})$ *such that* $\mathcal{C} \subseteq \mathcal{A}$ *gets mapped to the direct image* $f(\mathcal{C}) \subseteq \mathcal{B}$.

Here the **Poset** *denotes the category of partially ordered sets and monotone functions.*

The fact that this is indeed a functor follows trivially from the definition.
To each C^*-algebra $\mathcal{A}$, one associates the poset of commuting subalgebras $\mathcal{C}(\mathcal{A})$ ordered by inclusion. Then one considers the collection of functor $\mathcal{C}(\mathcal{A}) \rightarrow \mathbf{Sets}$ for each such poset of commuting subalgebras. These collection forms a topos $\mathbf{Sets}^{\mathcal{C}(\mathcal{A})=[\mathcal{C}(\mathcal{A}),\mathbf{Sets}]}$ and it is the topos associated to $\mathcal{A}$. Hence, by describing a quantum system in terms of a particular C^*-algebra one can associate to it its corresponding topos $\mathbf{Sets}^{\mathcal{C}(\mathcal{A})}$. The association of a topos to a quantum system represented by a C^*-algebra is functorial as it is shown by the following theorem:

Theorem 11.1.2 *The map*

$$\tau : \mathbf{CStar} \rightarrow \mathbf{Topos}$$
$$\mathcal{A} \mapsto \mathbf{Sets}^{\mathcal{C}(\mathcal{A})}$$

is a functor. Here **Topos** *is the category whose objects are topos and whose morphisms are geometric morphisms.*

Proof As a first step we need to define the action on morphisms. In particular, given a morphism $f : \mathcal{A} \rightarrow \mathcal{B}$, we know from Theorem 11.1.1 that there is a map $\mathcal{C}(f) : \mathcal{C}(\mathcal{A}) \rightarrow \mathcal{C}(\mathcal{B})$ in **Topos**. This is a map between categories and, as such, it induces a geometric morphism $f : \mathbf{Sets}^{\mathcal{C}(\mathcal{A})} \rightarrow \mathbf{Sets}^{\mathcal{C}(\mathcal{B})}$ with direct image part

$f_* : \mathbf{Sets}^{\mathcal{C}(\mathcal{A})} \to \mathbf{Sets}^{\mathcal{C}(\mathcal{B})}$ and inverse image part $f^* : \mathbf{Sets}^{\mathcal{C}(\mathcal{A})} \to \mathbf{Sets}^{\mathcal{C}(\mathcal{B})}$. The latter preserves finite limits.

We have shown that, for each morphism $f : \mathcal{A} \to \mathcal{B}$, τ assigns the geometric morphism $f : \mathbf{Sets}^{\mathcal{C}(\mathcal{A})} \to \mathbf{Sets}^{\mathcal{C}(\mathcal{B})}$, Clearly such an assignment is functorial. □

We have just defined the ambient topos for a particular quantum system represented by an algebra $\mathcal{A}$. The next step is to define all the quantum object internally to $\mathbf{Sets}^{\mathcal{C}(\mathcal{A})}$. The first object we will define is the internal C^*-algebra $\overline{\mathcal{A}}$.

Definition 11.1.1 The co-presheaf $\overline{\mathcal{A}}$ is defined on

1. objects: for each $C \in \mathcal{C}(\mathcal{A})$, then $\overline{\mathcal{A}}(C) = C$.
2. Morphisms: given a map $i_{C'C} : C' \subseteq C$ the corresponding co-presheaf map $\overline{\mathcal{A}}(i_{C'C}) : \overline{\mathcal{A}}(C') \to \overline{\mathcal{A}}(C)$ is simply the inclusion map.

In [42, 43] it was show that this functor is an internal commutative C^*-algebra for $\mathbf{Sets}^{\mathcal{C}(\mathcal{A})}$.

Theorem 11.1.3 *The functor $\overline{\mathcal{A}}$ is a vector space over the complex numbers $\overline{\mathbb{Q}}[i]$ with respect to the following operations:*

$$0 : \overline{1} \to \overline{\mathcal{A}}$$
$$* \mapsto 0;$$

$$+ : \overline{\mathcal{A}} \times \overline{\mathcal{A}} \to \overline{\mathcal{A}}$$
$$< a, b > \mapsto a + b;$$

$$\cdot : \overline{\mathbb{Q}}[i] \times \overline{\mathcal{A}} \to \overline{\mathcal{A}}$$
$$< z, a > \mapsto z \cdot a.$$

The algebra operation and involution operation are given respectively by

$$\cdot : \overline{\mathcal{A}} \times \overline{\mathcal{A}} \to \overline{\mathcal{A}}$$
$$< a, b > \mapsto a \cdot b$$

and

$$(-)^* : \overline{\mathcal{A}} \times \overline{\mathcal{A}} \to \overline{\mathcal{A}}$$
$$a \mapsto a^*.$$

The norm relation is instead given by

$$\overline{N} : \overline{\mathcal{A}} \times \overline{\mathbb{Q}}^+ \to \overline{\Omega} \quad \overline{N}(a, q) \text{ iff } ||a|| < q,$$

where the sub-object classifier $\overline{\Omega}$ *in* $\mathbf{Sets}^{\mathcal{C}(\mathcal{A})}$ *is given by the functor* $\overline{\Omega} : \mathcal{C}(\mathcal{A}) \to$ **Sets** *such that* $\overline{\Omega}(C) = \{S \subseteq \uparrow C | S \text{ is an upper set}\}$.

The proof of the fact that $\overline{\mathcal{A}}$ is a commutative C^*-algebra rests on the fact that, for each context $C \in \mathcal{C}(\mathcal{A})$, $\overline{\mathcal{A}}(C) = C$ is itself a commutative C^*-algebra. We have, therefore, a collection of 'local' commutative C^*-algebras. In particular, since the definition of commutative pre-semi-C^*-algebra consists only of geometric formulae, we can apply Lemma 10.3.1 which implies that $\overline{\mathcal{A}}$ is a pre-semi-C^*-algebra over $\overline{\mathbb{Q}}[i]$, since each $\overline{\mathcal{A}}(C)$ $(C \in \mathcal{C})$ is a pre-semi-C^*-algebra over $\mathbb{Q}[i]$. In order to show completeness and that the semi-norm is actually a norm, one uses the internal sheaf semantics. For a detailed proof the reader should refer to [42, 74]. This completes the proof that $\overline{\mathcal{A}}$ is an internal commutative C^*-algebra. One can generalise the above theorem as follows [74]:

Theorem 11.1.4 *An object* $\overline{\mathcal{A}}$ *is a* C^**-algebra internal to* $[\mathcal{C}, \mathbf{Sets}]$ *iff it is defined via a functor* $\overline{\mathcal{A}} : \mathcal{C} \to$ **C*alg** *where* **C*alg** *is the category of* C^**-algebras and *-homomorphisms in* **Sets**. $\overline{\mathcal{A}}$ *is commutative iff each* $\overline{\mathcal{A}}(C)$ *is commutative and unital iff each* $\overline{\mathcal{A}}(C)$ *is unital and for each* $f : C \to C'$ *in* $\mathcal{C}$ *the corresponding *-homomorphism* $\overline{\mathcal{A}}(f) : \overline{C} \to \overline{C'}$ *preserves the unit.*

Summarising what has been done so far: one starts with a non-commutative C^*-algebra A representing some quantum system. This algebra gets then internalised in the topos $[\mathcal{C}(\mathcal{A}), \mathbf{Sets}]$ obtaining the commutative C^*-algebra $\overline{\mathcal{A}}$. By internalising the algebra one goes from a non-commutative algebra to a commutative one and the multiplication of two non-commutative operators is no longer defined, since they belong to different commutative subalgebras, i.e. different contexts.

11.2 State-Space

Now that we have defined the internal C^*-algebra $\overline{\mathcal{A}}$ we would like to define its spectrum. To this end we apply the internal Gelfand duality mentioned in [5], which associates to each internal commutative C^*-algebra $\overline{\mathcal{A}}$ a completely regular compact locale $\overline{\Sigma}_{\overline{\mathcal{A}}}$, which represents its spectrum.

As shown in [5], in order to explicitly construct the state-space $\overline{\Sigma}_{\overline{\mathcal{A}}}$, one has to construct the propositional geometric theory $\mathbb{T}$ of multiplicative linear functionals on $\overline{\mathcal{A}}$, obtained by adapting that of linear functionals of norm ≤ 1 on the seminormed space $\overline{\mathcal{A}}$. In particular, $\mathbb{T}$ is constructed as follows: for each $a \in \overline{\mathcal{A}}$ and each open rectangle (r, s) in $\mathbb{C}$ we construct the basic formulae

$$a \in (r, s).$$

Any other formulae is obtained by application of the logical connective to the basic formulae. The set of sequents/axioms of the theory are:

1. $\top \vdash 0 \in (r, s)$ if $0 \in (r, s)$ and $0 \in (r, s) \vdash \bot$ otherwise.

2. $a \in (r, a) \vdash ta \in (tr, ts)$ for any number $t > 0$ and $a \in (r, a) \vdash ia \in i(r, s)$ where i represents the imaginary unit.
3. $a \in (r, s) \wedge a' \in (r', s') \vdash a + a' \in (r + r', s + s')$.
4. $\top \vdash a \in D(1)$ if $a \in D(1)$ where $D(1)$ represents the open disc of radius 1 centred at the origin in the complex plane.
5. $a \in (r, s) \vdash a \in (p.q) \vee a \in (p', q')$ if $(r, s) \lhd (p, q) \vee (p', q')$.
6. $a \in (r, s) \vdash \bigvee_{(r', s') \lhd (r, s)} a \in (r', s')$.
7. $\bigvee_{(r', s') \lhd (r, s)} a \in (r', s') \vdash a \in (r, s)$.

Given the theory $\mathbb{T}$ defined above, one can construct the associated locale obtained via the Lindenbaum algebra, i.e the locale whose objects are all formulae of $\mathbb{T}$ modulo provable equivalence and ordered by provable entailment. We will denote such a locale by $\underline{\Sigma}_{\underline{\mathcal{A}}}$ in anticipation of the identification of such a locale with the Gelfand spectrum of $\underline{\mathcal{A}}$.

The axioms of the above theory imply that any representation of the theory, hence any point of the locale $\underline{\Sigma}_{\underline{\mathcal{A}}}$, represents a linear functional of norm ≤ 1 on the semi-normed space $\underline{\mathcal{A}}$. To also account for the linear functional to be multiplicative we need to add a few extra axioms, namely:

1. $\top \vdash 1 \in (r, s)$ if $1 \in (r, s)$ and $1 \in (r, s) \vdash \bot$ otherwise.
2. $a \in (r, a) \vdash a^* \overline{(r, s)}$.
3. $aa' \in (r, s) \vdash \bigvee_i a \in (p_i, q_i) \wedge a' \in (p'_i, q'_i)$ if $\bigvee_i (p_i, q_i) \times (p'_i, q'_i) = \mu^*(r, s)$ where $\mu : \mathbb{C} \times \mathbb{C} \to \mathbb{C}$ denotes multiplication in the locale of complex numbers, and $\mu^*(r, s)$ is the inverse image of the open rectangle.

The locale obtained by adding the above axioms of the theory is the spectrum of the commutative C^*-algebra $\underline{\mathcal{A}}$, which we again denote by $\underline{\Sigma}_{\underline{\mathcal{A}}}$. By construction the points of $\underline{\Sigma}_{\underline{\mathcal{A}}}$ (models of the theory) are the multiplicative linear functionals of $\underline{\mathcal{A}}$.

In [5] it was shown that:

Theorem 11.2.1 *Given any commutative C^*-algebra A in a Grothendieck topos τ, the spectrum $\underline{\Sigma}_A$ is a compact, completely regular locale.*

Now that we have defined the spectrum of the internal C^*-algebra $\underline{\mathcal{A}}$ we can define the Gelfand transform. Here we will only briefly state what the Gelfand transform is. For a detailed derivation the reader should refer to [5, 42].

Given an internal C^*-algebra $\underline{\mathcal{A}}$ we consider the self-adjoint part, $\underline{\mathcal{A}}_{sa}$ whose elements are the self-adjoint elements in $\underline{\mathcal{A}}$. Then the Gelfand transform associates, to each $a \in \underline{\mathcal{A}}_{sa}$, the locale map

$$\hat{a} : \underline{\Sigma}_{\underline{\mathcal{A}}} \to \underline{\mathbb{R}}$$

where $\underline{\mathbb{R}}$ is the locale of internal Dedekind reals (see Appendix A.1). The associated frame map is

$$\hat{a}^{-1} : \mathcal{O}(\underline{\mathbb{R}}) \to \mathcal{O}(\underline{\Sigma}_{\underline{\mathcal{A}}}).$$

This definition of the Gelfand transform will be used to prove the following important theorem [42]:

Theorem 11.2.2 *The Kochen-Specker theorem is equivalent to the statement that the locale $\overline{\Sigma}_{\overline{\mathcal{A}}}$ has no points.*

Proof Given a point $p : \overline{1} \to \overline{\Sigma}_{\overline{\mathcal{A}}}$ (see Definition 8.1.2) of the locale $\overline{\Sigma}_{\overline{\mathcal{A}}}$ and the Gelfand transform $\hat{a} : \overline{\Sigma}_{\overline{\mathcal{A}}} \to \overline{\mathbb{R}}$ associated to $a \in \overline{\mathcal{A}}_{sa}$, one can construct the point $\hat{a} \circ p : \overline{1} \to \overline{\mathbb{R}}$ of the locale $\overline{\mathbb{R}}$. Since this can be done for each $a \in \overline{\mathcal{A}}_{sa}$, effectively we obtain a natural transformation $V_p : \overline{\mathcal{A}}_{sa} \to Pt(\overline{\mathbb{R}})$, which is an internal multiplicative functional. For each $C \in \mathcal{C}$ the components of V_p are $V_p(C) : \overline{\mathcal{A}}_{sa}(C) \to Pt(\overline{\mathbb{R}})(C)$ which are equivalent to $V_p(C) : C_{sa} \to \mathbb{R}$. From the naturality condition one obtains that for $C' \subseteq C$, $V_p(C)|_{C'} = V_p(C')$. Since each $V_p(C)$ are multiplicative functionals, the condition $V_p(C)|_{C'} = V_p(C')$ turns V_p into a valuation on $\mathcal{B}(\mathcal{H})$ which can not exist because of the Kochen-Specker theorem. □

Note that a similar theorem holds in the contravariant approach to topos quantum theory [26].

11.2.1 Relation Between Contravariant and Covariant State-Space

In [75], the author analysed the possible relation between the covariant state-space $\overline{\Sigma}_{\overline{\mathcal{A}}}$ and the contravariant state-space $\underline{\Sigma}$ [26, Def. 9.7]. This was done by internalising $\underline{\Sigma}$ and analysing the properties of the resulting internal local. As it turns out that such a locale is not regular, hence it cannot be identified with the spectrum of an internal C^*-algebra, however, the implications of such a result still need to be analysed in details.

In the following we will explain how the state-space $\underline{\Sigma}$ can be internalised and show that the resulting internal locale is not regular [75]. To this end we recall that the contravariant state-space was identified with the spectral presheaf $\underline{\Sigma}$ on $\mathcal{V}(\mathcal{H})$ (see Definition 9.7 in [26]), where $\mathcal{V}(\mathcal{A})$ represents the category of abelian von Neumann subalgebras of the von Neumann algebra $\mathcal{A}$ associated to some quantum system. Without loss of generality we can safely consider $\mathcal{A}$ to be a C^*-algebra associated to a quantum system and $\mathcal{C}(\mathcal{A})$ to be the category of unital abelian subalgebras of $\mathcal{A}$. Given the downwards Alexandroff topology[2] on $\mathcal{V}(A)$ (equivalently on $\mathcal{C}(\mathcal{A})$), $\underline{\Sigma}$ becomes a sheaf on $\mathcal{V}(\mathcal{A})$ whose corresponding etalé bundle has as bundle space

$$\Sigma = \coprod_{C \in \mathcal{V}(\mathcal{A})} \underline{\Sigma}_C$$

[2]We recall that the downwards Alexandroff topology on $\mathcal{V}(\mathcal{A})$ is the topology for which a subset $U \subseteq \mathcal{V}(\mathcal{A})$ is open if it is a downwards closed set, i.e. $U = \{C' \subseteq C | C \in \mathcal{V}(\mathcal{A})\}$.

where $\underline{\Sigma}_C$ is the Gelfand spectrum associated to the algebra C. The locale homomorphism (or bundle map) is defined as follows:

$$\pi : \Sigma \to \mathcal{V}(\mathcal{A})$$
$$\lambda_C \to C.$$

The topology on Σ is generated by the basis

$$W = \{S_{C,\lambda} | C \in \mathcal{V}(\mathcal{A}), \lambda \in \underline{\Sigma}_C\}, \quad S_C = \{\lambda|_D | D \subseteq C\}.$$

Given the fact that both Σ and $\mathcal{V}(\mathcal{A})$ can be seen as internal locales in **Sets**, we can identify the map $\pi : \Sigma \to \mathcal{V}(\mathcal{A})$ as a map of locales. Utilising the method for constructing internal locals explained in Sect. 9.3 the map $\pi : \Sigma \to \mathcal{V}(\mathcal{A})$ induces an internal locale in $Sh(\mathcal{V}(\mathcal{A})$, whose associate frame is given by [75]

$$\mathcal{O}\underline{\Sigma}_C = \mathcal{O}\underline{\Sigma}|_{\downarrow C}.$$

However, in order to better compare the internal locale associated to the spectral presheaf $\underline{\Sigma}$ to the locale $\overline{\Sigma}_{\overline{\mathcal{A}}}$, it is useful to equip the etalé space Σ with a topology generated by clopen subsets of $\underline{\Sigma}$. In particular, given the etalé bundle $\pi : \Sigma \to \mathcal{V}(\mathcal{A})$, a subset $U \in \Sigma$ is clopen if it is open in the etalé topology on Σ and, for each $C \in \mathcal{V}(\mathcal{A})$, the set $U_C := U \cap \underline{\Sigma}_C$ is clopen in $\underline{\Sigma}_C$, equipped with the spectral topology. When dealing with von Neumann algebras $C \in \mathcal{V}(\mathcal{A})$ each $\underline{\Sigma}_C$ has as basis clopen subsets. The collection of these etalé clopens form a basis for a topology. The resulting topological space is denoted [75] by $\Sigma_{\downarrow}$ and the rigorous definition is given by

Definition 11.2.1 The space $\Sigma_{\downarrow}$ is identified with the set $\Sigma = \coprod_{C \in \mathcal{V}(\mathcal{A})} \underline{\Sigma}_C$ where $U \in \Sigma_{\downarrow}$ is open if

1. If $\lambda \in U_C$ and $C' \subseteq C$, then $\lambda|_{C'} \in U_{C'}$.
2. $\forall C \in \mathcal{V}(\mathcal{A})$, U_C is open in $\underline{\Sigma}_C$.

With respect to the above topology the map $\pi : \Sigma_{\downarrow} \to \mathcal{V}(\mathcal{A})$ is no longer a homomorphism but it is continuous, therefore, it induces an internal locale in $[\mathcal{V}(\mathcal{A})^{\text{op}}, \mathbf{Sets}]$, which we will denote by $\underline{\Sigma}_{\downarrow}$. The following result shown in [75] is of particular importance:

Theorem 11.2.3 *Given a von Neumann algebra $\mathcal{A}$ such that $C(\mathcal{A}) \neq \{\mathbb{C} \cdot 1\}$ then the locale $\underline{\Sigma}_{\downarrow}$ in $[\mathcal{V}(\mathcal{A})^{\text{op}}, \mathbf{Sets}]$ is not regular.*

In order to prove the above theorem we need the following lemma [48]:

Lemma 11.2.1 *Consider two Locales, X, Y and a continuous map $f : X \to Y$. This gives rise to the geometric morphism $f : Sh(X) \to Sh(Y)$ whose associated direct image part is $f_* : Sh(X) \to Sh(Y)$. In this setting the locale $f_*(\Omega_X)$ is said to be regular iff, given any open $U \subseteq X$ and any element $x \in X$, there exists a* *neighbourhood N of f(x) in Y and open sets V, W in S such that*

$$x \in V, \quad V \cap W = \emptyset$$

$$f^{-1}(N) \subseteq U \cup W .$$

In this context a regular locale is defined as follows

Definition 11.2.2 Given a Locale L, this is said to be regular if every element $a \in L$ satisfies the following relations

$$a = \bigvee \{b \in L | b \,\overline{<}\, a\}.$$

The symbol $b \,\overline{<}\, a$ indicates that there exists a $c \in L$ such that $b \wedge c = 0$ and $a \vee c = 1$.

We are now able to prove Theorem 11.2.3.

Proof In what follows we will report the proof given in [75]. To this end consider the set

$$B_{C,U} = \{\lambda|_D | D \subseteq C, \lambda \in U\}$$

where $U \in \mathcal{O}\underline{\Sigma}_C$ and $C \in \mathcal{V}(\mathcal{A})$. Clearly $B_{C,U}$ satisfies condition 2 in Definition 11.2.1. To see that also condition 1 in Definition 11.2.1 is satisfied one has to recall that the presheaf maps $\underline{\Sigma}(i_{C'C}) : \underline{\Sigma}_C \to \underline{\Sigma}_{C'}$ $(C' \subseteq C)$, given by restriction, are open [26]. From Lemma 11.2.1, the locale $\underline{\Sigma}_{\downarrow}$ is regular iff for any $U \in \mathcal{O}\underline{\Sigma}_{\downarrow}$ and any $\lambda \in U$, there exists a pair of opens $V, W \in \mathcal{O}\underline{\Sigma}_{\downarrow}$ such that $\lambda \in V$, $V \cap W = 0$ and $B_{C,U} \subseteq U \cup W$. Now since we have assumed that $C(\mathcal{A}) \neq \{\mathbb{C} \cdot 1\}$, there exists some $C \in \mathcal{V}(\mathcal{A})$ such that $\underline{\Sigma}_C$ has at least two elements (Gelfand-Mazur theorem). Take two distinct elements $\lambda_1, \lambda_2 \in \underline{\Sigma}_C$, then we define the open $U := B_{C, \underline{\Sigma}_C \{\lambda_2\}}$ where clearly $\lambda_1 \in U$. If $\mathcal{O}\underline{\Sigma}_{\downarrow}$ were to be regular, then there would exist two opens $V, W \in \mathcal{O}\underline{\Sigma}_{\downarrow}$ such that $\lambda_1 \in V$, $\lambda_2 \in W$ and $V \cap W = 0$. Since such conditions have to be preserved by the presheaf maps, it follows that for each $C' \subseteq C$ then $\lambda_1|_{C'} \neq \lambda_2|_{C'}$. However, if we include the trivial subalgebra $\mathbb{C}$ as a context, it follows that the condition $\lambda_1|_{C'} \neq \lambda_2|_{C'}$ can not be satisfied, hence the locale $\underline{\Sigma}_{\downarrow}$ is not regular. □

The importance of the above theorem is given by the fact that if the locale $\underline{\Sigma}_{\downarrow}$ is not regular, then it can not be identified with the spectrum of an internal unital commutative C^* algebra. It does however satisfy the weaker property of being sober.

If we excluded the trivial context then in some case the locale $\underline{\Sigma}_{\downarrow}$ would be regular, for example for $A = M_2(\mathbb{C})$ the associated locale is regular, however for $n > 2$ the locale associated to $A = M_n(\mathbb{C})$ is not regular. In general it is safe to say that the locale $\underline{\Sigma}_{\downarrow}$ is not regular.

11.3 States

In this section we will describe how states are defined in the covariant approach. This was first introduced in [42]. The main result of this paper regarding states is given by Theorem 14 [42] in which it is shown that there exists an isomorphism between (quasi-)states on a C^*-algebra $\mathcal{A}$ and, either probability integrals on $\overline{\mathcal{A}}_{sa}$, or probability valuations on the Gelfand spectrum $\Sigma_{\overline{\mathcal{A}}}$. In what follows we will explain how this result is obtained. To this end, we first of all need to introduce the notion of a *state* and a *quasi-state*.

Definition 11.3.1 Given a unital C^*-algebra $\mathcal{A}$, a state on $\mathcal{A}$ is a linear functional $\rho : \mathcal{A} \to \mathbb{C}$ such that:

1. ρ is positive: $\rho(a^*a) \geq 0$ for all $a \in \mathcal{A}$.
2. $\rho(1) = 1$.

A state is said to be pure if it is not a mixture of two distinct other states. The collection of states on a C^*-algebra form a compact convex set where the extremal points represent pure states. The notion of a quasi-state is a relaxed version of the notion of a state, namely, the former is only linear on commuting operators. This weakened notion of a state suffices as far as quantum theory is concerned since only commuting observables are physically meaningful. The rigorous definition of a quasi-state is as follows:

Definition 11.3.2 Given a unital C^*-algebra $\mathcal{A}$, a functional $\rho : \mathcal{A} \to \mathbb{C}$ is called a *quasi-state* on $\mathcal{A}$ if the following conditions hold:

1. ρ is quasi-linear: it is linear on all commuting subalgebras and for all self-adjoint operators $\hat{A}, \hat{B}$ then $\rho(\hat{A} + i\hat{B}) = \rho(\hat{A}) + i\rho(\hat{B})$.
2. ρ is positive: $\rho(\hat{A}^*\hat{A}) \geq 0$.
3. $\rho(\hat{1}) = 1$.

The main result relating quasi-states on a unital C^*-algebra and states in the covariant approach to topos quantum theory is given by the following theorem:

Theorem 11.3.1 ([42]) *Given a unital C^*-algebra $\mathcal{A}$, quasi-sates on $\mathcal{A}$ are in bijective correspondence with both probability integrals on $\overline{\mathcal{A}}_{sa}$ and probability valuations on $\Sigma_{\overline{\mathcal{A}}}$.*

Before being able to prove this theorem we need to understand what probability integrals and probability valuations are in this context.

Definition 11.3.3 Given the self-adjoint part $\mathcal{A}_{sa}$ of a unital C^*-algebra $\mathcal{A}$, a *probability integral* on $\mathcal{A}_{sa}$ is a linear functional $I : \mathcal{A}_{sa} \to \mathbb{R}$ such that

i) if $\hat{A} \in \mathcal{A}_{sa}$ is such that $\hat{A} \geq \hat{0}$, then $I(\hat{A}) \geq 0$;
ii) $I(\hat{1}) = 1$.

An integral is *faithful* if its kernel is $\{0\}$, i.e. $\hat{A} \geq \hat{0}$ ($\hat{A} \in \mathcal{A}$) and $I(\hat{A}) = 0$ imply that $\hat{A} = \hat{0}$.

The collection of all probability integrals over $\mathcal{A}_{sa}$ forms a locale which we denote by $\mathcal{I}(\mathcal{A}_{sa})$.

Definition 11.3.4 Given a locale X and the set $[0, 1]_l$ of lower reals between 0 and 1, a *probability valuation* on X is a monotone map $\mu : \mathcal{O}(X) \to [0, 1]_l$ such that, given any $U, V \in \mathcal{O}(X)$ and a directed set $\{U_i\}_{i \in I}$, then the following conditions hold:

1. If $U \leq V$ then $\mu(U) \leq \mu(V)$, i.e. μ is monotone.
2. $\mu(\bot) = 0$ and $\mu(\top) = 1$.
3. $\mu(U) + \mu(V) = \mu(U + \wedge V) + \mu(U \vee V)$
4. $\mu(\bigvee_{i \in I} U_i) = \bigvee_{i \in I} \mu(U_i)$.

Essentially a probability valuation on a locale can be thought of as the same thing as a probability measure, but defined only on opens.

We are now ready to prove Theorem 11.3.1, [42].

Proof We will start by proving the correspondence between quasi-states on the algebra $\mathcal{A}$ and probability integrals on $\underline{\mathcal{A}}_{sa}$. In particular, consider a quasi-state ρ on $\mathcal{A}$. This gives rise to a natural transformation[3] $I : \underline{\mathcal{A}}_{sa} \to \underline{\mathbb{R}}$ as follows: for each $C \subseteq A$ we define the component

$$(I_\rho)_C := \rho|_{C_{sa}} : C_{sa} \to \mathbb{R}. \tag{11.3.1}$$

The collection of all such components can be easily seen to give rise to a natural transformation. Now consider an element $a\hat{A} \in \mathcal{A}_{sa}$ such that $\hat{A} \geq \hat{0}$. Given a subalgebra $C \subseteq \mathcal{A}$ such that $\hat{A} \in C$, from the definition (11.3.1) it follows that $(I_\rho)_C(\hat{A}) = \rho|_{C_{sa}}(\hat{A}) \geq 0$. Moreover $(I_\rho)_C(\hat{1}) = \rho|_{C_{sa}}(\hat{1}) = 1$ for all $C \subseteq \mathcal{A}$. Therefore, as defined above, I is indeed a probability integral on $\underline{\mathcal{A}}_{sa}$.

On the other hand, given a probability integral $I : \underline{\mathcal{A}}_{sa} \to \underline{\mathbb{R}}$ we have defined a quasi-state on $\mathcal{A}$ as follows

$$\begin{aligned} \rho : \mathcal{A}_{sa} &\to \mathbb{R} \\ \hat{A} &\mapsto \rho(\hat{A}) := I_{C^*(\hat{A})}(\hat{A}), \end{aligned}$$

where $C^*(\hat{A})$ is the sub-C^*-algebra generated by $\hat{A}$. Clearly $C^*(\hat{A} + \hat{B}) \subseteq C^*(\hat{A}, \hat{B})$ and $C^*(\hat{A}) \cup C^*(\hat{B}) \subseteq C^*(\hat{A}, \hat{B})$. Since I is a natural transformation and it is locally linear we obtain

$$\begin{aligned} \rho(\hat{A} + \hat{B}) &= I_{C^*(\hat{A}+\hat{B})}(\hat{A} + \hat{B}) \\ &= I_{C^*(\hat{A},\hat{B}}(\hat{A} + \hat{B}) \end{aligned}$$

[3]Here $\underline{\mathbb{R}}$ represents the internal locale of Dedekind reals.

$$= I_{C^*(\hat{A},\hat{B})}(\hat{A}) + I_{C^*(\hat{A},\hat{B})}(\hat{B})$$
$$= I_{C^*(\hat{A})}(\hat{A}) + I_{C^*(\hat{B})}(\hat{B})$$
$$= \rho(\hat{A}) + \rho(\hat{B})$$

Moreover $\rho(\hat{1}) = I_{C^*(\hat{1})}(\hat{1}) = 1$, while positivity of ρ follows from the fact that I is locally positive. Clearly the above assignments are inverse of each other.

This shows that there exists a bijective correspondence between quasi-states on $\mathcal{A}$ and probability integrals on $\overline{\mathcal{A}}_{sa}$. Now we want to show that there also exists a bijective correspondence between states on $\mathcal{A}$ and probability valuations on $\Sigma_{\overline{\mathcal{A}}}$. This bijection can be shown by proving that the locale of probability integrals on $\overline{\mathcal{A}}_{sa}$ is homeomorphic to the locale of probability valuation on $\Sigma_{\overline{\mathcal{A}}}$. This result was show in [10, 42] and it is essentially a generalisation of the Riesz-Markov theorem. The reader should refer to [10, 42] for a complete statement and proof of the theorem. □

The above discussion shows that to each quasi-state ρ on $\mathcal{A}$ there is associated to it a probability valuation $\overline{\mu}_\rho$ on $\Sigma_{\overline{\mathcal{A}}}$, given by the map of internal locals $\overline{\mu}_\rho : \mathcal{O}\Sigma_{\overline{\mathcal{A}}} \to \overline{\mathbb{R}}_l$. Here $\overline{\mathbb{R}}_l$ represents the lower reals in $[\mathcal{C}(\mathcal{A}), \mathbf{Sets}]$ (see Definition 11.5.1). Next we define the map $\overline{1}_l : \mathcal{O}\Sigma_{\overline{\mathcal{A}}} \to \overline{\mathbb{R}}_l$, such that for each context $C \in \mathcal{C}(\mathcal{A})$ we have

$$\overline{1}_{l,C} : \mathcal{O}\Sigma_{\overline{\mathcal{A}}}(C) \to \overline{\mathbb{R}}_l(C)$$
$$U \mapsto (\overline{1}_{l,C})(U) = 1_C :\uparrow C \to [0,1]$$

where $1_C :\uparrow C \to [0,1]$ is the map that is constantly 1.
The terms $\overline{\mu}_\rho$ and $\overline{1}_l$ are terms of type $\overline{\mathbb{R}}_l$ with free variables of type $\mathcal{O}\Sigma_{\overline{\mathcal{A}}}$, hence we can then construct the formula $[\underline{\mu}_\rho = \overline{1}_l]$ of type $\overline{\Omega}$ as follows

$$[\overline{\mu}_\rho = \overline{1}_l] : \mathcal{O}\Sigma_{\overline{\mathcal{A}}} \times \mathcal{O}\Sigma_{\overline{\mathcal{A}}} \to \overline{\Omega}$$

Using the alternative definition of state-space given in Definition 11.4.1 we can re-write the above formula as

$$[\overline{\mu}_\rho = \overline{1}_l] : \mathcal{O}\overline{\Sigma}_\uparrow \times \mathcal{O}\overline{\Sigma}_\uparrow \to \overline{\Omega}$$

Since any open $U \in \mathcal{O}\Sigma_\uparrow$ can be seen as a point of $\mathcal{O}\overline{\Sigma}_\uparrow$ we obtain $\overline{U} : \overline{1} \to \mathcal{O}\overline{\Sigma}_\uparrow$. As we will see in detail in the next section such points represent propositions in the contravariant topos approach. Hence, given a probability valuation $\overline{\mu}_\rho$ on $\mathcal{O}\overline{\Sigma}_\uparrow$ and a proposition $\overline{U}$, the formula

$$[\overline{\mu}_\rho = \overline{1}_l](U) = [\overline{\mu}_\rho(U) = \overline{1}_l] = [\overline{\mu}_\rho = \overline{1}_l] \circ \overline{U} : \overline{1} \to \overline{\Omega}$$

represents the truth value of the proposition $\overline{U}$ given the state $\overline{\mu}_\rho$.

11.4 Propositions

In this section we will investigate how proposition are defined in the covariant topos quantum theory approach. To this end we need to first introduce the concept of daseinisation and daseinisation map. The covariant version of such a map was first introduces in [42], but subsequently a simplified version was put forward in [75]. In the present context we will adopt this simplified version. However, we will need to slightly change the topos we work with, in particular we will consider the topos $[\mathcal{V}(\mathcal{A}), \mathbf{Sets}]$, where $\mathcal{V}(\mathcal{A})$ represents the category of abelian von Neumann subalgebras of $\mathcal{A}$. The reason for such a change is that we would like to define the covariant daseinisation map in terms of the contravariant one, to do this we need contexts which contain enough projections while general C^* algebras do not. Before describing the covariant daseinisation map we need to introduce an alternative description of the state-space $\underline{\Sigma}_{\overline{\mathcal{A}}}$ put forward in [74]

Definition 11.4.1 We define the space $\Sigma_{\uparrow}$ to be the set $\Sigma = \coprod_{C \in \mathcal{C}(\mathcal{A})} \Sigma_C$ such that $U \in \mathcal{O}\Sigma_{\uparrow}$ iff the following conditions hold:

1. given $\lambda \in U_C$, $C \subseteq C'$ and $\lambda' \in \Sigma_{C'}$ is such that $\lambda'|_C = \lambda$, then $\lambda' \in U_{C'}$.
2. $\forall C \in \mathcal{C}(\mathcal{A})$, $U_C \in \mathcal{O}\Sigma_C$.

We know from Sect. 9.3 that, given a locale map $f : Y \to X$, this gives rise to an internal locale $L(Y)$ in $Sh(X)$. For the case at hand we consider the space $\mathcal{C}(\mathcal{A})$ equipped with the upper Alexandroff topology and define the continuous projection map

$$\pi : \Sigma_{\uparrow} \to \mathcal{C}(\mathcal{A}) \tag{11.4.1}$$

$$\lambda \mapsto C. \tag{11.4.2}$$

This map between locales gives rise to the internal locale $\underline{\Sigma}_{\uparrow}$ in $[\mathcal{C}(\mathcal{A}), \mathbf{Sets}]$. The frame associated to $\underline{\Sigma}_{\uparrow}$ is given by

$$\mathcal{O}\underline{\Sigma}_{\uparrow} : \mathcal{C}(\mathcal{A}) \to \mathbf{Sets}$$
$$C \mapsto \mathcal{O}\underline{\Sigma}_{\uparrow}(C) := \mathcal{O}\Sigma|_{\uparrow C} = \{U \in \mathcal{O}\Sigma_{\uparrow} | U \subseteq \prod_{C' \in (\uparrow C)} \Sigma_{C'}\}.$$

For $C \subseteq C'$, the corresponding presheaf maps are given by

$$\mathcal{O}\underline{\Sigma}_{\uparrow}(C) \to \mathcal{O}\underline{\Sigma}_{\uparrow}(C')$$
$$U \mapsto \coprod_{C'' \in (\uparrow C')} U_{C''}.$$

In [74][Th. 2.2.2] it was shown that, up to isomorphism, $\overline{\Sigma}_{\uparrow}$ is the internal spectrum of the algebra $\overline{\mathcal{A}}$.
Given this new characterisation of the state-space in terms of $\overline{\Sigma}_{\uparrow}$ we can define the covariant daseinisation map in terms of the inner and outer daseinisation of the contravariant approach [26].

Definition 11.4.2 Consider the internal (to $[\mathcal{V}(\mathcal{A}), \mathbf{Sets}]$) locales $\underline{\Sigma}_{\uparrow}$ and $\overline{\mathbb{IR}}$,[4] whose associated frames are $\mathcal{O}\underline{\Sigma}_{\uparrow}$ and $\mathcal{O}\overline{\mathbb{IR}}$ respectively. For each $\hat{A} \in \mathcal{A}_{sa}$ we define the covariant daseinisation map as the arrow $\overline{\delta}(\hat{A})^{-1} : \mathcal{O}\overline{\mathbb{IR}} \to \mathcal{O}\underline{\Sigma}_{\uparrow}$ such that for each context $C \in \mathcal{V}(\mathcal{A})$ we have:

$$\overline{\delta}(\hat{A})^{-1}(\uparrow C' \times (p,q)_S) = \delta(\hat{A})^{-1}(p,q)_S \cap \Sigma_{\uparrow}|_{\uparrow C'}$$

where

$$\delta : \mathcal{A}_{sa} \to C(\Sigma_{\uparrow}, \mathbb{IR})$$
$$\hat{A} \mapsto \delta(\hat{A}) := \lambda \mapsto [\lambda(\delta^i(\hat{A})_C), \lambda(\delta^o(\hat{A})_C)].$$

Here δ^i and δ^o represent the inner and outer daseinisation introduced in the contravariant approach (see [26] Section 13.4, eq. 13.4.6). Moreover, $\Sigma_{\uparrow}|_{\uparrow C'} = \coprod_{C'' \in \uparrow C'} \Sigma_{C''}$ and $\uparrow C' \times (p,q)_S = \{(C, [r,s]| C \in \mathcal{C}(\mathcal{A}), C' \subseteq C, p < r \leq s < q\}$ is a basic open subset in $\mathcal{C}(\mathcal{A}) \times \mathbb{IR}$.

We now need to check that indeed the daseinisation map is well defined. In particular we need to check that $\delta(\hat{A}) : \Sigma_{\uparrow} \to \mathbb{IR}$ is continuous and that $\overline{\delta}^{-1}$, as defined above, is a frame map. To show that $\delta(\hat{A})$ is continuous we need to show that the inverse image of an open in $\mathbb{IR}$ is open in $\Sigma_{\uparrow}$. To this end consider an open $(p,q)_S$ in $\mathbb{IR}$. The inverse image, at a context $C \in \mathcal{C}(\mathcal{A})$, is given by

$$(\delta(\hat{A})^{-1}(p,q)_S)_C = \{\lambda \in \Sigma_C | \lambda(\delta^i(\hat{A})_C) > p\} \cap \{\lambda \in \Sigma_C | \lambda(\delta^o(\hat{A})_C) < q\}.$$

In order to determine whether $(\delta(\hat{A})^{-1}(p,q)_S)_C$ is open we need to show that it satisfies the conditions in Definition 11.4.1. We will start with the second condition, that is, we need to show that for all $C \in \mathcal{C}(\mathcal{A})$, $U \cap \Sigma_C \in \mathcal{O}(\Sigma_C)$. Since each Σ_C is equipped with the Gelfand topology, open sets are of the form $U = \{\lambda \in \Sigma_C | \lambda(\hat{A}) - z| < \sigma\}$ for $\hat{A} \in \mathcal{A}$, $z \in \mathbb{C}$ and $\epsilon > 0$ hence, it follow that both $\{\lambda \in \Sigma_C | \lambda(\delta^i(\hat{A})_C) > p\}$ and $\{\lambda \in \Sigma_C | \lambda(\delta^o(\hat{A})_C) < q\}$ are open, and the second condition is satisfied.

Next we need to show that "openness" is preserved by the presheaf map. In particular we need to show that if $\lambda \in U_C$ then, for $C \subseteq C'$, if $\lambda' \in \Sigma_{C'}$ is such

[4] $\overline{\mathbb{IR}}$ represents the internal Scotts interval domain (see Appendix A.2).

that $\lambda'|_C = \lambda$, then $\lambda' \in U_{C'}$. Let us assume that $\lambda \in (\delta(\hat{A})^{-1}(p,q)_S)_C \cap \Sigma_C$ and for $C \subseteq C'$, $\lambda' \in \Sigma_{C'}$ is such that $\lambda'|_C = \lambda$. We then consider the open $(\delta(\hat{A})^{-1}(p,q)_S)_{C'} = \{\lambda \in \Sigma_{C'}|\lambda(\delta^i(\hat{A})_{C'}) > p\} \cap \{\lambda \in \Sigma_{C'}|\lambda(\delta^o(\hat{A})_{C'} < q\}$. It was shown in [26] that, for $C \subseteq C'$, inner and outer daseinisation follow the following relation

$$\delta^i(\hat{A})_C \leq \delta^i(\hat{A})_{C'} \leq \hat{A} \leq \delta^o(\hat{A})_{C'} \leq \delta^o(\hat{A})_C.$$

It then follows that $p < \lambda'(\delta^i(\hat{A})_C) \leq \lambda'(\delta^i(\hat{A})_{C'})$ and $\lambda'(\delta^o(\hat{A})_{C'}) \leq \lambda(\delta^o(\hat{A})_C)) < q$. hence $\lambda' \in (\delta(\hat{A})^{-1}(p,q)_S)_{C'} \cap \Sigma_{C'}$.

The above discussion showed that the map, $\delta(\hat{A}) : \Sigma_\uparrow \to \mathbb{IR}$, as defined above is continuous. Next we need to show that the induced map $\overline{\delta}(\hat{A})$ is indeed a map of internal locales. To this end we recall that $\mathbf{Loc}(Sh(\mathcal{C}(\mathcal{A}))) \simeq \mathbf{Loc}/\mathcal{C}(\mathcal{A})$, hence to define a map between internal locales all we need to do is to determine the corresponding map in $\mathbf{Loc}/\mathcal{C}(\mathcal{A})$. For the case at hand such a map is given by the following commutative diagram:

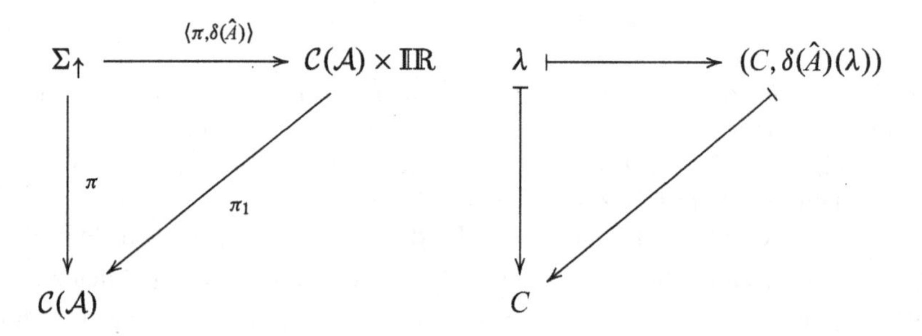

Having defined the covariant daseinisation map we can now describe how propositions are expressed in the covariant approach [42, 75].

Definition 11.4.3 Given $\hat{A} \in \mathcal{A}_{sa}$ and $(p,q) \in \mathcal{O}(\mathbb{R})$ the proposition $\hat{A} \in (p,q)$ is represented by the element in $\mathcal{O}\Sigma_\uparrow$ defined by

$$[\hat{A} \in (p,q)]_1 = \delta(\hat{A})^{-1}(p,q)_S \tag{11.4.3}$$

$$= \coprod_{C \in \mathcal{C}(\mathcal{A})} \{\lambda \in \Sigma_C | [\lambda(\delta^i(\hat{A})_C), \lambda(\delta^o(\hat{A})_C] \in (p,q)_S\}. \tag{11.4.4}$$

At the level of the spectrum $\underline{\Sigma}_\downarrow$ we have that

$$\overline{[\hat{A} \in (p,q)]} : \overline{1} \to \mathcal{O}\overline{\Sigma}_\uparrow$$

such that, for each context $C \in \mathcal{C}(\mathcal{A})$, the components are defined by

$$\overline{[\hat{A} \in (p,q)]}_C : \overline{1}_C \to \mathcal{O}\overline{\Sigma}_{\uparrow C}$$
$$\mapsto \overline{[\hat{A} \in (p,q)]}_C(*) = \coprod_{C' \uparrow C} [\hat{A} \in (p,q)]_{C'}.$$

It is also possible to write $\overline{[\hat{A} \in (p,q)]}$ directly in terms of the map $\overline{\delta}(\hat{A})^{-1}$. In particular we first define the map $\overline{(p,q)} : \overline{1} \to \mathcal{O}\overline{\mathbb{IR}}$ such that at each context $C \in \mathcal{C}(\mathcal{A})$ we have

$$\overline{(p,q)}_C : \overline{1}_C \to \mathcal{O}\overline{\mathbb{IR}}_C$$
$$* \mapsto \overline{(p.q)}_C(*) = \uparrow C \times (p,q)_S.$$

Combining this with $\overline{\delta}(\hat{A})^{-1}$ we obtain

$$\overline{[\hat{A} \in (p,q)]} = \overline{\delta}(\hat{A})^{-1} \circ \overline{(p.q)} : \overline{1} \to \mathcal{O}\overline{\Sigma}_{\uparrow}.$$

11.4.1 Relation Between Covariant Propositions and Contravariant Propositions

In this section we would like to analyse the relation between the covariant definition of propositions as elucidated in the previous section and the contravariant description given in [24, 26]. Recall from Definition 2.2.2 that in the contravariant approach propositions are identified with clopen sub-objects of the state-space by the process of outer daseinisation. In particular, given a proposition "$A \in \Delta$", which is represented by a projection operator $\hat{P}$, then the corresponding topos proposition is given by

$$\underline{\delta}^o(\hat{P}) := \Big(\mathfrak{S}_V(\delta^o(\hat{P})_V)\Big)_{V \in \mathcal{V}(\mathcal{H})} \tag{11.4.5}$$

where $\underline{\delta}^o(\hat{P})_V := \bigwedge\{\hat{R} \in P(V) | \hat{R} \geq \hat{P}\}$ and $\mathfrak{S}_V(\delta^o(\hat{P})_V) = \{\lambda \in \underline{\Sigma}_V | \lambda(\delta^o(\hat{P})_V) = 1\}$.

Let us consider a general C^*-algebra $\mathcal{A}$ and its collection of abelian subalgebras $\mathcal{V}(\mathcal{A})$, given a self-adjoint operator $\hat{A} \in \mathcal{A}_{sa}$, using definition (11.4.5), one can write the proposition $\hat{A} \in (p,q)$ as

$$\underline{\delta}^o(\hat{P}) = \coprod_{C \in \mathcal{V}(\mathcal{A})} \{\lambda \in \Sigma_C | \lambda(\delta^o \chi_{(p,q)}(\hat{A}))_C = 1\}, \tag{11.4.6}$$

where $\chi_{(p,q)}(\hat{A})$ denotes the spectral projection operator associated to the proposition $\hat{A} \in (p,q)$.

This way of expressing contravariant propositions better lends itself for a comparison with the covariant approach. In fact one can already see that on the one hand the definition of covariant propositions (11.4.3) relays on the approximation of the operator $\hat{A} \in \mathcal{A}_{sa}$, on the other hand, in the contravariant propositions expressed by (11.4.6) the entire projection associated to the proposition gets approximated.

In [75] it was shown that it is possible to express covariant propositions in terms of the inner daseinisation of the contravariant approach. In this new definition, similarly as is the case of the contravariant approach, the entire proposition gets approximated, albeit via the inner daseinisation. In particular we have that, given a proposition $[\hat{A} \in (p,q)]$, this can be expressed in the covariant approach as

$$[\hat{A} \in (p,q)]_2 = \coprod_{C \in \mathcal{V}(\mathcal{A})} \{\lambda \in \Sigma_C | \lambda(\delta^i(\chi_{(p,q)}(\hat{A}))_C) = 1\}. \tag{11.4.7}$$

In [75] it was shown that the two ways of expressing a proposition given in (11.4.6) and (11.4.7) are equivalent.

Lemma 11.4.1 *Consider a self-adjoint operator $\hat{A} \in \mathcal{A}_{sa}$ and an element $(p,q) \in \mathcal{O}(\mathbb{IR})$. For any $C \in \mathcal{V}(\mathcal{A})$ and $\lambda \in \Sigma_C$ it follows that:*

1. $[\hat{A} \in (p,q)]_1 \subseteq [\hat{A} \in (p,q)]_2$.
2. *If $\lambda \in [\hat{A} \in (p,q)]_2$ for some $\lambda \in \Sigma_C$, then $\lambda(\delta^i(\hat{A})_C) \geq p$ and $\lambda(\delta^o(\hat{A})_C) \leq q$.*

In order to prove the above Lemma we need to introduce the notion of *Antonymous functions* and *observable functions* [18]. These are used to describe the Gel'fand transforms of the operators associated to the inner daseinisation and outer daseinisation of self-adjoint operators. This is done in Sect. 4.1.

Proof Recall that the spectral resolution of a self-adjoint operator $\hat{A}$ is given by $(\hat{E}^{\hat{A}}_r)_{r \in \mathbb{R}}$ where $\hat{E}^{\hat{A}}_r = \chi_{(-\infty,r]}(\hat{A})$. We can now re-write Eqs. (4.1.14) and (4.1.15) as [74]:

$$\lambda(\delta^i(\hat{A})_C) = \sup\{r \in \mathbb{R} | \hat{1} - \hat{E}^{\hat{A}}_r \in C_{\mathcal{N}}(F_\lambda)\} \tag{11.4.8}$$

$$= \sup\{r \in \mathbb{R} | \exists \hat{P} \in P(C), \lambda(\hat{P}) = 1, \hat{P} \leq \chi_{[r,\infty)}(\hat{A})\} \tag{11.4.9}$$

and

$$\lambda(\delta^o(\hat{A})_C) = \inf\{r \in \mathbb{R} | \hat{E}^{\hat{A}}_r \in C_{\mathcal{N}}(F_\lambda)\} \tag{11.4.10}$$

$$= \sup\{r \in \mathbb{R} | \exists \hat{P} \in P(C), \lambda(\hat{P}) = 1, \hat{P} \leq \chi_{(-\infty,r)}(\hat{A})\}, \tag{11.4.11}$$

respectively. This will allow us to prove the claims of the above lemma.

1. Assume that $[\hat{A} \in (p,q)]_1 \subseteq [\hat{A} \in (p,q)]_2$. This is equivalent to the statement that if $\lambda(\delta^i(\hat{A})_C) > p$ and $\lambda(\delta^o(\hat{A})_C) < q$ hold, then $\lambda(\delta^i(\chi_{(p,q)}(\hat{A}))_C) = 1$. However, from Eq. (11.4.8), if $\lambda(\delta^i(\hat{A})_C) > p$ then, for $\epsilon > 0$, there exists a projection operator $\hat{P} \in P(C)$ such that $\lambda(\hat{P}) = 1$ and $\hat{P} \leq \chi_{[p+\epsilon,\infty)}(\hat{A})$. Similarly, from (11.4.10), if $\lambda(\delta^o(\hat{A})_C) < q$, then there exists a projection operator $\hat{Q} \in P(C)$ such that $\lambda(\hat{Q}) = 1$ and $\hat{Q} \leq \chi_{(-\infty,q)}(\hat{A})$. If we define $\hat{R} := \hat{P}\hat{Q}$, then $\hat{R} \in P(C)$, $\lambda(\hat{R}) = 1$ and $\hat{R} \leq \chi_{[r+\epsilon,s)}(\hat{A}) \leq \chi_{(p,q)}(\hat{A})$. Therefore $\hat{R} \leq \delta^i(\chi_{(p,q)}(\hat{A}))_C$ and $\lambda(\delta^i(\chi_{(p,q)}(\hat{A}))_C) = 1$.
2. Consider any $\lambda \in \Sigma_C$ such that $\lambda \in [\hat{A} \in (p,q)]_2$, then, $\lambda(\delta^i(\chi_{(p,q)}(\hat{A}))_C) = 1$. However, since $\delta^i(\chi_{(p,q)}(\hat{A}))_C \leq \chi_{(-\infty,q)}(\hat{A})$, it follows from (11.4.10) that $\lambda(\delta^o(\hat{A})_C) \leq q$. Similarly, since $\delta^i(\chi_{(p,q)}(\hat{A}))_C \leq \chi_{[p,\infty)}(\hat{A})$, (11.4.8) implies that $\lambda(\delta^o(\hat{A})_C) \geq p$.

□

11.5 Physical Quantities

In this section we will explain how physical quantities are defined in the covariant approach. Similarly as for the contravariant approach, even in the covariant approach physical quantities are defined as maps from the state-space to the quantity value object. Since both the state space $\underline{\Sigma}_{\uparrow}$ and the quantity value object $\underline{\mathbb{IR}}$ are internal topological spaces, physical quantities will be identified by continuous maps between internal topological space. It is worth, therefore, to recall how a continuous map between internal topological space is defined. Consider the topos $Sh(T)$ for some topological space T, and two internal topological spaces $\overline{X}$ and $\overline{Y}$ whose external description is given by the bundles $p : X \rightarrow T$ and $q : Y \rightarrow T$, respectively. By equipping $\overline{X}$ and $\overline{Y}$ with the topologies $\mathcal{O}(\overline{X})$ and $\mathcal{O}(\overline{Y})$, respectively we can define a continuous map $f : (\overline{X}, \mathcal{O}(\overline{X})) \rightarrow (\overline{Y}, \mathcal{O}(\overline{Y}))$ as a sheaf morphisms

$$f : \overline{X} \rightarrow \overline{Y}$$

such that, for all $U \in P(\overline{Y})$, then[5]

$$U \in \mathcal{O}(\overline{Y}) \Rightarrow f^{-1}(U) \in \mathcal{O}(\overline{X})$$

where $f^{-1} = P(f) : P(\overline{Y}) \rightarrow P(\overline{X})$. Here P can be seen as a contravariant functor $P : Sh(T) \rightarrow Sh(T)$ which sends each object B to its power object $P(B)$ and each morphism $f : B \rightarrow A$ to the map $P(f) = P(A) \rightarrow P(B)$, which is defined as the

[5]Here $P(\overline{Y})$ denotes the power object of $\overline{Y}$.

only arrow making the following diagram commute

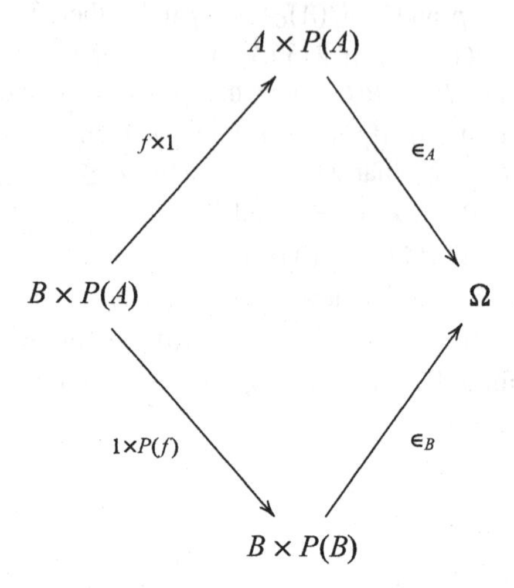

The map $\in_B$: $B \times PB \to \Omega$ is defined to be the arrow such that, given any map $f : B \times A \to \Omega$, there exists a unique map $g : A \to PB$ which makes the following diagram commute:

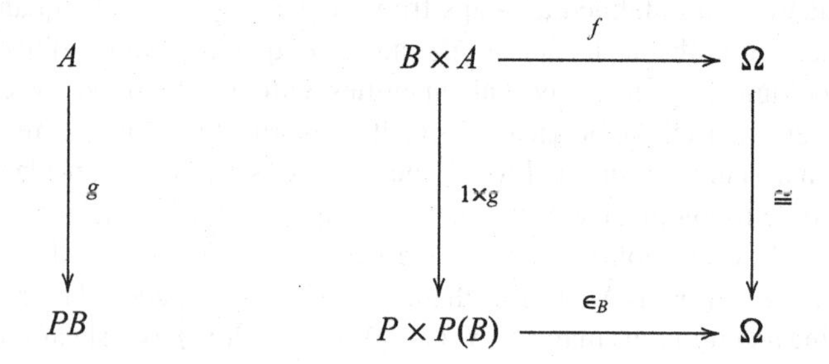

By considering sheaves over T as etalé bundles over T, then the map $f : \overline{X} \to \overline{Y}$ corresponds to the commuting triangle

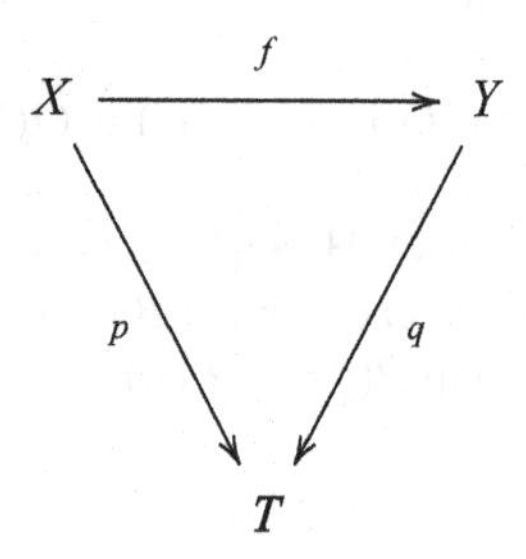

where f is continuous with respect to the etalé topology. If f is also continuous with respect to the coarser topologies on X and Y coming from the internal topologies of $\mathcal{O}(\overline{X})$ and $\mathcal{O}(\overline{Y})$, then f corresponds to an internal continuous map $f : (\overline{X}, \mathcal{O}(\overline{X})) \rightarrow (\overline{Y}, \mathcal{O}(\overline{Y}))$. This ends our small revision of continuous maps between internal topological spaces. Next we will define physical quantities as maps from the state space to the quantity value object however, in order to do so, we need to introduce the functors representing lower and upper reals.

Definition 11.5.1 In $[\mathcal{C}(\mathcal{A}), \mathbf{Sets}]$ lower reals are given by the functor $\overline{\mathbb{R}}_l : \mathcal{C}(\mathcal{A}) \rightarrow \mathbf{Sets}$ defined on:

1. objects: for each $C \in \mathcal{C}(\mathcal{A})$ the functor $\overline{\mathbb{R}}_l$ assigns the set $\overline{\mathbb{R}}_l(C) := OP((\uparrow C), \mathbb{R})$ of order preserving functions $\mu :\uparrow C \rightarrow \mathbb{R}$;
2. morphisms: given a morphism $D \hookrightarrow C$ the corresponding morphism is

$$OP((\uparrow D), \mathbb{R}) \rightarrow OP((\uparrow C), \mathbb{R})$$
$$\mu \mapsto \mu|_{\uparrow C} .$$

Similarly, we can define the functor of upper reals as follows:

Definition 11.5.2 In $[\mathcal{C}(\mathcal{A}), \mathbf{Sets}]$ upper reals are given by the functor $\overline{\mathbb{R}}_u : \mathcal{C}(\mathcal{A}) \rightarrow \mathbf{Sets}$ defined on:

1. objects: for each $C \in \mathcal{C}(\mathcal{A})$ the functor $\overline{\mathbb{R}}_u$ assigns the set $\overline{\mathbb{R}}_u(C) := OR((\uparrow C), \mathbb{R})$ of order reversing functions $\nu :\uparrow C \rightarrow \mathbb{R}$;
2. morphisms: given a morphism $D \hookrightarrow C$ the corresponding morphism is

$$OR((\uparrow D), \mathbb{R}) \rightarrow OR((\uparrow C), \mathbb{R})$$
$$\nu \mapsto \nu|_{\uparrow C} .$$

Having defined the functors of lower and upper reals we can now characterise both outer and inner daseinisation in terms of them.

Theorem 11.5.1 ([74]) *Given a self-adjoint operator $\hat{A} \in \mathcal{A}_{sa}$, outer daseinisation is identified with the internal locale map $\overline{\delta}^o(\hat{A}) : \Sigma_{\overline{\mathcal{A}}} \rightarrow \overline{\mathbb{R}}_u$. Since* $\mathbf{Loc}/\mathcal{C}(\mathcal{A}) \cong \mathbf{Loc}(Sh(\mathcal{C}(\mathcal{A})))$, *outer daseinisation can be also identified in terms of the following commuting diagram of continuous maps*[6]*:*

[6]In the following the set $\mathbb{R}_u$ denotes the set of upper reals. The topology on $\mathbb{R}_u$ is generated by the lower half open intervals $[-\infty, y)$, $y \in \mathbb{R}$.

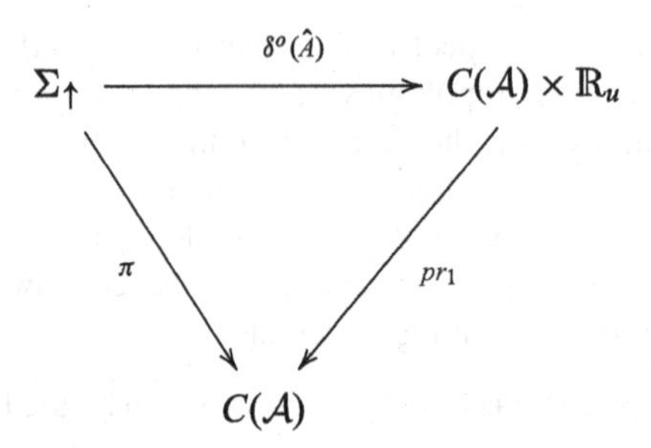

where $\Sigma_\uparrow$ was defined in Definition 11.4.1 while π was defined in (11.4.1).

Proof Consider the internal topological space of the upper reals $\overline{\mathbb{R}}_u$. In terms of etalé bundles this can be expressed by the map

$$\pi_u : \mathcal{R}_u \to C(\mathcal{A})$$

where the space $\mathcal{R}_u$ is given by

$$\mathcal{R}_u = \{s|s \in OR((\uparrow C), \mathbb{R})\}\,.$$

$\mathcal{R}_u$ is equipped with the topology generated by the etalé opens: for each $C \in C(\mathcal{A})$ and $c \in \mathbb{R}$ we defined the open

$$U_{x,C} = \{s \in \mathcal{R}_u | D \in \uparrow C, s(D) < x\}\,.$$

We can now express the diagram in Theorem 11.5.1 in terms of maps of etalé bundles as follows:

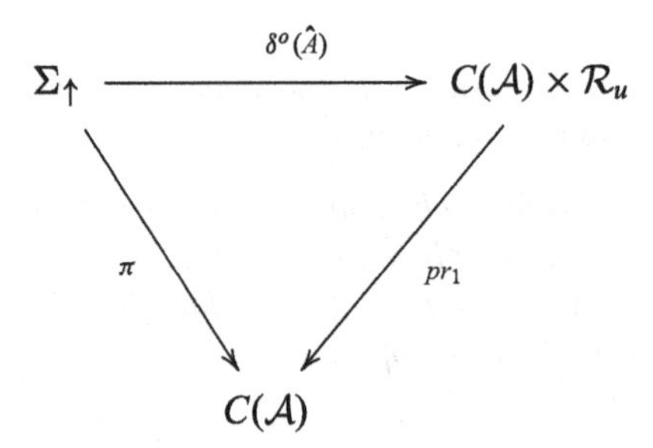

where, for each $\lambda \in (\Sigma_\uparrow)_C$ $C \in C(\mathcal{A})$, we obtain

$$\delta^o(\hat{A})(\lambda) = D \mapsto \lambda(\delta^o(\hat{A})_D)\,.$$

We now need to check that $\delta^o(\hat{A})$ is continuous with respect to the topology generated by the opens $U_{x,C}$. In particular we obtain

$$\delta^o(\hat{A})^{-1}(U_{x,C})_D = \begin{cases} \delta^o(\hat{A})_D^{-1}(-\infty, x) & \text{if } C \subseteq D \\ \emptyset & \text{if } D \nsubseteq C \end{cases}$$

where $\delta^o(\hat{A})_D : \Sigma_D \to \mathbb{R}$ is given by the Gelfand duality. Clearly, for each $D \in \mathcal{C}(\mathcal{A})$ the set $\delta^o(\hat{A})^{-1}(U_{x,C})_D$ is open in Σ_D. Moreover, given $\lambda \in \Sigma_D$ and $C \subseteq D$ such that $\lambda|_C \in \delta^o(\hat{A})^{-1}(U_{x,C})$ then, since

$$\lambda(\delta^o(\hat{A})_D) \leq \lambda|_C(\delta^o(\hat{A})_C) < x\,,$$

it follows that $\lambda \in \delta^o(\hat{A})^{-1}(U_{x,C})_D$. Therefore $\delta^o(\hat{A})^{-1}(U_{x,C})$ is open in $\Sigma_\uparrow$ since it satisfies the conditions of Definition 11.4.1. □

A similar theorem holds for inner daseinisation.

Theorem 11.5.2 *[74] Given a self-adjoint operator $a \in \mathcal{A}_{sa}$, inner daseinisation is identified with the internal locale map $\overline{\delta}^i(\hat{A}) : \underline{\Sigma}_{\overline{\mathcal{A}}} \to \overline{\mathbb{R}}_l$. Since* **Loc**$/\mathcal{C}(\mathcal{A}) \cong$ **Loc**$(Sh(\mathcal{C}(\mathcal{A})))$, *inner daseinisation can be also identified in terms of the following commuting diagram of continuous maps*[7]*:*

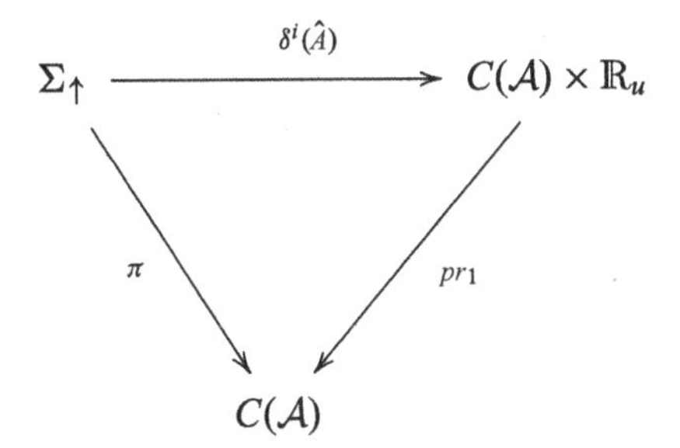

Proof Consider the internal topological space of the lower reals $\overline{\mathbb{R}}_l$. In terms of etalé bundles this can be expressed by the map

$$\pi_l : \mathcal{R}_l \to \mathcal{C}(\mathcal{A})$$

where the space $\mathcal{R}_l$ is given by

$$\mathcal{R}_l = \{s | s \in OP((\uparrow C), \mathbb{R})\}\,.$$

[7]In the following the set $\mathbb{R}_l$ denotes the set of lower reals. The topology on $\mathbb{R}_l$ is generated by the upper half intervals $(y, +\infty]$, $y \in \mathbb{R}$.

$\mathcal{R}_l$ is equipped with the topology generated by the etalé opens: for each $C \in \mathcal{C}(\mathcal{A})$ and $c \in \mathbb{R}$ we defined the open

$$U_{x,C} = \{s \in \mathcal{R}_l | D \in \uparrow C, s(D) > x\} .$$

We can now express the diagram in Theorem 11.5.1 in terms of maps of etalé bundles as follows:

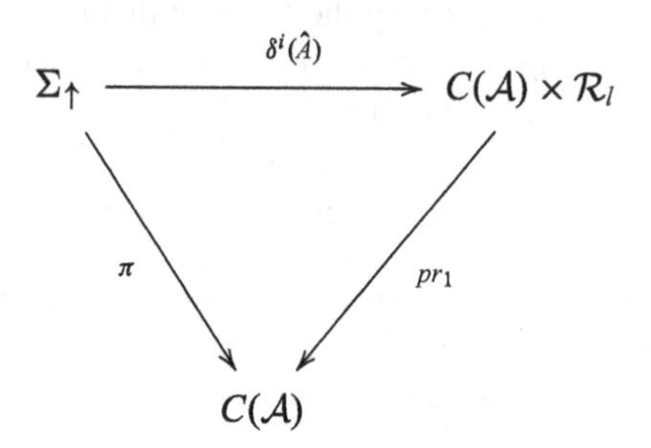

where, for each $\lambda \in (\Sigma_{\uparrow})_C$ $C \in \mathcal{C}(\mathcal{A})$, we obtain

$$\breve{\delta}^i(\hat{A})(\lambda) = D \mapsto \lambda(\delta^i(\hat{A})_D) .$$

We now need to check that $\breve{\delta}^i(\hat{A})$ is continuous with respect to the topology generated by the opens $U_{x,C}$. In particular we obtain

$$\breve{\delta}^i(\hat{A})^{-1}(U_{x,C})_D = \begin{cases} \breve{\delta}^i(\hat{A})_D^{-1}(x, +\infty) & \text{if } C \subseteq D \\ \emptyset & \text{if } D \nsubseteq C \end{cases}$$

where $\breve{\delta}^i(\hat{A})_D : \Sigma_D \to \mathbb{R}$ is given by the Gelfand duality. Clearly, for each $D \in \mathcal{C}(\mathcal{A})$ the set $\breve{\delta}^i(\hat{A})^{-1}(U_{x,C})_D$ is open in Σ_D. Moreover, given $\lambda \in \Sigma_D$ and $C \subseteq D$ such that $\lambda|_C \in \breve{\delta}^i(\hat{A})^{-1}(U_{x,C})$ then, since

$$\lambda(\delta^i(\hat{A})_D) \geq \lambda|_C(\delta^i(\hat{A})_C) > x ,$$

it follows that $\lambda \in \breve{\delta}^i(\hat{A})^{-1}(U_{x,C})_D$. Therefore $\breve{\delta}^i(\hat{A})^{-1}(U_{x,C})$ is open in $\Sigma_{\uparrow}$ since it satisfies the conditions of Definition 11.4.1. □

If we then combine the two above maps together we obtain the locale map

$$\overline{\delta}(\hat{A}) = \langle \delta^i(\hat{A}), \delta^o(\hat{A}) \rangle : \underline{\Sigma}_{\overline{\mathcal{A}}} \to \overline{\mathbb{R}}_l \times \overline{\mathbb{R}} ,$$

which externally is defined by the following diagram of continuous maps:

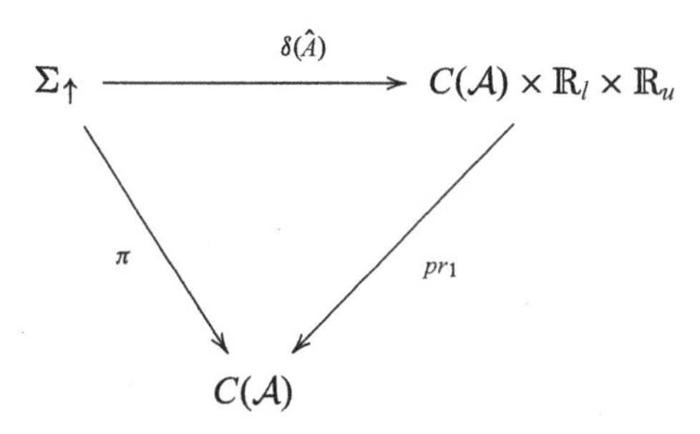

At the start of this section we claimed that physical quantities are maps from the state space $\underline{\Sigma}_{\uparrow}$ to the quantity value object $\underline{\mathbb{IR}}$, therefore we need to understand the relation between $\underline{\mathbb{IR}}$ and $\underline{\mathbb{R}}_l \times \underline{\mathbb{R}}_u$. We will start by considering the relation of the corresponding sets. In particular we will define an injective map [74]

$$
\begin{aligned}
j : \mathbb{IR} &\rightarrow \mathbb{R}_l \times \mathbb{R}_u \\
[x, y] &\mapsto (x, y)\,.
\end{aligned}
$$

The fact that j is injective follows trivially from the definition, moreover j is also continuous. In fact, since $\mathbb{R}_l \times \mathbb{R}_u$ comes equipped with the product topology, an open will be of the form $((r, +\infty] \times [-\infty, s))$, therefore the inverse image $j^{-1}\left((r, +\infty] \times [-\infty, s)\right)$ is

$$
j^{-1}((r, +\infty] \times [-\infty, s)) = \begin{cases} (r, s) \in \mathcal{O}(\mathbb{IR}) \text{ if } r < s \\ \emptyset \text{ if } s < r \end{cases}
$$

hence j is continuous.

From the definitions of inner and outer daseinisation given in previous sections we know that, for each context $C \in \mathcal{C}(\mathcal{A})$, $\delta^i(\hat{A})_C \leq \delta^o(\hat{A})_C$, therefore $\lambda(\delta^i(\hat{A})_C) \leq \lambda(\delta^o(\hat{A})_C)$ and hence $(\lambda(\delta^i(\hat{A})_C), \lambda(\delta^o(\hat{A})_C)) \in \mathcal{O}(\mathbb{IR})$. This implies that the daseinisation map can be factored through the map j. This is expressed by the following commuting diagram in $\mathbf{Loc}/C(\mathcal{A})$:

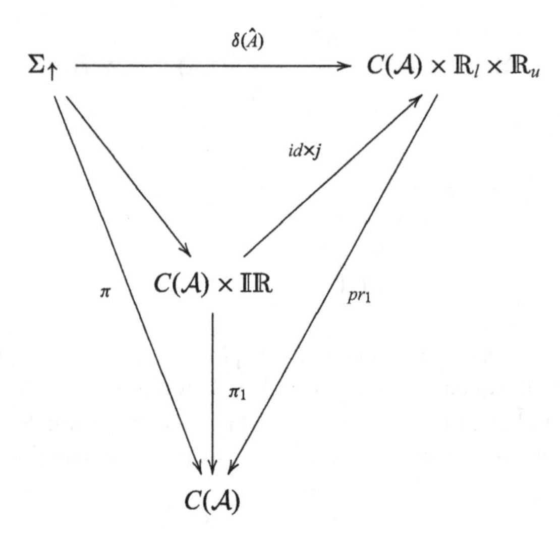

In the above, the map $\pi_1 : C(\mathcal{A}) \times \mathbb{IR} \to C(\mathcal{A})$ represents the external description of the interval domain $\overline{\mathbb{IR}}$ in $[C(\mathcal{A}), \mathbf{Sets}]$, while the factorisation map $\Sigma_\uparrow \to C(\mathcal{A}) \times \mathbb{IR}$ represents the external description of the daseinisation map defined in Definition 11.4.2. It is precisely in this sense that physical quantities are defined by maps from the state space to the quantity value object.

Chapter 12
Space Time in Topos Quantum Theory

One of the main challenges in theoretical physics in the past 50 years has been to define a theory of quantum gravity, i.e., a theory which consistently combines general relativity and quantum theory in order to define a theory of space-time itself seen as a fluctuating field (with respect to the connection and the metric). Therefore, a definition of space-time is of paramount importance but, it is precisely the attainment of such a definition which is one of the main stumbling blocks in quantum gravity.

The reason for such a difficulty is the seemingly incompatible roles of space-time put forward by general relativity and quantum theory. In fact, on the one hand, in general relativity, although the presence of both the 4-dimensional metric and the connection is assumed ab initio, they are both considered to be dynamical quantities and there is no preferred foliation of space-time.

On the other hand, quantum theory assumes a fixed (with respect to its differentiable structure and metric) space-time, implied by the mathematical formalism of the theory. For example the Schrodinger's equation describing unitary evolution between measurements is $i\hbar\frac{\partial\psi}{\partial t} = H\psi$ where the notion of time is fixed, while the measurement process "reduces" the state vector via the second law of evolution: $\psi \rightarrow \frac{\hat{P}\psi}{||\hat{P}\psi||}$ where $\hat{P}$ is the projection on the outcome of the measurement. Therefore the state gets projected on a spacelike surface. Clearly, a fixed geometry of spacetime is needed to define both the time t in the Schrodinger equation and the spacelike surface on which the state vector is reduced.

In classical physics the statement "the particle x has position y" makes sense, however in quantum theory, in order to have meaning, this statement should be changed to "if a measurement is performed on the position of the particle x, then it will have a certain probability to give outcome y". The difference in these statements reveals the discrepancy that exists between any classical theory and quantum theory. Although the full implications of this will be analysed in details later on, for now we want to emphasise that the concept of measurement is essential for any statement regarding quantum systems to make sense. This, in turn, implies that by necessity,

C. Flori, *A Second Course in Topos Quantum Theory*,
Lecture Notes in Physics 944, https://doi.org/10.1007/978-3-319-71108-9_12

a fixed space-time background has to be assumed in which the measurement takes place.

However, such a notion of a fixed background structure seems hard to accommodate in a theory of quantum gravity, where the varying field is space-time itself. In fact, by adopting quantum theory as it stands, a possible theory of quantum gravity would have to make sense of statements of the form "if a measurement is made of property x of the space-time field, then the outcome y will have probability z", where the notion of measurement requires a fixed space-time background.

But what can be said about the "structure" of such a space-time inherent in quantum theory? To answer this question one has to go back and analyse how exactly quantum theories come about. This is generally done through the process of "quantization", by which a classical theory is 'transformed' into a quantum theory. Now in a classical theory, in general, the configuration space of the system is mathematically represented by a differentiable manifold $\mathcal{M}$, while the phase space is represented by its cotangent bundle $T^*\mathcal{M}$. When quantising a classical theory, this concept of a phase space is inherited by the quantum theory. For example, if the classical configuration is $Q \simeq G/H$ for some Lie groups G and H, then a quantization of such a system would define the quantum states to be sections of a vector bundle over Q, whose fibres carry a representation of H. By definition, the domain of these sections would be the continuum Q. If quantization is so defined, and this begs the question as to why this is the case, then the space of values of quantum states is modelled by the continuum. This mathematical description of space-time agrees with that given by general relativity, which models space-time by a differentiable manifold $\mathcal{M}$, whose elements are interpreted as space-time points and the gravitational field is given by the curvature tensor of the pseudo-Riemannian metric on $\mathcal{M}$.

Therefore, although mathematically space-time is treated in an analogous way in both quantum theory and general relativity, its role in these two theories is very different. However, when defining a theory of quantum gravity, the very definition of space-time as a differentiable manifold is put into discussion. In fact, it is believed that at microscopic scales, space-time ceases to be continuous but acquires a discrete nature. Therefore, the continuum structure of space-time suggested by the two main ingredients of quantum gravity seems to be refuted by quantum gravity itself. This might seem an odd predicament, but it might also suggest that the mathematical description of space-time required for quantum gravity should be radically different from the continuum picture put forward by the two ingredient theories.

A candidate for an alternative description of space-time is given by the topos approach. In this approach the notion of a space-time point is replaced by the notion of a space-time region. Such regions should be interpreted as defining regions which are occupied by "extended" objects.

The interesting feature is that the collection of such "regions" carry a Heyting algebra structure, which is a generalised Boolean algebra where the law of excluded middle does not hold. This mathematical description of space-time in terms of what is technically called a locale fits well with the discrete notion of space-time put forward by quantum gravity.

12.1 A Lesson from Quantum Gravity?

As mentioned in the introduction, to date there is still no agreement upon *the* theory of quantum gravity. It is still possible and useful to analyse the various proposals to see if anything can be learned from them. In particular, we would like to get a feeling for how a theory of quantum gravity should address three main topics which we believe are directly related with the ensuing concept of space-time. These topics are:

(1) the use of the continuum;
(2) the role of the Planck length;
(3) the relation to the instrumentalist interpretation of quantum theory.

Use of the Continuum Both in classical physics and in quantum theory the continuum appears in three main areas:

1. To model configuration space.
2. As values of physical quantities.
3. As modelling the space of probabilities.

We will now analyse why the continuum is used in these areas and how its use in quantum gravity doesn't seem to be justified, other than as an a priori assumption. [7, 16, 45].

1. *To model configuration space.* If we consider a particle moving in three-dimensional space, we define its configuration to be $\mathbb{R}^3$. This is a consequence of the fact that physical space is modelled by a differentiable manifold. This comes as no surprise since we are used to measuring macroscopic objects with pointers and rulers. So it seems that the choice of a differentiable manifold to represent a classical configuration space is a consequence of modelling physical space by a manifold. As discussed briefly in the introduction, such a conception of configuration space is carried over to the quantum regime via the process of quantization. Thus, by necessity, the mathematical description of quantum theory is determined by a priori assumptions on the nature of space-time. The question is then if such a priori assumptions are still justified in the context of quantum gravity.
2. *Values of physical quantities*. In classical physics, quantities are real-valued. This is a consequence of the fact that such values are defined in terms of measurements carried out in the physical space, i.e. pointers and rulers which measure quantities living in the classical physical space. Thus, by using the continuum to model space-time, we also adopt the view that values of quantities should be real. Similarly in quantum theory, eigenvalues are expected to be real numbers. If we now consider quantum gravity, space-time is no longer a smooth manifold, thus it seems harder to justify the fact that physical quantities are real-valued.
3. *Probabilities*. In both classical physics and quantum theory probabilities are defined using the relative frequency interpretation. Namely, to obtain a probability of a certain outcome x_i, one has to repeat the measurement a large number

of times (N) and divide this number by the number of times the measurement gave the desired result (N_i), thus obtaining the probability of outcome x_i to be N/N_i. By necessity this has to be a rational number which lies between 1 (always obtain x_i) and 0 (never obtain x_i). If one takes the number of measurements to be infinite, then $N/N_i \in [0, 1]$. In this way quantum probabilities take their values in the closed interval $[0, 1]$. However, to make sense of such an interpretation one has to assume a continuum background space-time in which the measurements take place. However, if there is no classical spatio-temporal background in which observations can be made or if the relative frequency interpretation results meaningless for that background, then how can one justify the fact that probabilities should lie in the closed interval $[0, 1]$? There have been other notions of probabilities put forward, namely as propensities, measures of believes and possibilities. With respect to such interpretations it turns out that it is harder to justify the closed interval $[0, 1]$ as the space of probability values. In fact, all that is really required from such a space is that it be a partially ordered set with top and bottom element and equipped with a semi-additive structure so that probabilities of disjoint events can be added. However, other than this, there is no reason why $[0, 1]$ should be chosen.

Planck Scale We know by now that a theory of quantum gravity aims at reconciling quantum theory with general relativity. One of the striking features of quantum gravity is that, although both general relativity and quantum theory treat space-time as a 4-dimensional manifold equipped with a metric, quantum gravity would suggest that, at the microscopic scale, space-time is somewhat discrete. For example, approaches such as loop quantum gravity, spin foams, dynamical triangulation and causal set, all suggest that at the fundamental level space-time is discrete. In fact, it is believed that at the Planck scale classical space-time concepts cease to apply and a new way of viewing space-time is needed, which is not based on the notion of a continuum. In particular, the notion of point is not regarded as fundamental any more but it is replaced by the notion of a 'region' which physical objects occupy. For example in loop quantum gravity space-time is considered to be discrete, therefore there exist 'regions' of space-time with the smallest area and volume which can't be divided into smaller regions. Such a geometry of space time is described by spin foams [56]. Another example is given by Causal Dynamical Triangulations (CDT)[2], in which geometry is decomposed into triangular chunks or their higher-dimensional versions, respectively.

Therefore the continuum structure of space-time suggested by the two main ingredients of quantum gravity seems to be thrown into discussion by quantum gravity itself. This seems quite an odd predicament but it might suggest that perhaps a different mathematical structure, other than a smooth manifold, should model space-time. Clearly if such a new mathematical structure is defined, the questions to answer are

(a) Does it relate to the conception of space-time of both general relativity and quantum theory?
(b) Is the manifold structure emergent?

Relation to Instrumentalist Interpretation Most proponents of a theory of quantum gravity seem to adopt quantum theory without changing any of the mathematical formalism. This implies that they also inherit the conceptual interpretation of the theory which is an instrumentalist interpretation. This interpretation heavily relies on the distinction of observed system and observer and ascribes to the process of measurement an almost ontological status. However, such an interpretation seems to be problematic in the context of quantum gravity. In fact, if we consider quantum gravity as a theory of the entire universe, which is a closed system, then the observer and observed system distinction cannot be applied.[1]

Moreover, the notion of measurement in quantum theory requires a fixed space-time background in which the measurement takes place. However, quantum gravity is a theory of space-time itself seen as a dynamical quantity, hence the problem arises as to how one could measure space-time properties? Where would this measurement take place? Failing to clearly define measurements of space-time properties jeopardises the entire edifice of quantum theory, since any prediction of the theory is defined in terms of repeated sets of measurements. Hence the instrumentalist interpretation of quantum theory seems to be inconsistent with a theory of quantum gravity. However, if we analyse the mathematics which induces such an interpretation, we quickly arrive at the conclusion that it is a consequence of the Hilbert space formalism, which 'comes equipped' with the Born rule for probabilities. This reflection would suggest that maybe a theory of quantum gravity should make do without the Hilbert space formalism of quantum theory.

12.2 Modelling Space-Time as a Locale

Usually space-time is modelled in terms of the continuum, i.e. locally space-time is simply defined in terms of $\mathbb{R}^4$. However, as discussed in the introduction, in quantum gravity there seems to be no a priori justification for the adoption of the continuum to model space time. Moreover, it seems that the notion of a point in space time becomes secondary to the notion of a region. If one thinks about it, the notion of a space-time point is hardly justifiable. In fact, when talking about objects as being in space-time, these tend to occupy regions of space-time rather than points. This would suggest that perhaps a description of space-time in terms of extended regions might be more appropriate [11, 46]. As mentioned in the introduction, such extended regions would have to satisfy a certain algebra, whose operations would be given in terms of union, intersection etc.

[1]There have been attempts to reconcile the distinction between observed-system and observer by considering appropriate sub-systems [64], however, we feel that if quantum theory should be regarded as the ultimate theory, then even the interactions of such systems should undergo quantum laws, and the definition of a subsystem being an observer versus an observed system seems very hard to accommodate and be made precise.

With these reflections in mind, it seems reasonable to try to utilise some kind of locale (see Sect. 8.1) do describe space-time. A locale is the same thing as a frame which carries a Heyting algebra structure, and we have seen that the latter is the internal algebra of a topos. Hence, when considering topos theory as the mathematical framework of quantum theory, one facilitates a discrete notion of space-time.

In particular, in topos quantum theory space-time is modelled in terms of a locale $\mathcal{R}$ called the quantity value object. This object plays the same role as the Reals do in classical physics, i.e. it assigns values to quantities. These values, however, are not real numbers, but are related to them in some way. By using the locale $\mathcal{R}$, the concept of a point becomes secondary while the concept of a "region" of space becomes primary. Moreover, since a locale is equivalent to a complete Heyting algebra, the collection of space-time regions undergoes an intuitionistic rather than a classical (Boolean) logic. If one equates extended objects with the space-time regions they occupy, then, at the Planck scale, modelling space-time in terms of a locale would imply that statements of the form "an object ψ occupies region A or it does not" would be neither true nor false. This is another way of stating the fact that, in standard quantum theory, given a vector $\psi \in \mathcal{H}$ and a subspace W of $\mathcal{H}$, ψ can have non zero components in both W and $W^{\perp}$. So it would seem that modelling space-time, at least at the quantum level, as a locale, is in agreement with known facts about quantum theory. Motivated by this, let us try and make the locale definition of space-time more rigorous. As a first step we will introduce the rigorous definition of the quantity value object and show how this can be interpreted as a locale.

12.3 Topos Definition of the Quantity Value Object

We will now introduce the representation of the quantity value object $\mathcal{R}$ in $\mathbf{Sets}^{\mathcal{V}(\mathcal{H})^{op}}$. In classical theory, the quantity value object is simply the real numbers since each quantity takes on, as its value, an element of the reals. Similarly, in canonical quantum theory, we have the reals as the quantity value object. However, in topos quantum theory, the quantity value object is an object which has the same role as the reals have in standard quantum theory, but its elements will not be numbers. Clearly, each element of the quantity value object will be related to the real numbers in some way, but it will not be a real number itself.

It should be noted that in any topos there is an object which represents the real numbers, in fact there are several of them [50, 55]. However, the quantity value object we use for topos quantum theory is not one of them. As we will see, the motivations for defining the quantity value object in topos quantum theory come from physics requirements.

In the topos $\mathbf{Sets}^{\mathcal{V}(\mathcal{H})^{op}}$, the representation of the quantity value object $\mathcal{R}$ is given by the following presheaf [24, 28]:

Definition 12.3.1 The presheaf $\underline{\mathbb{R}^{\leftrightarrow}}$ acts as follows:

(i) Objects[2]:

$$\underline{\mathbb{R}^{\leftrightarrow}}_V := \{(\mu, \nu)|\mu, \nu :\downarrow V \to \mathbb{R},$$
$$\mu \text{ is order-preserving, } \nu \text{ is order-reversing} ; \mu \leq \nu\}.$$

(ii) Arrows: given two contexts $V' \subseteq V$ the corresponding morphisms is

$$\underline{\mathbb{R}^{\leftrightarrow}}_{V,V'} : \underline{\mathbb{R}^{\leftrightarrow}}_V \to \underline{\mathbb{R}^{\leftrightarrow}}_{V'}, \quad (\mu, \nu) \mapsto (\mu_{|V'}, \nu_{|V'}).$$

This presheaf is where physical quantities take their values, thus it has the same role as the reals in classical physics.

The reason why the quantity value object is defined in terms of order-reversing and order-preserving functions is because, in general, in quantum theory one can only give approximate values to the quantities.

Let us analyse the presheaf $\underline{\mathbb{R}^{\leftrightarrow}}$ in more depth. To this end, we assume that we want to define the value of a physical quantity A, given a state ψ. If ψ is an eigenstate of A, then we would get a sharp value of the quantity A, say a. If ψ is not an eigenstate, then we would get a certain range Δ of values for A.

Let us assume that $\Delta = [a, b]$. Then what the presheaf $\underline{\mathbb{R}^{\leftrightarrow}}$ does is to single out the extreme points a and b, so as to give a range (unsharp) of values for the physical quantity A. Obviously, since we are in the topos of presheaves, we have to define each object contextually, i.e. for each context $V \in \mathcal{V}(\mathcal{H})$. It is precisely to accommodate this fact that the pair of order-reversing and order-preserving functions was chosen to define the extreme values of our intervals.

To understand this we consider a context V, such that the self-adjoint operator $\hat{A}$, which represents the physical quantity A, does belong to V and such that the range of values of A at V is $[a, b]$. If we then consider the context $V' \subseteq V$, such that $\hat{A} \notin V$, we will have to approximate $\hat{A}$ so as to fit V'. The precise way in which self-adjoint operators are approximated is defined in [24, 28]. However, such an approximation will inevitably coarse-grain $\hat{A}$, i.e. it will deform it.

It follows that the range of possible values of such an approximated operator, which we denote by $\delta\hat{A}$, will be bigger. Therefore the range of values of $\delta\hat{A}$ at V' will be $[c, d] \supseteq [a, b]$, where $c \leq a$ and $d \geq b$. These relations between the extremal points can be achieved by the presheaf $\underline{\mathbb{R}^{\leftrightarrow}}$ through the order-reversing and order-preserving functions. Specifically, given that $a := \mu(V), b := \nu(V), V' \subseteq V$ implies that $c := \mu(V') \leq \mu(V)$ (μ being order-preserving) and $d := \nu(V') \geq \nu(V)$ (ν being order-reversing). Moreover, the fact that $\mu(V) \leq \nu(V)$ by definition, implies that as one goes to smaller and smaller contexts V', the intervals $(\mu(V'), \nu(V'))$ keep getting bigger or stay the same.

[2]A map $\mu :\downarrow V \to \mathbb{R}$ is said to be order-preserving if $V' \subseteq V$ implies that $\mu(V') \leq \mu(V)$. A map $\nu :\downarrow V \to \mathbb{R}$ is order-reversing if $V' \subseteq V$ implies that $\nu(V') \supseteq \nu(V)$.

An example of the quantity value object can be found in [17].

This object $\underline{\mathbb{R}^{\leftrightarrow}}$ can be given the structure of a locale and hence, that of a complete Heyting algebra.

12.4 Quantity Value Object as a Locale

We would like to apply the above mentioned procedure to the sheaf $\underline{\mathbb{R}^{\leftrightarrow}}$ so as to construct the associated locale in $Sh(\mathcal{V}(\mathcal{H}))$. As a first step, we will need to define a continuous map relating $\underline{\mathbb{R}^{\leftrightarrow}}$ to $\mathcal{V}(\mathcal{H})$ equipped with the Alexandroff topology. This can be done in various ways depending on what topology $\underline{\mathbb{R}^{\leftrightarrow}}$ is considered to have.

12.4.1 Locale Associated to $\underline{\mathbb{R}^{\leftrightarrow}}$ Part I

As shown in [75], it is possible to view $\underline{\mathbb{R}^{\leftrightarrow}}$ as a sub-object of a locale in $Sh(\mathcal{V}(\mathcal{H}))$. In order to understand how this is indeed possible we, first of all, have to consider the product $\mathcal{V}(\mathcal{H}) \times \mathbb{IR}$, where the set $\mathbb{IR}$ is the interval domain consisting of all compact intervals $[a, b]$ with $a, b \in \mathbb{R}$ and $a \leq b$ (see Appendix A.2). Normally, the interval domain comes equipped with the Scott topology (Appendix A.2), however, we will not consider such a topology when looking at the product $\mathcal{V}(\mathcal{H}) \times \mathbb{IR}$ since, instead of the product topology, we will consider a different topology. In particular, any algebra $V \in \mathcal{V}(\mathcal{H})$ and $(\mu, \nu) \in \underline{\mathbb{R}^{\leftrightarrow}}(V)$ define a basic open $U[V, (\mu, \nu)]$ with $(V', [a, b]) \in U[V, (\mu, \nu)]$ if $V' \subseteq V$ and $\mu(V') < a \leq b < \nu(V')$. It is straightforward to check that such opens form a basis for a topology on $\mathcal{V}(\mathcal{H}) \times \mathbb{IR}$. In terms of such a topology, the projection map $\pi_1 : \mathcal{V}(\mathcal{H}) \times \mathbb{IR} \to \mathcal{V}(\mathcal{H})$ is continuous, hence a map between locales. Since π_1 is an element of $\mathbf{Loc}/\mathcal{V}(\mathcal{H})$, we can now apply the previously defined procedure to obtain an internal locale $\underline{\mathrm{Loc}(\mathcal{V}(\mathcal{H}) \times \mathbb{IR})}$, whose associated frame is $\underline{\mathcal{O}(\mathcal{V}(\mathcal{H}) \times \mathbb{IR})}$. The sheaf $\underline{\mathcal{O}(\mathcal{V}(\mathcal{H}) \times \mathbb{IR})}$, for all $V \in \mathcal{V}(\mathcal{H})$, is given by:

$$\underline{\mathcal{O}(\mathcal{V}(\mathcal{H}) \times \mathbb{IR})}(\downarrow V) = \mathcal{O}(\mathcal{V}(\mathcal{H}) \times \mathbb{IR})|_{\downarrow V \times \mathbb{IR}}.$$

The associated presheaf $\underline{\mathbb{R}^{\leftrightarrow}}$ is a subpresheaf of $\underline{\mathcal{O}(\mathcal{V}(\mathcal{H}) \times \mathbb{IR})}$ given by

$$\begin{aligned} \phi_V : \underline{\mathbb{R}^{\leftrightarrow}}(V) &\to \underline{\mathcal{O}(\mathcal{V}(\mathcal{H}) \times \mathbb{IR})}(V) \\ (\mu, \nu) &\mapsto U[V, (\mu, \nu)]. \end{aligned}$$

We would now like to utilise this map to show that $\underline{\mathbb{R}^{\leftrightarrow}}$ is a sublocale of $\underline{\mathrm{Loc}(\mathcal{V}(\mathcal{H}) \times \mathbb{IR})}$. However, as a first step we need to define the locale structure on $\underline{\mathbb{R}^{\leftrightarrow}}$. The first step is to put an ordering on $\underline{\mathbb{R}^{\leftrightarrow}}(V)$ for each $V \in \mathcal{V}(\mathcal{H})$. We define $(\mu, \nu) \leq (\mu', \nu')$ iff for all $V' \subseteq V$, $\mu(V') \geq \mu'(V')$ and $\nu'(V') \geq \nu(V')$ The operations of join and meet are then defined as follows:

for all $V' \subseteq V$ we have

$$((\mu, \nu) \wedge (\mu', \nu'))(V') = ((\mu \wedge \mu')(V'), (\nu \wedge \nu')(V'))$$
$$:= (\max\{\mu(V'), \mu'(V')\}, \min\{\nu(V'), \nu'(V')\})$$

and

$$((\mu, \nu) \vee (\mu', \nu'))(V') = ((\mu \vee \mu')(V'), (\nu \vee \nu')(V'))$$
$$:= (\min\{\mu(V'), \mu'(V')\}, \max\{\nu(V'), \nu'(V')\}),$$

where the right hand side of both equations is defined with respect to the total ordering on $\mathbb{R}$. It is now straight forward to show that (8.1.1) holds.

Theorem 12.4.1 $\underline{\mathbb{R}^{\leftrightarrow}}$ *is a sublocale of* $\underline{\mathrm{Loc}(\mathcal{V}(\mathcal{H}) \times \mathbb{IR})}$.

Proof In this proof we will utilise the fact that a sublocale is given by the nucleus of the underlying frame (see Definitions 8.3.2(2) and 8.1.1). In this case, we will show that for each $V \in \mathcal{V}(\mathcal{H})$, ϕ_V defined above satisfies conditions (8.1.4). We will first show that ϕ_V preserves meets, i.e. $\phi_V((\mu, \nu) \wedge (\mu', \nu')) = \phi_V(\mu, \nu) \wedge \phi_V(\mu', \nu')$. Considering $\phi_V(\mu, \nu) \wedge \phi_V(\mu', \nu') = U[V, (\mu, \nu)] \wedge U[V, (\mu', \nu')]$ from the definition of the opens $U[V, (\mu, \nu)]$ it follows that $U[V, (\mu, \nu)] \wedge U[V, (\mu, \nu)] := U[V, (\mu \wedge \mu', \nu \wedge \nu')]$ where $\mu \wedge \mu'$ is defined for each context $V' \subseteq V$ as $(\mu \wedge \mu')(V') := \max\{\mu(V'), \mu'(V')\}$ and, similarly, $(\nu \wedge \nu')(V') := \min\{\nu(V'), \nu'(V)\}$. Applying the definitions we then get

$$\phi_V((\mu, \nu) \wedge (\mu', \nu')) = \phi_V((\mu \wedge \mu'), (\nu \wedge \nu')) = U[V, (\mu \wedge \mu', \nu \wedge \nu')].$$

It now remains to show that ϕ_V satisfies (B.2) and (B.3). This, however, is equivalent to the requirement that the image of $\underline{\mathbb{R}^{\leftrightarrow}}$ under ϕ is closed upon taking the pseudo-complement, i.e. for all $(\mu, \nu) \in \underline{\mathbb{R}^{\leftrightarrow}}(V)$ and for all $W \in \mathcal{O}(\mathcal{V}(\mathcal{H}) \times \mathbb{IR})(V)$, $(W \rightarrow U[V, (\mu, \nu)]) \in \phi_V(\underline{\mathbb{R}^{\leftrightarrow}}(V))$. Here $U[V, (\mu, \nu)] = \bigvee\{\overline{U'}|U' \wedge W \leq U[V, (\mu, \nu)]\}$. Since the opens $U[V, (\mu, \nu)]$ are a basis of the topology on $\mathcal{V}(\mathcal{H}) \times \mathbb{IR}$, the result of the theorem follows. □

The reason why the above topology was chosen for $\mathcal{V}(\mathcal{H}) \times \mathbb{IR}$ is because the author in [75] wanted to analyse the connection between the topos approach to quantum theory put forward by Isham and Doering and described above with an alternative formulation by Heunen, Landsman and Spitters in [42]. This latter formulation has as a starting point a C^*-algebra $\mathcal{A}$ and an ambient topos $\mathbf{Sets}^{C(\mathcal{A})}$ where $C(\mathcal{A})$ is the category of abelian subalgebras of $\mathcal{A}$. They then promote $\mathcal{A}$ to an internal C^*-algebra $\underline{\mathcal{A}}$ in $\mathbf{Sets}^{C(\mathcal{A})}$ and define the spectrum $\underline{\Sigma}$ as an internal locale. In this context, the quantity value object is identified with the internal locale of interval domain $\overline{\mathbb{IR}}$ (See Appendix A.2).

Through the above construction it transpires that the quantity value object $\underline{\mathbb{R}^{\leftrightarrow}}$ is intimately related to $\overline{\mathbb{IR}}$.

12.4.2 Locale Associated to $\underline{\mathbb{R}^{\leftrightarrow}}$ Part II

We now would like to analyse an alternative topology for the quantity value object $\underline{\mathbb{R}^{\leftrightarrow}}$.

Given such a presheaf, we define the set $R := \coprod_{V\in\mathcal{V}(\mathcal{H})} \underline{\mathbb{R}^{\leftrightarrow}}_V$ with associated map $p_R : R \to \mathcal{V}(\mathcal{H})$ such that $p_R(\mu, \nu) = V$ for $(\mu, \nu) \in \underline{\mathbb{R}^{\leftrightarrow}}_V$. We would like to define a topology on R such that the map p_R be continuous.

A possibility would be to define the discrete topology on each fibre $p_R^{-1}(V) = \underline{\mathbb{R}^{\leftrightarrow}}_V$g In or Sign Up https://www.faceb which would accommodate for the fact that $p_R : R \to \mathcal{V}(\mathcal{H})$ is an étale bundle. We could then define the disjoint union topology, but this would not account for the 'horizontal' topology on the base category $\mathcal{V}(\mathcal{H})$ given by the Alexandroff topology.

Another possibility would be to consider as a basis for the topology on R the collection of all open sub-objects. Thus a basis set would be of the form $S = \coprod_{V\in\mathcal{V}(\mathcal{H})} \underline{S}_V$ such that $\underline{S}_V$ is open in $\underline{\mathbb{R}^{\leftrightarrow}}_V$, which is equipped with the discrete topology. In such a setting, the 'horizontal' topology would be accounted for by the presheaf maps.

Since each $\underline{\mathbb{R}^{\leftrightarrow}}_V$ is equipped with the discrete topology, the topology on the entire set R would essentially be the discrete topology in which all sub-objects of R are open.

Obviously, with respect to such a topology, the bundle map p_R would be continuous since for each $\downarrow V$, $p^{-1}(\downarrow V) = \coprod_{V'\in\downarrow V} \underline{\mathbb{R}^{\leftrightarrow}}_{V'}$ will represent the open sub-object whose value is $\underline{\mathbb{R}^{\leftrightarrow}}_{V'}$ for all $V' \in\downarrow V$ and $\emptyset$ everywhere else.

We now have at our disposal the continuous map $p_R : R \to \mathcal{V}(\mathcal{H})$ which is a locale map. This gives rise to the geometric morphism $p_R : Sh(R) \to Sh(\mathcal{V}(\mathcal{H}))$. At this point we can apply the procedure defined in the previous section to construct the internal locale $\mathcal{O}(R) = p_{R*}(\underline{\Omega}^R)$ such that, for any $V \in \mathcal{V}(\mathcal{H})$, we obtain

$$(p_{R*}(\underline{\Omega}^R))_V = \underline{\Omega}^R(p_R^{-1}(V)) = \underline{\Omega}^R(\coprod_{V'\in\downarrow V} \underline{\mathbb{R}^{\leftrightarrow}}_{V'})$$

where

$$\underline{\Omega}^R(\coprod_{V'\in\downarrow V} \underline{\mathbb{R}^{\leftrightarrow}}_{V'}) = \{\underline{S} \in \mathcal{O}(R) | S \subseteq \coprod_{V'\in\downarrow V} \underline{\mathbb{R}^{\leftrightarrow}}_{V'}\}.$$

We recall that

$$\underline{S} := \coprod_{V'\in\downarrow V} \underline{S}_{V'} \tag{12.4.1}$$

where $\underline{S}_{V'}$ is an open subset of $\underline{\mathbb{R}^{\leftrightarrow}}_{V'}$. Since the latter has a discrete topology, the basis will be formed of singletons. These are defined, for each $V' \subseteq V$ as a pair $(\mu, \nu) :\downarrow V' \to \mathbb{R}$ of order preserving and order reversing functions taking values in $\mathbb{R}$, which describe varying intervals of real numbers. Conceptually what this means is the following: for each context $V \in \mathcal{V}(\mathcal{H})$, which represents a classical snapshot of the quantum system, we obtain a "locale" space-time seen as a collection of unions of varying intervals of real numbers. Such a space-time has the property that when considering two contexts V' and V such that $V' \subseteq V$, then the intervals of real numbers describing the space-time associated to the context V' are "bigger", i.e. are less precise and thus have less information than the intervals of the space-time associated to the bigger context. Since these intervals are interpreted as the regions of space-time which physical objects occupy, what the above result signifies is that when going to a smaller contexts V', which contains less information due to coarse-graining, then the precision with which one is able to determine the position of physical objects decreases.

12.4.3 Locale Associated to $\underline{\breve{\mathbb{R}}}$

An alternative way of associating to the quantity value object a locale is by considering the definition of the quantity value object put forward in [27]. This new definition was the result of introducing the notion of a group and a group action in the topos quantum theory framework. To achieve this, the topos utilised to define quantum theory had to be slightly changed so that now the base category, although still remaining $\mathcal{V}(\mathcal{H})$, is considered to be invariant under any group transformation, i.e. the group acts trivially. This slightly different category is denoted by $\mathcal{V}_f(\mathcal{H})$ where f stands for fixed. All the group actions were relegated to an intermediate category, resulting in a construction of sheaves over $\mathcal{V}_f(\mathcal{H})$ as sheaves over this intermediate category and, then, "pushed down" to sheaves on $\mathcal{V}_f(\mathcal{H})$ in the appropriate way. We will not go into the details of how this is done, but the interested reader should refer to [27]. All that we will do in the present context is to state how the resulting new quantity value object is defined.

Definition 12.4.1 The quantity value object $\underline{\breve{\mathbb{R}}}^{\leftrightarrow}$ is a presheaf of order-preserving and order-reversing functions on $\mathcal{V}_f(\mathcal{H})$ defined as follows:

– On objects $V \in \mathcal{V}_f(\mathcal{H})$ we have

$$\underline{\breve{\mathbb{R}}}^{\leftrightarrow}_V := \coprod_{\phi_g \in Hom(\downarrow V, \mathcal{V}(\mathcal{H}))} \underline{\mathbb{R}^{\leftrightarrow}}_{\phi_g(V)} \tag{12.4.2}$$

where each[3]

$$\underline{\mathbb{R}^{\leftrightarrow}}_{\phi_g(V)} := \{(\mu,\nu)|\mu \in OP(\downarrow \phi_g(V), \mathbb{R}) \ , \ \mu \in OR(\downarrow \phi_g(V), \mathbb{R}), \ \mu \leq \nu\}. \tag{12.4.3}$$

Here the maps $\phi_g : \mathcal{V}_f(\mathcal{H}) \rightarrow \mathcal{V}(\mathcal{H})$ defined by $\phi_g(V) = \hat{U}_g V \hat{U}_g^{-1}$ for $g \in G$, represent faithful representations of a given group G. The downward set $\downarrow \phi_g(V)$ comprises all the sub-algebras $V' \subseteq \phi_g(V)$. The condition $\mu \leq \nu$ implies that for all $V' \in \downarrow \phi_g(V)$, $\mu(V') \leq \nu(V')$.

- On morphisms $i_{V'V} : V' \rightarrow V$ $(V' \subseteq V)$, we get:

$$\underline{\breve{\mathbb{R}}}^{\leftrightarrow}(i_{V'V}) : \underline{\breve{\mathbb{R}}}^{\leftrightarrow}_V \rightarrow \underline{\breve{\mathbb{R}}}^{\leftrightarrow}_{V'} \tag{12.4.4}$$

$$\coprod_{\phi_g \in Hom(\downarrow V, \mathcal{V}(\mathcal{H}))} \underline{\mathbb{R}^{\leftrightarrow}}_{\phi_g(V)} \rightarrow \coprod_{\phi_{g'} \in Hom(\downarrow V', \mathcal{V}(\mathcal{H}))} \underline{\mathbb{R}^{\leftrightarrow}}_{\phi_{g'}(V')} \tag{12.4.5}$$

where for each element $(\mu, \nu) \in \underline{\mathbb{R}^{\leftrightarrow}}_{\phi_g(V)}$ we obtain

$$\underline{\breve{\mathbb{R}}}^{\leftrightarrow}(i_{V'V})(\mu,\nu) := \underline{\mathbb{R}^{\leftrightarrow}}(i_{\phi_g(V), \phi_{g'}(V')})(\mu,\nu) \tag{12.4.6}$$

$$= (\mu_{|\phi_g(V')}, \nu_{|\phi_{g'}(V')}) \tag{12.4.7}$$

where $\mu_{|\phi_g(V')}$ denotes the restriction of μ to $\downarrow \phi_{g'}(V') \subseteq \downarrow \phi_g(V)$, and analogously for $\nu_{|\phi_{g'}(V')}$.

We are now interested in defining a topology for $\underline{\breve{\mathbb{R}}}^{\leftrightarrow}$. This was done in [27]. As a first step we define the set

$$\mathcal{R} = \coprod_{V \in \mathcal{V}_f(\mathcal{H})} \underline{\breve{\mathbb{R}}}^{\leftrightarrow}_V = \bigcup_{V \in \mathcal{V}_f(\mathcal{H})} \{V\} \times \underline{\breve{\mathbb{R}}}^{\leftrightarrow}_V \tag{12.4.8}$$

where each $\underline{\breve{\mathbb{R}}}^{\leftrightarrow}_V := \coprod_{\phi_g \in Hom(\downarrow V, \mathcal{V}(\mathcal{H}))} \underline{\mathbb{R}^{\leftrightarrow}}_{\phi_g(V)}$.

The above represents a bundle over $\mathcal{V}_f(\mathcal{H})$ with bundle map $p_{\mathcal{R}} : \mathcal{R} \rightarrow \mathcal{V}_f(\mathcal{H})$ defined by $p_{\mathcal{R}}(\mu, \nu) = V$, where V is the context such that $(\mu, \nu) \in \underline{\mathbb{R}^{\leftrightarrow}}_{\phi_g(V)}$. In this setting $p_{\mathcal{R}}^{-1}(V) = \underline{\breve{\mathbb{R}}}^{\leftrightarrow}_V$ are the fibres of the map $p_{\mathcal{R}}$.

We would like to define a topology on $\mathcal{R}$ with the minimal requirement that the map $p_{\mathcal{R}}$ is continuous. We know that the category $\mathcal{V}_f(\mathcal{H})$ has the Alexandroff topology whose basis open sets are of the form $\downarrow V$ for some $V \in \mathcal{V}_f(\mathcal{H})$. Thus we are looking for a topology such that the pullback $p_{\mathcal{R}}^{-1}(\downarrow V) := \coprod_{V' \in \downarrow V} \underline{\breve{\mathbb{R}}}_{V'}$ is open in $\mathcal{R}$.

[3]Here OP stands for *order-preserving* while OR stands for *order-reversing*.

Following the discussion at the end of [27, Section 2.1], we know that each $\underline{\mathbb{R}^{\leftrightarrow}}$ is equipped with the discrete topology in which all sub-objects are open (in particular each $\underline{\mathbb{R}^{\leftrightarrow}}_V$ has the discrete topology).

Therefore we define a subsheaf $\underline{\breve{Q}}$ of $\underline{\breve{\mathbb{R}}}^{\leftrightarrow}$ to be *open* if for each $V \in \mathcal{V}_f(\mathcal{H})$ the set $\underline{\breve{Q}}_V \subseteq \underline{\breve{\mathbb{R}}}_V$ is open, i.e., each $\underline{Q}_{\phi_g(V)} \subseteq \underline{\mathbb{R}^{\leftrightarrow}}_{\phi_g(V)}$ is open in the discrete topology on $\underline{\mathbb{R}^{\leftrightarrow}}_{\phi_g(V)}$. It follows that the sheaf $\underline{\breve{\mathbb{R}}}^{\leftrightarrow}$ gets induced the discrete topology in which all sub-objects are open. In this setting the 'horizontal' topology on the base category $\mathcal{V}_f(\mathcal{H})$ would be accounted for by the sheaf maps.

For each $\downarrow V$ we then obtain the open set $p_{\mathcal{R}}^{-1}(\downarrow V)$ which has value $\underline{\breve{\mathbb{R}}}_{V'}$ at contexts $V' \in \downarrow V$ and $\emptyset$ everywhere else.

Given the continuous map $p_{\mathcal{R}}$, this can be seen as an element in $\mathbf{Loc}/\mathcal{V}_f(\mathcal{H})$, allowing us to construct the corresponding internal locale in $Sh(\mathcal{V}_f(\mathcal{H}))$. In particular, we consider the induced geometric morphism $p_{\mathcal{R}} : Sh(\mathcal{R}) \to Sh(\mathcal{V}_f(\mathcal{H}))$. The internal locale we are looking for is then given by $\underline{\mathcal{O}(\mathcal{R})} = p_{\mathcal{R}*}(\Omega^{\mathcal{R}})$, therefore, for any open $\downarrow V \in \mathcal{V}_f(\mathcal{H})$ we obtain

$$\underline{\mathcal{O}(\mathcal{R})}(\downarrow V) = \underline{\Omega}^{\mathcal{R}}(p_{\mathcal{R}}^{-1}(\downarrow V)) = \underline{\Omega}^{\mathcal{R}}\Big(\coprod_{V' \in \downarrow V} \underline{\breve{\mathbb{R}}}_{V'}\Big)$$

where

$$\underline{\Omega}^{\mathcal{R}}\Big(\coprod_{V' \in \downarrow V} \underline{\breve{\mathbb{R}}}_{V'}\Big) = \{\underline{U} \in \mathcal{O}(\mathcal{R}) | U \subseteq \coprod_{V' \in \downarrow V} \underline{\breve{\mathbb{R}}}_{V'}\} = \mathcal{O}(\mathcal{R})|_{\downarrow V}$$

and

$$\underline{U} := \coprod_{V' \in \downarrow V} \ \coprod_{\phi_g \in Hom(\downarrow V', \mathcal{V}(\mathcal{H}))} \underline{U}_{\phi_g(V')}. \tag{12.4.9}$$

In this situation, for each context V the "locale" space-time is a collection of unions of equivalence classes of varying intervals of real numbers where such equivalence is defined with respect to a group G.

A tentative interpretation is that these space-time regions represent diffeomorphic regions of space-time. Since such a locale is a sheaf, one could interpret a global section as a particular choice of space-time. Each such global section (choice of space-time) would then be related to each other by space-time diffeomorphisms.

Summary In this section we have defined three different locales representing the quantity value object. The first two of these locales are very similar and only differ in the type of topology used to construct such a locale. The third locale, on the other hand, carries also information regarding group transformations and, as such, it can be seen as a covariant description of R.

12.5 Modelling Space-Time as a Locale

In the previous section We have seen that the quantity value object of topos quantum theory can be interpreted as an internal locale $\mathcal{R}$ and, hence, as a complete Heyting algebra.

If we then modelled space-time in terms of the locale $\mathcal{R}$, the concept of a point would be secondary while the concept of a "region" of space would be primary. Moreover, since a locale is equivalent to a complete Heyting algebra, the collection of space-time regions would undergo an intuitionistic rather than a classical (Boolean) logic. If one equates extended objects with the space-time regions they occupy, then, at the Planck scale, modelling space-time in terms of a locale would imply that statements of the form "an object ψ occupies region A or it does not" would be neither true nor false. This is another way of stating the fact that in standard quantum theory, given a vector $\psi \in \mathcal{H}$ and a subspace W of $\mathcal{H}$, ψ can have non zero components in both W and $W^{\perp}$. So it would seem that modelling space-time, at least at the quantum level, as a locale is in agreement with known facts about quantum theory. Motivated by this, let us try and make the locale definition of space-time more rigorous.

Since we would like to somehow retrieve the classical concept of space-time in the appropriate limit, we take quantum space-time to be constructed in a similar way as in classical physics, i.e. as the fourth power of the quantity value object. In order to achieve this we need to introduce the notion of tensor product of locales. We will do this in terms of frames.

Definition 12.5.1 ([73]) Given two frames A and B, the tensor product $A \otimes B$ is defined to be the frame represented by the following presentation:

$$\mathcal{T}\langle a \otimes b, a \in A \text{ and } b \in B |$$

$$\bigwedge_i (a_i \otimes b_i) = \Big(\bigwedge_i a_i\Big) \otimes \Big(\bigwedge_i b_i\Big) \tag{12.5.1}$$

$$\bigvee_i (a_i \otimes b) = \Big(\bigvee_i a_i\Big) \otimes b \tag{12.5.2}$$

$$\bigvee_i (a \otimes b_i) = a \otimes \Big(\bigvee_i b_i\Big). \tag{12.5.3}$$

In other words, we form the formal products, $a \otimes b$, of elements $a \in A$, $b \in B$ and WE subject them to the relations in Eqs. (12.5.1)–(12.5.3). We note that there are injective maps

$$\begin{aligned} i : A &\to A \otimes B \\ a &\mapsto a \otimes \text{true} \end{aligned} \tag{12.5.4}$$

and

$$j : B \to A \otimes B$$
$$b \mapsto \text{true} \otimes b. \tag{12.5.5}$$

Alternatively, following [73, Proposition 6.4.2], it is possible to define the tensor product of frames in a more categorical way as coproducts[4] of frames.

Lemma 12.5.1 *Given three frames A, B, C and frame homomorphisms $f : A \to C$ and $g : B \to C$, there exists a unique frame homomorphism $h : A \otimes B \to C$ such that $f = h \circ i$ and $g = h \circ j$.*

An immediate corollary of the above is:

Corollary 12.5.1 $\text{pt}(A \otimes B) \simeq \text{pt}(A) \times \text{pt}(B)$

Proof The points $\text{pt}(A \otimes B)$ correspond to frame maps $p^{-1} : A \otimes B \to \Omega$. By the property of coproducts, such maps are uniquely defined as the product maps $[p_A^{-1}, p_B^{-1}]$ for the frame maps $p_A^{-1} : A \to \Omega$ and $p_B^{-1} : B \to \Omega$. □

Equipped with this definition, we can now define space-time to be the locale $\mathcal{R} \otimes \mathcal{R} \otimes \mathcal{R} \otimes \mathcal{R} =: \mathcal{R}^4$ constructed by iterations of Definition 12.5.1. Because of Lemma 12.5.1, the object $\mathcal{R}^4$ is defined as a coproduct in $Sh(\mathcal{V}(\mathcal{H}))$.

In this setting we then interpret space-time as the internal locale $\mathcal{R}^4$ in $Sh(\mathcal{V}(\mathcal{H}))$. Clearly according to which of the above defined locales we consider $\mathcal{R}$ to be, we will obtain a slightly different interpretation of space-time, however all agree on the fact that the basic notions are now given by extended regions rather than points. It is not clear at this point which locale would be more suitable to represent space-time. We leave the answer to this question as a topic for a subsequent study.

The implications of adopting a localic description of space-time and the detailed analysis of the physics it might ensue is beyond the scope of the present chapter. In the present instance we are only interested in describing possible alternative mathematical descriptions of space-time which do not rely on the notion of the continuum.

A natural question to ask is if the notion of a point can be defined within these locales. In particular, can the notion of a sheaf of points of the locales be defined? Making use of Corollary 12.5.1, we know that the presheaf $\text{pt}(\mathcal{R}^4) \simeq \times_4 \text{pt}(\mathcal{R})$, therefore it suffice to analyse the single presheaf $\text{pt}(\mathcal{R})$. We will now analyse such an object for each of the locales defined in the previous section.

[4]Coproducts are the categorical generalisation of disjoint unions in **Sets**.

12.5.1 Can We Retrieve the Notion of Space-Time Points?

It is natural to ask the question of how the points of the locales used to model space-time are represented assuming they exists. It turns out that it is possible to retrieve the notion of a point for all three of the locales defined in the previous section. The "collection" of such points will itself be a sheaf in $Sh(X)$. In order to define the points of internal locales we will make use of the functor $IPt : \text{Loc}(Sh(X)) \to Top(Sh(X))$, where $\text{Loc}(Sh(X))$ is the category of internal locales in $Sh(X)$, while $Top(Sh(X))$ is the category of internal topological spaces in $Sh(X)$. This functor is the internal analogue of the functor defined in (12.5.1). IPt is the right adjoint of $ILoc : Top(Sh(X)) \to \text{Loc}(Sh(X))$ which is the internal analogue of (8.1.7).

Internal topological spaces are defined as follows:

Definition 12.5.2 ([68]) A topological space object in a topos τ consists of a pair (A, T_A) where $T_A \subseteq PA$ such that the following conditions are satisfied

1. $\emptyset \in T_A$ and $A \in T_A$.
2. For all $B, B' \in PA$ if $B \in T_A$ and $B' \in T_A$ then $B \cap B' \in T_A$.
3. For all $S \in P(PA)$, if $S \subseteq T_A$ then $\bigcup S \in T_A$.

Where $\bigcup : P(PA) \to PA$ is the exponential adjoint of the characteristic morphism of the sub-object of $P(PA) \times A$ consisting of those elements (S, a), such that there exists a $B \in PA$ for which $a \in B$ and $B \in S$. Similarly $\bigcap : P(PA) \to PA$ is the exponential adjoint of the characteristic morphism of the sub-object of $P(PA) \times A$ consisting of those elements (S, a) such that for all $B \in PA$ if $B \in S$, then $a \in S$.
In the particular case in which the topos τ is $Sh(X)$ for some topological space X, the above definition reduces to the following

Definition 12.5.3 ([69]) Consider a sheaf $\underline{A}$ in $Sh(X)$ for some topological space X with topology T_X and its associated etalé bundle $A \xrightarrow{p} X$. A topological structure on $\underline{A}$ consists of a second topology on A, T_A, which is courser than the etalé topology but which still makes the map p continuous.
In the present situation, the topological space we are considering is $\mathcal{V}(\mathcal{H})$ equipped with the Alexandrov topology. In this case $Sh(\mathcal{V}(\mathcal{H})) \simeq \mathbf{Sets}^{\mathcal{V}(\mathcal{H})^{\text{op}}}$, therefore also an internal topological space will be a presheaf $\underline{Y}$. This implies that for each $V \in \mathcal{V}(\mathcal{H})$, $\underline{Y}_V$ will be a topological space in **Sets** with the appropriate topology. The corresponding etalé space would then be $Y = \coprod_{V \in \mathcal{V}(\mathcal{H}} \underline{Y}_V$ and the finer topology T_Y will be identified with the appropriate disjoint union topology. The definition of $ILoc : Top(Sh(X)) \to \text{Loc}(Sh(X))$ is now straightforward: given an internal topological space $\underline{Y}$ then $\text{Loc}(\underline{Y})$ is a presheaf such that, for each $V \in \mathcal{V}(\mathcal{H})$, $ILoc(\underline{Y})_V := \{U | U \text{is open in } \underline{Y}_V\}$ and it is the locale associated to the topological space $\underline{Y}_V$ in **Sets**. Therefore, for each $V \in \mathcal{V}(\mathcal{H})$ we have $ILoc(\underline{Y})_V = \text{Loc}(\underline{Y}_V)$.

In the same way $IPt : \text{Loc}(Sh(X)) \to Top(Sh(X))$ is such that, given a locale $\underline{L}$ then $I\text{pt}(\underline{L})_V := \text{pt}(\underline{L}_V)$. The adjointness relation $Loc \dashv Pt$ induces the adjunction $ILoc \dashv IPt$.

In the previous sections we have seen that, for any topological space X, we have the following equivalence of categories $\text{Loc}(Sh(X)) \simeq \mathbf{Loc}/X$, hence another way of defining a point of an internal locale $\underline{L}$ in $Sh(X)$ is as a continuous cross section $\pi : \underline{L} \to X$.

In the following we will consider each locale defined in the previous section and for each of them construct its points. For notational simplicity we will denote *IPt* simply as *Pt*, keeping in mind that we are now considering internal locales and topological spaces.

Point of the Locale $\underline{\text{Loc}(\mathcal{V}(\mathcal{H}) \times \mathbb{IR})}$ As briefly discussed in [42], the points of the locale $\underline{\text{Loc}(\mathcal{V}(\mathcal{H}) \times \mathbb{IR})}$ are given by the presheaf $\text{pt}(\underline{\text{Loc}(\mathcal{V}(\mathcal{H}) \times \mathbb{IR})})$ such that for each $V \in \mathcal{V}(\mathcal{H})$ we obtain the set given by a collection of pairs of sub-objects of $\underline{\mathbb{Q}}$, as follows:

$$\begin{aligned} \text{pt}(\underline{\text{Loc}(\mathcal{V}(\mathcal{H}) \times \mathbb{IR})})(V) = \{(\underline{L}, \underline{U})_{|\downarrow V} | \; \forall\; V' \subseteq V, \\ (L(V'), U(V')) \text{ is an element of the locale } \mathbb{IR}\}, \end{aligned} \tag{12.5.6}$$

where both $\underline{L}$ and $\underline{U}$ are sub-objects of the constant presheaf $\underline{\mathbb{Q}}$ which assigns to each $V \in \mathcal{V}(\mathcal{H})$ the rationals $\mathbb{Q}$. Therefore, for each $V' \subseteq V$, the pair $(\underline{L}(V'), \underline{U}(V'))$ satisfies Definition A.1.1.

However, since we are interested in the sublocale $\underline{\mathbb{R}^{\leftrightarrow}}$, we will only consider a subpresheaf of $\text{pt}(\underline{\text{Loc}(\mathcal{V}(\mathcal{H}) \times \mathbb{IR})})$. In order to understand how the points in $\underline{\mathbb{R}^{\leftrightarrow}}$ are defined, let us go back to the locale $\underline{\text{Loc}(\mathcal{V}(\mathcal{H}) \times \mathbb{IR})}$ with associated frame $\underline{\mathcal{O}(\mathcal{V}(\mathcal{H}) \times \mathbb{IR})}$, where $\underline{\mathcal{O}(\mathcal{V}(\mathcal{H}) \times \mathbb{IR})}(V) = \mathcal{O}(\downarrow V \times \mathbb{IR})$. Each of these sets is isomorphic to the set of order-preserving functions $OP(\downarrow V, \mathcal{O}(\mathbb{IR}))$. In particular, given a topological space X, there is a standard bijection $\mathcal{O}(X) \simeq C(X, S)$ where S is the Sierpinski space $\{0, 1\}$ whose only non-trivial open set is $\{1\}$. This bijection is defined as $f : U \mapsto \chi_U$ and $f^{-1} : g \mapsto g^{-1}(\{1\})$. Applying this result to the case at hand, we obtain that $\mathcal{O}(\downarrow V \times \mathbb{IR}) \simeq C(\downarrow V \times \mathbb{IR}, S)$. However, as shown in [42], because of lambda-abstraction $C(\downarrow V \times \mathbb{IR}, S) \simeq C(\downarrow V, S^{\mathbb{IR}})$. Moreover, $C(\mathbb{IR}, S) \simeq \mathcal{O}(\mathbb{IR})$, therefore $\mathcal{O}(\downarrow V \times \mathbb{IR}) \simeq C(\downarrow V, \mathcal{O}(\mathbb{IR}))$. In this setting continuity is given by monotonicity, therefore we replace $C(\downarrow V, \mathcal{O}(\mathbb{IR}))$ by $OP(\downarrow V, \mathcal{O}(\mathbb{IR}))$ obtaining the following isomorphisms:

$$\begin{array}{llll} \mathcal{O}(\downarrow V \times \mathbb{IR}) \to & C(\downarrow V \times \mathbb{IR}, S) \to & C(\downarrow V, S^{\mathbb{RR}}) \to & OP(\downarrow V, \mathcal{O}(\mathbb{IR})) \\ U & \mapsto \chi_U & \mapsto h_{\chi_U} & \mapsto g_{h_{\chi_U}} \end{array}$$

where

$$\chi_U(V', [a, b]) = \begin{cases} 1 & \text{iff } (V', [a, b]) \in U \\ 0 & \text{otherwise} \end{cases}, \qquad h_{\chi_U} :' V \mapsto \chi_U(V', -)$$

and

$$g_{h_{\chi_U}} : V' \mapsto \Big((\chi_U(V', -))^{-1}(\{1\}) = \{[a,b] | \chi_U(V', [a,b]) = 1\} \\ = \{[a,b] | (V', [a,b]) \in U\} \Big). \tag{12.5.7}$$

We will now utilise the isomorphism $\mathcal{O}(\downarrow V \times \mathbb{IR}) \simeq OP(\downarrow, \mathcal{O}(\mathbb{IR}))$ to allow us to relate points of $\underline{\text{Loc}(\mathcal{V}(\mathcal{H}) \times \mathbb{IR})}$, defined in (12.5.6) to order-reversing maps $\downarrow V \to \mathcal{O}(\mathbb{IR})$. In particular, to each element of $\mathbb{IR}$ there corresponds an open $U \in \mathcal{O}(\mathbb{IR})$ via the equivalence between $p : 1 \to \mathbb{IR}$ and $p^{-1} : \mathcal{O}(\mathbb{IR}) \to \Omega$ and, to each such open U, there corresponds a map $g_{h_{\chi_U}}$. Now, since we are in **Sets**, the element $(L(V'), U(V'))$ defines a compact interval $[\sup(L(V')), \inf(U(V'))]$ so that $\text{pt}(\mathbb{IR})$ can be identified with the classical Scott interval domain (See Appendix). Therefore, to each pair $(L(V'), U(V'))$, there corresponds an open $W \in \mathcal{O}(\mathbb{IR})$ such that the interval $[sup(L(V')), inf(U(V'))] \in W$. However, since each such pair is defined for all $\downarrow V$, then each $(L, U)|_{\downarrow V}$ defines a map $g_{h_{\chi_U}} : \downarrow V \to \mathcal{O}(\mathbb{IR})$ which, in turn, identifies an open $U \in \mathcal{O}(\mathcal{V}(\mathcal{H}) \times \mathbb{IR})$. In this way, the set $\text{pt}(\underline{\text{Loc}(\mathcal{V}(\mathcal{H}) \times \mathbb{IR})})(V)$ can be written as

$$\text{pt}(\underline{\text{Loc}(\mathcal{V}(\mathcal{H}) \times \mathbb{IR})})(V) = \{(\underline{L}, \underline{U})_{|\downarrow V} | \; \forall \; V' \subseteq V, \\ (V', [\sup(L(V')), \inf(U(V')]) \in \mathcal{O}(\downarrow V \times \mathbb{IR})\} \\ \simeq \{g_{h_{\chi_U}} : \downarrow V \to \mathcal{O}(\mathbb{IR})\}.$$

However, we are only interested in elements of the sublocale $\underline{\mathbb{R}^{\leftrightarrow}}$. Recalling that the sublocale map takes elements (μ, ν) and maps them to base opens $U[V, (\mu, \nu)]$, then

$$\text{pt}(\underline{\mathbb{R}^{\leftrightarrow}})(V) = \{(\underline{L}, \underline{U})_{|\downarrow V} | \exists \; (\mu, \nu) \in \underline{\mathbb{R}^{\leftrightarrow}}_V \text{ s.t. } \forall \; V' \subseteq V, \\ \mu(V') < \sup(L(V')) \leq \inf(U(V')) < \nu(V')\}.$$

Recall that the compact intervals $[a, b]$ seen as elements of $\mathbb{IR}$ are ordered by reverse inclusion. Therefore, the condition $\mu(V') < \sup(L(V')) \leq \inf(U(V')) < \nu(V')$ implies that the interval $[\mu(V'), \nu(V')]$ is the biggest, i.e. the one with least amount of information.

In this setting, $\text{pt}(\underline{\mathbb{R}^{\leftrightarrow}})$ can be seen as a presheaf such that for each $V \in \mathcal{V}(\mathcal{H})$, the set $\text{pt}(\underline{\mathbb{R}^{\leftrightarrow}})(V)$ is a collection of assignments to each $V' \subseteq V$ of an interval domain $[\sup(L(V')), \inf(U(V'))]$ associated to a global element $(\mu, \nu) \in \Gamma(\underline{\mathbb{R}^{\leftrightarrow}})$. This interval domain represents a refinement of the information contained in the interval $[\mu(V'), \nu(V')]$.

Points of the Locale R We are now interested in analysing the points of the locale R with associated frame $\underline{\mathcal{O}(R)}$. In particular, the correspondence between **Loc**$(Sh(X))$ and **Loc**$/X$ for some topological space X implies that the points of the locale R

correspond to continuous cross-sections of $\pi : R \to \mathcal{V}(\mathcal{H})$. These are the locale maps:

$$\phi : \mathcal{V}(\mathcal{H}) \to R$$
$$V \mapsto (V, \overline{\phi}(V))$$

where $\overline{\phi}(V) \in \underline{\mathbb{R}^{\leftrightarrow}}(V)$. At the level of frames we then have

$$\phi^{-1} : \mathcal{O}(R) \to \mathcal{O}(\mathcal{V}(\mathcal{H}))$$
$$U_{V,(\mu,\nu)} \mapsto \{V' \in \mathcal{V}(\mathcal{H}) | \overline{\phi}(V') = (\mu, \nu)|_{V'}\}$$

where $U_{V,(\mu,\nu)} = \{(V', (\mu, \nu)|_{V'} | V' \subseteq V\}$ is a basic open set for the topology on R. Since ϕ is continuous, the set $\phi^{-1}(U_{V,(\mu,\nu)})$ is open in the Alexandroff topology of $\mathcal{V}(\mathcal{H})$. This implies that if $\overline{\phi}(V) = (\mu, \nu)$ and $V' \subseteq V$ then $\overline{\phi}(V') = (\mu, \nu)|_{V'}$. Hence a point of the locale R corresponds to a global section of $\underline{\mathbb{R}^{\leftrightarrow}}$, i.e. $\text{pt}(R) \simeq \Gamma\underline{\mathbb{R}^{\leftrightarrow}}$.

In [19, Proposition 4.2] it was shown that $\Gamma\underline{\mathbb{R}^{\leftrightarrow}} \simeq OP(\mathcal{V}(\mathcal{H}), \mathbb{IR})$. As a first step the authors notices that each order-preserving map $f :\downarrow V \to \mathbb{IR}$ can be decomposed into two maps $f_+, f_- :\downarrow V \to \mathbb{R}$ which pick out the end points of the interval, i.e. $f(V) := [f_-(V), f_+(V)]$. This implies that $f_- \leq f_+$ and f_- is order-preserving while, f_+ is order-reversing.

On the other hand, each pair $(\mu, \nu) \in \underline{\mathbb{R}^{\leftrightarrow}}$ gives rise to an order-preserving map $f :\downarrow V \to \mathbb{IR}$ such that $f(V') = [\mu(V'), \nu(V')]$ for all $V' \subseteq V$. This correspondence implies that we can now characterise the presheaf $\underline{\mathbb{R}^{\leftrightarrow}}$ in terms of the interval domain as follows:

Theorem 12.5.1 *The quantity value object $\underline{\mathbb{R}^{\leftrightarrow}}$ acts on*

- *objects: for all $V \in \mathcal{V}(\mathcal{H})$,*

$$\underline{\mathbb{R}^{\leftrightarrow}}(V) = \{f :\downarrow V \to \mathbb{IR} | f \textit{ is order-preserving}\}$$

- *Morphisms: for all $i_{V'V} : V' \subseteq V$*

$$\underline{\mathbb{R}^{\leftrightarrow}}(i_{V'V}) : \underline{\mathbb{R}^{\leftrightarrow}}(V) \to \underline{\mathbb{R}^{\leftrightarrow}}(V'); \qquad f \mapsto f|_{\downarrow V'}.$$

Given this new characterization, one can show that $\Gamma\underline{\mathbb{R}^{\leftrightarrow}} \simeq OP(\mathcal{V}(\mathcal{H}), \mathbb{IR})$ as follows:

Consider a global element $\gamma : \underline{1} \to \underline{\mathbb{R}^{\leftrightarrow}}$. This, for each $V \in \mathcal{V}(\mathcal{H})$, assigns out an element $p_V := \gamma_V(\{*\}) :\downarrow V \to \mathbb{IR}$. Since the global element is a natural transformation, its naturality implies that $p_{V'} = \underline{\mathbb{R}^{\leftrightarrow}}(i_{V'V})(p_V) = p_V|_{V'}$, therefore each global element γ gives rise to a map:

$$\rho_\gamma : \mathcal{V}(\mathcal{H}) \to \mathbb{IR}$$
$$V' \mapsto p_V|_{\downarrow V'}$$

where $V' \subseteq V$. This is well-defined since for any other V'', such that $V \subseteq V''$, we have $p_{V''}|_{V'} = p_{V'} = p_V|_{V'}$. The fact that it is also order-preserving follows from naturality.

On the other hand, given an order-preserving map $\rho : \mathcal{V}(\mathcal{H}) \to \mathbb{IR}$, we can define a global element by setting $p_V := \rho|_{\downarrow V}$ for all $V \in \mathcal{V}(\mathcal{H})$.

This result implies that

$$\text{pt}(R) \simeq OP(\mathcal{V}(\mathcal{H}), \mathbb{IR}).$$

Therefore the points in the locale utilised to eventually model space-time are order-preserving functions from the context category to the interval domain. They describe varying intervals of real numbers.

Points of the Locale $\mathcal{R}$ We will now define the points of the last locale which represents a possible candidate for modelling space-time. Clearly such a locale is intimately connected to the locale R and, as a consequence, also the characterization of its points will be very similar. In fact, the locale $\mathcal{R}$ can be written out as $\mathcal{R} = \coprod_{V\in\mathcal{V}(\mathcal{H})} \coprod_{\phi:\downarrow V\to\mathcal{V}(\mathcal{H})} \underline{\mathbb{R}^{\leftrightarrow}}(\phi(V))$. Therefore, for each $V \in \mathcal{V}(\mathcal{H})$, we obtain

$$\text{pt}(\mathcal{R})(V) \simeq \coprod_{\phi:\downarrow V\to\mathcal{V}(\mathcal{H})} \text{pt}(R)(\phi(V)) \simeq \coprod_{\phi:\downarrow V\to\mathcal{V}(\mathcal{H})} OR(\downarrow \phi(V), \mathbb{IR}).$$

What this implies is that the presheaf $\text{pt}(\mathcal{R})$ assigns to each $V \in \mathcal{V}(\mathcal{H})$ a collection of intervals $[\mu(V), \nu(V)]$ all related by a group transformation. Conceptually this might be interpreted as stating that the points of the locale $\mathcal{R}$ represent equivalence classes of intervals under some group transformation. Therefore, if we consider space-time to be modelled by such a locale, then space-time points become equivalence classes of regions under a symmetry transformation. So for a point $p \in \text{pt}(\mathcal{R}(V))$, we will write $p = \coprod_{\phi:\downarrow V\to\mathcal{V}(\mathcal{H})} p_\phi :\downarrow \phi(V) \to \mathbb{IR}$.

12.6 Conclusions

Most theories of quantum gravity seem to suggest that, at the fundamental level, space-time has a discrete structure. As discussed in the introduction, this view of space-time seems to contradict the way in which it is described, both in quantum theory and in general relativity. In fact, in both these two theories, space-time is seen as a continuum modelled on the real numbers. This discrepancy of the description of space-time between quantum theory and general relativity on the one hand and quantum gravity on the other seems rather surprising, since quantum gravity is supposed to combine general relativity with quantum theory in a coherent framework. This odd feature has motivated some researchers to question the fact that space-time should be modelled by a continuum and has suggested that, also at the level of quantum theory and general relativity, a new mathematical model for space-time should be created. In this chapter we proposed an alternative definition of space-time put forward by topos quantum theory. Such a definition consists of modelling space-time in terms of a locale where, now, the fundamental space-time building blocks are regions, not points. This idea reflects the fact that space-time points are not physically meaningful since real objects occupy space-time regions.

A locale is equivalent to a complete Heyting algebra. Therefore, modelling space-time in terms of a locale can be interpreted, roughly, as modelling space-time via an algebra of open regions of space-time where, the algebraic operations are interpreted as defining unions and intersections of space-time regions. The reason why open regions are preferred is to account for quantum indeterminacies given by the generalised uncertainty principle. Therefore, modelling space-time via a locale seems to resonate more accurately with our common-sense interpretation of space-time.

In this chapter, we propose various candidate locales which are all ultimate related to one another. Each of these locales could be adopted for modelling space-time, however a discussion on which one would be the best suited is left as a topic for future research. To derive the above mentioned locales, we started with the quantity value object of topos quantum theory which plays the same role as the Reals in classical physics. This object being a sheaf, we applied the standard technique to 'transform' sheaves into a locale internal to the appropriate topos. When considering the locale associated to the quantity value object $\underline{\mathbb{R}}^{\leftrightarrow}$, we discovered that, for each context, the elements of such a locale were identified with collection of unions of pairs of order-preserving and order-reversing functions, from the context category to the interval domain. These describe unions of varying intervals of real numbers. Conceptually what this means is the following: for each context $V \in \mathcal{V}(\mathcal{H})$ which represents a classical snapshot of the quantum system, we obtain a "locale" space-time seen as a collection of unions of varying intervals of real numbers. Such a space-time has the property that, when considering two contexts V' and V such that $V' \subseteq V$, then the intervals of real numbers describing the space-time associated to the context V' are "bigger", i.e. are less precise and, thus, have less information, than the intervals of the space-time associated to the bigger context. Since these intervals are to be interpreted as the regions of space-time which physical objects occupy, what the above result signifies is that when going to a smaller contexts V', which contains less information due to coarse-graining, then the precision with which one is able to determine the position of physical objects decreases. Moreover, when considering the locale associated to the 'covariant' quantity value object $\breve{\underline{\mathbb{R}}}$, we discovered that for each context V the "locale" space-time was considered to be a collection of unions of equivalence classes of varying intervals of real numbers, where such equivalence was defined with respect to a group G. A tentative interpretation is that these space-time regions represent diffeomorphic regions of space-time. Since such a locale is a sheaf, one could interpret a global section as a particular choice of space-time. Each such global section (choice of space-time) would then be related to each other by space-time diffeomorphisms. Clearly this is a very speculative idea and a much more thorough analysis is needed to determine whether such an interpretation is physically reasonable. As previously stated, in this chapter we only want to elucidate possible candidates for an interpretation of space-time which makes no fundamental use of the continuum. This is a first step towards defining a mathematical model of space-time. In fact, although the discreteness of space-time in quantum theory is almost universally accepted, no mathematical model of discrete space-time has been constructed so far. We hope that this article will provide ideas for possible such models by offering various candidates. Which, if any, of these candidates would be the most appropriate choice for representing space time is a question that still remains to be answered and will be the topic of a subsequent studies.

Chapter 13
Extending the Topos Quantum Theory Approach

As it has been developed so far, the mathematical formalism of topos quantum theory only allows for taking into consideration one physical system at a time. This has clearly some limitations, in particular when trying to consider composite systems. Hence it comes natural to try and enlarge the mathematical formalism so that is possible to take into consideration various physical systems at the same time. This implies considering a topos somewhat "larger" than the topos $\mathbf{Sets}^{\mathcal{V}(\mathcal{H})^{\text{op}}}$. In particular what needs to be "enlarged" is the category $\mathcal{V}(\mathcal{H})$. In fact this category only refers to the physical system with associated von Neumann algebra $\mathcal{N}$, whose category of abelian subalgebras is given by $\mathcal{V}(\mathcal{H})$. However we would like to consider all physical systems, each of which, has associated to it a different von Neumann algebra. To account for this, one possibility would be to construct a category in which each element is itself a topos which represents the mathematical formalism of a physical system. Then one would have to construct a mapping which associates to each physical system its associated topos. The aim would be to turn this map into a geometric morphism of some sort between topoi, such that it possesses nice properties which would help to better understand composite systems. In the following we will present all work done so far in this direction, which is an exposition of the results obtained in [30]. As it will be clear in due course, there are still many open problems to be addressed.

C*-algebra theory is a blend of algebra and analysis which turns out to be much more than the sum of its parts, as already illustrated by its fundamental results of Gelfand duality and the GNS representation theorem. Nevertheless, the C*-algebra axioms seem somewhat mysterious, and it may not be very clear what they mean or where they actually 'come from'. To see the point, consider the axioms of groups for comparison: these have a clear meaning in terms of symmetries and the composition of symmetries, and this provides adequate motivation for these axioms. Do C*-algebras also have an interpretation which motivates their axioms in a similar manner?

C. Flori, *A Second Course in Topos Quantum Theory*,
Lecture Notes in Physics 944, https://doi.org/10.1007/978-3-319-71108-9_13

A plausible answer to this question would be in terms of applications of C*-algebras to areas outside of pure mathematics. The most evident application of C*-algebras is to quantum mechanics and quantum field theory [35, 52, 70]. However, also in this context the C*-algebra axioms do not seem well-motivated. In fact, not even the multiplication, which results in the algebra structure, does have a clear physical meaning. This is in stark contrast to other physical theories, such as relativity: especially in special relativity, the mathematical structures that come up are derived from physical considerations and principles, often via the use of thought experiments. A similar derivation of C*-algebraic quantum mechanics does not seem to be known.

Hence it seems pertinent to try and reformulate the C*-algebra axioms in a more satisfactory manner that would allow for a clear interpretation.

13.1 C*-Algebras as Functors CHaus → Sets

In this section, we explain how to regard a C*-algebra as a functor CHaus → **Sets**, and how this encodes the usual functional calculus for normal elements in a C*-algebra, as well as its multivariate generalization.

The Yoneda embedding realizes a C*-algebra A as the hom-functor

$$\mathsf{C^*alg}_1(-, A) \; : \; \mathsf{C^*alg}_1^{\mathrm{op}} \to \mathbf{Sets}.$$

We are interested in studying this hom-functor on the commutative C*-algebras, meaning that we consider its restriction to a functor $\mathsf{cC^*alg}_1^{\mathrm{op}} \to \mathbf{Sets}$. Applying Gelfand duality, we can equivalently consider it as a functor

$$-(A) \; : \; \mathsf{CHaus} \to \mathbf{Sets},$$

assigning to every compact Hausdorff space $X \in \mathsf{CHaus}$ a set $X(A)$, which is the set of all $*$-homomorphisms $C(X) \to A$. Our notation $X(A)$ suggests thinking of it as the set of *generalized A-points* of X.

Example 13.1.1 If X is finite, a $*$-homomorphism $C(X) \to A$ or generalized A-point in X corresponds to a *partition of unity* in A indexed by X, i.e. a family of pairwise orthogonal projections summing up to 1.

Example 13.1.2 If A is a W*-algebra, the spectral theorem [31, Theorem 1.44] implies that $X(A)$ is precisely the collection of all regular projection-valued measures on X with values in A.

Remark 13.1.1 In terms of algebraic quantum mechanics, where a physical system is described by a C*-algebra A of observables [52, 70], we interpret a $*$-homomorphism $\alpha \; : \; C(X) \to A$ as a projective measurement with values in X, described in the Heisenberg picture. So the physical meaning of our $X(A)$ is as the collection of all measurements with outcomes in the space X.

Those $*$-homomorphisms $C(X) \to A$, whose image is in the center of A, are called *$C(X)$-algebras*, and they correspond exactly to upper semicontinuous C*-bundles over X [58].[1]

At the level of morphisms, every $f : X \to Y$ acts by composing a $*$-homomorphism $\alpha : C(X) \to A$ with $C(f)$ to $\alpha \circ C(f) : C(Y) \to A$, so that

$$\begin{aligned} f(A) : X(A) &\longrightarrow Y(A) \\ \alpha &\longmapsto \alpha \circ C(f) \end{aligned} \tag{13.1.1}$$

is the action of f on generalized A-points.

Remark 13.1.2 The physical interpretation of $f(A)$ is as a *post-processing* or *coarse-graining* of measurements. Under $f(A)$, a measurement $\alpha : C(X) \to A$ with values in X becomes a measurement $\alpha \circ C(f) : C(Y) \to A$ with values in Y, implemented by first conducting the original measurement α and then processing the outcome via application of the function f. Since we work in the Heisenberg picture, the order of composition is reversed, so that $C(f)$ happens first.

This construction is also functorial in A: for any $*$-homomorphism $\zeta : A \to B$ and $X \in \mathsf{CHaus}$, we have $X(\zeta) : X(A) \to X(B)$. Furthermore, for any $f : X \to Y$ there is the evident naturality diagram

$$\begin{array}{ccc} X(A) & \xrightarrow{f(A)} & Y(A) \\ {\scriptstyle X(\zeta)}\downarrow & & \downarrow{\scriptstyle Y(\zeta)} \\ X(B) & \xrightarrow[f(B)]{} & Y(B) \end{array}$$

which expresses the bifunctoriality of the hom-functor $\mathsf{C^*alg}_1(-,-)$ in our setup.

Before proceeding with technical developments, it is worthwhile pondering on how these considerations relate to functional calculus.

13.1.1 Functoriality Captures the 'Commutative Part' of the C-Algebra Structure*

In a somewhat informal sense, the functor $-(A)$ captures the entire 'commutative part' of the structure of a C*-algebra A. We will obtain a precise result along these lines as Theorem 13.3.1. Here, we perform some simple preparations.

[1] We thank Klaas Landsman for pointing this out to us.

Lemma 13.1.1 *For any compact set $S \subseteq \mathbb{C}$, evaluating an $\alpha : C(S) \to A$ on $\mathrm{id}_S : S \to \mathbb{C}$,*

$$\alpha \longmapsto \alpha(\mathrm{id}_S), \tag{13.1.2}$$

is a bijection between $S(A)$ and the normal elements[2] *of A with spectrum in S.*

Proof If $\alpha, \beta : C(S) \to A$ coincide on id_S, then they must coincide on the *-algebra generated by id_S. Since id_S separates points, this *-algebra is dense in $C(S)$ by the Stone-Weierstrass theorem, so that $\alpha = \beta$ by continuity. This establishes injectivity of (13.1.2).

Concerning surjectivity, applying functional calculus to a given normal element with spectrum in S results in a $*$-homomorphism $C(S) \to A$ which realizes the given element via (13.1.2). □

Due to this correspondence, we will not distinguish notationally between a $*$-homomorphism $\alpha : C(S) \to A$ and its associated normal element, i.e. we also denote the latter simply by $\alpha \in A$. Moreover, we can also think of a $*$-homomorphism $C(X) \to A$ for arbitrary $X \in \mathsf{CHaus}$, as a sort of 'generalized normal element' of A.

For any two compact objects $S, T \subseteq \mathbb{C}$ and $f : S \to T$, functional calculus—in the sense of applying f to normal elements with spectrum in S—is encoded in two ways:

- in evaluating an $\alpha : C(S) \to A$ on $f : S \to \mathbb{C}$, as in the proof of Lemma 13.1.1;
- in the functoriality $f(A) : S(A) \to T(A)$, since applying this functorial action to α results in the same normal element of A,

$$f(A)(\alpha)(\mathrm{id}_T) \overset{(13.1.1)}{=} (\alpha \circ C(f))(\mathrm{id}_T) = \alpha(C(f)(\mathrm{id}_T)) = \alpha(\mathrm{id}_T \circ f) = \alpha(f). \tag{13.1.3}$$

From now on, what we mean by 'functional calculus' is the functoriality, i.e. the second formulation.

Writing $\bigcirc \subseteq \mathbb{C}$ for the unit disk, the normal elements of norm ≤ 1 are identified with the $*$-homomorphisms $\alpha : C(\bigcirc) \to A$. For every $r \in [0, 1]$, we have the multiplication map $r\cdot : \bigcirc \to \bigcirc$, so that $(r\cdot)(A) : \bigcirc(A) \to \bigcirc(A)$ represents scalar multiplication of normal elements by r. Based on this, we can recover the norm of a normal element $\alpha \in \bigcirc(A)$ as the largest r for which α factors through $C(r)$,

$$||\alpha|| = \max \{ r \in [0, 1] \mid \alpha \in \mathrm{im}((r\cdot)(A)) \} .$$

As we will see next, the functoriality also captures part of the binary operations of a C*-algebra.

[2]Given a C^*-algebra A, a normal element $a \in A$ is an element such that $aa^* = a^*a$.

Lemma 13.1.2 *For $S, T \subseteq \mathbb{C}$, applying functoriality to the product projections*

$$p_S : S \times T \longrightarrow S, \qquad p_T : S \times T \longrightarrow T \tag{13.1.4}$$

establishes a bijection between $(S \times T)(A)$ and pairs of commuting *normal elements $(\alpha, \beta) \in A \times A$ with $\mathrm{sp}(\alpha) \subseteq S$ and $\mathrm{sp}(\beta) \subseteq T$.*

This generalizes Lemma 13.1.1 to commuting pairs of normal elements. Of course, there are analogous statements for tuples of any size (finite or even infinite), and this encodes multivariate functional calculus.

Proof We need to show that the map

$$(p_S(A), p_T(A)) \; : \; (S \times T)(A) \longrightarrow S(A) \times T(A)$$

is injective, and that its image consists of precisely the pairs (α, β) with $\alpha : C(S) \to A$ and $\beta : C(T) \to A$, that have commuting ranges. Injectivity holds because $p_S : S \times T \to \mathbb{C}$ and $p_T : S \times T \to \mathbb{C}$ separate points, so that the same argument as in the proof of Lemma 13.1.1 applies. For surjectivity, let α and β be given. Since their ranges commute, we can find a commutative subalgebra $C(X) \subseteq A$ that contains both, so that the pair (α, β) has a preimage in the upper right corner of the diagram

$$\begin{array}{ccc} (S \times T)(C(X)) & \longrightarrow & S(C(X)) \times T(C(X)) \\ \downarrow & & \downarrow \\ (S \times T)(A) & \longrightarrow & S(A) \times T(A) \end{array}$$

Now the upper row is equal to the canonical map $\mathsf{CHaus}(X, S \times T) \to \mathsf{CHaus}(X, S) \times \mathsf{CHaus}(X, T)$, which is a bijection due to the universal property of $S \times T$. Hence we can find a preimage of (α, β) also in the upper left corner, and then also in the lower left corner by commutativity of the diagram. □

In the physical interpretation, the elements of $(S \times T)(A)$ are measurements that have outcomes in $S \times T$ (Remark 13.1.1). Lemma 13.1.2 now shows that such a measurement corresponds to a pair of *compatible* measurements taking values in S and T, respectively, and one obtains these measurements by coarse-graining along the product projections (13.1.4), i.e. by forgetting the other outcome.

As part of bivariate functional calculus, we can now consider the addition map

$$S \times T \longrightarrow S + T, \qquad (x, y) \longmapsto x + y, \tag{13.1.5}$$

where $S + T$ is the Minkowski sum

$$S + T = \{\, x + y \mid x \in S, \; y \in T \,\},$$

again considered as a compact subset of $\mathbb{C}$. Under the identifications of Lemmas 13.1.1 and 13.1.2, the addition map

$$+(A) \,:\, (S \times T)(A) \longrightarrow (S+T)(A). \tag{13.1.6}$$

takes a pair of commuting normal elements with spectra in S and T and takes it to a normal element with spectrum in $S+T$.

Lemma 13.1.3 *On commuting normal elements, this recovers the usual addition in A.*

Proof By Lemma 13.1.2, it is enough to take a $\gamma \in (S \times T)(A)$ and to compute the resulting normal element that one obtains by applying $+(A)$ in a manner analogous to (13.1.3),

$$\begin{aligned}
(+(A))(\gamma)(\mathrm{id}_{S+T}) &\overset{(13.1.1)}{=} (\gamma \circ C(+))(\mathrm{id}_{S+T}) = \gamma(\mathrm{id}_{S+T} \circ +) \\
&= \gamma(\mathrm{id}_S \circ p_S + \mathrm{id}_T \circ p_T) \\
&= \gamma(\mathrm{id}_S \circ p_S) + \gamma(\mathrm{id}_T \circ p_T) \\
&= (\gamma \circ C(p_S))(\mathrm{id}_S) + (\gamma \circ C(p_T))(\mathrm{id}_T) \\
&\overset{(13.1.1)}{=} (p_S(A))(\gamma)(\mathrm{id}_S) + (p_T(A))(\gamma)(\mathrm{id}_T),
\end{aligned}$$

where the crucial assumption of additivity of γ has been used to obtain the expression in the third line. □

In the analogous manner, one can show that the multiplication map

$$S \times T \longrightarrow ST, \qquad (x, y) \longmapsto xy. \tag{13.1.7}$$

lets us recover the product of two commuting normal elements in A. More generally, we can recover any polynomial or continuous function of any number of commuting normal elements.

Summarising, we think of the functor $-(A) : \mathsf{CHaus} \to \mathbf{Sets}$ associated to $A \in \mathsf{C^*alg}_1$ as a generalization of functional calculus, which remembers the entire 'commutative structure' of A. The generalization is from applying functions to individual normal elements—as in the conventional picture of functional calculus—to applying functions to 'generalized' normal elements in the guise of $*$-homomorphisms of the form $C(X) \to A$. In particular, the C*-algebra operations acting on commuting normal elements are encoded in the functoriality. In the remainder of this chapter, we will always have this point of view in mind, together with its physical interpretation:

functoriality = generalized functional calculus = post-processing of measurements.

Remark 13.1.3 In Sect. 13.2.2, we will also consider functors $F : \mathsf{CHaus} \to \mathbf{Sets}$ that do not necessarily arise from a C*-algebra in this way. In terms of the physical interpretation, this means that, instead of modelling physical systems in terms of their algebras of observables, as the primary structure, we model them in terms of a functor F, as the most fundamental structure that describes physics. This is motivated by the fact that the C*-algebra structure of the observables is (a priori) not physically well-motivated, as discussed in the introduction. Thanks to Remarks 13.1.1 and 13.1.2, our functors F : CHaus → **Sets** do have a meaningful operational interpretation in terms of measurements: $F(X)$ is the set of (projective) measurements with outcomes in X, and the action of F on morphisms is the post-processing. This bare-bones structure turns out to carry a surprising amount of information about the algebra of observables. We will try to equip F with additional properties and structure such as to uniquely specify the algebra of observables.

In spirit, this approach is similar to the existing reconstructions of quantum mechanics from operational axioms [34]. In recent years, a wide range of reconstruction theorems, with a large variety of choices for the axioms, have been derived, as pioneered by Hardy [39, 40]. In these theorems, 'quantum mechanics' refers to the Hilbert space formulation in finite dimensions, and the reconstruction theorems recover the Hilbert space structure within the framework of general probabilistic theories. In contrast to this, our work focuses on the C*-algebraic formulation of quantum mechanics and it is not limited to a finite-dimensional setting. Moreover, we do not make use of the possibility of taking stochastic mixtures since we are (currently) only dealing with projective measurements, therefore, taking stochastic mixtures is not possible in our setup.

13.2 C*-Algebras as Sheaves CHaus → Sets

Functional calculus let us apply functions to operators or, more generally, to $*$-homomorphisms $C(X) \to A$ as in the previous section. In some situations, one can also go the other way: for certain families of functions $\{f_i : X \to Y_i\}_{i \in I}$ with common domain, a collection of $*$-homomorphisms $\{\beta_i : C(Y_i) \to A\}_{i \in I}$ arises from a unique $*$-homomorphism $\alpha : C(X) \to A$ by functoriality along the f_i, if and only if the β_i satisfy a simple compatibility requirement. This property is a *sheaf condition*, and it turns our functors $-(A)$ into sheaves on the category CHaus.

We would like to emphasis that the sheaf conditions that we consider do not arise from a Grothendieck topology (on $\mathsf{CHaus}^{\mathrm{op}}$), since the axiom of stability under pullback fails to hold. Moreover, sheaf conditions are typically formulated for contravariant functors (i.e. presheaves), instead our sheaves live in a covariant setting. To emphasize this distinction we could have named our sheaves 'cosheaves', however this term usually refers to dualizing the standard notion of sheaf on the codomain category, while we dualize on the domain category, hence we did not find the name appropriate.

A good way of talking about sheaf conditions on large categories is not in terms of sieves or cosieves—which would usually have to be large—but in terms of cocones or cones [65]:

Definition 13.2.1 A *cone* in CHaus is any small family of morphisms $\{f_i : X \to Y_i\}_{i\in I}$ with common domain.

Definition 13.2.2 A functor F : CHaus $\to$ **Sets** satisfies the *sheaf condition* on a cone $\{f_i : X \to Y_i\}_{i\in I}$ if the $F(f_i)$ implement a bijection between the sections $\alpha \in F(X)$ and the families of sections $\{\beta_i\}_{i\in I}$ with $\beta_i \in F(Y_i)$ that are *compatible* in the following sense: for any $i, j \in I$ and any diagram

$$\begin{array}{ccc} X & \xrightarrow{f_i} & Y_i \\ {\scriptstyle f_j}\downarrow & & \downarrow{\scriptstyle g} \\ Y_j & \xrightarrow[h]{} & Z \end{array} \tag{13.2.1}$$

we have $F(g)(\beta_i) = F(h)(\beta_j)$.

Since CHaus has pushouts, the compatibility condition holds if and only if it holds on every pushout diagram

$$\begin{array}{ccc} X & \xrightarrow{f_i} & Y_i \\ {\scriptstyle f_j}\downarrow & & \downarrow \\ Y_j & \longrightarrow & Y_i \,{}_{f_i}\!\amalg_{f_j} Y_j \end{array}$$

Hence the sheaf condition holds on $\{f_i\}$ if and only if the diagram

$$F(X) \longrightarrow \prod_{i\in I} F(Y_i) \rightrightarrows \prod_{i,j\in I} F(Y_i \,{}_{f_i}\!\amalg_{f_j} Y_j),$$

is an equalizer in **Sets**, where the arrows are the canonical ones [55, p. 123]. At times it is convenient to apply the compatibility condition as in (13.2.1) instead of considering the pushout, while at other times it is necessary to work, explicitly, with the pushout.

13.2.1 Effective-Monic Cones in CHaus

Since we are interested in sheaf conditions satisfied by a functor of the form $-(A)$: CHaus $\to$ **Sets** for $A \in$ C*alg$_1$, it makes sense to first consider the commutative case. Then our functor takes the form $-(C(W))$, which is isomorphic to the hom-functor CHaus$(W, -)$.

Definition 13.2.3 (e.g. [65, Definition 2.22]) A cone $\{f_i : X \to Y_i\}_{i \in Y}$ in CHaus is *effective-monic* if every representable functor $\mathsf{CHaus}(W, -)$ satisfies the sheaf condition on it.

Hence $\{f_i\}$ is effective-monic if and only if X is the equalizer in the diagram

$$X \longrightarrow \prod_{i \in I} Y_i \rightrightarrows \prod_{i,j \in I} (Y_i \, {}_{f_i}\amalg_{f_j} \, Y_j),$$

or, equivalently, the limit in the diagram

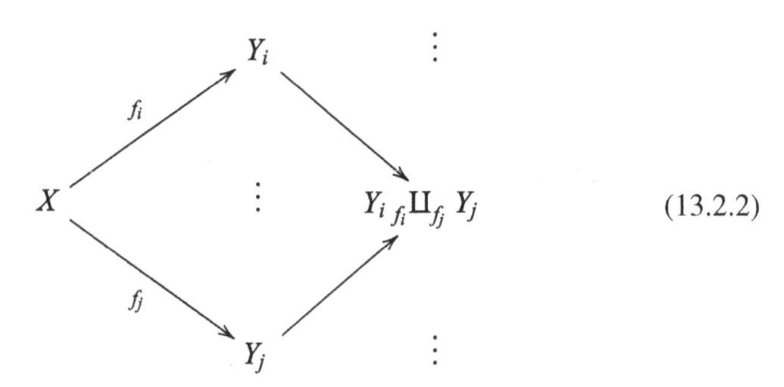

Before giving some examples we recall the definition of a representable functor.

Definition 13.2.4 Given a locally small category C, a functor $F : C \to \mathbf{Sets}$ is said to be representable if it is naturally isomorphic to the Hom-functor $Hom(A, -) : C \to \mathbf{Sets}$ for some object $A \in C$. A representation of F is a pair (A, Φ) where

$$\Phi : Hom(A, -) \to F$$

is a natural isomorphism.

Example 13.2.1 Let Λ be a small category and $L : \Lambda \to \mathsf{CHaus}$ a functor of which we consider the limit $\lim_\Lambda L \in \mathsf{CHaus}$. The limit projections $p_\lambda : \lim_\Lambda L \to L(\lambda)$ assemble into a cone $\{p_\lambda\}_{\lambda \in \Lambda}$, which is effective-monic.

Fortunately, it is not necessary to consider arbitrary W in Definition 13.2.3:

Lemma 13.2.1 *A cone $\{f_i\}$ is effective-monic if and only if $\mathsf{CHaus}(\mathbf{1}, -)$ satisfies the sheaf condition on it.*

Proof CHaus is well-known to be monadic over **Sets**, with the forgetful functor being, precisely, the functor of points $\mathsf{CHaus}(\mathbf{1}, -) : \mathsf{CHaus} \to \mathbf{Sets}$. In particular, this functor creates limits. □

The above discussion implies that X must be the subspace of the product space $\prod_{i \in I} Y_i$ consisting of all those families of points $\{y_i\}_{i \in I}$, such that the image of $y_i \in Y_i$ coincides with the image of $y_j \in Y_j$ in the pushout space $Y_i \, {}_{f_i}\amalg_{f_j} \, Y_j$. This condition also applies for $j = i$, in which case it is equivalent to $y_i \in \mathrm{im}(f_i)$.

Remark 13.2.1 For a given Y, the cone of all functions $\{f : X \to Y\}_{f:X\to Y}$ is effective-monic for every X if and only if Y is codense.

While these categorical considerations have been extremely general, we will now analyse CHaus in more details. We write $\square := [0,1] \times [0,1]$ for the unit square, and consider it as embedded in $\square \subseteq \mathbb{R}^2 = \mathbb{C}$, where the unit interval $[0,1] \subseteq \mathbb{R}$ is an edge of $\square$.

Lemma 13.2.2 *For every $X \in$ CHaus, the cone $\{f : X \to \square\}_{f:X\to\square}$ consisting of all functions $f : X \to \square$ is effective-monic.*

By Remark 13.2.1, this is a restatement of the known fact that $\square$ is codense in CHaus [44].

Differently from the 'conventional' sheaf condition, which states that a function is uniquely determined by a compatible assignment of values to all (local neighbourhoods of) points, the above sheaf condition states that a point is uniquely determined by a compatible assignment of values to all functions.

Proof We need to show that the diagram

$$X \longrightarrow \prod_{f:X\to\square} \square \rightrightarrows \prod_{g,h:X\to\square} (\square\, {}_g\amalg_h\, \square)$$

is an equalizer. Since functions $X \to \square$ separate points in X, it is clear that the map $X \to \prod_f \square$ is injective.

Surjectivity is more difficult. Suppose that $v \in \prod_{f:X\to\square} \square$ is a compatible family of sections. Then in particular, we have

$$v(hf) = h(v(f)) \quad \text{for all} \quad h : \square \to \square \tag{13.2.3}$$

as an instance of the compatibility condition, since the square

$$\begin{array}{ccc} X & \xrightarrow{hf} & \square \\ {\scriptstyle f}\downarrow & & \| \\ \square & \xrightarrow[h]{} & \square \end{array} \tag{13.2.4}$$

commutes.

We have to show that there exists a point $x \in X$ with $v(f) = f(x)$ for all $f : X \to \square$. This set of equations is equivalent to $x \in \bigcap_f f^{-1}(v(f))$. Hence it is enough to show that $\bigcap_f f^{-1}(v(f))$ is non-empty. By compactness, it is sufficient to prove that any finite intersection

$$f_1^{-1}(v(f_1)) \cap \ldots \cap f_n^{-1}(v(f_n))$$

for a finite set of functions $f_1, \ldots, f_n : X \to \square$ is non-empty. Using induction on n, the induction step is obvious if for given f_1, f_2 we can exhibit $g : X \to \square$, such that

$$g^{-1}(v(g)) = f_1^{-1}(v(f_1)) \cap f_2^{-1}(v(f_2)).$$

First, by (13.2.3), we can assume that both f_1 and f_2 actually take values in $[0, 1]$, e.g. by considering

$$h_1 : \square \longrightarrow [0, 1], \qquad t \longmapsto |t - v(f_1)|$$

and replacing f_1 by $h_1 f_1$, which results in

$$(h_1 f_1)^{-1}(v(h_1 f_1)) \overset{(13.2.3)}{=} f_1^{-1}(h_1^{-1}(h_1(v(f_1)))) = f_1^{-1}(h_1^{-1}(0)) = f_1^{-1}(v(f_1)),$$

and similarly for f_2. After this replacement, we can take $g(t) := (f_1(t), f_2(t))$, and the induction step is complete upon applying (13.2.3) to the two coordinate projections.

Finally, we need to show that any individual set $f^{-1}(v(f))$ is non-empty as the base of the induction. To this end, given a $s \in [0, 1] \setminus \mathrm{im}(f)$, choose h such that $h(\mathrm{im}(f)) = \{0\}$ and $h(s) = 1$ by the Tietze extension theorem. Then

$$0 = v(0) = v(hf) = h(v(f)),$$

and hence $v(f) \neq s$. Therefore $v(f) \in \mathrm{im}(f)$, as it was to be shown. □

The effective-monic cone $\{f : X \to \square\}_{f:X\to\square}$ will be of particular importance on subsequent sections. To better understand effective-monic cone and shed some light on their general behaviour we will consider some other examples which arise in CHaus. However, before doing so we should point out that, as shown in the counterexample given in the proof of [44, Theorem 2.6], Lemma 13.2.2 does not hold with $[0, 1]$ in place of $\square$. However, if X is extremally disconnected[3] Lemma 13.2.2 is an immediate consequence of the following result:

Lemma 13.2.3 *If X is extremally disconnected, then $\{f : X \to \mathbf{4}\}$ is effective-monic.*

Here, we write $\mathbf{4} := \{0, 1, 2, 3\}$, and the proof uses indicator functions $\chi_Y : X \to \mathbf{4}$ of clopen sets $Y \subseteq X$.

Proof Since the clopen sets separate points, the injectivity is again clear and the burden of the proof is in the surjectivity. To this end let $v : \mathbf{4}^X \to \mathbf{4}$ be a compatible family of sections.

[3] A topological space is said to be *extremally disconnected* if the closure of every open set in it is open.

As in the proof of Lemma 13.2.2, we show that the intersection

$$\bigcap_{Y \text{ clopen}, \, v(\chi_Y)=1} Y$$

is non-empty. From compactness and an induction argument similar as the one given in the proof of Lemma 13.2.2, it is enough to show that, for any clopen $Y_1, Y_2 \subseteq X$ with $v(\chi_{Y_1}) = 1$ and $v(\chi_{Y_2}) = 1$, we also have $v(\chi_{Y_1 \cap Y_2}) = 1$. To see this, we consider the function

$$f := \chi_{Y_1} + 2\chi_{Y_2},$$

and then apply the compatibility condition in the form (13.2.3) for various h. If we choose h such that $0, 2 \mapsto 0$ and $1, 3 \mapsto 1$ we obtain that $hf = \chi_{Y_1}$, hence $v(f) \in \{1, 3\}$. Similarly, if we choose h such that $0, 1 \mapsto 0$ and $2, 3 \mapsto 1$ we obtain $hf = \chi_{Y_2}$, therefore $v(f) \in \{2, 3\}$. Overall, we obtain $v(f) = 3$, and apply h with $0, 1, 2 \mapsto 0$ and $3 \mapsto 1$ to conclude $v(\chi_{Y_1 \cap Y_2}) = 1$ from $hf = \chi_{Y_1 \cap Y_2}$.

So there is at least one point $x_0 \in X$ such that $v(\chi_Y) = 1$ implies $x_0 \in Y$ for all clopen $Y \subseteq X$. We then claim that $v(f) = f(x_0)$ for all $f : X \to \mathbf{4}$. This follows from writing

$$f = 0\chi_{Y_0} + 1\chi_{Y_1} + 2\chi_{Y_2} + 3\chi_{Y_3}$$

for a partition of X by clopens $Y_0, Y_1, Y_2, Y_3 \subseteq X$, and applying (13.2.3) with h, such that $v(f) \mapsto 1$, while the other three integers map to 0. □

A singleton cone $\{f : X \to Y\}$ is effective-monic if and only if f is injective. For cones consisting of exactly two functions, the necessary and sufficient criterion is as follows:

Lemma 13.2.4 *A cone $\{f : X \to Y, g : X \to Z\}$ consisting of exactly two functions is effective-monic if and only if the pairing $(f, g) : X \to Y \times Z$ is a Mal'cev relation, meaning that f and g are jointly injective and their joint image*

$$R := \mathrm{im}((f, g)) \subseteq Y \times Z$$

satisfies the implication

$$\Big((y, z) \in R, \quad (y', z) \in R, \quad (y, z') \in R\Big) \quad \Longrightarrow \quad (y', z') \in R. \tag{13.2.5}$$

For the notion of Mal'cev relation, see [32].

Proof We use the criterion of Lemma 13.2.1. The injectivity part of the sheaf condition is equivalent to injectivity of $(f, g) : X \to Y \times Z$. Assuming that this holds, we identify X with the joint image $R \subseteq Y \times Z$.

Now if $\{f, g\}$ is effective-monic and we have $y, y' \in Y$ and $z, z' \in Z$ as in (13.2.5), then each of the three pairs (y, z), (y', z) and (y, z') represents a point of X. Therefore, since (y, z) is in particular a compatible pair of sections, in $Y\,{}_f\!\amalg_g Z$ the image of y coincides with the image of z. If we apply the same reasoning to (y', z) and (y, z'), then also y' and z', respectively get mapped to the same point in $Y\,{}_f\!\amalg_g Z$. Hence also (y', z') is a compatible pair of sections, which must correspond to a point of X due to the sheaf condition.

Conversely, suppose that (13.2.5) holds. The pushout $Y\,{}_f\!\amalg_g Z$ is the quotient of the coproduct $Y \amalg Z$ by the closed equivalence relation generated by $f(x) \sim g(x)$ for all $x \in X$, i.e. by $y \sim z$ for all $(y, z) \in R$. In terms of relational composition, it is straightforward to check that

$$\mathrm{id}_{Y\amalg Z} \cup R \cup R^{\mathrm{op}} \cup (R \circ R^{\mathrm{op}}) \cup (R^{\mathrm{op}} \circ R)$$

is already an equivalence relation thanks to (13.2.5). As a finite union of closed sets, it is also closed, hence two points in $Y \amalg Z$ get identified in $Y\,{}_f\!\amalg_g Z$ if and only if they satisfy this relation. In particular, $y \in Y$ and $z \in Z$ map to the same point in $Y\,{}_f\!\amalg_g Z$ if and only if $(y, z) \in R$. □

In general, the pushout of an effective-monic cone along an arbitrary function is not effective-monic again. The following example shows that the effective-monic cones on CHaus do not form a coverage (see Definition 13.2.5); an even more drastic example can be found in the proof of Proposition 13.2.2.

Example 13.2.2 Take $X := \mathbf{4} = \{0, 1, 2, 3\}$, and consider two maps to spaces with 3 points,

$$f : \{0, 1, 2, 3\} \longrightarrow \{01, 2, 3\}, \qquad g : \{0, 1, 2, 3\} \longrightarrow \{0, 1, 23\},$$

as illustrated by the projection maps in Fig. 13.1. By Lemma 13.2.4, this cone is effective-monic. However, taking the pushout along the identification map

$$h : \{0, 1, 2, 3\} \longrightarrow \{0, 12, 3\}$$

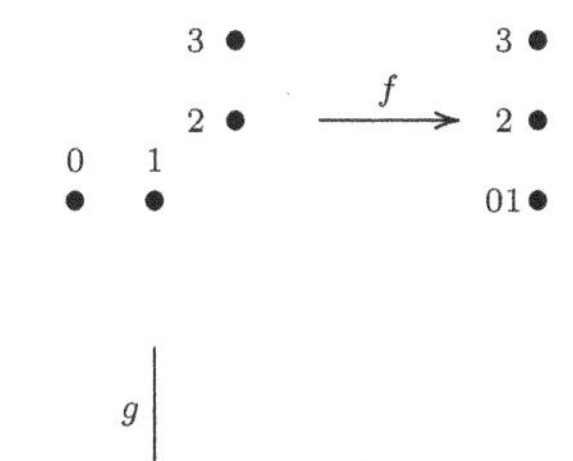

Fig. 13.1 Illustration of the cone $\{f, g\}$ of Example 13.2.2

results in a cone consisting of $f' : \{0, 12, 3\} \to \{012, 3\}$ and $g' : \{0, 12, 3\} \to \{0, 123\}$. Since the criterion of Lemma 13.2.4 fails, the cone $\{f', g'\}$ is not effective-monic. In particular, the pushout of an effective-monic cone is not necessarily effective-monic again. Worse, the collection of all effective-monic cones is not a coverage (see Definition 13.2.5), in fact, for our original $\{f, g\}$, there does not exist any effective-monic cone $\{k_i : \{0, 12, 3\} \to Y_i\}_{i\in I}$ such that every $k_i h$ would factor through f or g,

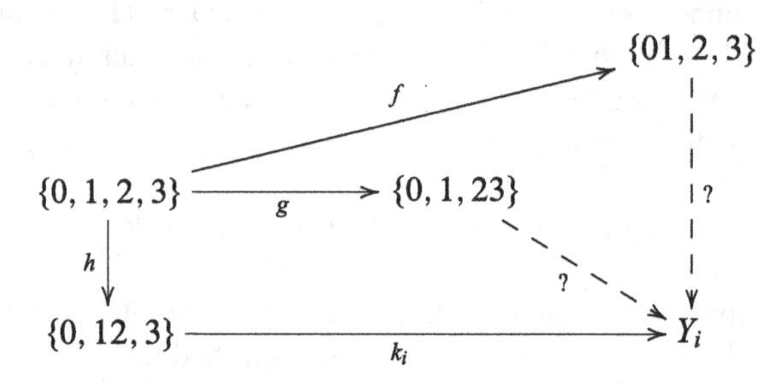

The reason is as follows: for every $i \in I$, we would need to have $k_i(0) = k_i(12)$ or $k_i(12) = k_i(3)$. If the former happens, consider the point $y_i := k_i(3) \in Y_i$, while if the latter happens take $y_i := k_i(0)$. (If both cases apply, these two prescriptions result in the same point $y_i = k_i(0) = k_i(3)$.) It is easy to check that the resulting family of points $\{y_i\}_{i\in I}$ is compatible. However, it does not arise from a point of $\{0, 12, 3\}$. In fact, since the k_i must separate points, there must be i with $k_i(0) = k_i(12) \neq k_i(3)$, and another i with $k_i(0) \neq k_i(12) = k_i(3)$. Hence neither of $x \in \{0, 12, 3\}$ results in the given compatible family, and the cone $\{k_i\}$ is not effective-monic.

Incidentally, the cone $\{f', g'\}$ from above is arguably the simplest example of a cone that separates points (it is jointly injective) without being effective-monic.

We recall the definition of a coverage to be

Definition 13.2.5 Given a category C, a coverage on C consists of a function assigning to each object $U \in C$ a collection of families of morphisms $\{f_i : U_i \to U\}_{i\in I}$ called covering families, such that given any morphism $g : V \to U$, then there exists a covering family $\{h_j : V_j \to V\}$ such that each composite $g \circ h_j$ factors though some f_i, i.e.

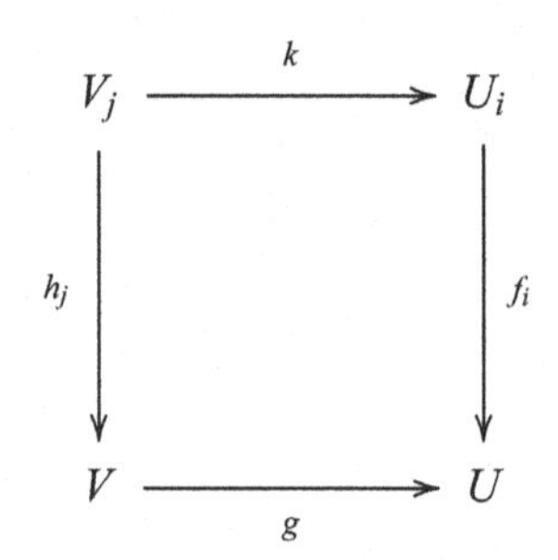

The previous example can also be understood in terms of effectus theory [47, Assumption 1]: the relevant pushout square is of the form

$$\begin{array}{ccc} W+Y & \xrightarrow{\mathrm{id}+f} & W+Z \\ {\scriptstyle g+\mathrm{id}}\downarrow & & \downarrow{\scriptstyle g+\mathrm{id}} \\ X+Y & \xrightarrow[\mathrm{id}+f]{} & X+Z \end{array}$$

where '+' is the coproduct in CHaus and both f and g are the unique map $\mathbf{2} \to \mathbf{1}$. In general, any cone consisting of $\mathrm{id}+f : W+Y \to W+Z$ and $g+\mathrm{id} : W+Y \to X+Y$ is effective-monic by Lemma 13.2.4.

It is conceivable that there are deeper connections with effectus theory than just at the level of examples, but so far we have not explored this theme in depth.

Going back to C*-algebras, we record one more statement about cones for further use.

Lemma 13.2.5 *A cone $\{f_i : X \to Y_i\}$ separates points if and only if the ranges of the $C(f_i) : C(Y_i) \to C(X)$ generate $C(X)$ as a C*-algebra.*

Proof By the Stone-Weierstrass theorem, the C*-subalgebra generated by the ranges of the $C(f_i)$ equals $C(X)$ if and only if it separates points (as a subalgebra). This C*-subalgebra is generated by the elements $g_i \circ f_i \in C(X)$, where $g_i : Y_i \to [0, 1]$ ranges over all functions, and hence the subalgebra separates points if and only if these functions separate points. This in turn is equivalent to the f_i separating points, since the $g_i : Y_i \to [0, 1]$ also separate points. □

13.2.2 How to Guarantee Commutativity?

The previous subsection was concerned with sheaf conditions satisfied by the functors $-(A)$ for commutative A. Now, we want to investigate which of these sheaf conditions hold for general A.

Definition 13.2.6 An effective-monic cone $\{f_i : X \to Y_i\}_{i\in I}$ in CHaus is *guaranteed commutative* if every functor $-(A)$ satisfies the sheaf condition on it.

In detail, $-(A)$ satisfies the sheaf condition on $\{f_i\}$ if and only if restricting a $*$-homomorphism $\alpha : C(X) \to A$ along all $C(f_i) : C(Y_i) \to C(X)$ to families $\beta_i : C(Y_i) \to A$, that are compatible in the sense that $\beta_i \circ C(g) = \beta_j \circ C(h)$ for every diagram of the form (13.2.1),

$$\begin{array}{ccc} X & \xrightarrow{f_i} & Y_i \\ {\scriptstyle f_j}\downarrow & & \downarrow{\scriptstyle g} \\ Y_j & \xrightarrow[h]{} & Z \end{array}$$

results in a bijection. In terms of the functor $C : \mathsf{CHaus}^{\mathrm{op}} \to \mathsf{C^*alg}_1$, this holds if and only if the diagram

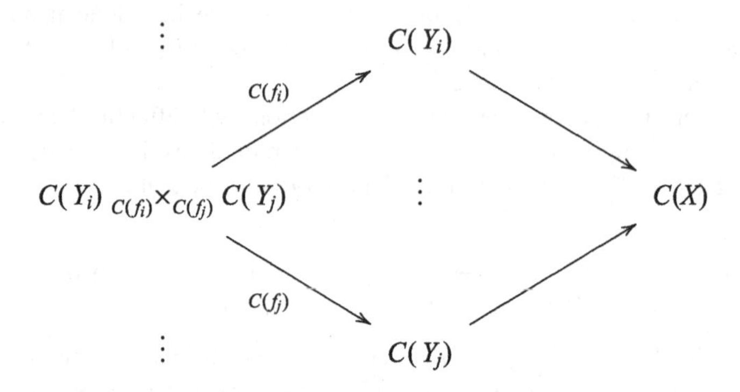

which is the image of (13.2.2) under C, is a colimit in $\mathsf{C^*alg}_1$. Here, we have used the canonical isomorphism $C(Y_i \;{}_{f_i}\!\amalg_{f_j} Y_j) \cong C(Y_i) \;{}_{C(f_i)}\!\times_{C(f_j)} C(Y_j)$, which holds because C is a right adjoint. So we are dealing with an instance of the question: *which limits does C turn into colimits?*

Remark 13.2.2 In terms of the physical interpretation of Remarks 13.1.1 and 13.1.3, the sheaf condition on a cone $\{f_i : X \to Y_i\}$ states that every compatible family of measurements with outcomes in the Y_i corresponds to a unique measurement with values in X, which coarse-grains to the given measurements via the f_i.

The terminology of Definition 13.2.6 is motivated by the following observation:

Lemma 13.2.6 *An effective-monic cone $\{f_i : X \to Y_i\}_{i \in I}$ is guaranteed commutative if and only if, for every $A \in \mathsf{C^*alg}_1$ and compatible family $\beta_i : C(Y_i) \to A$, the ranges of the β_i commute.*

Proof Suppose that the criterion holds. For $A \in \mathsf{C^*alg}_1$, we show that restricting a $*$-homomorphism $C(X) \to A$ to a compatible family of $*$-homomorphisms $C(Y_i) \to A$ is a bijection. We first show injectivity. To this end let $\alpha, \alpha' : C(X) \to A$ be such that the resulting families coincide, i.e. $\beta_i = \beta_i'$. In particular, this means that the range of each β_i coincides with the range of β_i', and hence $\mathrm{im}(\alpha) = \mathrm{im}(\alpha')$ by Lemma 13.2.5. We are, then, back in the commutative case, where Gelfand duality and the effective-monic assumption apply.

For surjectivity, let a compatible family $\beta_i : C(Y_i) \to A$ be given. By assumption, there is some commutative subalgebra $B \subseteq A$ which contains the ranges of all β_i, and it is sufficient to prove the sheaf condition with B in place of A. The claim then follows from Gelfand duality together with the assumption that $\{f_i\}$ is effective-monic.

Conversely, if the sheaf condition holds on a functor $-(A)$, then the $\beta_i : C(Y_i) \to A$ all arise from restricting some $\alpha : C(X) \to A$ along $C(f_i) : C(Y_i) \to C(X)$. In particular, the range of every β_i is contained in the range of α, which is a commutative C*-subalgebra. □

The crucial ingredient here is the fact that commutativity is a pairwise property, in the sense that if any family of elements in a C*-algebra commute pairwise, then they generate a commutative C*-subalgebra. We will meet this property again in Definition 3.1.5.

In the sense of Lemma 13.2.6, the question is under what conditions an effective-monic cone 'guarantees commutativity' of the ranges of a compatible family.

Example 13.2.3 The effective-monic cone of Example 13.2.2 is guaranteed commutative. In terms of indicator functions of individual points, the compatibility assumption on a pair of $*$-homomorphisms $\beta_f : C(\{01, 2, 3\}) \to A$ and $\beta_g : C(\{0, 1, 23\}) \to A$ is that

$$\beta_f(\chi_{01}) = \beta_g(\chi_0) + \beta_g(\chi_1), \qquad \beta_g(\chi_{23}) = \beta_f(\chi_2) + \beta_f(\chi_3).$$

So $\beta_g(\chi_0)$ is a projection below $\beta_f(\chi_{01})$, and in particular orthogonal to $\beta_f(\chi_2)$ and $\beta_f(\chi_3)$, so that it commutes with every element in the range of β_f. Proceeding like this proves that the ranges of β_f and β_g commute entirely.

Example 13.2.4 Let $\mathbb{T} \subseteq \mathbb{C}$ be the unit circle, and $p_{\mathfrak{R}}, p_{\mathfrak{I}} : \mathbb{T} \to [-1, +1]$ the two coordinate projections. Then the cone $\{p_{\mathfrak{R}}, p_{\mathfrak{I}}\}$ is effective-monic. This can be seen by either applying Lemma 13.2.4 or, alternatively, noting that applying $p_{\mathfrak{R}}$ and $p_{\mathfrak{I}}$ establishes a bijection between points of x and pairs of numbers $y_{\mathfrak{R}}, y_{\mathfrak{I}} \in [-1, +1]$ with $y_{\mathfrak{R}}^2 + y_{\mathfrak{I}}^2 = 1$. Hence compatible families $\{\beta_{\mathfrak{R}}, \beta_{\mathfrak{I}}\}$ are $*$-homomorphisms $\beta_{\mathfrak{R}} : C([-1, +1]) \to A$ and $\beta_{\mathfrak{I}} : C([-1, +1]) \to A$ that correspond to self-adjoint elements $\beta_{\mathfrak{R}}(\mathrm{id}), \beta_{\mathfrak{I}}(\mathrm{id}) \in [-1, +1](A)$ with $\beta_{\mathfrak{R}}(\mathrm{id})^2 + \beta_{\mathfrak{I}}(\mathrm{id})^2 = 1$. Functional calculus tells us that such a pair of self-adjoints arises from a unitary if and only if they commute. As a counter example consider an A with non-commuting symmetries $s_{\mathfrak{R}}$ and $s_{\mathfrak{I}}$, this gives rise to a compatible family upon defining $\beta_{\mathfrak{R}} := s_{\mathfrak{R}}/\sqrt{2}$ and $\beta_{\mathfrak{I}} := s_{\mathfrak{I}}/\sqrt{2}$. However, such a family does not arise in the way described above. Therefore $\{p_{\mathfrak{R}}, p_{\mathfrak{I}}\}$ is not guaranteed commutative.

So far, we know of one powerful sufficient condition for guaranteeing commutativity:

Definition 13.2.7 An effective-monic cone $\{f_i : X \to Y_i\}_{i \in I}$ in CHaus is *directed* if for every $i \in I$ there is a cone $\{g_i^j : Y_i \to Z_i^j\}_{j \in J_i}$ which separates points, and it is such that for every $i, i' \in I$ and $j \in J_i, j' \in J_{i'}$ there is $k \in I$ and a diagram

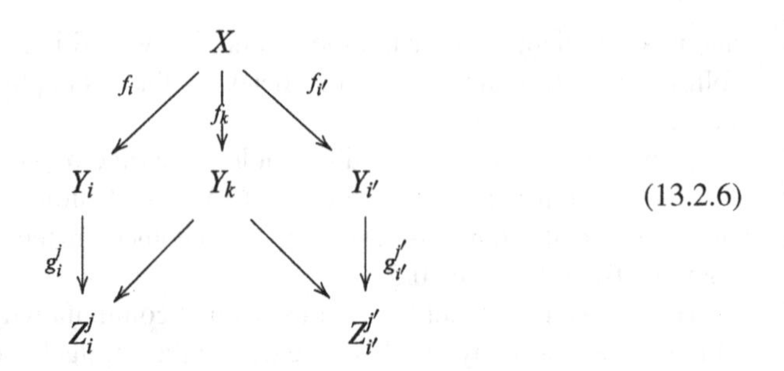

(13.2.6)

Note that this definition can be considered in principle in any category.

Proposition 13.2.1 *If* $\{f_i\}$ *is effective-monic and directed, then it is also guaranteed commutative.*

Proof By Lemma 13.2.6, it is enough to show that the ranges of a compatible family $\{\beta_i : C(Y_i) \to A\}$ commute. By Lemma 13.2.5, it is enough to prove that the range of $\beta_i \circ C(g_i^j) : C(Z_i^j) \to A$ commutes with the range of $\beta_{i'} \circ C(g_{i'}^{j'}) : C(Z_{i'}^{j'}) \to A$ for any $i, i' \in I$ and $j \in J_i, j' \in J_{i'}$. Thanks to (13.2.6) and the compatibility, both of these ranges are contained in the range of $\beta_k : C(Y_k) \to A$, which is commutative.

□

Example 13.2.5 Let $\mathbf{2}^{\mathbb{N}}$ be the Cantor space, with projections $p_n : \mathbf{2}^{\mathbb{N}} \to \mathbf{2}^n$ for every $n \in \mathbb{N}$. Then the cone $\{p_n\}_{n\in\mathbb{N}}$ is effective-monic and directed. Therefore it is also guaranteed commutative.

More generally, let Λ be a small cofiltered category[4] and $L : \Lambda \to \mathsf{CHaus}$ a functor of which we consider the limit $\lim_\Lambda L \in \mathsf{CHaus}$. The cone of limit projections $\{p_\lambda : \lim_\Lambda L \to L(\lambda)\}$ is effective-monic (Example 13.2.1). With the trivial cones $\{\mathrm{id}\}$ on the codomains $L(\lambda)$, the cofilteredness implies that the cone is also directed, and therefore guaranteed commutative. What we have shown hereby in a roundabout manner is that a filtered colimit of commutative C*-algebras is again commutative.

Unfortunately, the converse to Proposition 13.2.1 is not true:

Example 13.2.6 The effective-monic cone $\{f, g\}$ of Examples 13.2.2 and 13.2.3 is not directed, despite being guaranteed commutative. The reason is that the additional cones, as in Definition 13.2.7, would have to contain some $h : \{12, 3, 4\} \to Z_{12}$ with $h(3) \neq h(4)$, and, similarly, some $k : \{1, 2, 34\} \to Z_{34}$ with $k(1) \neq k(2)$. By (13.2.6), this would mean that the cone $\{f, g\}$ would have to contain a function that separates both 1 from 2 and 3 from 4, which is not the case.

[4] A (finitely) cofiltered category is a category $\mathcal{C}$ in which every finite diagram has a cone. The dual notion is that of a filtered category. A diagram $F : \mathcal{C} \to \mathcal{D}$ for $\mathcal{C}$ a cofiltered category is called a cofiltered diagram. A limit of a cofiltered diagram is called a cofiltered colimit. Again, dual notions apply.

While Proposition 13.2.1 is sufficiently powerful for the results of this chapter, it remains an open question to find a necessary and sufficient condition for guaranteeing commutativity.

Lemma 13.2.7 *For any $X \in$ CHaus, the cone $\{f : X \to \square\}$ of all functions $f : X \to \square$ is directed.*

By Lemma 13.2.2, we already know that this cone is effective-monic. By Proposition 13.2.1, we can now conclude that it is also guaranteed commutative.

Proof In Definition 13.2.7, take every $\{g_i^j\}_{j\in J_i}$ to be the cone consisting of all functions $\square \to [0,1]$. Since the pairing of any two functions $X \to [0,1]$ is a function $X \to \square$, the cone $\{f : X \to \square\}$ is directed. □

In terms of Remark 13.2.2, Lemma 13.2.7 'explains' why physical measurements are numerical. In fact, for every conceivable measurement with values in some arbitrary space X, conducting that measurement and recording the outcome in X is equivalent to conducting a sufficient number of measurements with values in $\square$ and recording their outcomes, which are now plain (complex) numbers.

Lemma 13.2.8 *If two cones $\{f_i : W \to Y_i\}_{i\in I}$ and $\{g_j : X \to Z_i\}_{j\in J}$ are effective-monic and directed, then so is the product cone*

$$\{f_i \times g_j : W \times X \to Y_i \times Z_j\}_{(i,j)\in I\times J}.$$

Proof Let $\{h_i^k : Y_i \to U_i^k\}_{k\in K_i}$ and $\{k_j^l : Z_j \to V_j^l\}_{l\in L_j}$ be the families of additional cones that witness the directedness. Then for $(i,j) \in I \times J$, consider the cone at $Y_i \times Z_j$ given by

$$\{h_i^k p_{Y_i} : Y_i \times Z_j \to U_i^k\} \cup \{k_j^l p_{Z_j} : Y_i \times Z_j \to V_j^l\} \tag{13.2.7}$$

with index set $K_i \amalg L_j$. This cone separates the points of $X_i \times Y_j$, since any two different points differ in at least one coordinate. To check that the condition of Definition 13.2.7 is satisfied, one needs to distinguish the cases of the left and the right morphism in (13.2.6) belonging to either part of (13.2.7). The only interesting case that comes up is when one considers a $h_i^k p_{Y_i} : Y_i \times Z_{j'} \to U_i^k$ together with a $k_j^l p_{Z_j} : Y_{i'} \times Z_j \to \times V_j^l$, resulting in a diagram of the form

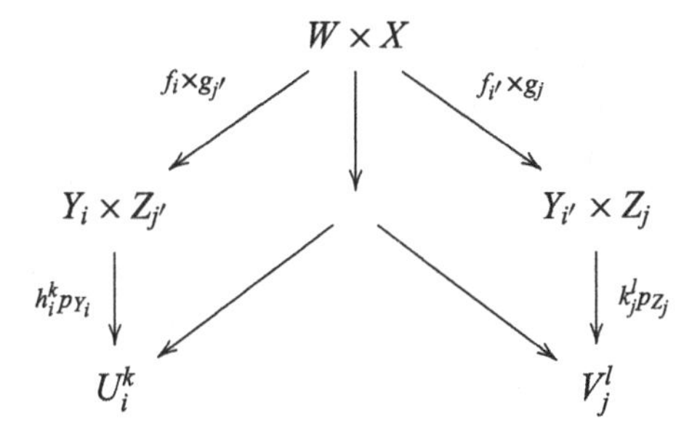

where indeed the central vertical arrow can be taken to be $f_i \times g_j$. □

In combination with Lemma 13.2.7, we therefore obtain:

Corollary 13.2.1 *For any* $X, Y \in \mathsf{CHaus}$, *the cone* $\{f \times g : X \times Y \to \Box \times \Box\}$ *indexed by all functions* $f : X \to \Box$ *and* $g : Y \to \Box$ *is directed.*

Another simple class of examples is as follows:

Lemma 13.2.9 *Let* $\{f_i : X \to Y_i\}_{i \in I}$ *be an effective-monic cone on* $X \in \mathsf{CHaus}$. *Then the cone*

$$\{(f_{i_1}, \ldots, f_{i_n}) : X \to Y_{i_1} \times \ldots \times Y_{i_n}\}$$

consisting of all finite tuplings of the f_i *is effective-monic and directed.*

Alternatively, we could phrase this as saying that, if an effective-monic cone is closed under pairing, then it is directed.

Proof Mapping points of X to compatible families of points in all finite products $\prod_{m=1}^{n} Y_{i_m}$ is trivially injective, since it is already so on single-factor products due to the effective-monic assumption. Concerning surjectivity, the compatibility assumption guarantees that the component $(y_1, \ldots, y_n) \in Y_{i_1} \times \ldots \times Y_{i_n}$ is uniquely determined by the components in every individual y_i, since this is precisely the compatibility condition on diagrams of the form

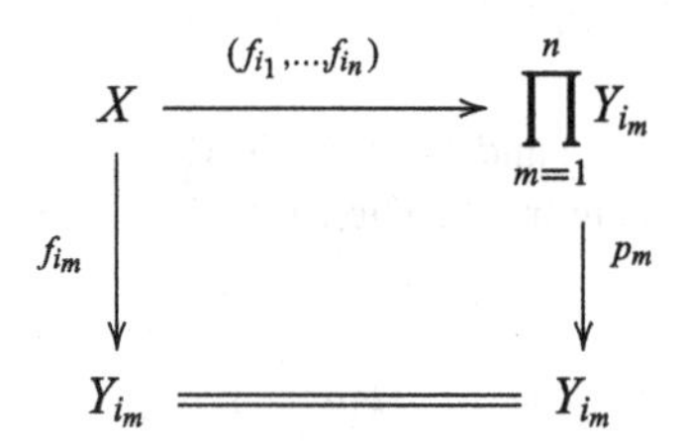

Hence the new cone is also effective-monic.

The condition of Definition 13.2.7 holds by construction, with the trivial cone $\{\mathrm{id}\}$ on the codomains. □

Next, we briefly investigate the collection of directed effective-monic cones in its entirety.

Proposition 13.2.2 *The collection of all directed effective-monic cones on* CHaus *is not a coverage (see Definition 13.2.5).*

Results along the lines of [61, Theorem 1.1] indicate that this is not due to the potential inadequacy of our definitions, but rather due to fundamental obstructions related to the noncommutativity.

Proof Consider $X := \{0, 1\}^3$ together with the three product projections $p_1, p_2, p_3 : \{0, 1\}^3 \to \{0, 1\}$. Applying a similar reasoning as that utilised in the proof of Lemma 13.2.2 it can be deduced that their three pairings

$$\{\,(p_1, p_2),\ (p_1, p_3),\ (p_2, p_3)\ :\ \{0, 1\}^3 \longrightarrow \{0, 1\}^2\,\} \tag{13.2.8}$$

form an effective-monic cone. From the proof of Lemma 13.2.7 it follows that this cone is directed.

Now consider the function $f : \{0, 1\}^3 \to \mathbf{4}$ defined by mapping every element of $\{0, 1\}^3$ to the sum of its digits. In any square of the form

$$\begin{array}{ccc} \{0,1\}^3 & \xrightarrow{(p_1,p_2)} & \{0,1\}^2 \\ {\scriptstyle f}\downarrow & & \downarrow{\scriptstyle g} \\ \mathbf{4} & \xrightarrow[h]{} & Z \end{array}$$

we necessarily have

$$\underbrace{h(f(000))}_{=h(0)} = g(00) = h(f(001)) = h(f(010)) = g(01)$$

$$= \underbrace{h(f(010))}_{=h(1)} = \ldots = \underbrace{h(f(110))}_{=h(2)} = \ldots = \underbrace{h(f(111))}_{=h(3)},$$

and therefore h must be constant. By symmetry, the same must hold with (p_1, p_3) or (p_2, p_3) in place of (p_1, p_2). Hence any cone on $\mathbf{4}$ that factors through (13.2.8) must identify *all* points of $\mathbf{4}$. In particular, no such cone can be effective-monic, let alone directed. □

We close this subsection with another potential criterion for guaranteeing commutativity.

Lemma 13.2.10 *The following conditions on a cone $\{f_i : X \to Y_i\}$ in* CHaus *are equivalent:*

1. *For every $x \in X$ and neighbourhood $U \ni x$ there exists $i \in I$ with*

$$f_i^{-1}(f_i(x)) \subseteq U.$$

2. *For every $x \in X$ and neighbourhood $U \ni x$ there exist $i \in I$ and a neighbourhood $V \ni f_i(x)$ with*

$$f_i^{-1}(V) \subseteq U.$$

3. *The sets of the form $f_i^{-1}(V)$ for open $V \subseteq Y_i$ form a basis for the topology on X.*

Proof

1⇒2: Since $X \setminus U$ is compact, $f_i(X \setminus U)$ is a closed set, and disjoint from $\{x\}$ by assumption. Now take V to be any open neighbourhood of $f_i(x)$ disjoint from $f_i(X \setminus U)$.

2⇒3: Suppose $x \in f_i^{-1}(V_i) \cap f_j^{-1}(V_j)$. Then by assumption, there is k and an open $V_k \subseteq Y_k$ with $f_k(x) \in V_k$ such that

$$f_k^{-1}(V_k) \subseteq f_i^{-1}(V_i) \cap f_j^{-1}(V_j).$$

3⇒1: There must be a basic open $f_i^{-1}(V_i)$ with $x \in f_i^{-1}(V_i) \subseteq U$. □

Definition 13.2.8 If the above conditions hold, we say that the cone $\{f_i\}$ is *locally injective*.

Clearly, a locally injective cone separates points. However, it is not necessarily effective-monic:

Example 13.2.7 The cone consisting of all three surjective functions $\mathbf{3} \to \mathbf{2}$ is locally injective. However, it is not effective-monic since the pushout of any two different maps $\mathbf{3} \to \mathbf{2}$ is trivial, and hence there are 2^3 compatible families of points in the cone, but only 3 points in X.

Example 13.2.8 The cone $\{p_{\mathfrak{R}}, p_{\mathfrak{I}}\}$ from Example 13.2.4 is not locally injective. In fact, for any angle $0 < \varphi < \pi/2$, the point $(\cos\varphi, \sin\varphi) \in \mathbb{T}$ cannot be distinguished from $(\cos\varphi, -\sin\varphi) \in \mathbb{T}$ under $p_{\mathfrak{R}}$, and not from $(-\cos\varphi, \sin\varphi)$ under $p_{\mathfrak{I}}$.

Conjecture 13.2.1 An effective monic cone $\{f_i : X \to Y_i\}$ that is locally injective is also guaranteed commutative.

Since the cone of all functions $X \to \square$ is an effective-monic and a locally injective cone, proving this conjecture would again show that $\{f : X \to \square\}$ is guaranteed commutative. Furthermore, this would detect some cones as guaranteed commutative that are not detected as such by Proposition 13.2.1. For example, the effective-monic cone of Examples 13.2.2 and 13.2.3.

Example 13.2.9 In the setting of Example 13.2.5, the topology of $\lim_\Lambda L$ is generated by the preimages of opens in all the $L(\lambda)$. The cofilteredness assumption implies that these opens form a basis: for $U_\lambda \subseteq L(\lambda)$ and $U_{\lambda'} \subseteq L(\lambda')$, we have $\hat{\lambda}$ and morphisms $f : \hat{\lambda} \to \lambda$ and $f' : \hat{\lambda} \to \lambda'$ such that

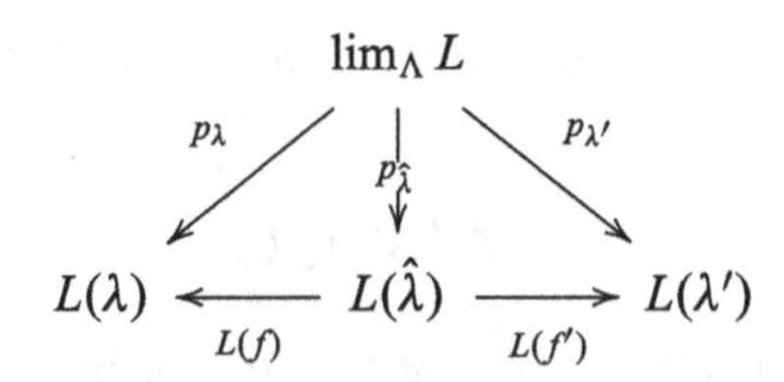

commutes. In particular, $f^{-1}(U_\lambda) \cap f'^{-1}(U_{\lambda'})$ is an open in $L(\hat{\lambda})$ whose preimage in $\lim_\Lambda L$ is exactly the intersection of the preimages of U_λ and $U_{\lambda'}$. Hence the limit cone $\{p_\lambda\}$ is also locally injective. By Example 13.2.5, this is in accordance with Conjecture 13.2.1.

As already seen in Proposition 13.2.1, being locally injective is also not a necessary condition for guaranteeing commutativity of effective-monic cones.

Example 13.2.10 There are effective-monic cones that are directed and hence guaranteed commutative, but not locally injective. For example with $⧉ := [0, 1]^3$ the unit cube, the three face projections $p_1, p_2, p_3 : ⧉ \to \square$ form a cone $\{p_1, p_2, p_3\}$ that is effective-monic but not locally injective. Nevertheless, if one considers copies of the cone $\{p_{\Re}, p_{\Im} : \square \to [0, 1]\}$, Definition 13.2.7 shows that the cone is directed, and hence guaranteed commutative. In particular, the converse to Conjecture 13.2.1 is false.

13.2.3 The Category of Sheaves and Its Smallness Properties

Now that we have some idea of which sheaf conditions are satisfied by C*-algebras, we investigate completely general functors CHaus → **Sets** satisfying some of these sheaf conditions.

Definition 13.2.9 A functor F : CHaus → **Sets** is a *sheaf* if it satisfies the sheaf condition on all effective-monic cones that are directed.

We write Sh(CHaus) for the resulting category of sheaves, which is a full sub-category of $\mathbf{Sets}^{\mathsf{CHaus}}$. Due to Proposition 13.2.2, the sheaf conditions are *not* those of a (large) site. Nevertheless, we expect that Sh(CHaus) is an instance of a category of sheaves on a *quasi-pretopology* or on a *Q-category*, whose categories of sheaves were investigated by Kontsevich and Rosenberg in the context of noncommutative algebraic geometry [62, 63].[5]

A priori, Sh(CHaus) may seem rather unwieldy, and it is not clear whether it is locally small.

Lemma 13.2.11 *Let* $F, G \in$ Sh(CHaus). *Evaluating natural transformations on* $\square$ *is injective,*

$$\mathsf{Sh(CHaus)}(F, G) \hookrightarrow \mathbf{Sets}(F(\square), G(\square)).$$

Proof Since F and G satisfy the sheaf condition on $\{f : X \to \square\}$ by Corollary 13.2.7, the canonical map

$$F(X) \longrightarrow \prod_{f:X\to\square} F(\square)$$

[5]It is natural to suspect that the reason why Grothendieck topologies do not apply is in both cases due to the noncommutativity, as it has been formally proven in [61]. However, so far, we have not explored the relation to the work of Kontsevich and Rosenberg any further.

is injective. Hence for any $\eta : F \to G$, the naturality diagram

$$\begin{array}{ccc} F(X) & \longrightarrow & \prod_{f:X\to\square} F(\square) \\ \downarrow{\scriptstyle \eta_X} & & \downarrow{\scriptstyle \prod_f \eta_\square} \\ G(X) & \longrightarrow & \prod_{f:X\to\square} G(\square) \end{array}$$

shows that every component η_X is uniquely determined by $\eta_\square$. $\square$

Corollary 13.2.2 $\mathrm{Sh}(\mathsf{CHaus})$ *is locally small.*

Proof Lemma 13.2.11 provides an upper bound on the size of each hom-set. $\square$

With functors $-(A)$ and $-(B)$ for $A, B \in \mathsf{C^*alg}_1$ in place of F and G, Lemma 13.2.11 also follows from the Yoneda lemma (Lemma A.7.2) and the fact that $C(\square)$ is a separator in $\mathsf{C^*alg}_1$. The latter is true more generally:

Corollary 13.2.3 $-(C(\square))$ *is a separator (see Definition 13.2.10) in* $\mathrm{Sh}(\mathsf{CHaus})$.

Recall that as functors $-(C(\square)), \mathsf{CHaus}(\square, -) : \mathsf{CHaus} \to \mathbf{Sets}$ are such that $-(C(\square)) \cong \mathsf{CHaus}(\square, -)$.

Proof By the Yoneda lemma (Lemma A.7.2),

$$\mathrm{Sh}(\mathsf{CHaus})(-(C(\square)), F) = \mathbf{Sets}^{\mathsf{CHaus}}(\mathsf{CHaus}(\square, -), F) = F(\square), \tag{13.2.9}$$

and hence the claim follows from Lemma 13.2.11. $\square$

We recall the definition of a separator to be the following:

Definition 13.2.10 Consider a category C, an object $S \in C$ is a separator if given any parallel morphism $f, g: X \to Y$ in C, if $f \circ e = g \circ e$ for every morphism $e: S \to X$, then $f = g$.

If C is locally small category, S is a separator if the functor $Hom(S, -): C \to \mathbf{Sets}$ is faithful, i.e. for all $X, Y \in C$ then

$$Hom(S, -) : C(X, Y) \to \mathbf{Sets}(Hom(S, X), Hom(S, Y))$$

is injective.

The following stronger injectivity property will play a role in the next section:

Lemma 13.2.12 *For* $F \in \mathrm{Sh}(\mathsf{CHaus})$, *the following are equivalent:*

1. *The canonical map*

$$(F(p_1), F(p_2)) \; : \; F(\square \times \square) \longrightarrow F(\square) \times F(\square) \tag{13.2.10}$$

   is injective.

2. *For every* $X \in \mathsf{CHaus}$ *and effective-monic* $\{f_i : X \to Y_i\}$*, the canonical map*

$$F(X) \longrightarrow \prod_{i \in I} F(Y_i)$$

is injective.

In 2, the point is that the cone may not be directed so, generically, F does not satisfy the sheaf condition on it. The intuition behind the lemma is that when these (equivalent) conditions hold in the C*-algebra case, then the image of (13.2.10) consists of precisely the pairs of commuting normal elements. In terms of the interpretation as measurements on a physical system, this image consists of the pairs of measurements (with values in $\square$) that are jointly measurable.

In the proof, we can start to put the seemingly haphazard lemmas of the previous subsection to some use.

Proof Since the cone $\{p_1, p_2 : \square \times \square \to \square\}$ is effective-monic, condition 1 is a special case of 2.

In the other direction, we first show that for every $X, Y \in \mathsf{CHaus}$, the canonical map $F(X \times Y) \to F(X) \times F(Y)$ is injective. By Corollary 13.2.1, the left vertical arrow in

$$\begin{array}{ccc} F(X \times Y) & \longrightarrow & F(X) \times F(Y) \\ \downarrow & & \downarrow \\ \prod_{f:X \to \square, g:Y \to \square} F(\square \times \square) & \longrightarrow & \left(\prod_{f:X \to \square} F(\square)\right) \times \left(\prod_{g:Y \to \square} F(\square)\right) \end{array}$$

is injective. Since the lower horizontal arrow is injective by assumption, it follows that the upper horizontal arrow is also injective. By induction, we then obtain that $F(\prod_{j=1}^n X_j) \to \prod_{j=1}^n F(X_j)$ is injective for any finite product.

Now let $\{f_i\}$ be an arbitrary effective-monic cone on X. By Lemma 13.2.9, F satisfies the sheaf condition on the cone consisting of all the finite tuplings $(f_{i_1}, \ldots, f_{i_n})$. Hence we have the diagram

$$\begin{array}{ccc} F(X) & \longrightarrow & \prod_{i \in I} F(Y_i) \\ \downarrow & & \downarrow \\ \prod_{n \in \mathbb{N}} \prod_{i_1, \ldots, i_n \in I} F\left(\prod_{m=1}^n Y_{i_m}\right) & \longrightarrow & \prod_{n \in \mathbb{N}} \prod_{i_1, \ldots, i_n \in I} \prod_{m=1}^n F(Y_{i_m}) \end{array}$$

where the left vertical arrow is injective due to the sheaf condition, and the lower horizontal one due to the first part of the proof. Hence also the upper horizontal arrow is injective. □

So far, we do not know of any sheaf CHaus → **Sets** that would *not* have the property characterized by the above Lemma.

By Gelfand duality, the commutative C*-algebras are precisely the representable functors $\mathsf{CHaus}(W, -) : \mathsf{CHaus} \to \mathbf{Sets}$ (see Definition 13.2.4). These are characterized in terms of a condition similar to the previous lemma:

Lemma 13.2.13 *For $F \in \mathrm{Sh}(\mathsf{CHaus})$, the following are equivalent:*

1. *The canonical map*

$$(F(p_1), F(p_2)) : F(\square \times \square) \longrightarrow F(\square) \times F(\square)$$

   is bijective.
2. *F satisfies the sheaf condition on every effective-monic cone $\{f_i : X \to Y_i\}$ in* CHaus.
3. *F is representable.*

Proof By the definition of effective-monic, 3 trivially implies 2. Also if 2 holds, then it is easy to show 1: the empty cone is effective-monic on $1 \in \mathsf{CHaus}$, which implies $F(1) \cong 1$. With this in mind, 1 is the sheaf condition on the effective-monic cone $\{p_1, p_2 : \square \times \square \to \square\}$.

The burden of the proof is the implication from 1 to 3. By the representable functor theorem [54, p. 130] and the generation of limits by products and equalizers, it is enough to show that F preserves products and equalizers, which we do in several steps. As a first step we note that, since the functor $- \times Y : \mathsf{CGHaus} \to \mathsf{CGHaus}$ is a left adjoint [54, Theorem VII.8.3], it preserves colimits, in particular preserves pushouts for any $Y \in \mathsf{CHaus}$. Moreover the inclusion functor CHaus → CGHaus also preserves finite colimits, since it preserves finite coproducts and coequalizers (the latter by the automatic compactness of quotients of compact spaces).

As a second step, we prove that the canonical map $F(X \times \square) \longrightarrow F(X) \times F(\square)$ is a bijection for every $X \in \mathsf{CHaus}$. To this end, we consider the effective-monic cone $\{f \times \mathrm{id}_\square : X \times \square \to \square \times \square\}$ indexed by $f : X \to \square$. We know that this cone is directed by Lemmas 13.2.7 and 13.2.8. This entails that $F(X \times \square)$ is equal to the set of compatible families $\{\beta_f\}_{f:X\to\square}$ of elements of $\prod_{f:X\to\square} F(\square \times \square)$. Since $- \times \square$ preserves pushouts as per the first observation, the compatibility condition is the one associated to the squares of the form

$$\begin{array}{ccc} X \times \square & \xrightarrow{f\times \mathrm{id}_\square} & \square \times \square \\ {\scriptstyle g\times \mathrm{id}_\square}\downarrow & & \downarrow \\ \square \times \square & \longrightarrow & (\square \,{}_{f\times\mathrm{id}}\amalg_{g\times\mathrm{id}}\, \square) \times \square \end{array}$$

By using the fact that the maps $\Box \,{}_{f\times\mathrm{id}}\amalg_{g\times\mathrm{id}}\, \Box \longrightarrow \Box$ separate points, it is sufficient to postulate the compatibility on all commuting squares of the form

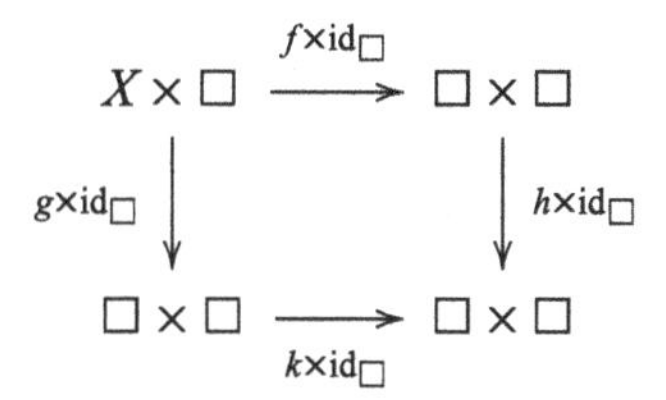

Upon decomposing $\beta_f = (\beta_f^1, \beta_f^2)$ via $F(\Box\times\Box) = F(\Box)\times F(\Box)$, the compatibility condition is equivalent to the fact that $F(h)(\beta_f^1) = F(k)(\beta_g^1)$ and that $\beta_f^2 = \beta_g^2$ for all $h, k : \Box \to \Box$ with $hf = kg$. Since the family of first components corresponds precisely to an element of $F(X)$, we conclude that the canonical map $F(X\times\Box) \longrightarrow F(X)\times F(\Box)$ is an isomorphism.

We now use this result to show that $F(X\times Y) \longrightarrow F(X)\times F(Y)$ is an isomorphism for all $X, Y \in$ CHaus. The proof is the same as above, just with $-\times\Box$ replaced by $-\times Y$. The case of finite products $F(\prod_{i=1}^n X_i) \cong \prod_{i=1}^n F(X_i)$ can then be proven by induction, while the case of infinite products can be proven using the sheaf condition.

We now show that F preserves equalizers. Since every monomorphism $f : X \to Y$ in CHaus is regular, the singleton cone $\{f\}$ is effective-monic. The fact that this cone is trivially directed implies that F satisfies the sheaf condition on it, which entails that $F(f) : F(X) \to F(Y)$ must be injective.

Recall that a diagram

$$E \overset{e}{\longrightarrow} X \underset{g}{\overset{f}{\rightrightarrows}} Y$$

is an equalizer if and only if

$$\begin{array}{ccc} E & \overset{e}{\longrightarrow} & X \\ {\scriptstyle e}\downarrow & & \downarrow{\scriptstyle (\mathrm{id}_X, f)} \\ X & \underset{(\mathrm{id}_X, g)}{\longrightarrow} & X\times Y \end{array}$$

is a pullback. By constructing the pushout $X \,{}_e\amalg_e\, X$ as a quotient of $X \amalg X$ and doing a case analysis on pairs of points in $X \,{}_e\amalg_e\, X$, the induced arrow k in

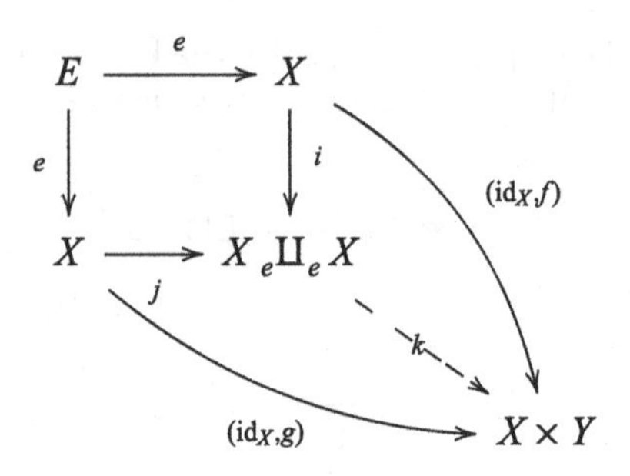

is seen to be a monomorphism and, therefore, so is $F(k)$. If $\beta \in F(X)$ is such that $F(f)(\beta) = F(g)(\beta)$, then also $F(i)(\beta) = F(j)(\beta)$. But by the sheaf condition on the singleton cone $\{e\}$, this means that β is in the image of $F(e)$, as it was to be shown. □

For $F \in \mathrm{Sh}(\mathsf{CHaus})$, any $W \in \mathsf{CHaus}$ and any $\alpha \in F(W)$, let $F_\alpha : \mathsf{CHaus} \to$ **Sets** be the subfunctor of F generated by α. Concretely, over every $X \in \mathsf{CHaus}$, the set $F_\alpha(X)$ consists of all the images $F(f)(\alpha)$ for $f : W \to X$.

Proposition 13.2.3 *If the canonical map*

$$F(\square \times \square) \longrightarrow F(\square) \times F(\square) \tag{13.2.11}$$

is injective, then such an F_α is representable.

Proof It is straightforward to verify that F_α is also a sheaf. If we are able to show that every pair of elements $(\beta_1, \beta_2) \in F_\alpha(\square) \times F_\alpha(\square)$ actually comes from an element of $F_\alpha(\square \times \square)$, then Lemma 13.2.13 and the injectivity assumption on F complete the proof.

To this end, we write $\beta_1 = F(f_1)(\alpha)$ and $\beta_2 = F(f_2)(\alpha)$ for certain f_1, f_2 : $W \to \square$. Now considering α transported along the pairing $(f_1, f_2) : W \to \square \times \square$ results in an element of $F(\square \times \square)$ that reproduces (β_1, β_2). □

Here is another smallness result:

Proposition 13.2.4 $\mathrm{Sh}(\mathsf{CHaus})$ *is well-powered.*

Before beginning the proof we recall the definition of a category being well-powered.

Definition 13.2.11 Given a category C, we say that C is well-powered if every object $X \in C$ has a small poset of subobjects, seen as an equivalence class of monomorphisms with codomain X.

In other words, for every object $X \in C$ the (generally large) preordered set of monomorphisms with codomain X is equivalent to a small poset.

For example Every Grothendieck topos is well-powered by the existence of a subobject classifier and the smallness of hom-sets.

We are now ready to prove Proposition 13.2.4.

Proof Let $\eta : F \to G$ be a monomorphism in Sh(CHaus). Then upon composing morphisms of the form $-(C(\square)) \longrightarrow F$ with η, the Yoneda lemma (13.2.9) shows that the component $\eta_\square : F(\square) \to G(\square)$ is injective, since the diagram

$$\begin{array}{ccc} \mathrm{Sh(CHaus)}(-(C(\square)), F) & \xrightarrow[\cong]{(13.2.9)} & F(\square) \\ \downarrow{\scriptstyle \eta\circ -} & & \downarrow{\scriptstyle \eta_\square} \\ \mathrm{Sh(CHaus)}(-(C(\square)), G) & \xrightarrow[\cong]{(13.2.9)} & G(\square) \end{array}$$

commutes.

Again, using the sheaf condition on all functions $X \to \square$ and the fact that $\square$ is a coseparator in CHaus, we can identify the $\alpha \in F(X)$ with the families $\{\beta_f\}$ with $\beta_f \in F(\square)$ that are indexed by $f : X \to \square$ and satisfy the compatibility condition that $F(h)(\beta_f) = \beta_{hf}$ for all f and $h : \square \to \square$. Hence we have the diagram

$$\begin{array}{ccccc} F(X) & \longrightarrow & \prod_{f:X\to\square} F(\square) & \rightrightarrows & \prod_{f:X\to\square,h:\square\to\square} F(\square) \\ \downarrow{\scriptstyle \eta_X} & & \downarrow{\scriptstyle \prod_f \eta_\square} & & \downarrow{\scriptstyle \prod_{f,h} \eta_\square} \\ G(X) & \longrightarrow & \prod_{f:X\to\square} G(\square) & \rightrightarrows & \prod_{f:X\to\square,h:\square\to\square} G(\square) \end{array}$$

in which both rows are equalizers. Therefore, for fixed G, the set $F(X)$ is determined by the inclusion map $\eta_\square : F(\square) \to G(\square)$. Hence the number of sub-objects of G is bounded by $2^{|G(\square)|}$. □

Corollary 13.2.4 *Every sheaf* F : CHaus → **Sets** *for which* (13.2.11) *is injective is a (small) colimit in* Sh(CHaus) *of representable functors.*

Proof We show that F is the colimit in Sh(CHaus) of the subfunctors of the form F_α from Proposition 13.2.3, as ordered by inclusion. Thanks to Proposition 13.2.4, this colimit is equivalent to a small colimit.

To show the required universal property suppose, first, that $\eta, \eta' \in$ Sh(CHaus)(F, G) coincide upon restriction to all F_α. Then in particular, $\eta_\square(\alpha) = \eta'_\square(\alpha)$ for all $\alpha \in F(\square)$, and hence $\eta = \eta'$ by the previous results. Conversely, let $\{\phi^\alpha\}_\alpha$ be a family of natural transformations $\phi^\alpha : F_\alpha \to G$ that are compatible in the sense that if $F_\beta \subseteq F_\alpha$, then $\phi^\alpha|_{F_\beta} = \phi^\beta$. Now define the component $\eta_X : F(X) \to G(X)$ on every $\alpha \in F(X)$ as

$$\eta_X(\alpha) := \phi^\alpha_X(\alpha).$$

The commutativity of the naturality square

$$\begin{array}{ccc} F(X) & \xrightarrow{\eta_X} & G(X) \\ {\scriptstyle F(f)}\downarrow & & \downarrow{\scriptstyle G(f)} \\ F(Y) & \xrightarrow[\eta_Y]{} & G(Y) \end{array}$$

on some $\alpha \in F(X)$ follows from

$$G(f)(\phi_X^\alpha(\alpha)) = \phi_Y^\alpha(F(f)(\alpha)) = \phi_Y^{F(f)(\alpha)}(F(f)(\alpha)),$$

where the first equation is naturality of ϕ^α and the second one is the assumed compatibility.

To see that η restricts to ϕ^α on every F_α we show that the components coincide, i.e. $\eta_Y = \phi_Y^\alpha$ for all $Y \in \mathsf{CHaus}$ and $\beta \in F_\alpha(Y)$ where $\beta = F(f)(\alpha)$ for suitable $f : X \to Y$. To this end consider the diagram

$$\begin{array}{ccccc} F_\alpha(X) & \longrightarrow & F(X) & \xrightarrow{\eta_X} & G(X) \\ \downarrow & & {\scriptstyle F(f)}\downarrow & & \downarrow{\scriptstyle G(f)} \\ F_\alpha(Y) & \longrightarrow & F(Y) & \xrightarrow[\eta_Y]{} & G(Y) \end{array}$$

Starting with α in the upper left, we have $\phi_X^\alpha(\alpha)$ in the upper right, and hence

$$G(f)(\phi_X^\alpha(\alpha)) = \phi_Y^\alpha(F(f)(\alpha)) = \phi_Y^\alpha(\beta)$$

in the lower right, where the first equation is as above. Since we also have β in the lower left, we obtain the desired $\eta_Y(\beta) = \phi_Y^\alpha(\beta)$. □

In light of the upcoming Theorem 13.3.1, this result is closely related to [72, Theorem 5]. The only potential difference is that our colimit is taken in Sh(CHaus), while van den Berg and Heunen consider it in $\mathsf{pC^*alg_1}$, and it is not clear whether these two definitions of colimits are equivalent.

Since it is currently unclear whether Definition 13.2.9 is the most adequate collection of sheaf conditions that one can postulate, we will not investigate any further the categorical properties of Sh(CHaus).

13.3 Piecewise C*-Algebras as Sheaves CHaus → Sets

In this section, we will establish that Sh(CHaus) contains the category of *piecewise C*-algebras* introduced by van den Berg and Heunen [72] as a full sub-category. The definitions of partial algebras and piecewise *-homomorphism was given in Definitions 3.1.5 and 3.1.6 respectively.

The discussion of Sect. 13.1 extends canonically to piecewise C*-algebras. That is to say, Gelfand duality still implements an equivalence of $\mathsf{CHaus}^{\mathrm{op}}$ with a full sub-category of $\mathsf{pC^*alg}_1$, so that for every $A \in \mathsf{pC^*alg}_1$ we can restrict the hom-functor

$$\mathsf{pC^*alg}_1(-,A)\ :\ \mathsf{pC^*alg}_1^{\mathrm{op}} \to \mathbf{Sets}$$

to a functor $\mathsf{CHaus} \to \mathbf{Sets}$, which maps $X \in \mathsf{CHaus}$ to the set of piecewise $*$-homomorphisms $C(X) \to A$. For any $A \in \mathsf{C^*alg}_1$, this results precisely in the functor $\mathsf{CHaus} \to \mathbf{Sets}$ that we already know from Sect. 13.1, since in this case $\mathsf{pC^*alg}_1(C(X),A) = \mathsf{C^*alg}_1(C(X),A)$. In other words, we have a diagram of functors

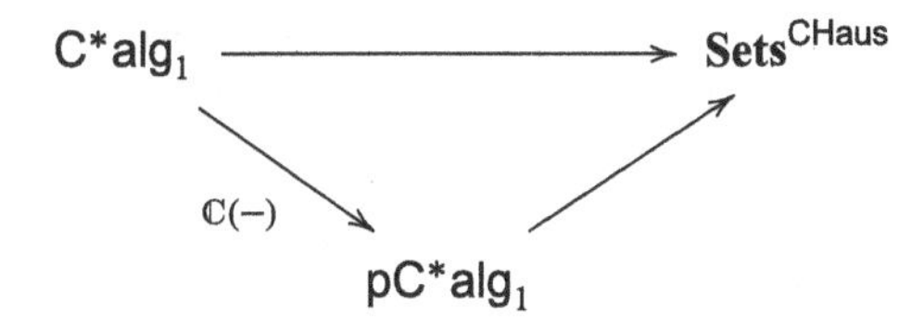

In fact, the proof of Proposition 13.2.1 still holds for piecewise C*-algebras. Hence, also the image of the functor $\mathsf{pC^*alg}_1 \to \mathbf{Sets}^{\mathsf{CHaus}}$ resides in the full sub-category Sh(CHaus), and the commutative triangle of functors can be taken to be

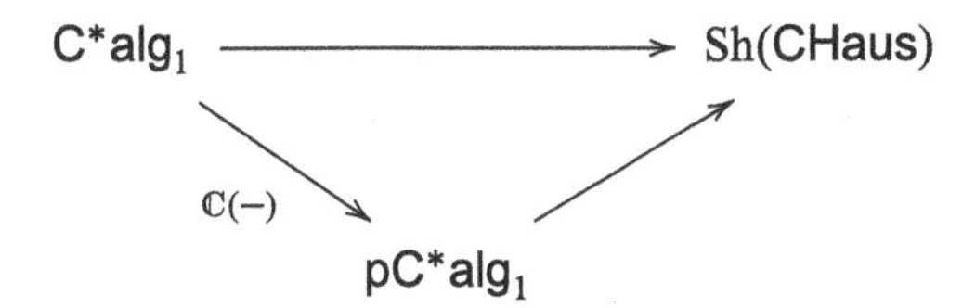

We now investigate a bit further the functor on the right, finding that it is close to being an equivalence. In the following, we use the unit disk $\bigcirc \subseteq \mathbb{C}$. Since it is homeomorphic to the unit square $\square$ we have been working with until now, all previous statements apply likewise with $\square$ replaced by $\bigcirc$.

Theorem 13.3.1 *The functor* $\mathsf{pC^*alg}_1 \longrightarrow \mathrm{Sh}(\mathsf{CHaus})$ *is fully faithful, with essential image given by all those* $F \in \mathrm{Sh}(\mathsf{CHaus})$ *for which the canonical map*

$$F(\bigcirc \times \bigcirc) \longrightarrow F(\bigcirc) \times F(\bigcirc) \tag{13.3.1}$$

is injective.

In other words, this functor forgets at most 'property', namely the property of injectivity of (13.3.1) as investigated in Lemma 13.2.12. This property is equivalent to F being separated (in the presheaf sense) on the effective-monic cones. It seems natural to suspect that not every sheaf on CHaus is separated in this sense, but this remains an open question. So it is also conceivable that $\mathsf{pC^*alg_1} \to \mathsf{Sh(CHaus)}$ actually is an equivalence of categories (see Definition 8.1.5).

In particular, this shows that $\mathsf{cC^*alg_1}$ is dense in $\mathsf{pC^*alg_1}$, i.e. that the canonical functor $\mathsf{pC^*alg_1} \to \mathbf{Sets}^{\mathsf{cC^*alg_1^{op}}}$ is fully faithful. For a potentially related result of a similar flavour, see [66, Corollary 8].

For the sake of completeness we will recall the definition of a fully faithful functor and that of an essential image of a functor.

Definition 13.3.1 Given a functor $F: C \to D$ from a category C to a category D, then F is said to be full and faithful if for each pair of objects $x, y \in C$, the function

$$F: C(x, y) \to D(F(x), F(y))$$

between hom-sets is both surjective (full condition) and injective (faithful condition).

Definition 13.3.2 Given a functor $F : C \to D$ between two categories C and D, the essential image of F is the smallest subcategory S of D such that:

1. S contains the image of F.
2. Given any object $x \in S$ and any isomorphism $f: x \cong y$ in D, both y and f are also in S. This condition means that S is a replete subcategory of D.

We will now prove the Theorem 13.3.1.

Proof A piecewise $*$-homomorphism $\zeta : A \to B$ is determined by its action on the unit ball, which is the set of elements with spectrum in $\bigcirc$. In particular, ζ is uniquely determined by the associated transformation $-(\zeta) : -(A) \to -(B)$, so that the functor under consideration is faithful.

Concerning fullness, let $\eta : -(A) \to -(B)$ be a natural transformation. Its component at $\bigcirc$ is a map $\eta_\bigcirc : \bigcirc(A) \to \bigcirc(B)$. The pairs of commuting elements $\alpha, \beta \in \bigcirc(A)$ are precisely those that are in the image of the canonical map

$$(\bigcirc \times \bigcirc)(A) \longrightarrow \bigcirc(A) \times \bigcirc(A),$$

and hence the requirements (3.1.5) follow from naturality and the consideration of functions like (13.1.5) and (13.1.7). The other axioms are likewise simple consequences of naturality. This exhibits a piecewise $*$-homomorphism $\zeta : A \to B$ such that $\eta_\bigcirc$ coincides with $\bigcirc(\zeta)$. Then by Lemma 13.2.11, we have $\eta = -(\zeta)$.

Finally, we show that every $F \in \mathsf{Sh(CHaus)}$ for which (13.3.1) is injective is isomorphic to $-(A)$ for some $A \in \mathsf{pC^*alg_1}$. Concretely, we construct a piecewise C*-algebra A by first defining its unit ball to be

$$\bigcirc(A) := F(\bigcirc).$$

This set comes equipped with a commutation relation, namely, $\alpha \perp\!\!\!\perp \beta$ is declared to hold for $\alpha, \beta \in A$ precisely when (α, β) is in the image of (13.3.1). In this case, we can define the sum $\alpha + \beta$ and the product $\alpha\beta$ using the functoriality on maps such as (13.1.5) and (13.1.7). Likewise there is a scalar multiplication by numbers $z \in \bigcirc$ and an involution arising from functoriality on the complex conjugation map $\bigcirc \to \bigcirc$.

If we define A to consist of pairs $(\alpha, z) \in F(\bigcirc) \times \mathbb{R}_{>0}$, modulo the equivalence $(\alpha, z) \sim (sa, sz)$ for all $s \in (0, 1)$, we would obtain a piecewise C*-algebra. In fact, the relevant structure of Definition 3.1.5 extends canonically from $\bigcirc(A)$ to all of A. Moreover, we also claim that any set $\{\gamma_i\}_{i\in I} \subseteq \bigcirc(A)$ of pairwise commuting elements is contained in a commutative C*-subalgebra. We write this family as a single element of the I-fold product,

$$\gamma \in F(\bigcirc)^I.$$

The cone $\{(p_i, p_j) : \bigcirc^I \to \bigcirc \times \bigcirc\}_{i,j\in I}$ consisting of all pairings of projections $p_i : \bigcirc^I \to \bigcirc$ is effective-monic and directed. By the commutativity assumption on γ, the pair $(\gamma_i, \gamma_j) \in F(\bigcirc) \times F(\bigcirc)$ comes from an element of $F(\bigcirc \times \bigcirc)$. Hence, by the sheaf condition, γ is actually the image of an element $\gamma' \in F\left(\bigcirc^I\right)$ under the canonical map. The subfunctor $F_{\gamma'} \subseteq F$, as in Proposition 13.2.3, is representable and it corresponds to the commutative C*-subalgebra generated by the γ_i. □

The following criterion—due to Heunen and Reyes—describes the image of the functor $\mathbb{C}(-) : \mathsf{C^*alg}_1 \to \mathsf{pC^*alg}_1$ at the level of morphisms.

Lemma 13.3.1 ([41, Proposition 4.13]) *For $A, B \in \mathsf{C^*alg}_1$, a piecewise $*$-homomorphism $\zeta : \mathbb{C}(A) \to \mathbb{C}(B)$ extends to a $*$-homomorphism $A \to B$ if and only if it is additive on self-adjoints and multiplicative on unitaries.*

By faithfulness of $\mathsf{C^*alg}_1 \to \mathsf{pC^*alg}_1$, we already know such an extension, if it exists, to be unique.

Proof The 'only if' part is clear, so we focus on the 'if' direction. Every element of A is of the form $a+ib$ for $a, b \in \mathbb{R}(A)$, and linearity forces us to define the candidate extension of f by

$$\hat{\zeta}(a + ib) := \zeta(a) + i\zeta(b).$$

In this way, $\hat{\zeta}$ becomes linear due to the first assumption, and it is evidently involutive and unital. On a unitary ν, we have $\hat{\zeta}(\nu) = \zeta(\nu)$, since

$$\begin{aligned}\hat{\zeta}(\nu) &= \hat{\zeta}\left(\frac{\nu + \nu^*}{2} + i\frac{\nu - \nu^*}{2}\right) = \frac{1}{2}\zeta(\nu + \nu^*) + \frac{1}{2}\zeta(\nu - \nu^*)\\ &= \frac{1}{2}\zeta(\nu) + \frac{1}{2}\zeta(\nu^*) + \frac{1}{2}\zeta(\nu) - \frac{1}{2}\zeta(\nu^*) = \zeta(\nu),\end{aligned}$$

where the third step uses $\nu \perp\!\!\!\perp \nu^*$.

We finish the proof by arguing that $\hat{\zeta}$ is multiplicative on two arbitrary elements $\alpha, \beta \in [-1, +1](A)$, which is enough to prove multiplicativity, and hence to show that $\hat{\zeta}$ is indeed a $*$-homomorphism. By functional calculus, we can find unitaries $\nu, \tau \in \mathbb{T}(A)$ such that $\alpha = \nu + \nu^*$ and $\beta = \tau + \tau^*$. Then

$$\begin{aligned}
\hat{\zeta}(\alpha\beta) = \hat{\zeta}\left((\nu + \nu^*)(\tau + \tau^*)\right) &= \hat{\zeta}\left(\nu\tau + \nu\tau^* + \nu^*\tau + \nu^*\tau^*\right) \\
&= \hat{\zeta}(\nu\tau) + \hat{\zeta}(\nu\tau^*) + \hat{\zeta}(\nu^*\tau) + \hat{\zeta}(\nu^*\tau^*) \\
&= \zeta(\nu\tau) + \zeta(\nu\tau^*) + \zeta(\nu^*\tau) + \zeta(\nu^*\tau^*) \\
&= \zeta(\nu)\zeta(\tau) + \zeta(\nu)\zeta(\tau^*) + \zeta(\nu^*)\zeta(\tau) + \zeta(\nu^*)\zeta(\tau^*) \\
&= (\zeta(\nu) + \zeta(\nu^*))(\zeta(\tau) + \zeta(\tau^*)) \\
&= (\hat{\zeta}(\nu) + \hat{\zeta}(\nu^*))(\hat{\zeta}(\tau) + \hat{\zeta}(\tau^*)) \\
&= \hat{\zeta}(\nu + \nu^*)\hat{\zeta}(\tau + \tau^*) = \hat{\zeta}(\alpha)\hat{\zeta}(\beta),
\end{aligned}$$

where we have used that $\hat{\zeta}$ coincides with ζ on unitaries (third and sixth line) and the assumption of multiplicativity on unitaries (fourth line). □

In fact, this result can be improved upon:

Proposition 13.3.1 *A piecewise $*$-homomorphism $\zeta : \mathbb{C}(A) \to \mathbb{C}(B)$ extends to a $*$-homomorphism $A \to B$ if and only if it is multiplicative on unitaries.*

Proof By Lemma 13.3.1, it is enough to prove that such a ζ is additive on self-adjoints.

To this end we use the following fact, which follows from the exponential series: for every $\alpha, \beta \in \mathbb{R}(A)$ and real parameter $t \in \mathbb{R}$, the unitary

$$e^{it(\alpha+\beta)}e^{-it\alpha}e^{-it\beta}$$

differs from 1 by at most $O(t^2)$ as $t \to 0$. Since ζ preserves the spectrum of unitaries, we conclude that also

$$\zeta\left(e^{it(\alpha+\beta)}e^{-it\alpha}e^{-it\beta}\right) = e^{it\zeta(\alpha+\beta)}e^{-it\zeta(\alpha)}e^{-it\zeta(\beta)}$$

is a unitary that differs from 1 by at most $O(t^2)$. By the same argument as above, this implies that $\zeta(\alpha + \beta) = \zeta(\alpha) + \zeta(\beta)$, as it was to be shown. □

As the proof shows, we actually need only multiplicativity on products of exponentials, i.e. on the connected component of the identity $1 \in \mathbb{T}(A)$. The method of proof also suggests a relation to the Baker-Campbell-Hausdorff formula, which may be worth exploring further.

Finally, it is worth noting that a piecewise $*$-homomorphism $\zeta : \mathbb{C}(A) \to \mathbb{C}(B)$ is additive on self-adjoints if and only if it is a Jordan homomorphism. In fact, the condition $\zeta(\alpha^2) = \zeta(\alpha)^2$ for $\alpha \in \mathbb{R}(A)$ is automatic since ζ preserves functional calculus.

Let us end this section by stating an important open problem:

Problem 13.3.1 Is the functor $\mathsf{pC^*alg_1} \to \mathsf{Sh(CHaus)}$ an equivalence of categories, i.e. does *every* sheaf on CHaus satisfy the injectivity condition of Lemma 13.2.12?

13.4 Almost C*-Algebras as Piecewise C*-Algebras with Self-Action

What we have learnt so far is that considering a C*-algebra A as a sheaf $-(A) : \mathsf{CHaus} \to \mathbf{Sets}$, or equivalently as a piecewise C*-algebra, recovers the entire 'commutative part' of the C*-algebra structure of A. Nevertheless, the functor $\mathsf{C^*alg_1} \to \mathsf{pC^*alg_1}$ is not full, which indicates that part of the relevant structure is lost. For example, a C*-algebra A is in general not isomorphic to A^{op} [60], although the two are canonically isomorphic as piecewise C*-algebras. This raises the question: *which natural piece of additional structure on a sheaf* $\mathsf{CHaus} \to \mathbf{Sets}$ *or piecewise C*-algebra would let us recover the missing information?*

Of course, what kind of additional structure counts as 'natural' is a subjective matter. However, we can take inspiration from quantum physics and re-express the question as follows: *which additional structure would have a clear physical interpretation?* Our proposal to answer such a question is based on a central feature of quantum mechanics, namely, that observables generate dynamics, in the sense that to every observable (self-adjoint operator) $\alpha \in \mathbb{R}(A)$, one associates the one-parameter group of inner automorphisms given by

$$\mathbb{R} \times A \longrightarrow A, \qquad (t, \beta) \longmapsto e^{i\alpha t}\beta e^{-i\alpha t}. \tag{13.4.1}$$

For example, if α is energy, then the resulting one-parameter family of automorphisms is given precisely by time translations, i.e. by the inherent dynamics of the system under consideration. If α is a component of angular momentum, then the resulting family of automorphisms are the rotations around that axis. As it is obvious from (13.4.1), this natural way in which A acts on itself by inner automorphisms is a purely noncommutative feature, in that it becomes trivial in the commutative case.

More formally, the construction of (13.4.1) really consists of two parts:

1. First, for every $t \in \mathbb{R}$, one forms the unitary $\nu := e^{-i\alpha t}$; since this is functional calculus, it is captured by the functoriality $\mathsf{CHaus} \to \mathbf{Sets}$.
2. Second, one lets ν act on A via conjugation, as $\beta \mapsto \nu^*\beta\nu$.

The last part is not captured by what we have discussed so far, and hence we axiomatize it as an additional piece of structure. Our definition is similar in spirit to the 'active lattices' of Heunen and Reyes [41] and also seems related to [6, Section VI].

Definition 13.4.1 An *almost C*-algebra* is a pair $(A, \mathfrak{a})$ consisting of a piecewise C*-algebra $A \in \mathsf{pC^*alg}_1$ and a *self-action* of A, which is a map

$$\mathfrak{a} : \mathbb{T}(A) \longrightarrow \mathsf{pC^*alg}_1(A, A)$$

assigning to every unitary $\nu \in \mathbb{T}(A)$ a piecewise automorphism $\mathfrak{a}(\nu) : A \to A$ such that

- ν commutes with $\tau \in \mathbb{T}(A)$ if and only if $\mathfrak{a}(\nu)(\tau) = \tau$;
- in this case, $\mathfrak{a}(\nu\tau) = \mathfrak{a}(\nu)\mathfrak{a}(\tau)$.

So $\mathfrak{a}$ must satisfy two equations on commuting unitaries. The first equation implies that a commutative C*-algebra, considered as a piecewise C*-algebra, can act on itself only trivially. Conversely, if the self-action is trivial, in the sense that every $\mathfrak{a}(\nu)$ is the identity, then A must be commutative. The second equation implies that if ν and τ commute, then also their actions commute:

$$\mathfrak{a}(\nu)\mathfrak{a}(\tau) = \mathfrak{a}(\nu\tau) = \mathfrak{a}(\tau\nu) = \mathfrak{a}(\tau)\mathfrak{a}(\nu).$$

Introducing a self-action $\mathfrak{a} : \mathbb{T}(A) \longrightarrow \mathsf{pC^*alg}_1(A, A)$ was motivated by physics considerations discussion above and we expect the appearance of $\mathbb{T}$ to be related to Pontryagin duality. The physical interpretation of the first axiom could instead be related to Noether's theorem.

Almost C*-algebras form a category denoted $\mathsf{aC^*alg}_1$ as follows:

Definition 13.4.2 An *almost $*$-homomorphism* $\zeta : (A, \mathfrak{a}) \to (B, \mathfrak{b})$ is a piecewise $*$-homomorphism $\zeta : A \to B$ which preserves the self-actions in the sense that

$$\mathfrak{b}(\zeta(\nu))(\zeta(\alpha)) = \zeta(\mathfrak{a}(\nu)(\alpha)). \tag{13.4.2}$$

The forgetful functor $\mathsf{C^*alg}_1 \to \mathsf{pC^*alg}_1$ factors through $\mathsf{aC^*alg}_1$ by associating to every C*-algebra A and unitary $\nu \in \mathbb{T}(A)$ its conjugation action,

$$\mathfrak{a}(\nu)(\alpha) := \nu^*\alpha\nu.$$

Every $*$-homomorphism $\zeta : A \to B$ is compatible with the resulting self-actions. The condition (13.4.2) becomes simply

$$\zeta(\nu)^*\zeta(\alpha)\zeta(\nu) = \zeta(\nu^*\alpha\nu). \tag{13.4.3}$$

Our main question is whether the additional structure of a self-action, that is present in an almost C*-algebras, is sufficient to recover the entire C*-algebra structure:

Problem 13.4.1 Is the forgetful functor $\mathsf{C^*alg}_1 \to \mathsf{aC^*alg}_1$ an equivalence of categories?

In order for this to be the case, one would have to show that the functor is both fully faithful and essentially surjective. While the latter question is wide open, it is clear that the functor is faithful, since the forgetful functor $\mathsf{C^*alg_1} \to \mathsf{pC^*alg_1}$ is. We can also prove fullness in a W*-algebra setting:

Theorem 13.4.1 $\mathsf{C^*alg_1} \to \mathsf{aC^*alg_1}$ *is fully faithful on morphisms out of any W*-algebra.*

This result is similar to [41, Theorem 4.11], but does not directly follow from it.[6]

Proof We need to show surjectivity, i.e. if $\zeta : \mathbb{C}(A) \to \mathbb{C}(B)$ for a W*-algebra A is a piecewise $*$-homomorphism which satisfies (13.4.3), then ζ extends to a $*$-homomorphism $A \to B$. Let us first consider the case that A contains no direct summand of type I_2. Then for every state $\phi : B \to \mathbb{C}$, the map

$$\alpha + i\beta \longmapsto \phi(\zeta(\alpha) + i\zeta(\beta)) \tag{13.4.4}$$

for $\alpha, \beta \in \mathbb{R}(A)$ is a quasi-linear functional on A in the sense of [36, Definition 5.2.5] and, therefore, it is uniquely determined by its values on the projections $\mathbf{2}(A)$ [36, Proposition 5.2.6]. On the other hand, by the generalized Gleason theorem [36, Theorem 5.2.4], this map $\mathbf{2}(A) \to \mathbb{R}$ uniquely extends to a state $A \to \mathbb{R}$. In conclusion, composition with ζ takes states on B to states on A, and hence $\mathbb{R}(\zeta) : \mathbb{R}(A) \to \mathbb{R}(B)$ is linear.

Furthermore, on $\mathbb{R}(A)$ we have $\zeta(\alpha^2) = \zeta(\alpha)^2$, which makes ζ into a Jordan homomorphism. By a deep result of Størmer [67, Theorem 3.3], this means that there exists a projection $\pi \in \mathbf{2}(B)$, commuting with the range of ζ, such that $\alpha \mapsto \pi\zeta(\alpha)$ uniquely extends to a (generally nonunital) $*$-homomorphism, and similarly $\alpha \mapsto (1 - \pi)\zeta(\alpha)$ uniquely extends to a (generally nonunital) $*$-anti-homomorphism. In other words, ζ decomposes into the sum of the restriction (to normal elements) of a $*$-homomorphism and a $*$-anti-homomorphism. So far, we have only made use of the assumption that ζ is a piecewise $*$-homomorphism.

In order to complete the proof in the case of A without type I_2 summand we will work with the corner $(1 - \pi)A(1 - \pi)$ in place of A itself. This then shows that it is enough to consider the case $\pi = 0$, i.e. when ζ is the restriction of a $*$-anti-homomorphism. In particular,

$$\zeta(\nu)^*\zeta(\alpha)\zeta(\nu) \overset{(13.4.3)}{=} \zeta(\nu^*\alpha\nu) = \zeta(\nu)\zeta(\alpha)\zeta(\nu)^*,$$

and therefore $\zeta(\alpha)\zeta(\nu^2) = \zeta(\nu^2)\zeta(\alpha)$ for all $\nu \in \mathbb{T}(A)$ and $\alpha \in \mathbb{C}(A)$. Since every exponential unitary $e^{i\beta}$ is the square of another unitary, we know that $\zeta(\alpha)$ commutes with every exponential unitary. Since every element of A is a linear combination of

[6] This is because the notion of 'active lattice' of [41] includes a group that acts on the lattice and a morphism of active lattices. In particular it is *assumed* to be a homomorphism of the corresponding groups. If we had assumed something analogous in our definition of almost C*-algebra, the fullness of the forgetful functor would simply follow from Proposition 13.3.1.

exponential unitaries, we conclude that $\zeta(\alpha)$ commutes with $\zeta(\beta)$ for every $\beta \in \mathbb{C}(A)$. Hence the range of ζ is commutative. In particular, ζ is also the restriction of a $*$-homomorphism, which completes the proof for the case at hand.

Now consider the case of an almost $*$-homomorphism $\zeta : \mathbb{C}(M_2) \to \mathbb{C}(B)$. Due to the isomorphism $M_2 \cong \mathrm{Cl}(\mathbb{R}^2) \otimes \mathbb{C}$ with a complexified Clifford algebra, M_2 is freely generated as a C*-algebra by two self-adjoints σ_x and σ_y subject to the relations

$$\sigma_x^2 = \sigma_y^2 = 1, \qquad \sigma_x\sigma_y + \sigma_y\sigma_x = 0.$$

Since ζ commutes with functional calculus, the first two equations are clearly preserved by ζ in the sense that $\zeta(\sigma_x)^2 = \zeta(\sigma_y)^2 = 1$. Concerning the third equation, we know

$$-\zeta(\sigma_x) = \zeta(-\sigma_x) = \zeta(\sigma_y\sigma_x\sigma_y) \overset{(13.4.3)}{=} \zeta(\sigma_y)\zeta(\sigma_x)\zeta(\sigma_y).$$

Hence $\zeta(\sigma_x)\zeta(\sigma_y) + \zeta(\sigma_y)\zeta(\sigma_x) = 0$ due to $\zeta(\sigma_y)^2 = 1$. Therefore the values $\zeta(\sigma_x)$ and $\zeta(\sigma_y)$ extend uniquely to a $*$-homomorphism $\hat{\zeta} : M_2 \to B$. The problem is now to show that this coincides with the original ζ on normal elements. Since any symmetry $\nu \in \{-1, +1\}(M_2)$ is conjugate to σ_x, we certainly have $\hat{\zeta}(\nu) = \zeta(\nu)$ by (13.4.3) and the assumption $\hat{\zeta}(\sigma_x) = \zeta(\sigma_x)$. However, because in the special case of M_2, every normal element can be obtained from a symmetry by functional calculus, and both ζ and $\hat{\zeta}$ preserve functional calculus, it is sufficient to show that $\hat{\zeta} = \zeta$ on normal elements. This finishes off the case $A = M_2$.

A general W*-algebra of type I_2 is of the form $A \cong L^\infty(\Omega, \mu, M_2)$ for a suitable measure space (Ω, μ). Let $\zeta : \mathbb{C}(A) \to \mathbb{C}(B)$ be an almost $*$-homomorphism, we will first show that ζ uniquely extends to a bounded $*$-homomorphism on the *-subalgebra of simple functions. For a measurable set $\Gamma \subseteq \Omega$, let $\chi_\Gamma : \Omega \to \{0, 1\}$ be the associated indicator function. For non-empty Γ, the algebra elements of the form $\alpha\chi_\Gamma$ for $\alpha \in M_2$ form a C*-subalgebra isomorphic to M_2 itself (with different unit). By the previous discussion, we know that ζ uniquely extends to a $*$-homomorphism on this subalgebra. Furthermore, ζ behaves as expected on a simple function $\sum_{i=1}^n \alpha_i\chi_{\Gamma_i}$. Assuming that the Γ_i's form a partition of Ω, we have $\alpha_i\chi_{\Gamma_i} \cdot \alpha_j\chi_{\Gamma_j} = 0$ for $i \neq j$, and hence ζ is additive on the sum which implies

$$\zeta\left(\sum_{i=1}^n \alpha_i\chi_{\Gamma_i}\right) = \sum_{i=1}^n \zeta(\alpha_i)\zeta(\chi_{\Gamma_i}). \tag{13.4.5}$$

We show that ζ is linear on the sum of two self-adjoint simple functions. By choosing a common refinement, it is enough to consider the case that the two partitions are the same. Additivity then follows from (13.4.5) and additivity on M_2. Multiplicativity on unitary simple functions is analogous. Since the proof of Lemma 13.3.1 still goes through in the present situation (where the *-algebra of simple functions is generally not a C*-algebra), we conclude that ζ extends uniquely to a $*$-homomorphism on the simple functions. By construction this $*$-homomorphism is bounded. Therefore it uniquely extends to a $*$-homomorphism $\hat{\zeta} : A \to B$, which coincides with ζ on the normal simple functions. It remains to be shown that $\hat{\zeta}(\alpha) = \zeta(\alpha)$ for all $\alpha \in \mathbb{C}(A)$.

To obtain this for a given $\alpha \in \mathbb{C}(A)$, we distinguish those points $x \in \Omega$ for which $\alpha(x)$ is degenerate from those for which it is not. Since degeneracy is detected by the vanishing of the discriminant $\mathrm{tr}^2 - 4\det$, the relevant set is

$$\Delta := \{\, x \in \Omega \mid \mathrm{tr}(\alpha(x))^2 - 4\det(\alpha(x)) = 0 \,\}.$$

This set is measurable since both trace and determinant are measurable functions $M_2 \to \mathbb{C}$. For every $x \in \Omega \setminus \Delta$ there is a unique unitary $\nu(x) \in \mathbb{T}(M_2)$ such that $\nu(x)^*\alpha(x)\nu(x)$ is diagonal. Since the eigenbasis of a nondegenerate self-adjoint matrix depends continuously on the matrix, it follows that the function $x \mapsto \nu(x)$ is also measurable. By arbitrarily choosing $\nu(x) := 1$ on $x \in \Delta$, we have constructed a unitary $\nu \in \mathbb{T}(L^\infty(\Omega, \mu, M_2))$, such that $\nu^*\alpha\nu$ is pointwise diagonal. Thanks to (13.4.3), it is therefore sufficient to prove the desired identity $\hat{\zeta}(\alpha) = \zeta(\alpha)$ on diagonal α only. However, since these diagonal elements generate a commutative C*-subalgebra, which contains a dense *-subalgebra of simple functions on which $\hat{\zeta}$ and ζ are known to coincide, the proof is complete because both $\hat{\zeta}$ and ζ are $*$-homomorphisms on this commutative subalgebra.

Now a general W*-algebra A is a direct sum of a W*-algebra without I_2 summand and one that is of type I_2 [71, Theorems 1.19 & 1.31]. Again, by considering corners, it is straightforward to check that if the fullness property holds on almost $*$-homomorphisms out of $A, B \in \mathsf{C^*alg}_1$, then it also holds on almost $*$-homomorphisms out of $A \oplus B$. □

In general, the problem of fullness is related to the cohomology of the unitary group $\mathbb{T}(A)$ as follows: Let $\zeta : \mathbb{C}(A) \to \mathbb{C}(B)$ be an almost $*$-homomorphism between C*-algebras. We can assume, without loss of generality, that $\mathrm{im}(\zeta)$ generates B as a C*-algebra. For unitaries $\nu, \tau \in \mathbb{T}(A)$ and any $\alpha \in \bigcirc(A)$, we have

$$\begin{aligned} \zeta(\alpha) &= \zeta\left(\tau^*\nu^*(\nu\tau)\alpha(\nu\tau)^*\nu\tau\right) \\ &\overset{(13.4.3)}{=} \zeta(\tau)^*\zeta(\nu)^*\zeta(\nu\tau)\zeta(\alpha)\zeta(\nu\tau)^*\zeta(\nu)\zeta(\tau) \\ &= \left(\zeta(\nu\tau)^*\zeta(\nu)\zeta(\tau)\right)^* \ \zeta(\alpha)\ \left(\zeta(\nu\tau)^*\zeta(\nu)\zeta(\tau)\right). \end{aligned}$$

Hence the unitary $\zeta(\nu\tau)^*\zeta(\nu)\zeta(\tau)$ commutes with $\zeta(\alpha)$. By the assumption that $\mathrm{im}(\zeta)$ generates B, this means that there exists $c(\nu, \tau)$ in the centre of $\mathbb{T}(B)$, such that

$$\zeta(\nu\tau) = c(\nu, \tau)\zeta(\nu)\zeta(\tau).$$

As in the theory of projective representations of groups, we can use this relation to evaluate ζ on a product of three unitaries $\nu, \tau, \chi \in \mathbb{T}(A)$, resulting in

$$c(\nu\tau, \chi)c(\nu, \tau)\zeta(\nu)\zeta(\tau)\zeta(\chi) = \zeta(\nu\tau\chi) = c(\nu, \tau\chi)c(\tau, \chi)\zeta(\nu)\zeta(\tau)\zeta(\chi).$$

This establishes the cocycle equation

$$c(\tau, \chi)c(\nu\tau, \chi)^* c(\nu, \tau\chi)c(\nu, \tau)^* = 1,$$

showing that c is a 2-cocycle on $\mathbb{T}(A)$ with values in the centre of $\mathbb{T}(B)$, which is equal to the unitary group of the centre of B. Unfortunately, we do not know whether this can be used to show that $\mathbb{T}(\zeta) : \mathbb{T}(A) \to \mathbb{T}(B)$ is a group homomorphism, which would be enough to prove fullness in general by Proposition 13.3.1.

Let us now restate the remaining part of Problem 13.4.1:

Problem 13.4.2 Is the functor $\mathsf{C^*alg}_1 \to \mathsf{aC^*alg}_1$ full in general? If so, could it even be essentially surjective?

13.5 Groups as Piecewise Groups with Self-Action

In order to get a better intuition for the relation between C*-algebras and almost C*-algebras, it is instructive to perform analogous considerations for other mathematical structures. In this section, we investigate the case of groups, which may also be of interest in its own right.

By analogy with piecewise C*-algebras, we have:

Definition 13.5.1 ([41]) A *piecewise group* is a set G equipped with the following pieces of structure:

1. a reflexive and symmetric relation $\perp\!\!\!\perp \subseteq G \times G$. If $x \perp\!\!\!\perp y$, we say that x and y *commute*;
2. a binary operation $\cdot : \perp\!\!\!\perp \to G$;
3. a distinguished element $1 \in G$;

such that every subset $C \subseteq G$ of pairwise commuting elements is contained in some subset $\bar{C} \subseteq G$ of pairwise commuting elements, which is an abelian group with respect to the data above.

Abelian groups are precisely those piecewise groups for which the commutativity relation $\perp\!\!\!\perp$ is total. Piecewise groups form a category pGrp in the obvious way:

Definition 13.5.2 Given piecewise groups G and H, a *piecewise group homomorphism* is a function $\zeta : G \to H$ such that if $g \perp\!\!\!\perp h$ in G, then

$$\zeta(g) \perp\!\!\!\perp \zeta(h), \qquad \zeta(gh) = \zeta(g)\zeta(h). \tag{13.5.1}$$

It is straightforward to show that a piecewise group homomorphism satisfies $\zeta(1) = 1$.

Every group can be considered as a piecewise group. This gives rise to a forgetful functor $\mathsf{Grp} \to \mathsf{pGrp}$, which is faithful and reflects isomorphisms. Since it is not full (taking inverses $g \mapsto g^{-1}$ is a piecewise group homomorphism for every G, but a group homomorphism only if G is abelian), this functor forgets some of the structure that groups have. By analogy with Definition 13.4.1, we try to recover this structure by equipping a piecewise group with a notion of inner automorphisms:

Definition 13.5.3 An *almost group* is a pair $(G, \mathfrak{a})$ consisting of $G \in \mathsf{pGrp}$ and a *self-action* on G, which is a map

$$\mathfrak{a} : G \longrightarrow \mathsf{pGrp}(G, G)$$

assigning to every element $g \in G$ a piecewise automorphism $\mathfrak{a}(g) : G \to G$ such that:

- g commutes with h if and only if $\mathfrak{a}(g)(h) = h$;
- in this case, $\mathfrak{a}(gh) = \mathfrak{a}(g)\mathfrak{a}(h)$.

Almost groups form a category denoted aGrp as follows:

Definition 13.5.4 An *almost group homomorphism* $\zeta : (G, \mathfrak{a}) \to (H, \beta)$ is a piecewise group homomorphism $\zeta : A \to B$ such that

$$\mathfrak{a}(\zeta(g))(\zeta(h)) = \zeta(\mathfrak{a}(g)(h)). \tag{13.5.2}$$

The forgetful functor $\mathsf{Grp} \to \mathsf{pGrp}$ factors through aGrp by associating to every group G and to every element $g \in G$, the conjugation action

$$\mathfrak{a}(g)(h) := g^{-1}hg.$$

Every group homomorphism $\zeta : G \to H$ respects the resulting self-actions and the condition (13.4.2) simplifies to:

$$\zeta(g)^{-1}\zeta(h)\zeta(g) = \zeta(g^{-1}hg). \tag{13.5.3}$$

One can ask whether this forgetful functor $\mathsf{Grp} \to \mathsf{aGrp}$ is an equivalence of categories. In contrast to the discussion of Sect. 13.4, and in particular Theorem 13.4.1, here we know the answer to be negative:

Theorem 13.5.1 *The forgetful functor* $\mathsf{Grp} \to \mathsf{aGrp}$ *is not full.*

In general, going from a group to an almost group still constitutes a loss of structure.

Proof We provide an explicit example of an almost group homomorphism between groups that is not a group homomorphism.

Let $\mathbb{F}_2$ be the free group on two generators a and b. For any word $w \in \mathbb{F}_2$, let $\hat{w}$ be the cyclically reduced word associated to w. Then consider the map $\zeta : \mathbb{F}_2 \to \mathbb{Z}$ where $\zeta(w)$ is defined as the number of times that the generator a directly precedes the generator b in $\hat{w}$, minus the number of times that the generator b^{-1} directly precedes the generator a^{-1} in $\hat{w}$. By construction, this is invariant under conjugation and therefore satisfies (13.5.3). If $v, w \in \mathbb{F}_2$ commute, then they must be of the form $v = u^m$ and $w = u^n$ for some $u \in \mathbb{F}_2$ and $m, n \in \mathbb{Z}$ [53, Proposition 2.17]. Hence to verify that ζ is a piecewise group homomorphism, it is enough to show that $\zeta(u^k) = \zeta(u)^k$ for all $k \in \mathbb{Z}$. This is the case because we have $\widehat{u^k} = \hat{u}^k$ at the level of reduced cyclic words.

On the other hand, ζ is not a group homomorphism since $\zeta(a) = \zeta(b) = 0$, while $\zeta(ab) = 1$. □

As the second half of the proof indicates, part of the problem is that a free group has very few commuting elements. One can hope that the situation will be better for finite groups.

Problem 13.5.1 Is the restriction of the functor $\mathsf{Grp} \to \mathsf{aGrp}$ from finite groups to finite almost groups an equivalence of categories?

13.6 Open Problems: States and State-Space

As explained in previous chapters and in [26], topos quantum theory is formulated in terms of a topos that depends on the particular physical system under consideration, namely the category of presheaves on the poset of commutative subalgebras of the algebra of observables $\mathcal{A}$. Instead of working with commutative subalgebras only, in this chapter we consider all *-homomorphisms $C(X) \to \mathcal{A}$ for all commutative C*-algebras $C(X)$. Doing so means that $\mathcal{A}$ becomes a functor $\mathsf{CHaus} \to \mathbf{Sets}$. In this way, we can consider all physical systems as described by objects in the functor category $\mathbf{Sets}^{\mathsf{CHaus}}$ or the category of sheaves $\mathrm{Sh}(\mathsf{CHaus})$. This will allows us to consider multiple physical systems at once, shedding light on the problem of composite systems in topos quantum theory.

As a first step in this direction would be to determine the state-space for a quantum system in terms of the topos $\mathrm{Sh}(\mathsf{CHaus})$. This is still an open problem, however a few things can be said on that matter. It is clearly possible to construct a functor $P : \mathsf{CHaus} \to \mathbf{Sets}$ which assigns to each space $X \in \mathsf{CHaus}$ the set of regular probability measures $P(X)$. In this setting a state could be defined as a natural transformation

$$-(A) \to P$$

such that for every space $X \in \mathsf{CHaus}$ the component $X(A) \to P(X)$ assigned to each measurement α with outcome in X, the probability measure associated to that measurement. Each state $s \in A$ induces such a natural transformation, however it is not clear at this stage if the converse holds, namely, if each natural transformation $-(A) \to P$ arises from a state. If this would be the case we would obtain a bijective correspondence between states of the algebra A and natural transformations $-(A) \to P$. It can be argued that if A is any kind of algebra for which Gleason's theorem applies, then the natural transformations $-(A) \to P$ are *exactly* the states on A.

Given the above reasoning it would seem reasonable to define the state-space as some kind of exponential object. The question is then to figure out which category to use to define the exponential object. On the one hand we have sheaves $-(A) \in Sh(CHaus)$ for each $A \in pC^*Alg_1$, which represent potential physical systems, and on the other we have the presheaves $P : \mathbf{Sets}^{CHaus}$, which assigns probability measure to potential physical systems. Ideally the state-space would be the exponential object $P^{-(A)}$, however, it is still unclear in which category to take this object. A possible candidate would be $\mathbf{Sets}^{CHaus}$, since both functors live in there, but *CHaus* is not even locally small.

Solving this issue is an important open problem in the topos quantum theory formalism.

Chapter 14
Quantization in Topos Quantum Theory: An Open Problem

In this chapter we are interested in analysing how, if at all, different quantizations can be represented in Topos Quantum Theory. We already know from the work of [57] that it is indeed possible to define the concept of quantization within a topos. We would like to extend this program to incorporate all possible equivalent quantizations.

When talking about quantization we are faced by two situations:

(a) there is an existing underlining classical system;
(b) there is no underlining classical system.

In the first sections of this chapter we will perform a general analysis of both these situations. However, when utilising the quantization in a topos defined in [57] we will assume the existence of an underlining classical system as was done by the author.

The detailed development of the case in which there is no underlying classical system is left for future work.

14.1 Abstract Characterisation of Quantization

In this section we would like to give a general description of a possible way to define quantization in a topos. A possible approach for defining quantization would be to choose, as a base category, a collection of objects representing label of physical quantities with some structure, then quantization would involve associating these labels with specific operators in a concrete Hilbert space and, hence, with one of our topos structures.

If this approach is adopted, the first issue would be to define what sort of mathematical structure such a collection of labels should have. The choice will depend on whether or not we are considering an underlining classical system with

C. Flori, *A Second Course in Topos Quantum Theory*,
Lecture Notes in Physics 944, https://doi.org/10.1007/978-3-319-71108-9_14

a symplectic manifold, S as state space or not. We will deal first with the situation when there is an underlying classical theory.

In this case the base category can be chosen to be the poset of subalgebras of $C^\infty(S, \mathbb{R})$[1] that are abelian as computed with the Poisson bracket on S. The idea here is that each (see below) element of $C^\infty(S, \mathbb{R})$ is labelled by the physical quantity to which it corresponds, and that this labelling is fixed in concrete: i.e., one cannot start performing symplectic covariance transformations on the whole system.

Quantization will then involve defining various 'topos representation' of this base poset. Of particular importance will be the construction of sheaves/presheaves over the poset on which the Dirac covariance group, G, acts. However the following issues have to be addressed:

1. We know from the van Hove effect that the naive interpretation of quantization, namely associating commutators with Poisson brackets, is not possible. Thus it seems necessary to restrict $C^\infty_{Lie}(S, \mathbb{R})$ in some way. When S is finite[2] dimensional and admits a finite 1-dimensional transitive group G of symplectic transformations, then we restrict our attention to the subalgebra $L(G) \subseteq C^\infty_{Lie}(S, \mathbb{R})$ spanned by G. The corresponding base category/poset of commutative Lie subalgebras of $L(G)$ will be denoted[3] PG. However, it will generally be the case that one needs to add 'by hand' certain physical quantities (e.g., the Hamiltonian) which are not already contained in $L(G)$. Of course, for some very limited situation this will not be the case: for example in the SHO the Hamiltonian itself belongs to the Lie algebra.

   When there is no classical Lie algebra chosen, we will write the poset of labels as PS. To get the poset structure it is necessary to use the Lie bracket on one particular quantization, thus $PS \simeq \mathcal{V}(\mathcal{H})$ although it must be born in mind that the unitary group $U(\mathcal{H})$ does not act on PS even if it does act on $\mathcal{V}(\mathcal{H})$.
2. Should we place a topology on $C^\infty_{Lie}(S, \mathbb{R})$ and, if so, do the abelian subalgebras have to be closed subsets? Of course for finite-dimensional G this problem does not arise.
3. Is $C^\infty_{Lie}(S, \mathbb{R})$ general enough? The reason C^∞-functions are chosen is because the Poisson bracket between any two of them is well-defined and belongs to the same space. However, this clearly excludes certain functions on S that, arguably, are of physical interest, but which are not of this type. For example, the characteristic function, χ_A, of a (measurable) subset $A \subseteq S$ corresponds to a proposition. This is not C^∞ but, in the quantum theory, it is represented by a projection operator.
4. One might want to enlarge the classical space of observables to the space, $Meas(S, \mathbb{R})$, of measurable real-valued functions. The problem of course is

[1] When we think of $C^\infty(S, \mathbb{R})$ as a Lie algebra we will use the notation $C^\infty_{Lie}(S, \mathbb{R})$.

[2] There is always an infinite-dimensional transitive group, namely the group of symplectic transformations of S.

[3] Note that the use of PG raises the interesting question as to the extent to which the non-commutative structure of the Lie algebra $L(G)$ can be recovered from knowing the poset structure of its abelian Lie subalgebras.

that Poisson brackets are not defined on such functions (unless, one admits distributional results?).

We will now set aside the above issues and analyse various possible quantization strategies using the poset, PG, of commutative Lie subalgebras of $L(G)$. To do this we will pick one particular representation on a Hilbert space $\mathcal{H}$ although, in principal, one could consider unitarily inequivalent quantizations for systems where the Stone von Neumann theorem does not hold. In this setting we define a quantization on $\mathcal{H}$ to be any poset morphism, $\phi : PG \rightarrow \mathcal{V}(\mathcal{H})$, from the poset PG to the poset $\mathcal{V}(\mathcal{H})$. Although any quantization can be seen as such a morphism, the converse is not true.

Denoting the collection of all poset morphisms by $Hom_{poset}(PG, \mathcal{V}(\mathcal{H}))$ we can define a group action on them by the group[4] $G \subseteq \mathcal{U}(\mathcal{H})$ as follows:

$$(l_g\phi)L := l_g(\phi(L)) = \hat{U}_g\phi(L)\hat{U}_{g^{-1}} \tag{14.1.1}$$

for each $L \in PG$ and $\phi \in Hom_{poset}(PG, \mathcal{V}(\mathcal{H}))$. Note that we are assuming no group action on $Hom_{poset}(PG, \mathcal{V}(\mathcal{H}))$.[5]

This definition of representation is clearly very naive, since all that each poset morphisms ϕ does is to associate, to each abelian subalgebra L of PG, an abelian von Neumann subalgebra, $\phi(L) \in \mathcal{V}(\mathcal{H})$ in such a way that the ordering is preserved, i.e. if $L_1 \subseteq L_2$ then $\phi(L_1) \subseteq (L_2)$. However this is not enough to characterise a well defined quantization, in particular we would want that the maps ϕ respect the vector space structure on each $L \in PG$. In order to solve this problem we will introduce a new presheaf later on, but for now we will stick with our simplified model and see what we can learn from it.

14.1.1 Quantization Presheaf

We would now like to define a presheaf over PG which, in a way, represents all possible quantization of PG. A possibility is to construct a presheaf using local poset homomorphisms as follows:

[4] Clearly to define an action of G on these poset morphisms we are using a representation of G on $\mathcal{H}$. By factoring this representation by its Kernel, we can assume that it is indeed faithful.

[5] One can think of this as the idea that the elements in each poset L represent labels of physical quantities and these are fixed once and for all.

Definition 14.1.1 We define the presheaf $\underline{R}$ on PG which has as

1. Objects: for each $L \in PG$ we have $\underline{R}_L := Hom_{poset}(\downarrow L, \mathcal{V}(\mathcal{H}))$.
2. Morphisms: given a morphism $i_{L_1,L_2} : L_1 \to L_2$ then the corresponding presheaf morphism is

$$\underline{R}(i_{L_1,L_2}) : Hom_{poset}(\downarrow L_2, \mathcal{V}(\mathcal{H})) \to Hom_{poset}(\downarrow L_1, \mathcal{V}(\mathcal{H})) \quad (14.1.2)$$

$$\phi \mapsto \phi_{|\downarrow L_1} . \quad (14.1.3)$$

We then have the following lemma[6]:

Lemma 14.1.1

$$\Gamma\underline{R} \simeq Hom_{poset}(PG, \mathcal{V}(\mathcal{H})) . \quad (14.1.4)$$

Proof In order to prove the above lemma we need to define a map between $Hom_{poset}(PG, \mathcal{V}(\mathcal{H}))$ and $\Gamma\underline{R}$ and then, show, that it is indeed the desired isomorphisms. We define the following:

$$i : Hom_{poset}(PG, \mathcal{V}(\mathcal{H})) \to \Gamma\underline{R} \quad (14.1.5)$$

such that for each $\phi \in Hom_{poset}(PG, \mathcal{V}(\mathcal{H}))$ and $L \in PG$ we have

$$i(\phi)(L) := \phi_{|\downarrow L} \quad (14.1.6)$$

where $\phi_{|\downarrow L} \in Hom_{poset}(\downarrow L, \mathcal{V}(\mathcal{H}))$. If we then consider $L_1 \subseteq L_2$ then $i(\phi)(L_1) := \phi_{|\downarrow L_1} = (\phi_{\downarrow L_2})_{\downarrow L_1} = (i(\phi)(L_2))_{\downarrow L_1} = \underline{R}(i_{L_1,L_2})(i(\phi)(L_2))$. Thus indeed $i(\phi)$, as defined above, is a global element.
We now define the inverse as follows:

$$j : \Gamma\underline{R} \to Hom_{poset}(PG, \mathcal{V}(\mathcal{H})) \quad (14.1.7)$$

by

$$(j(\gamma))(L) := \gamma(L)(L) \quad (14.1.8)$$

for all $L \in PG$ and $\gamma \in \Gamma\underline{R}$. Moreover given $L_1 \subseteq L_2$ then $j(\gamma)(L_1) := \gamma(L_1)(L_1) = \gamma(L_2)(L_1) \subseteq \gamma(L_2)(L_2) = (j(\gamma))L_2$. If follows that $j(\gamma)$, as defined above, is indeed a poset morphisms. It is easy to see that i and j are inverse of each other. □

The group action on $\underline{R}$ is defined similarly as in Eq. (14.1.1), namely

$$(l_g\phi)L_i := l_g(\phi(L_i)) \quad (14.1.9)$$

for all $\phi \in Hom_{poset}(\downarrow L, \mathcal{V}(\mathcal{H}))$, $L_i \in \downarrow L$ and $g \in U(\mathcal{H})$.

[6] Here $\Gamma\underline{R}$ denotes the global sections of $\underline{R}$.

Since both $\downarrow L$ and $\mathcal{V}(\mathcal{H})$ are equipped with the Alexandroff topology, we will make use of the following Lemma

Lemma 14.1.2 *Let $\alpha : P_1 \to P_2$ be a map between posets P_1 and P_2. Then α is order preserving if and only if for each lower set $L \subseteq P_2$, we have that $\alpha^{-1}(L)$ is a lower subset of P_1.*

Proof Let us assume that α is order preserving and let $L \subseteq P_2$ be lower. Now let $z \in \alpha^{-1}(L) \in P_1$, i.e., $\alpha(z) = l$ for some $l \in L$, and suppose $y \in P_1$ is such that $y \leq z$. Since α is order preserving we have $\alpha(y) \leq \alpha(z) = l \in L$, which, since L is lower, means that $\alpha(y) \in L$, i.e., $y \in \alpha^{-1}(L)$. Hence $\alpha^{-1}(L)$ is lower.

Conversely, suppose that for any lower set $L \in P_2$ we have that $\alpha^{-1}(L) \in P_1$ is lower, and consider a pair $x, y \in P_1$ such that $x \leq y$. Now $\downarrow (y)$ is lower in P_2 and hence $\alpha^{-1}(\downarrow \alpha(y))$ is a lower subset of P_1. However $\alpha(y) \in \downarrow \alpha(y)$ and hence $y \in \alpha^{-1}(\downarrow \alpha(y))$. Therefore, the fact that $x \leq y$ implies that $x \in \alpha^{-1}(\downarrow \alpha(y))$, i.e., $\alpha(x) \in \downarrow \alpha(y)$, which means that $\alpha(x) \leq \alpha(y)$. Therefore α is order preserving. □

Given the above Lemma we can write

$$\underline{R}_L = Hom_{poset}(\downarrow L, \mathcal{V}(\mathcal{H})) = C(\downarrow L, \mathcal{V}(\mathcal{H})) \tag{14.1.10}$$

where $C(\downarrow L, \mathcal{V}(\mathcal{H}))$ denotes the set of all continuous functions. Moreover, given two continuous functions $f, g : \downarrow L \to \mathcal{V}(\mathcal{H})$, these have the same germ at $L \in PG$ iff $f_{\downarrow L} = g_{\downarrow L}$. Thus, denoting the sheaf of germs (See Sect. A.4 in the Appendix) of continuous functions between any two topological spaces X and Y as $\underline{C}(X, Y)$, we can write

$$\underline{R} \simeq \underline{C}(PG, \mathcal{V}(\mathcal{H})), \tag{14.1.11}$$

such that for each $L \in PG$, the space of germs at L is

$$\underline{R}_L \simeq \underline{C}_L(PG, \mathcal{V}(\mathcal{H})). \tag{14.1.12}$$

As a consequence of Lemma 14.1.1 we have that

$$\Gamma\underline{R} \simeq C(PG, \mathcal{V}(\mathcal{H})). \tag{14.1.13}$$

14.1.2 Considering All Quantizations At Once

Since our aim is, eventually, to define a topos quantum theory which takes into consideration all possible quantization, we would like to consider these quantizations as elements of a new base category. This can indeed be done bydefining an ordering on the bundle space $\Lambda\underline{R}$ of the etalé bundle $p_R : \Lambda\underline{R} \to PG$, associated to the sheaf $\underline{R}$. Such an ordering is defined as follows: given two elements $\phi_1, \phi_2 \in \Lambda\underline{R}$ then

$$\phi_1 \leq \phi_2 \text{ iff } p_{\underline{R}}(\phi_1) \subseteq p_{\underline{R}}(\phi_2) \text{ and } \phi_1 = \phi_{2|p_{\underline{R}}(\phi_1)} . \tag{14.1.14}$$

Given such an ordering we then have the following theorem:

Theorem 14.1.1 *The Alexandrof topology on $\underline{R}$ defined via the above ordering is homeomorphic to the etalé topology of $\Lambda(\underline{R})$ associated with the etalé bundle $p_{\underline{R}}$: $\Lambda\underline{R} \to PG$.*

Before proving the above theorem we will briefly recall the definition of an etalé bundle.

Definition 14.1.2 Given a topological space X, a bundle $p_E : E \to X$ is said to be etalé iff p_A is a local homeomorphism. By this we mean that, for each $e \in E$ there exists an open set V with $e \in V \subseteq E$, such that pV is open in X and $p_{|V}$ is a homeomorphism $V \to pV$.

We now will prove the above theorem.

Proof Let us consider an open set U in the etalé topology of $\Lambda(\underline{R})$. Since $p_{\underline{R}}$: $\Lambda(\underline{R}) \to PG$ is a local homeomorphism,[7] then $p_{\underline{R}}(U)$ is open in PG, i.e., it is a lower set in the Alexandroff topology. However, by the definition of the poset structure on $\Lambda\underline{R}$, p is order preserving, thus $p_{\underline{R}}^{-1} \circ p_{\underline{R}}(U)$ is a lower set in $\Lambda\underline{R}$. Moreover since $p_{\underline{R}}$ is a local homeomorphism then $p_{\underline{R}}^{-1} \circ p_{\underline{R}}(U) = U$ is a lower set in $\Lambda\underline{R}$.

Conversely, let U be an open set in the Alexandroff topology on $\Lambda\underline{R}$. Since $p_{\underline{R}}$ is order preserving then $p_{\underline{R}}(U)$ is a lower set in PG. Now since $p_{\underline{R}} : \Lambda\underline{R} \to PG$ is an etalé bundle we know that $p_{\underline{R}}$ is a local homeomorphism in the etalé topology. Thus, restricting only to open sets, we have that $p_{\underline{R}}^{-1}(p_{\underline{R}}(U))$ is an open set in the etalé topology. However $p_{\underline{R}}^{-1} \circ p_{\underline{R}}(U) = U$, i.e., U is open in the etalé topology. □

If we, indeed, do want to use $\Lambda\underline{R}$ as the base category we need a way of "pulling back" the presheaves we defined in $\mathcal{V}(\mathcal{H})$ to presheaves on $\Lambda\underline{R}$. This can be done with the aid of the following functor:

Theorem 14.1.2 *The map $J_0 : Sh(\mathcal{V}(\mathcal{H})) \to Sh(\Lambda\underline{R})$ defined on*

1. *Objects: given $\phi \in \Lambda\underline{R}$ we define*

$$(J_0(\underline{A}))_\phi := \underline{A}_{\phi(p_R(\phi))} = (\phi^*\underline{A})_{p_R(\phi)} \tag{14.1.15}$$

such that if $\phi_1 \leq \phi_2$ ($p_R(\phi_1) \subseteq p_R(\phi_2)$ and $\phi_1 = (\phi_2)_{p_R(\phi_1)}$) we define $J_0(\underline{A})(i_{\phi_1\phi_2}) : J_0(\underline{A})_{\phi_2} \to J_0(\underline{A})_{\phi_1}$ by

$$J_0(\underline{A})(i_{\phi_1\phi_2}) := \underline{A}(i_{\phi_1\phi_2}) : \underline{A}_{\phi_2(p_R(\phi_2))} \to \underline{A}_{\phi_1(p_R(\phi_1))} . \tag{14.1.16}$$

[7] In the sense that for each element $\phi \in (\Lambda\underline{R})_V$, given the open neighbourhood U, $p_R(U)$ is open in PG and p_R restricted to $U \ni \phi$ is a homomorphisms, i.e., $(p_R)|U : U \to p_{\underline{R}}(U)$ is a homomorphisms.

2. *Maps: given* $f : \underline{A} \to \underline{B}$ *in* $Sh(\mathcal{V}(\mathcal{H}))$ *we define* $J_0(f)_\phi : J_0(\underline{A})_\phi \to J_0(\underline{B})_\phi$ *as the maps* $f_{\phi(p_R(\phi))} : \underline{A}_{\phi(p_R(\phi))} \to \underline{B}_{\phi(p_R(\phi))}$.

Proof Consider an arrow $f : \underline{A} \to \underline{B}$ in $Sh(\mathcal{V}(\mathcal{H}))$ such that, for each $V \in \mathcal{V}(\mathcal{H})$, the local component is $f_V : \underline{A}_V \to \underline{B}_V$ with commutative diagram

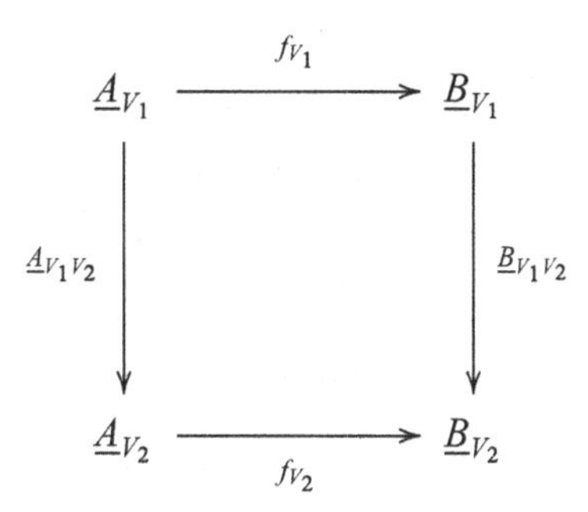

for all pairs V_1, V_2 with $V_2 \leq V_1$. Now suppose that $\phi_2 \leq \phi_1$, such that (i) $p_R(\phi_2) \subseteq p_R(\phi_1)$; and (ii) $\phi_2 = \phi_1|_{p_R(\phi_2)}$. We want to show that the action of the J_0 functor gives the commutative diagram

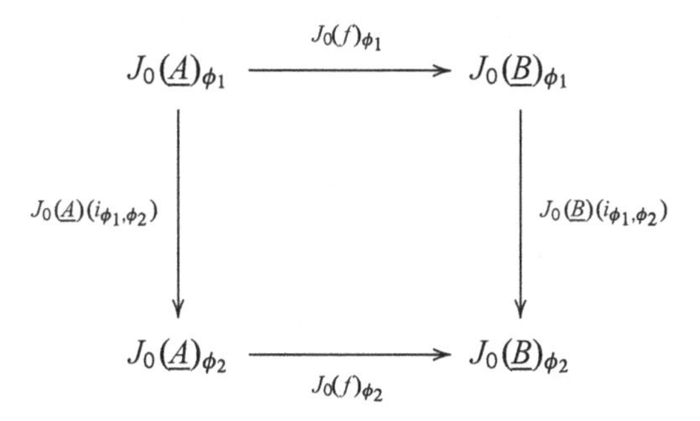

for all $V_2 \subseteq V_1$. By applying the definitions, we get

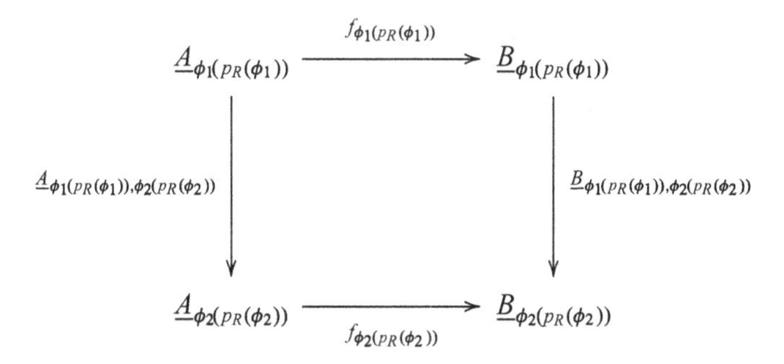

which is commutative. Therefore $J_0(f)$ is a well defined arrow in $Sh(\Lambda \underline{R})$ from $J_0(\underline{A})$ to $J_0(\underline{B})$.

Given two arrows f, g in $Sh(\mathcal{V}(\mathcal{H}))$ then it follows that:

$$J_0(f \circ g) = J_0(f) \circ J_0(g)\,. \tag{14.1.17}$$

This proves that the J_0 is a functor from $Sh(\mathcal{V}(\mathcal{H}))$ to $Sh(\Lambda\underline{R})$. The propertied of this □

Corollary 14.1.1 *The functor J_0 preserves monics.*

Proof Given a monic arrow $f : \underline{A} \to \underline{B}$ in $Sh(\mathcal{V}(\mathcal{H}))$ then by definition

$$J_0(f)_{w^g_{V_2}} : J_0(\underline{A})_{w^g_{V_2}} \to J_0(\underline{B})_{w^g_{V_2}} \tag{14.1.18}$$

$$f_{\phi^g_2(p_J(\phi^g_2))} : \underline{A}_{\phi^g_2(p_J(\phi^g_2))} \to \underline{B}_{\phi^g_2(p_J(\phi^g_2))} \tag{14.1.19}$$

The fact that such a map is monic is straightforward. □

Similarly we can show that

Corollary 14.1.2 *The functor J_0 preserves epic arrows.*

Proof Given an epic arrow $f : \underline{A} \to \underline{B}$ in $Sh(\mathcal{V}(\mathcal{H}))$ then by definition

$$J_0(f)_{w^g_{V_2}} : J_0(\underline{A})_{w^g_{V_2}} \to J_0(\underline{B})_{w^g_{V_2}} \tag{14.1.20}$$

$$f_{\phi^g_2(p_J(\phi^g_2))} : \underline{A}_{\phi^g_2(p_J(\phi^g_2))} \to \underline{B}_{\phi^g_2(p_J(\phi^g_2))} \tag{14.1.21}$$

The fact that such a map is epic is straightforward. □

We would now like to know how such a functor behaves with respect to the terminal object. To this end we define the following corollary:

Corollary 14.1.3 *The functor J_0 preserves the terminal object.*

Proof The terminal object in $Sh(\mathcal{V}(\mathcal{H}))$ is the objects $\underline{1}_{Sh(\mathcal{V}(\mathcal{H}))}$ such that to each element $V \in \mathcal{V}(\mathcal{H})$ it associates the singleton set $\{*\}$. We now apply the J_0 functor to such an object obtaining

$$J_0(\underline{1}_{Sh(\mathcal{V}(\mathcal{H}))})_{w^g_V} := (\underline{1}_{Sh(\mathcal{V}(\mathcal{H}))})_{\phi^g(p_J(\phi^g))} = \{*\} \tag{14.1.22}$$

where ϕ^g is the unique homeomorphism associated to the coset w^g_V.
Thus it follows that $J_0(\underline{1}_{Sh(\mathcal{V}(\mathcal{H}))}) = \underline{1}_{Sh(\Lambda(\underline{G/G_F}))}$ □

We now check whether J_0 preserves the initial object. We recall that the initial object in $Sh(\mathcal{V}(\mathcal{H}))$ is simply the sheaf $\underline{O}_{Sh(\mathcal{V}(\mathcal{H}))}$ which assigns to each element V the empty set $\{\emptyset\}$. We then have

$$J_0(\underline{O}_{Sh(\mathcal{V}(\mathcal{H}))})_{w^g_V} := (\underline{O}_{Sh(\mathcal{V}(\mathcal{H}))})_{\phi^g(p_J(\phi^g))} = \{\emptyset\} \tag{14.1.23}$$

where $\phi^g \in Hom(\downarrow V, \mathcal{V}(\mathcal{H}))$ is the unique homeomorphism associated with the coset w^g_V.

It follows that:

$$J_0(\underline{O}_{Sh(\mathcal{V}(\mathcal{H}))}) = \underline{O}_{Sh(\Lambda(\underline{G/G_F}))} \tag{14.1.24}$$

From the above proof it transpires that the reason the functor J_0 preserves monic, epic, terminal object, and initial object is mainly due to the fact that the action of J_0 is defined component-wise as $(J_0(\underline{A}))_\phi := \underline{A}_{\phi(V)}$ for $\phi \in Hom(\downarrow V, \mathcal{V}(\mathcal{H}))$. In particular, it can be shown that J_0 preserves all limits and colimits.

Theorem 14.1.3 *The functor J_0 preserves limits.*

In order to prove the above theorem we first of all have to recall some general results and definitions. To this end consider two categories $\mathcal{C}$ and $\mathcal{D}$, such that there exists a functor between them $F : \mathcal{C} \to \mathcal{D}$. For a small index category J, we consider diagrams of type J in both $\mathcal{C}$ and $\mathcal{D}$, i.e. elements in $\mathcal{C}^J$ and $\mathcal{D}^J$, respectively. The functor F then induces a functor between these diagrams as follows:

$$F^J : \mathcal{C}^J \to \mathcal{D}^J \tag{14.1.25}$$

$$A \mapsto F^J(A) \tag{14.1.26}$$

such that $(F^J(A))(j) := F(A(j))$. Therefore, if limits of type J exist in $\mathcal{C}$ and $\mathcal{D}$ we obtain the diagram

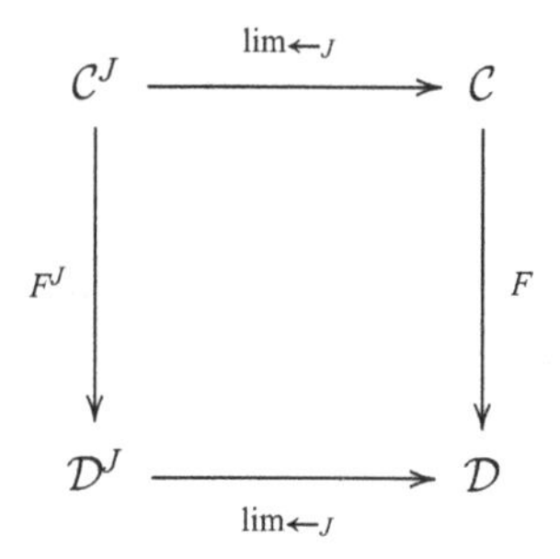

where the map

$$\lim_{\leftarrow J} : \mathcal{C}^J \to \mathcal{C} \tag{14.1.27}$$

$$A \mapsto \lim_{\leftarrow J}(A) \tag{14.1.28}$$

assigns, to each diagram A of type J in $\mathcal{C}$, its limit $\lim_{\leftarrow J}(A) \in \mathcal{C}$. By the universal properties of limits we obtain the *natural transformation*

$$\alpha_J : F \circ \lim_{\leftarrow J} \to \lim_{\leftarrow J} \circ F^J \tag{14.1.29}$$

We then say that F preserves limits if α_J is a *natural isomorphisms*.

For the case at hand, in order to show that the functor J_0 preserves limits we need to show that there exists a map

$$\alpha_J : J_0 \circ \lim_{\leftarrow J} \rightarrow \lim_{\leftarrow J} \circ J_0^J \tag{14.1.30}$$

which is a natural isomorphisms. Here J_0^J represents the map

$$J_0^J : (Sh(\mathcal{V}(\mathcal{H})))^J \rightarrow \left(Sh(\Lambda(\underline{G/G_F}))\right)^J \tag{14.1.31}$$

$$A \mapsto J_0^J(A) \tag{14.1.32}$$

where $(J_0^J(A)(j))_\phi := J_0(A(j))_\phi$.
The proof of α_J being a natural isomorphisms will utilise a result derived in [55] where it is shown that for any diagram $A : J \rightarrow \mathcal{C}^\mathcal{D}$ of type J in $\mathcal{C}^\mathcal{D}$ the following isomorphisms holds

$$\left(\lim_{\leftarrow J} A\right) D \simeq \lim_{\leftarrow J} A_D \;\forall\; D \in \mathcal{D} \tag{14.1.33}$$

where $A_D : J \rightarrow \mathcal{C}$ is a diagram in $\mathcal{D}$. With these results in mind we are now ready to prove Theorem 14.1.3

Proof Let us consider a diagram $A : J \rightarrow Sets^{\mathcal{V}(\mathcal{H})}$ of type J in $Sets^{\mathcal{V}(\mathcal{H})}$:

$$A : J \rightarrow Sets^{\mathcal{V}(\mathcal{H})} \tag{14.1.34}$$

$$j \mapsto A(j) \tag{14.1.35}$$

where $A(j)(V) := A_V(j)$ for $A_V : j \rightarrow Sets$ a diagram in $Sets$. Assume that L is a limit of type J for A, i.e. $L : \mathcal{V}(\mathcal{H}) \rightarrow Sets$ such that $\lim_{\leftarrow J} A = J$. We then construct the diagram

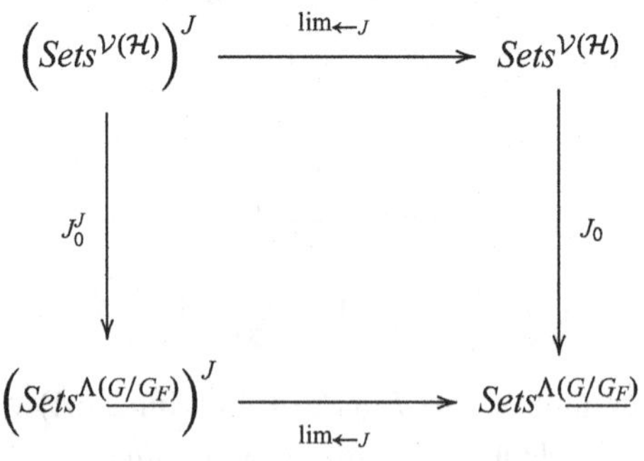

$$\begin{array}{ccc} \left(Sets^{\mathcal{V}(\mathcal{H})}\right)^J & \xrightarrow{\lim_{\leftarrow J}} & Sets^{\mathcal{V}(\mathcal{H})} \\ \Big\downarrow J_0^J & & \Big\downarrow J_0 \\ \left(Sets^{\Lambda(\underline{G/G_F})}\right)^J & \xrightarrow[\lim_{\leftarrow J}]{} & Sets^{\Lambda(\underline{G/G_F})} \end{array}$$

and the associated natural transformation

$$\alpha_J : J_0 \circ \lim_{\leftarrow J} \rightarrow \lim_{\leftarrow J} \circ J_0^J \tag{14.1.36}$$

For each diagram $A : J \to Sets^{\mathcal{V}(\mathcal{H})}$ and $\phi \in \Lambda(\underline{G/G_F})$ we obtain

$$\Big(J_0 \circ \lim_{\leftarrow J}(A)\Big)_\phi = \Big(J_0\big(\lim_{\leftarrow J} A\big)\Big)_\phi := \big(\lim_{\leftarrow J} A\big)_{\phi(V)} \simeq \lim_{\leftarrow J} A_{\phi(V)} \tag{14.1.37}$$

where $A_{\phi(V)} : J \to Sets$, such that $A_{\phi(V)}(j) = A(j)(\phi V)$.[8]

On the other hand

$$\Big(\big(\lim_{\leftarrow J} \circ J_0^J\big)A\Big)_\phi = \Big(\lim_{J}(J_0^J(A))\Big)_\phi \simeq \lim_{\leftarrow J}(J_0^J(A))_\phi = \lim_{\leftarrow J} A_{\phi(V)} \tag{14.1.38}$$

where

$$J_0^J(A) : J \to Sets^{\Lambda(\underline{G/G_F})} \tag{14.1.39}$$

$$j \mapsto J_0^J(A)(j) \tag{14.1.40}$$

such that for all $\phi \in \Lambda(\underline{G/G_F})$ we have $\big(J_0^J(A(j))\big)_\phi = \big(J_0(A(j))\big)_\phi = A(j)_{\phi(V)}$.
It follows that

$$J_0 \circ \lim_{\leftarrow J} \simeq \lim_{\leftarrow J} \circ J_0^J \tag{14.1.41}$$

□

Similarly one can show that

Theorem 14.1.4 *The functor J_0 preserves all colimits*

Since colimits are simply duals to the limits, the proof of this theorem is similar to the proof given above. However, for completeness sake we will, nonetheless, report it here.

Proof We first of all construct the analogue of the diagram above:

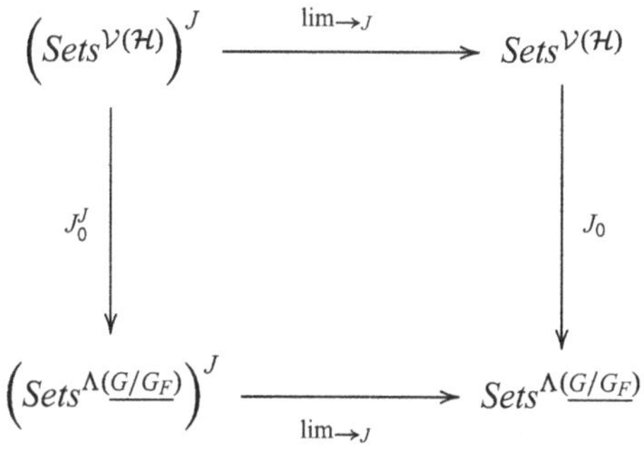

$$\begin{array}{ccc}
\left(Sets^{\mathcal{V}(\mathcal{H})}\right)^J & \xrightarrow{\lim_{\to J}} & Sets^{\mathcal{V}(\mathcal{H})} \\
\Big\downarrow{\scriptstyle J_0^J} & & \Big\downarrow{\scriptstyle J_0} \\
\left(Sets^{\Lambda(\underline{G/G_F})}\right)^J & \xrightarrow[\lim_{\to J}]{} & Sets^{\Lambda(\underline{G/G_F})}
\end{array}$$

[8] Recall that $A : J \to Sets^{\mathcal{V}(\mathcal{H})}$ is such that $A_V(j) = A(j)(V)$, therefore $\big(J_0(A(j))\big)_\phi := A(j)_{\phi(V)} = A_{\phi(V)}(j)$.

where $\lim_{\rightarrow J} : \left(Sets^{\mathcal{V}(\mathcal{H})}\right)_0^J \rightarrow Sets^{\mathcal{V}(\mathcal{H})}$ represents the map which assigns colimits to all diagrams in $\left(Sets^{\mathcal{V}(\mathcal{H})}\right)_0^J$.

We now need to show that the associated natural transformation

$$\beta_J : J_0 \circ \lim_{\rightarrow J} \rightarrow \lim_{\rightarrow J} \circ J_0^J \tag{14.1.42}$$

is a natural isomorphisms.

For any diagram $A \in \left(Sets^{\mathcal{V}(\mathcal{H})}\right)_0^J$ and $\phi \in \Lambda(\underline{G/G_F})$ we compute

$$\left(J_0 \circ \lim_{\rightarrow J}(A)\right)_\phi = \left(J_0(\lim_{\rightarrow J} A)\right)_\phi = \left(\lim_{\rightarrow J} A\right)_{\phi(V)} \simeq \lim_{\rightarrow J} A_{\phi(V)} \tag{14.1.43}$$

where $\left(\lim_{\rightarrow J} A\right)_{\phi(V)} \simeq \lim_{\rightarrow J} A_{\phi(V)}$ is the dual of (14.1.33). On the other hand

$$\left((\lim_{\rightarrow J} \circ J_0^J)(A)\right)_\phi = \left(\lim_{\rightarrow J}(J_0^J(A))\right)_\phi \simeq \lim_{\rightarrow J}(J_0^J(A))_\phi = \lim_{\rightarrow J} A_{\phi(V)} \tag{14.1.44}$$

It follows that indeed β_J is a natural isomorphisms. □

The Adjoint Functor Theorem as applied to Grothendieck toposes [50] states that any colimit preserving functor between Grothendieck topoi has a right adjoint and any limit preserving functor has a left adjoint, so J_0 has both adjoints. The construction of these functors is left as an exercise.

It is a standard result that, given a map $f : X \rightarrow Y$ between topological spaces X and Y, we obtain the following geometric morphisms:

$$f^* : Sh(Y) \rightarrow Sh(X) \tag{14.1.45}$$

$$f_* : Sh(X) \rightarrow Sh(Y) \tag{14.1.46}$$

and we know that $f^* \dashv f_*$, i.e., f^* is the left adjoint of f_*. If f is an etalé map, however, there also exists the left adjoint $f!$ to f^*, namely

$$f! : Sh(X) \rightarrow Sh(Y) \tag{14.1.47}$$

with $f! \dashv f^* \dashv f_*$.

It can be shown that:

$$f!(p_A : A \rightarrow X) = f \circ p_A : A \rightarrow Y \tag{14.1.48}$$

so that we combine the etalé bundle $p_A : A \rightarrow X$ with the etalé map $f : X \rightarrow Y$, to give the etalé bundle $f \circ p_A : A \rightarrow Y$.

Given a map $\alpha : A \rightarrow B$ of etalé bundles over X, we obtain the map $f!(\alpha) : f!(A) \rightarrow f!(B)$ which is defined as follows.

We start with the collection of fibre maps $\alpha_x : A_x \to B_x$, $x \in X$, where $A_x := p^{-1}A(\{x\})$, then, for each $y \in Y$ we want to define the maps $f!(\alpha)_y : f!(A)_y \to f!(B)_y$, i.e., $f!(\alpha)_y : p^{-1}\big(A(f^{-1}\{y\})\big) \to p^{-1}\big(B(f^{-1}\{y\})\big)$. These are defined as:

$$f!(\alpha)_y(a) := \alpha_{p_A(a)}(a) \tag{14.1.49}$$

for all $a \in f!(A)_y = p^{-1}\big(A(f^{-1}\{a\})\big)$.

For the case at hand, we will utilised the left adjoint functor

$$p_{\underline{R}}! : Sh(\Lambda \underline{R}) \to Sh(PG) \qquad p_{\underline{R}}! \vdash p^*_{\underline{R}}\,.$$

to map sheaves over $\Lambda \underline{R}$ to sheaves over the poset category PG.

Thus, given a sheaf $\underline{K} \in Sh(\Lambda \underline{R}$ the associated etalé bundle over $\Lambda \underline{R}$ is $p_K : \Lambda \underline{K} \to \Lambda \underline{R}$. Then[9] $p_{\underline{R}}! : Et(\Lambda \underline{R}) \to Et(PG)$ is defined by

$$p_{\underline{R}}!(p_{\underline{K}} : \Lambda \underline{K} \to \Lambda \underline{R}) := p_{\underline{R}} \circ p_{\underline{K}} : \Lambda \underline{K} \to PG\,.$$

The precise way in which the left adjoint functor $p_{\underline{R}}!$ is constructed is reported in the Sect. A.5 in the Appendix.

Given a sheaf $\underline{A} \in Sh(\mathcal{V}(\mathcal{H}))$ we are now able to define the corresponding sheaf over PG as $p_{\underline{R}} \circ J_0(\underline{A})$ where, now, the bundle space is

$$\Lambda(p_{\underline{R}} \circ J_0(\underline{A}))_L := \bigcup_{\phi \in Hom_{poset}(\downarrow L, \mathcal{V}(\mathcal{H}))} \phi^* = (\underline{A})_L\,. \tag{14.1.50}$$

It is interesting to note that, although the functor J_0 preserves the terminal object, the composition $p_{\underline{R}} \circ J_0$ doesn't (this was already shown in [27] but for different functors). In particular, given the terminal object $\underline{1}_{Sh(\mathcal{V}(\mathcal{H}))}$ then, for all $\phi \in \Lambda \underline{R}$ we have

$$(J_0(\underline{1}_{Sh(\mathcal{V}(\mathcal{H}))}))_\phi = (\underline{1}_{Sh(\mathcal{V}(\mathcal{H}))})_{\phi(p_{\underline{R}}(\phi))} = \{*\}\,, \tag{14.1.51}$$

hence

$$J_0(\underline{1}_{Sh(\mathcal{V}(\mathcal{H}))}) = \underline{1}_{Sh(\Lambda \underline{R})}\,. \tag{14.1.52}$$

However, if we then apply the functor $p_{\underline{R}}!$ we obtain

$$p_{\underline{R}}!(\underline{1}_{Sh(\Lambda \underline{R})}) = \underline{R} \neq \underline{1}_{Sh(PG)}\,. \tag{14.1.53}$$

[9]Here $Et(\Lambda \underline{R})$ indicates the etalé bundles over $\Lambda \underline{R}$.

14.1.3 Preservation of Linear Structure Through the Presheaf $\underline{I}$

The presheaf $\underline{R}$, although useful at a conceptual level it does not address the preservation of linear structure. A possible way to preserve such a structure is to construct a poset morphism $\phi :\downarrow L \rightarrow \mathcal{V}(\mathcal{H})$ ($L \in PG$) by first introducing a Lie algebra homomorphism, $f : L \rightarrow \mathcal{B}(\mathcal{H})$ and then defining

$$\phi_f(L) := f(L)'' \tag{14.1.54}$$

where $f(L)''$ is the double commutant of the abelian subalgebra $f(L) \subseteq \mathcal{B}(\mathcal{H})$. Clearly f extends to a poset morphisms from $\downarrow L$ to $\mathcal{V}(\mathcal{H})$ as $f(L_0) := \phi_{f|L_0}(L_0)''$ where $\phi_{f|L_0} : L_0 \rightarrow \mathcal{B}(\mathcal{H})$ is f restricted to the subalgebra $L_0 \subseteq L$.
Given this construction it is now possible to define the following presheaf:

Definition 14.1.3 The pre-quantization presheaf has:

- as objects

$$\underline{I}_L := Hom_{Lie}(\downarrow L, \mathcal{B}(\mathcal{H})) \tag{14.1.55}$$

  for $L \in PG$
- as Morphisms: given a map $i_{L_1L_2} : L_1 \subseteq L_2$, the presheaf maps are $\underline{I}(i_{L_1L_2}) : \underline{I}_{L_2} \rightarrow \underline{I}_{L_1}$ such that

$$\underline{I}(i_{L_1L_2})(f) := f_{|L_1} \tag{14.1.56}$$

  for all $f \in \underline{I}_{L_1} = Hom_{Lie}(L_1, \mathcal{B}(\mathcal{H}))$.

Since we want faithful representations, we restrict the homomorphisms to only include injective ones.

We could obviously restrict f to irreducible representations. Clearly each such irreducible representation will give a global element of $\underline{I}$, however it is not clear at this stage if the converse is true.

The bundle space $\Lambda\underline{I}$ can be given a poset structure in a similar way as it was done for $\Lambda\underline{R}$, namely: given two elements $f_1, f_2 \in \Delta I$ then

$$f_1 \leq f_2 \qquad \text{iff} \qquad p_I(f_1) \subseteq p_I(f_2)\,.$$

The group action is defined in the obvious way: for each $\hat{U} \in \mathcal{U}(\mathcal{H})$ and $f \in \underline{I}_L = Hom_{Lie}(L, \mathcal{B}(\mathcal{H}))$ then $l_{\hat{U}}(f) \in \underline{I}_L$ is

$$(l_{\hat{U}}f)(v) := \hat{U}f(v)\hat{U}^{-1}\,. \tag{14.1.57}$$

$l_{\hat{U}}(f)$ is indeed an algebra homomorphisms.

Similarly, as it was done for the $\underline{R}$ presheaf, we now utilise the poset $\Lambda\underline{I}$ as our base category and define a functor $I_0 : Sh(\mathcal{V}(\mathcal{H})) \to Sh(\Lambda\underline{I})$.

Theorem 14.1.5 *The functor I_0 is defined:*

1. *on objects: for all $f \in \Lambda\underline{I}$ we define*

$$(I_0(\underline{A}))(f) := \underline{A}_{f(p_I(f))''} \tag{14.1.58}$$

where $p_I : \Lambda\underline{I} \to PG$.
If $f_2 \leq f_1$ then $I_0\underline{A}_{f_1,f_2} : (I_0(\underline{A}))(f_1) \to (I_0(\underline{A}))(f_2)$ is defined as

$$I_0\underline{A}_{f_1,f_2} := \underline{A}_{f_1(p_I(f_1))''f_2(p_I(f_2))''} : \underline{A}_{f_1(p_I(f_1))''} \to \underline{A}_{f_2(p_I(f_2))''}\,. \tag{14.1.59}$$

2. *On arrows: for any $g : \underline{A} \to \underline{B}$, and for each $f \in \Lambda\underline{I}$ we define $\underline{I}(g)_f : \underline{A}_f \to \underline{B}_f$ as $g_{f(p_I(f))''} : \underline{A}_{f(p_I(f))''} \to \underline{B}_{f(p_I(f))''}$.*

Proof Consider an arrow $f : \underline{A} \to \underline{B}$ in $Sh(\mathcal{V}(\mathcal{H}))$, so that for each $V \in \mathcal{V}(\mathcal{H})$ the local component is $f_V : \underline{A}_V \to \underline{B}_V$ with commutative diagram

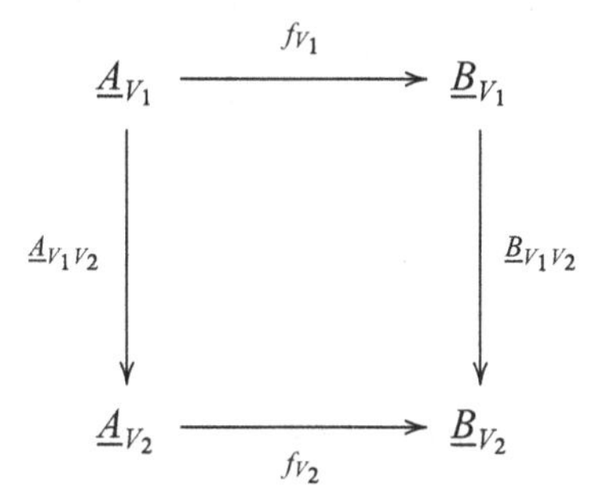

for all pairs V_1, V_2 with $V_2 \leq V_1$. Now suppose that $f_2 \leq f_1$, such that (i) $p_{\underline{I}}(f_2) \subseteq p_{\underline{I}}(f_1)$; and (ii) $f_2 = f_1|_{p_{\underline{I}}(f_2)}$, we want to show that, for all $V_2 \subseteq V_1$, the action of the I_0 functor gives the following commutative diagram

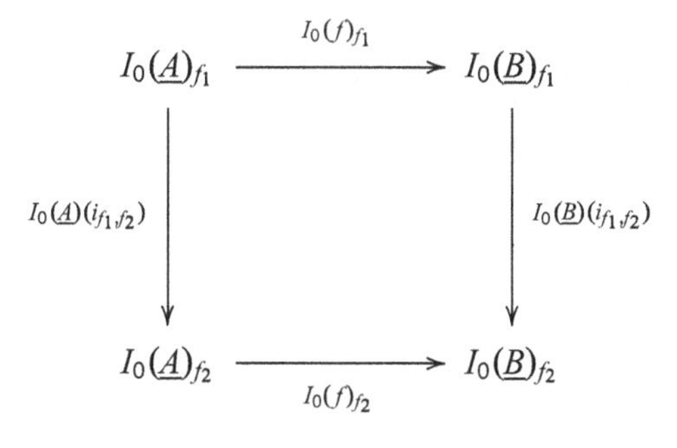

By applying the definitions we get

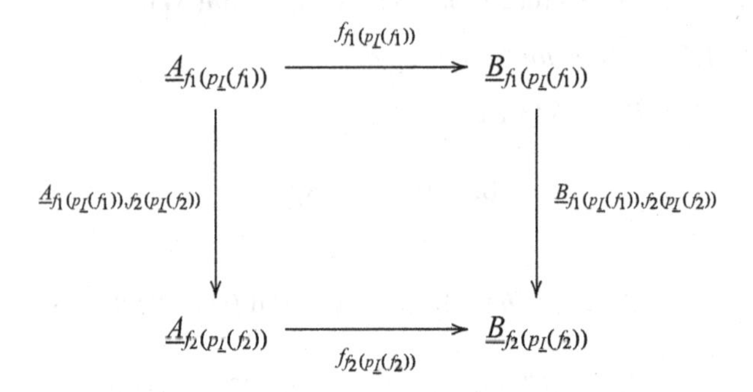

which is commutative. Therefore $I_0(f)$ is a well defined arrow in $Sh(\Lambda \underline{I})$ from $I(\underline{A})$ to $I(\underline{B})$. Moreover, given two arrows f, g in $Sh(\mathcal{V}(\mathcal{H}))$ it easy to see that

$$I_0(f \circ g) = I_0(f) \circ I(g)\,.$$

This proves that I is a functor from $Sh(\mathcal{V}(\mathcal{H}))$ to $Sh(\Lambda \underline{I})$. □

Given the etalé bundle map $p_{\underline{I}} : \Lambda \underline{I} \to PG$ we then utilise the left adjoint functor $p_{\underline{I}}!$ to map the sheaves over $\Lambda \underline{I}$ to sheaves over the poset PG, thus obtaining the composite functor

$$F := p_{\underline{I}}! \circ I_0\,. \tag{14.1.60}$$

14.1.4 Relation Between the Functors $\underline{I}$ and $\underline{R}$

Theorem 14.1.6 *The map $k : \underline{I} \to \underline{R}$ defined for each $L \in PG$ as*

$$k_L : Hom_{Lie}(L, \mathcal{B}(\mathcal{H})) \to Hom_{poset}(\downarrow L, \mathcal{V}(\mathcal{H})) \tag{14.1.61}$$

$$q \mapsto k_L(q) \tag{14.1.62}$$

such that $k_L(q)(L_0) := (q(L_0))''$ for all $L_0 \subseteq L$, is a functor.

Proof To show that the map k, as defined above, is indeed a functor we need to show that for all $L \in PG$ the following diagram commutes:

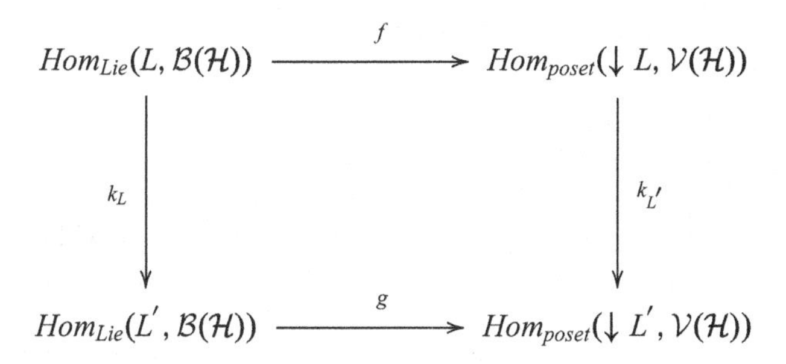

That is, we need to show that for all $q \in Hom_{Lie}(L, \mathcal{B}(\mathcal{H}))$, then

$$k_{L'} \circ f(q) = g \circ k_L(q) \,. \tag{14.1.63}$$

Now, for all $L_0 \subseteq L$ $(k_{L'} \circ f(q))(L_0) := (f(q(L_0))'' = (q'_L(L_0))''$. This is equivalent to $(q'_L(L_0))'' = (q(L_0))''$.

On the other hand $g \circ k_L(q) := (k_L(q))_{|\downarrow L'}$ therefore, for all $L_0 \subseteq L$, $(k_L(q))_{|\downarrow L'}(L_0) = k_{L'}(q)(L_0) = (q(L_0))''$. □

Now that we have the two functors I_0 and J_0 it is interesting to note that for each $\underline{A} \in Sh(\mathcal{V}(\mathcal{H}))$ we have, for $\phi \in Hom_{poset}(\downarrow L, \mathcal{V}(\mathcal{H}))$

$$J_0(\underline{A})_\phi := \underline{A}_{\phi(L)} = (\phi^* \underline{A})_L \tag{14.1.64}$$

and

$$I_0 \underline{A}_f = J_0(\underline{A})_{k_l(f)} = k_L^*(J_0(\underline{A}))_f \tag{14.1.65}$$

for $f \in Hom_{Lie}(L, \mathcal{B}(\mathcal{H}))$. Therefore we obtain that

$$I_0 \underline{A}_{f_1 f_2} := \underline{A}_{\phi_{f_1}(p_I(f_1))'' \phi_{f_2}(p_I(f_2))''} : \underline{A}_{\phi_{f_1}(p_I(f_1))''} \to \underline{A}_{\phi_{f_2}(p_I(f_2))''} \,. \tag{14.1.66}$$

From the definition it follows that this k functor can be seen as a context preserving functor. In particular, if we see it as a bundle map we would obtain the map $k : \Lambda \underline{I} \to \Lambda \underline{R}$ between the etalé bundle spaces. Consequently the pull back would be

$$[k^*(J_0(\underline{A}))]_f = J_0(\underline{A})_{k_L(f)} = \underline{A}_{f(L)''} = I_0(\underline{A})_f \tag{14.1.67}$$

for all $f \in \Lambda I$ and $p_{\underline{I}}(f) = L$.

14.2 Quantization by Nakayama

In [57] the author defines a concrete way of defining a quantization in a topos given an underlining classical system. This method turns out to be a particular application of the above abstract description of a quantization in a topos, although it is not presented in that way by the author. In the following we will first report the main ideas in the paper, then try to generalise it to incorporate unitary equivalent quantizations.

In [57] the author starts by considering a collection of so called pre-quantization categories which are basically posets of lie-abelian classical observables. In particular, let us consider the set $\mathcal{O}$ of all classical observables, then $\mathbf{C}_o$ is the collection of all subsets $C \subseteq \mathcal{O}$ such that for any $a, b \in C$ then $[a, b] = 0$. $\mathbf{C}_o$ forms a category under inclusion. A pre-quantization category is then defined as follows:

Definition 14.2.1 Any full subcategory of $\mathbf{C}_o$ is a pre-quantization category.

It is straightforward to see that the collection of pre-quantization categories forms itself a category under inclusion. We will denote such a category by $\mathfrak{C}$.
In what follows we will assume that $\mathbf{C}_o$ is invariant under any symplectic covariance transformation. Given such a category, it is possible, as shown in [57] to define a quantization functor. In particular, given any classical observable a we can define the quantization of a through the map[10]

$$\tilde{v} : a \mapsto e^{i\hat{a}} . \tag{14.2.1}$$

This map induces a quantization functor ϕ defined as follows:

$$\phi : \mathbf{C}_o \to \mathcal{V}(\mathcal{H}) \tag{14.2.2}$$
$$C \mapsto \Upsilon(C)'' \tag{14.2.3}$$

where $\Upsilon(C)'' = (\tilde{v}(C) \cup \tilde{v}(C)^*)''$ (here $''$ represents the double commutant operator). Thus $\Upsilon(C)''$ is the smallest abelian von Neumann algebra containing $\Upsilon(C)$. Since $\phi(C') \subseteq \phi(C)$ whenever $C' \subseteq C$, it follows that indeed ϕ is a functor. It is easy to 'extend' ϕ to a functor on $\mathfrak{C}$, where for all $C \in \mathfrak{C}$, then $\phi|_C : C \to \mathcal{V}(\mathcal{H})$ is the restriction of ϕ to C.

We are interested not only in a single quantization but in all possible unitary equivalent quantizations, therefore we need to define the action of $G \subseteq U(\mathcal{H})$ on ϕ. This will allow us to define the notion of unitary equivalent quantizations implementing the Dirac covariance of quantum theory. In particular, for each $g \in G$ and $C \in \mathbf{C}_o$ we define

$$l_g\phi(C) := l_g(\phi(C)) .$$

[10]The author in [57] claims that the map $\tilde{v}$ is faithful, however because of the periodicity of the exponential function it is not clear to us how he justifies his claim.

Having defined the action of G on the quantization functor we can define the *quantization presheaf* over $\mathbf{C}_o$ as follows:

Definition 14.2.2 The quantization presheaf $\underline{Q} : \mathbf{C}_0 \rightarrow Sets$ is defined on

1. Objects: for each $C \in \mathbf{C}_o$ we assign the collection of unitary equivalent quantization maps, i.e. $\underline{Q}(C) := \{l_g\phi : \downarrow C \rightarrow \mathcal{V}(\mathcal{H}) | g \in G\}$ where $l_g\phi(C) := l_g(\phi(C))$. We assume that there is no group action on $\mathbf{C}_o$.
2. Morphisms: given a map $i_{C_1,C_2} : C_1 \subseteq C_2$ the corresponding presheaf map is

$$\underline{Q}(i_{C_1,C_2}) : \underline{Q}(C_2) \rightarrow \underline{Q}(C_1) \tag{14.2.4}$$

$$\phi \mapsto \phi_{|C_1} \tag{14.2.5}$$

We would now like to consider all possible equivalent quantizations at the same time. To this end we will adopt a similar trick as the one adopted in [27] and utilise, as our new base category, the poset $\Lambda\underline{Q}$. This will be the topic of the next Section.

14.2.1 Sheaves over $\Lambda\underline{Q}$

Given the sheaf $\underline{Q}$, the associated étalé bundle is $p_Q : \Lambda\underline{Q} \rightarrow \mathbf{C}_o$ where ΛQ is the bundle space. We want to show that ΛQ is actually a poset.

Lemma 14.2.1 *Given two elements $\phi_1 \in \Lambda\underline{Q}$ and $\phi_2 \in \Lambda\underline{Q}$, then*

$$\phi_1 \leq \phi_2 \textit{ iff } p_Q(\phi_1) \subseteq p_Q(\phi_2) \textit{ and } \phi_1 = \phi_{2|p_Q(\phi_1)}$$

Proof

1. *Reflexivity*. Trivially $\phi \leq \phi$ since $p_Q(\phi) \subseteq p_Q(\phi)$ and $\phi = \phi$.
2. *Transitivity*. If $\phi_i \leq \phi_j$ and $\phi_j \leq \phi_k$, then $p_Q(\phi_i) \subseteq p_Q(\phi_j)$ and $p_Q(\phi_j) \subseteq p_Q(\phi_k)$, therefore $p_Q(\phi_i) \subseteq p_Q(\phi_k)$. Moreover we have that $\phi_i = \phi_j|_{p_Q(\phi_i)}$ and $\phi_j = \phi_k|_{p_Q(\phi_j)}$, therefore $\phi_i = \phi_k|_{p_Q(\phi_i)}$.
3. *Antisymmetry*. If $\phi_i \leq \phi_j$ and $\phi_j \leq \phi_i$ it implies that $p_Q(\phi_i) \subseteq p_Q(\phi_j)$ and $p_Q(\phi_j) \subseteq p_Q(\phi_i)$, thus $p_Q(\phi_i) = p_Q(\phi_j)$. Moreover we have that $\phi_i = \phi_j|_{p_Q(\phi_i)}$ and $\phi_j = \phi_i|_{p_Q(\phi_j)}$, therefore $\phi_i = \phi_j$.

□

We are now interested in 'transforming' all the physically relevant sheaves on $\mathcal{V}(\mathcal{H})$ to sheaves over $\Lambda\underline{Q}$ which, being a poset, is equipped with the Alexandroff topology. What this means is that we want to construct a functor

$$I : Sh(\mathcal{V}(\mathcal{H})) \rightarrow Sh(\Lambda\underline{Q}) \tag{14.2.6}$$

$$\underline{A} \mapsto I(\underline{A}) . \tag{14.2.7}$$

As a first attempt we define, for each context ϕ,

$$\left(I(\underline{A})\right)_{\phi} := \underline{A}_{\phi(C)} = \left((\phi)^{*}(\underline{A})\right)(C)$$

where $\phi :\downarrow C \rightarrow \mathcal{V}(\mathcal{H})$ is the quantization functor defined for $\downarrow C$.

Next we need to define the morphisms: given $i_{\phi_2,\phi_1} : \phi_2 \leq \phi_2$ ($\phi_1 \in Hom(\downarrow C_1, \mathcal{V}(\mathcal{H}))$ and $\phi_2 \in Hom(\downarrow C_2, \mathcal{V}(\mathcal{H}))$) we define the associated morphisms $I\underline{A}(i_{\phi_2,\phi_1}) : \left(I(\underline{A})\right)_{\phi_1} \rightarrow \left(I(\underline{A})\right)_{\phi_2}$ as

$$(I\underline{A}(i_{\phi_2,\phi_1}))(a) := \underline{A}_{\phi_1(C_1),\phi_2(C_2)}(a) \qquad \forall\, a \in \underline{A}_{\phi(C_1)} \,.$$

In the above equation $C_1 = p_Q(\phi_1)$ and $C_2 = p_Q(\phi_2)$.[11] Moreover, since $\phi_2 \leq \phi_1$, then $\phi_2(C_2) \subseteq \phi_1(C_1)$ and $\phi_2 = \phi_1|_{C_2}$.

Theorem 14.2.1 *The map $I : Sh(\mathcal{V}(\mathcal{H})) \rightarrow Sh(\Lambda\underline{Q})$ is a functor defined as follows:*

(i) Objects: $\left(I(\underline{A})\right)_{\phi_1} := \underline{A}_{\phi_1(C)} = \left((\phi)^{*}(\underline{A})\right)(C)$. If $\phi_2 \leq \phi_1$ ($\phi_1 \in Hom(\downarrow C_1, \mathcal{V}(\mathcal{H}))$ and $\phi_2 \in Hom(\downarrow C_2, \mathcal{V}(\mathcal{H}))$), then

$$(I\underline{A})(i_{\phi_2,\phi_1}) := \underline{A}_{\phi_1(C_1),\phi_2(C_2)} : \underline{A}_{\phi_1(C_1)} \rightarrow \underline{A}_{\phi_2(C_2)}$$

where $C_1 = p_Q(\phi_1)$ and $C_2 = p_Q(\phi_2)$.

(ii) Morphisms: if we have a morphisms $f : \underline{A} \rightarrow \underline{B}$ in $Sh(\mathcal{V}(\mathcal{H}))$ we, then, define the corresponding morphisms in $Sh(\Lambda(\underline{Q}))$ as

$$I(f)_{\phi_1} : I(\underline{A})_{\phi_1} \rightarrow I(\underline{B})_{\phi_1} \tag{14.2.8}$$

$$f_{\phi_1} : \underline{A}_{\phi_1(p_Q(\phi_1))} \rightarrow \underline{B}_{\phi_1(p_Q(\phi_1))} \,. \tag{14.2.9}$$

Proof Consider an arrow $f : \underline{A} \rightarrow \underline{B}$ in $Sh(\mathcal{V}(\mathcal{H}))$ so that, for each $V \in \mathcal{V}(\mathcal{H})$, the local component is $f_V : \underline{A}_V \rightarrow \underline{B}_V$ with commutative diagram

$$\begin{array}{ccc} \underline{A}_{V_1} & \xrightarrow{f_{V_1}} & \underline{B}_{V_1} \\ \downarrow{\scriptstyle \underline{A}_{V_1V_2}} & & \downarrow{\scriptstyle \underline{B}_{V_1V_2}} \\ \underline{A}_{V_2} & \xrightarrow{f_{V_2}} & \underline{B}_{V_2} \end{array}$$

[11] Recall that $p_Q : \Lambda\underline{Q} \rightarrow \mathbf{C}_o$.

for all pairs V_1, V_2 with $V_2 \leq V_1$. Now, suppose that $\phi_2 \leq \phi_1$, such that (i) $p_Q(\phi_2) \subseteq p_Q(\phi_1)$; and (ii) $\phi_2 = \phi_1|_{p_Q(\phi_2)}$. We now want to show that the action of the I functor gives the commutative diagram

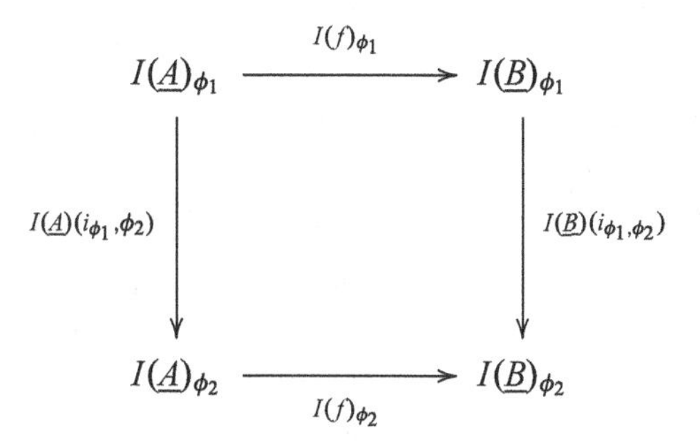

for all $V_2 \subseteq V_1$. By applying the definitions we get

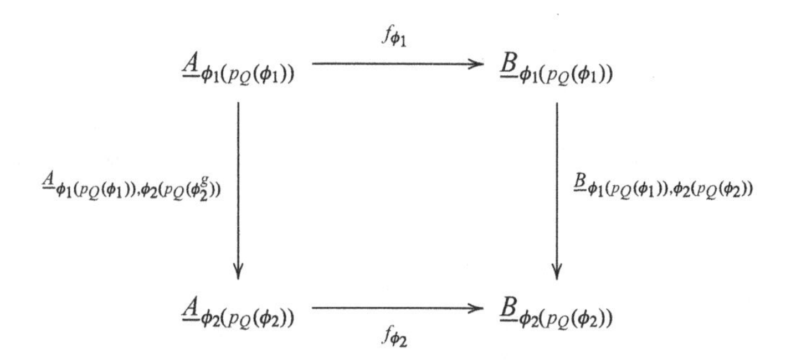

which is commutative. Therefore $I(f)$ is a well defined arrow in $Sh(\Lambda \underline{Q})$ from $I(\underline{A})$ to $I(\underline{B})$.

Given two arrows f, g in $Sh(\mathcal{V}(\mathcal{H}))$ then it follows that:

$$I(f \circ g) = I(f) \circ I(g)\,. \tag{14.2.10}$$

This proves that I is a functor from $Sh(\mathcal{V}(\mathcal{H}))$ to $Sh(\Lambda \underline{Q})$. □

By considering $\Lambda \underline{Q}$ as our new base category we are effectively considering a context to be an element of $\{\phi :\downarrow C \rightarrow \mathcal{V}(\mathcal{H})\}$ for some abelian Lie subalgebra C of classical observables. We can think of any such element as a local quantization in which classical observables are attached to specific self-adjoint operators in the Hilbert space $\mathcal{H}$, in a globally coherent way.

Given that

$$\Lambda \underline{Q} = \coprod_{C \in \mathbf{C}_0} \{\phi_i :\downarrow C \rightarrow \mathcal{V}(\mathcal{H})\} \tag{14.2.11}$$

we can then think of a context as a pair (ϕ, C) of a Lie algebra of commuting classical observables and a specific quantization. Given such pairs it is possible to give $\Lambda\underline{Q}$ a poset structure as follow:

$$(\phi_1, C_1) \leq (\phi_2, C_2) \Longleftrightarrow C_1 \subseteq C_2 \text{ and } \phi_1 = \phi_1|_{C_2} \tag{14.2.12}$$

Corollary 14.2.1 *The etalé topology on $\Lambda\underline{Q}$ is homeomorphic to the Alexandroff topology, given by the above ordering.*

To prove the above Corollary we will make use of the following Lemma:

Lemma 14.2.2 *Let $\alpha : P_1 \to P_2$ be a map between posets P_1 and P_2. Then α is order-preserving if and only if, for each lower set $L \subseteq P_2$ we have that $\alpha^{-1}(L)$ is a lower subset of P_1.*

Proof We now assume that α is order-preserving and $L \subseteq P_2$ is lower. Next let $z \in \alpha^{-1}(L) \in P_1$, i.e., $\alpha(z) = l$ for some $l \in L$, and suppose that $y \in P_1$ is such that $y \leq z$. Since α is order-preserving we have $\alpha(y) \leq \alpha(z) = l \in L$, which, since L is lower, it means that $\alpha(y) \in L$, i.e., $y \in \alpha^{-1}(L)$. Hence $\alpha^{-1}(L)$ is lower.

Conversely, suppose that for any lower set $L \in P_2$ we have that $\alpha^{-1}(L) \in P_1$ is lower. If we consider a pair $x, y \in P_1$ such that $x \leq y$, $\downarrow(y)$ is lower in P_2 and, hence, $\alpha^{-1}(\downarrow \alpha(y))$ is a lower subset of P_1. However $\alpha(y) \in \downarrow \alpha(y)$ and, hence, $y \in \alpha^{-1}(\downarrow \alpha(y))$. Therefore, the fact that $x \leq y$ implies that $x \in \alpha^{-1}(\downarrow \alpha(y))$, i.e., $\alpha(x) \in \downarrow \alpha(y)$, which means that $\alpha(x) \leq \alpha(y)$. We can now say that α is order-preserving. □

We can now prove Corollary 14.2.1.

Proof Let us consider an open set U in the étale topology of $\Lambda(\underline{Q})$. Since $p_Q : \Lambda\underline{Q} \to \mathbf{C}_o$ is a local homeomorphism[12] then $p_Q(U)$ is open in $\mathbf{C}_o$, i.e., is a lower set in the Alexandroff topology. However, by the definition of the poset structure on $\Lambda\underline{Q}$, p_Q is order-preserving, thus $p_Q^{-1} \circ p_Q(U)$ is a lower set in $\Lambda\underline{Q}$. Moreover since p_Q is a local homeomorphism then $p_Q^{-1} \circ p_Q(U) = U$ is a lower set in $\Lambda\underline{Q}$.
Conversely, let U be an open set in the Alexandroff topology on $\Lambda\underline{Q}$. Since p_Q is order-preserving then $p_Q(U)$ is a lower set in $\mathbf{C}_o$. Now, since $p_Q : \Lambda\underline{Q} \to \mathbf{C}_o$ is an étale bundle, we know that p_Q is a local homeomorphism in the étale topology. Thus, restricting only to open sets, we have that $p_Q^{-1}(p_Q(U))$ is an open set in the étale topology. However $p_Q^{-1} \circ p_Q(U) = U$, i.e., U is open in the étale topology. □

Given the map $p_Q : \Lambda\underline{Q} \to \mathbf{C}$ between topological space, we obtain the left adjoint functor $p_Q! : Sh(\Lambda\underline{Q}) \to Sh(\mathbf{C}_o)$ of $p_Q^* : Sh(\mathbf{C}_o) \to Sh(\Lambda\underline{Q})$. The existence of such a functor enables us to define the composite functor

$$F := p_Q! \circ I : Sh(\mathcal{V}(\mathcal{H})) \to Sh(\mathbf{C}_o)\,. \tag{14.2.13}$$

[12]In the sense that for each element $\phi \in (\Lambda\underline{Q})_C$, given the open neighbourhood $U, p_Q(U)$ is open in $\mathbf{C}_o$ and p_Q restricted to $U \ni \phi$ is a homomorphisms, i.e., $p_Q|_U : U \to p_Q(U)$ is a homomorphisms.

Such a functor sends all the original sheaves we had defined over $\mathcal{V}(\mathcal{H})$ to new sheaves over $\mathbf{C}_o$. Thus, denoting the sheaves over $\mathbf{C}_o$ as $\breve{\underline{A}}$ we have

$$\breve{\underline{\Sigma}} := F(\underline{\Sigma}) = p_Q! \circ I(\underline{\Sigma}) .$$

For each element $C \in \mathbf{C}_o$ we then obtain:

$$\coprod_{\phi \in (\Lambda \underline{Q})(C)} \underline{\Sigma}(\phi(C)) . \tag{14.2.14}$$

Conclusions In this chapter we considered an abstract characterization of quantization in a topos, both in the case when there is an underlining classical system and when such a system is absent. So far, to our knowledge, the only current application of this schema was done in [57]. In this paper the author considers an underlying classical system and defined quantization through the map $\tilde{v} : a \mapsto e^{i\hat{a}}$. However it is not clear to us how this map can be injective. Nevertheless, the work done in [57] is a very good start to tackling the problem of quantization in a topos. What needs to be done at this stage is to enlarge such a schema so that the issues mentioned in the introduction of this chapter can be addressed.

Appendix A

A.1 Dedekind Reals in a Topos

In Sect. 9.2 we gave the definition of the internal natural number object $\underline{\mathbb{Z}}_\tau$ and the internal rational number object $\underline{\mathbb{Q}}_\tau$. These will now be utilised to define the internal Dedekind reals $\overline{\mathbb{R}}$. To this end, let us consider the set $\mathbb{R}$ of ordered reals, each real number $r \in \mathbb{R}$ defines two disjoint subsets in $\mathbb{Q}$, namely

$$L = \{q \in \mathbb{Q} | q < r\}$$
$$U = \{q \in \mathbb{Q} | q > r\}.$$

These subsets have the following properties:

1. Each subset is non-empty.
2. L is a downwards closed set but has no largest element.
3. U is an upwards closed set but has no smallest element.
4. If x is a rational number then $L \cup U \subseteq \mathbb{Q}$, otherwise $L \cup U = \mathbb{Q}$.

Given the above we can now define the notion of a *Dedekind cut*.

Definition A.1.1 A Dedekind cut is a pair of disjoint subsets (L, U) of $\mathbb{Q}$ such that the following conditions hold:

1. Non-degenerate:

$$\exists\, p \in \mathbb{Q}\ (p \in L), \quad \exists\, q \in \mathbb{Q}\ (q \in U).$$

2. Inward-closed:

$$\forall\, p, q \in Q\ (p < q \wedge q \in L \Rightarrow p \in L);$$
$$\forall\, p, q \in Q\ (q < p \wedge q \in U \Rightarrow p \in U).$$

C. Flori, *A Second Course in Topos Quantum Theory*,
Lecture Notes in Physics 944, https://doi.org/10.1007/978-3-319-71108-9

3. Outward-open:

$$\forall q \in \mathbb{Q}\ (q \in L \Rightarrow \exists p \in \mathbb{Q}(p \in L \wedge q < p);$$
$$\forall q \in \mathbb{Q}\ (q \in U \Rightarrow \exists p \in \mathbb{Q}(p \in U \wedge p < q)\,.$$

4. Located:

$$\forall p, q \in \mathbb{Q}\ (q < p \Rightarrow (q \in L \vee p \in U)).$$

5. Mutually exclusive:

$$L \cap U = \emptyset\,.$$

It is possible to internalise the above definition of Dedekind cut in any topos with a natural number object. This is done by first defining the ordering relation $<$ in $\underline{\mathbb{Q}}_\tau$ as a sub-object of $\underline{\mathbb{Q}}_\tau \times \underline{\mathbb{Q}}_\tau$. In particular, given two elements $m/n, p/q \in \underline{\mathbb{Q}}_\tau$, then $<:= \{(m/n, p/q) \in \underline{\mathbb{Q}}_\tau \times \underline{\mathbb{Q}}_\tau | m \cdot q < p \cdot n\}$ where $m \cdot q < p \cdot n$ is the order relation on the integers. We can then re-write all of the above conditions in Definition A.1.1 in terms of the internal language of τ, by simply replacing $\mathbb{Q}$ by $\underline{\mathbb{Q}}_\tau$ and interpreting all the logical symbols in term of the internal language of τ.
The internal Dedekind reals $\overline{\mathbb{R}}$ is then defined as follows:

$$\overline{\mathbb{R}} = \{(L, U) \in P(\underline{\mathbb{Q}}_\tau) \times P(\underline{\mathbb{Q}}_\tau) | (L, U) \text{ is a Dedekind cut }\}\,.$$

It is also possible to define the internal Dedekind reals in terms of a geometric theory [50, D4.7.4]. In particular, we consider the geometrical theory $\mathbb{T}_{\mathbb{R}}$ generated by the symbols $(p, q) \in \mathbb{Q} \times \mathbb{Q}$ with $p < q$. These formal symbols undergo an ordering defined as follows: $(p, q) \leq (p', q')$ iff $p \leq p'$ and $q \leq q'$. The axioms of the theory $\mathbb{T}_{\mathbb{R}}$ are

1. $(p_1, q_1) \wedge (p_2, q_2) = \begin{cases} (\max\{p_1, p_2\},\ \min\{q_1, q_2\}) \text{ if } \max\{p_1, p_2\} \leq\ \min\{q_1, q_2\} \\ \bot \text{ otherwise }. \end{cases}$
2. $(p, q) = \bigvee\{(p', q') | p < p' < q' < q\}$.
3. $\top = \bigvee\{(p, q) | p < q\}$.
4. $(p, q) = (p, q_1) \vee (p_1, q)$ if $p \leq p_1 \leq q_1 \leq q$.

Given a topos τ, an interpretation of the theory $\mathbb{T}_{\mathbb{R}}$ in τ gives rise to a locale $\mathbb{R}_\tau$ with associated frame $\mathcal{O}(\mathbb{R}_\tau)$. The points of the locale $\mathbb{R}_\tau$, i.e. the maps $p^{-1} : \mathcal{O}(\mathbb{R}_\tau) \rightarrow \Omega_\tau$, are in bijective correspondence with Dedekind cuts defined in Definition A.1.1.

A.2 Scott's Interval Domain

We will start by giving a brief description of Scott's interval domain $\mathbb{IR}$ in **Sets**. In particular, as a set, $\mathbb{IR}$ consists of all compact subsets of the form $[a, b]$ with $a, b \in \mathbb{R}$ and $a \leq b$. In $\mathbb{IR}$ are included also the singletons $[a, a] = \{a\}$ for each $a \in \mathbb{R}$. $\mathbb{IR}$ is a poset under reverse inclusion. In this sense, elements of $\mathbb{IR}$ can be though of as approximations of real numbers. The smaller the subsets the better it approximates the corresponding real number. It is possible to equip $\mathbb{IR}$ with the so called Scott topology which is defined as follows:

Definition A.2.1 Given a subset $U \subseteq \mathbb{IR}$, we say that U is Scott open if the following conditions hold:

1. If $[a, b] \in U$ and $[a, b] \subseteq [a', b']$ then $[a', b'] \in U$. This means that U is upwards closed.
2. If all directed sets S with supremum in U have non-empty intersection with U, i.e. for any directed subset $S \subseteq \mathbb{IR}$ with supremum $\bigvee S$, if $\bigvee S \in U$, then there exists a $W \in S$ such that $W \in U$. This property means that U is inaccessible by directed joins.

Clearly the complement of a Scott open is Scott closed. These can be defined as follows:

Definition A.2.2 Given a subset $U \subseteq \mathbb{IR}$, we say that U is Scott closed if the following two condition hold:

1. U is a down set.
2. If S is a directed set contained in U and the supremum ($Sup(S)$) of S exists, then $Sup(S) \in U$.

Given the above definition, a basis for the Scott topology is given by the collection of the following subsets

$$(p, q)_S := \{[r, s] | p < r \leq s < q\}, \quad p, q \in \mathbb{Q} \text{ and } p < q.$$

We will denote the set $\mathbb{IR}$, equipped with the Scott topology, by $\mathcal{O}(\mathbb{IR})$.

Next we would like to internalise the object $\mathbb{IR}$ in the topos $[\mathcal{C}(\mathcal{A}), \mathbf{Sets}]$. This can be done utilising the technique elucidated in Sect. 9.3. In particular, we recall that the category $\mathcal{C}(\mathcal{A})$ is equipped with the upwards Alexandroff topology, such that the product $\mathcal{C}(\mathcal{A}) \times \mathbb{IR}$ is given the product topology. We then consider the continuous projection map $\pi : \mathcal{C}(\mathcal{A}) \times \mathbb{IR} \to \mathcal{C}(\mathcal{A})$, $(C, [a, b]) \mapsto C$. The associated frame map is then $\pi^{-1} : \mathcal{O}(\mathcal{C}(a)) \to \mathcal{O}(\mathcal{C}(\mathcal{A}) \times \mathbb{IR})$. Such a map describes the internal locale $\underline{\mathbb{IR}}$.

Similarly, as for the internal Dedekind reals, also the internal interval domain can be defined in terms of a geometric theory $\mathbb{T}_{\mathbb{IR}}$. In particular, the generating symbols for $\mathbb{T}_{\mathbb{IR}}$ are $(p, q) \in \mathbb{Q} \times \mathbb{Q}$ with $p < q$. These formal symbols undergo an ordering defined as follows: $(p, q) \leq (p', q')$ iff $p \leq p'$ and $q \leq q'$. The axioms of the theory $\mathbb{T}_{\mathbb{IR}}$ are

1. $(p_1, q_1) \wedge (p_2, q_2) = \begin{cases} (\max\{p_1, p_2\},\ \min\{q_1, q_2\}) \text{ if } \max\{p_1, p_2\} \leq \ \min\{q_1, q_2\} \\ \perp \text{ otherwise} \ . \end{cases}$
2. $(p, q) = \bigvee\{(p', q') | p < p' < q' < q\}$.
3. $\top = \bigvee\{(p, q) | p < q\}$.

Given a topos τ, an interpretation of the theory $\mathbb{T}_{\mathbb{IR}}$ in τ gives rise to a locale $\mathbb{IR}_\tau$ with associated frame $\mathcal{O}(\mathbb{IR}_\tau)$. The points of the locale $\mathbb{IR}_\tau$, i.e. the maps p^{-1} : $\mathcal{O}(\mathbb{IR}_\tau) \rightarrow \Omega_\tau$ are in bijective correspondence with elements of $\mathbb{IR}$ as defined above.

As one can see from the above definitions, the Scott interval domain is closely related to Dedekind cuts, in fact, only axiom (4.) in the definition of $\mathbb{T}_{\mathbb{R}}$ fails to hold for $\mathbb{T}_{\mathbb{IR}}$.

A.3 Properties of Daseinised Projections

In [26, Sec. 10.2] various properties of the daseinisation map where introduced. In this context it was also shown that sub-objects $\underline{\delta(\hat{P})}$ of the spectral presehaf $\underline{\Sigma}$ are 'special' in the sense that they are the only elements in $Sub_{cl}(\underline{\Sigma})$ for which the presheaf maps are surjective. In particular, in the definition of a sub-object of $\underline{\Sigma}$ we have the condition that, for each $V^{'} \subseteq V$, the respective presheaf map

$$\underline{\Sigma}(i_{V'V}) : \underline{\Sigma}_V \rightarrow \underline{\Sigma}_{V'} \tag{A.3.1}$$

is such that, for a given subset $\underline{S}_V \subseteq \underline{\Sigma}_V$, then

$$\underline{\Sigma}(i_{V'V})(\underline{S}_V) \subseteq \underline{S}_{V'} \subseteq \underline{\Sigma}_{V'} \ . \tag{A.3.2}$$

However, for subobjects of the form $\underline{\delta(\hat{P})}$ we obtain an equality in (A.3.2). This property is encoded in the following theorem:

Theorem A.3.1 *Given any projection operator $\hat{P}$ and any two contexts $V' \subseteq V$, then the following relation holds:*

$$S_{\underline{O}(i_{V'V})}\delta^o(\hat{P})_V = \underline{\Sigma}(i_{V'V})(S_{\delta^o(\hat{P})_V}) \ . \tag{A.3.3}$$

Proof As a first step we will show that the map

$$\underline{\Sigma}(i_{V'V}) : P(\underline{\Sigma}_V) \rightarrow P(\underline{\Sigma}_{V'}) \tag{A.3.4}$$
$$S \mapsto r_{V'V}S := \{\lambda_{|V'} | \lambda \in S\} \ ,$$

is continuous, closed and open. For notational simplicity we will write $r = \underline{\Sigma}(i_{V'V})$ and Eq. (A.3.3) becomes

$$r(S_{\delta^o(\hat{P})_V}) = S_{\mathcal{O}(i_{V'V})}(\delta^o(\hat{P})_V) = S_{\delta^o(\hat{P})_{V'}} . \tag{A.3.5}$$

Let us first show that such a map is continuous. Consider an open basis set[1] $R \in \underline{\Sigma}_{V'}$, we know that $\underline{\Sigma}_V := \{\lambda : V \to \mathbb{C}|\lambda(\hat{1}) = 1\}$ and, similarly, $\underline{\Sigma}_{V'} := \{\lambda : V' \to \mathbb{C}|\lambda(\hat{1}) = 1\}$. Moreover if $\lambda \in \underline{\Sigma}_V$ then from the definition of the presheaf maps it follows that $\lambda_{|V'} \in \underline{\Sigma}_{V'}$ when $V' \subseteq V$. We can then define, for any $R \in P(\underline{\Sigma}'_V)$ the following:

$$r^{-1}(R) := R \cap \underline{\Sigma}_V . \tag{A.3.6}$$

Since the intersection of open sets is open, $r^{-1}(R)$ is open.
Next we need to show that r is closed. Consider a closed subset $S \subseteq \underline{\Sigma}_V$. Since $\underline{\Sigma}_V$ is compact so is S and, since r is continuous, then $r(S)$ is compact in $\underline{\Sigma}_{V'}$. But $\underline{\Sigma}_{V'}$ is Hausdorff thus $r(S)$ is closed.
To show that r is open we note that since every $\lambda_{V'} \in \underline{\Sigma}_{V'}$ is of the form $\lambda_V|_{V'}$ for some $\lambda_V \in S \subseteq \underline{\Sigma}_V$, then

$$r(S) = \{\lambda_{V'}|\lambda \in S\} = S \cap \underline{\Sigma}_{V'} . \tag{A.3.7}$$

If S is open, then $S \cap \underline{\Sigma}_V$ is the intersection of two opens thus it is itself open.

Given the above properties of r, a clopen subset $S_{\delta^o(\hat{P})_V} \subseteq \underline{\Sigma}_V$ gets mapped to the clopen subset $r(S_{\delta^o(\hat{P})_V}) \in \underline{\Sigma}_{V'}$. Such a subset[2] is

$$r(S_{\delta^o(\hat{P})_V}) = int \bigcap \{S_{\hat{Q}} \in Sub_{cl}(\underline{\Sigma}_{V'})|r(S_{\delta^o(\hat{P})_V}) \subseteq S_{\hat{Q}}\} \tag{A.3.8}$$

thus $r(S_{\delta^o(\hat{P})_V}) \subseteq S_{\hat{Q}}$.

We now need to show that $\hat{Q} \geq \delta^o(\hat{P})_V$. We prove this by contradiction. Assume that $\delta^o(\hat{P})_V \geq \hat{Q}$ and define $\hat{R} := \delta(\hat{P})_V - \hat{Q} \in P(V)$, such that $\lambda \in S_{\hat{R}}$. It follows that $\lambda \in S_{\delta^o(\hat{P})_V}$ but $\lambda \notin S_{\hat{Q}} \subseteq \underline{\Sigma}_V$.

However if $\delta^o(\hat{P})_V = \delta^o(\hat{P})_{V'}$ then $r(S_{\delta^o(\hat{P})_V}) = S_{\delta^o(\hat{P})_{V'}}$. In fact, given an element $\lambda \in \underline{\delta(\hat{P})}_V = S_{\delta^o(\hat{P})_V}$ by definition $\lambda(\delta^o(\hat{P})_V) = 1$. Since $\delta^o(\hat{P})_{V'} \geq \delta^o(\hat{P})_V$ ($V' \subseteq$

[1]Note that for each $V \in \mathcal{V}(\mathcal{H})$, $\underline{\Sigma}_V$ has the spectral topology (being the spectrum of V) which is compact and Hausdorff. The details of such a topology are not necessary to prove continuity. It is worth saying, though, that it can be shown that a basis for this topology is the collection of clopen subsets. This renders the prof of continuity easier, however we will not use it here. On the other hand, when proving closeness of r we will use the fact that $\underline{\Sigma}_V$ is a Hausdorff compact space.

[2]Note that the *int* operation is needed for the subset to be clopen, otherwise it would only be closed.

V), then $\lambda_{|V'} \in S_{\delta^o(\hat{P})_{V'}}$. On the other hand, if $\lambda \notin S_{\delta^o(\hat{P})_V}$ then $\lambda(\delta^o(\hat{P})_V) = 0$. Since $\delta^o(\hat{P})_{V'} = \delta^o(\hat{P})_V$ then $\lambda_{|V'} \notin S_{\delta^o(\hat{P})_{V'}}$.

Given the fact that every $\lambda_{V'} \in \underline{\Sigma}_{V'}$ is of the form $\lambda_V|_{V'} = r(\lambda_V)$ for some $\lambda \in S_{\delta^o(\hat{P})_V}$, then $r(S_{\delta^o(\hat{P})_V}) = S_{\delta^o(\hat{P})_{V'}}$ and $r\big((S_{\delta^o(\hat{P})_V})^c\big) = S^c_{\delta^o(\hat{P})_{V'}} \in \underline{\Sigma}_{V'}$.

It follows that in our case we have $r(S_{\hat{Q}}) = S_{\hat{Q}}$ and $r\big((S_{\hat{Q}})^c\big) = (S_{\hat{Q}})^c$. We have shown that, $\lambda_{V'} \notin S_{\hat{Q}} \subseteq \underline{\Sigma}_V$ but $\lambda \in r(S_{\delta^o(\hat{P})_V})$, what this means is that

$$\text{if } \hat{Q} \leq \delta^o(\hat{P})_V \text{ then } r(S_{\delta^o(\hat{P})_V}) \nsubseteq S_{\hat{Q}}\,. \tag{A.3.9}$$

However, this is a contradiction, therefore it must be the case that $\hat{Q} \geq \delta^o(\hat{P})_V$. We can now write $r(S_{\delta^o(\hat{P})_V})$ as

$$r(S_{\delta^o(\hat{P})_V}) = int \bigcap \{S_{\hat{Q}} \in Sub_{cl}(\underline{\Sigma}_{V'}) | \hat{Q} \geq \delta^o(\hat{P})_V\} \tag{A.3.10}$$

$$= S_{\hat{Q} \in P(V') | \hat{Q} \geq \delta^o(\hat{P})_V} = S_{\mathcal{O}(i_{V'V})\delta^o(\hat{P})_V}, \tag{A.3.11}$$

therefore

$$\underline{\Sigma}(i_{V'V}) : S_{\delta^o(\hat{P})_V} \to S_{\delta^o(\hat{P})_{V'}}\,. \tag{A.3.12}$$

It follows that the clopen sub-objects of the form $\underline{\delta(\hat{P})}$ are such that the presheaf maps are also surjective. □

A.4 Connection Between Sheaves and Etalé Bundles

In this Section we will investigate the connection between sheaves and an etalé bundles is. To this end we need to introduce the notion of a *germ* of a function. Once we have introduced such a notion, it can be shown that each sheaf is a sheaf of cross sections of a suitable bundle. All this will become clear as we proceed. First of all: what is a *germ*? *Germs* represent constructions which define local properties of functions. In particular they indicate how similar two functions are locally. Because of this locality requirement, *germs* are generally defined on functions acting on topological spaces, such that the word local acquires meaning. For example, one can consider measure of 'locality' to be a power series expansion of a function around some fixed point. Thus, one can say that two holomorphic functions $f, g : U \to \mathbb{C}$ have the same germ at a point $a \in U$ iff the power series expansions around that point are the same. Thus f, g agree on some neighbourhood of a, i.e., with respect to that neighbourhood they "look" the same.

This definition obviously holds only if a power series expansion exists, however it is possible to generalise such a definition in a way that it only requires topological properties of the spaces involved. For example two functions $f, g : X \to E$ have the same germ at $x \in X$ if there exist some neighbourhood of x on which they agree. In this case we write[3] $germ_x f = germ_x g$ which implies that $f(x) = g(x)$. However the converse is not true.

How do we generalise such a definition of germs in the case of presheaves? Let us consider a presheaf $P : \mathcal{O}(X) \to Sets \in Sets^{\mathcal{O}(X)^{op}}$, where X is a topological space and $\mathcal{O}(X)^{op}$ is the category of open sets with reverse ordering to the inclusion ordering. Given a point $x \in X$ and two neighbourhoods U and V of x, the presheaf P assigns two sets $P(U)$ and $P(V)$. Now consider two points $t \in P(V)$ and $s \in P(U)$, we then say that t and s have the same germ at x iff there exists some open $W \subseteq U \cap V$, such that $x \in W$ and $s_{|W} = t_{|W} \in P(W)$.

The condition of having the same germ at x defines an equivalence class which is denoted as $germ_x s$. Thus $t \in germ_x s$ iff, given two opens $U, V \ni x$ then there exists some $W \subseteq U \cap V$ such that $x \in W$ and $t_{|W} = s_{|W} \in P(W)$, where $s \in P(U)$ and $t \in P(V)$. It follows that the set of all elements obtained through the P presheaf get 'quotient' through the equivalence relation of *"belonging to the same germ"*. Therefore, for each point $x \in X$ there will exist a collection of germs at x, i.e., a collection of equivalence classes:

$$P_x := \{germ_x s | s \in P(U), x \in U \text{ open in } X\} . \tag{A.4.1}$$

We can now collect all these set of germs for all points $x \in X$, defining

$$\Lambda_P = \coprod_{x \in X} P_x = \{\text{all } germ_x s | s \in X, s \in P(U)\} . \tag{A.4.2}$$

What we have done so far is, basically, to divide the presheaf space in equivalence classes. We can now define the map

$$\begin{aligned} p : \Lambda_P &\to X \\ germ_x s &\mapsto x \\ germ_y s &\mapsto y \end{aligned} \tag{A.4.3}$$

which sends each germ to the point in which it is taken. It follows that each $s \in P(U)$ defines a function

$$\begin{aligned} \dot{s} : U &\to \Lambda_P \\ x &\mapsto germ_x s . \end{aligned} \tag{A.4.4}$$

[3]This should be read as: the germ of f at x is the same as the germ of g at x.

It is straightforward to see that $\dot{s}$ is a section of $p : \Lambda_P \rightarrow X$. Since the assignments $s \rightarrow \dot{s}$ is unique, it is possible to replace each element s in the original presheaf with a section $\dot{s}$ on the set of germs Λ_P.

We now define a topology on Λ_P by considering as basis of open sets all the image sets $\dot{s}(U) \subseteq \Lambda_P$ for U open in X, i.e. open sets are unions of images of sections. Such a topology obviously makes p continuous. In fact, given an open set $U \subseteq X$ then $p^{-1}(U)$ is open by definition of the topology on Λ_P, since $p^{-1}(U) = \bigcup_{s_i \in P(U)} \dot{s}_i(U)$.

On the other hand it is also possible to show that the sections $\dot{s}$, as defined above, are continuous with respect to the topology on Λ_P. To understand this consider two elements $t \in P(V)$ and $s \in P(U)$ such that $\dot{t}(x) = \dot{s}(x)$, i.e. $germ_x(t) = germ_x(s)$ where $x \in V \cap U$. It then follows that there exists an open set $W \ni x$ such that $W \subseteq V \cap U$. If we consider all those elements $y \in V \cap U \subseteq X$ for which $\dot{s}(y) = \dot{x}(y)$, then all such elements will comprise the open set $W \subseteq V \cap U$. Given this reasoning we want to show that for any open $\mathcal{O} \in \Lambda_P$, then $\dot{s}^{-1}(\mathcal{O})$ is open in X. Without loss of generality we can choose $\mathcal{O}$ to be a basis set, i.e.

$$\dot{s}(W) = \{germ_x(s) | \forall x \in W\} . \tag{A.4.5}$$

Thus $\dot{s}^{-1}\dot{s}(W) = W$ consists of all those points x such that $\dot{s}(x) = \dot{t}(x)$ for $t, s \in germs_x(s)$. It follows that W is open.

One can also show that $\dot{s}$ is open and an injection. The property of being open follows directly from the definition of topology on Λ_p since the basis of open sets are all the image sets $\dot{s}(U) \subseteq \Lambda_P$ for U open in X. To show that it is injective we need to show that if $germ_x s = germ_y s$ then $x = y$. This follows from the definition of germs at a point. Putting all these results together we show that $\dot{s} : U \rightarrow \dot{s}(U)$ is a homeomorphism. So we have managed to construct a bundle $p : \Lambda_P \rightarrow X$ which is a local homeomorphism, since each point $germ_x(s) \in \Lambda_P$ has an open neighbourhood $\dot{s}(U)$ so that p, restricted to $\dot{s}(U)$, $p : \dot{s}(U) \rightarrow X$ has a two sided inverse $\dot{s} : U \rightarrow \dot{s}(U)$:

$$p \circ \dot{s} = id_X; \quad \dot{s} \circ p = id_{\Lambda_P} \tag{A.4.6}$$

hence p is a local homeomorphism.

The above reasoning shows how, given a presheaf P it is possible to construct an etalé bundle $p : \Lambda_P \rightarrow X$ out of it. Given such a bundle, it is then possible to construct a sheaf in terms of it. In particular we have the following theorem:

Theorem A.4.1 *The presheaf*

$$\begin{aligned} \Gamma(\Lambda_P) : \mathcal{O}^{op} &\rightarrow Sets \\ U &\mapsto \{\dot{s} | s \in P(U)\} \end{aligned} \tag{A.4.7}$$

is a sheaf.

Proof In the presheaf

$$\Gamma(\Lambda_P) : \mathcal{O}(X)^{op} \to Sets \qquad \text{(A.4.8)}$$
$$U \mapsto \{\dot{s}|s \in P(U)\}$$

the maps are defined by restriction, i.e. given $U_i \subseteq U$, then

$$\Gamma(\Lambda_P) : \mathcal{O}(X)^{op} \to Sets \qquad \text{(A.4.9)}$$
$$U_i \mapsto \{\dot{s}_i|s_i \in P(U_i)\}$$

where $\dot{s} \mapsto \dot{s}_i$ is defined via $\dot{s}_i = P(i_{U_iU})s$. Now since

$$\dot{s} : U \to \Lambda_p(U) \qquad \text{(A.4.10)}$$
$$x \mapsto germ_x s$$

while

$$\dot{s}_i : U_i \to \Lambda_p(U_i) \qquad \text{(A.4.11)}$$
$$y \mapsto germ_y s_i \, .$$

Since $U_i \subseteq U$, then

$$\dot{s} : U_i \to \Lambda_p(U_i) \qquad \text{(A.4.12)}$$
$$y \mapsto germ_y s$$

it follows that $\dot{s}_i = \dot{s}_{|U_i}$.

In order to show that the above is indeed a sheaf we need to show that the diagram

$$\Gamma(\Lambda_p(U)) \xrightarrow{e} \prod_i \Gamma(\Lambda_p(U_i)) \underset{q}{\overset{p}{\rightrightarrows}} \prod_{i,j} \Gamma(\Lambda_p(U_i \cap U_j))$$

is an equaliser. By applying the definition of the sheaf maps we obtain

$$e : \Gamma(\Lambda_p(U)) \to \prod_i \Gamma(\Lambda_p(U_i)) \qquad \text{(A.4.13)}$$
$$\dot{s} \to e(\dot{s}) = \{\dot{s}_{U_i}|i \in I\} = \{\dot{s}_i|i \in I\} \, .$$

On the other hand

$$p(\dot{s}_i) = \{s_i|_{U_i \cap U_j}\} = \{s_{|U_i \cap U_j}\} \qquad \text{(A.4.14)}$$

while

$$q(\dot{s}_j) = \{s_j|_{U_i \cap U_j}\} = \{s_{|U_i \cap U_j}\}\,. \tag{A.4.15}$$

□

$\Gamma(\Lambda_P)$ is called the sheaf of cross sections of the bundle $p : \Lambda_P \to X$. We can now define a map

$$\eta : P \to \Gamma \circ \Lambda_P \tag{A.4.16}$$

such that for each context $U \in \mathcal{O}(X)^{op}$ we obtain

$$\begin{aligned} \eta_U : P_U &\to \Gamma(\Lambda_P)(U) \\ s &\mapsto \dot{s}\,. \end{aligned} \tag{A.4.17}$$

Theorem A.4.2 *If P is a sheaf then η is an isomorphism.*

Proof We need to show that η is 1:1 and onto.

1. One to one:
   We want to show that if $\dot{s} = \dot{t}$ then $t = s$. Given $t, s \in P(U)$, $\dot{s} = \dot{t}$ means that $germ_x(s) = germ_x(t)$ for all $x \in U$. Therefore there exists opens $V_x \subseteq U$ such that $x \in V_x$ and $t_{|V_x} = s_{|V_x}$. The collection of these opens V_x for all $x \in U$ form a cover of U such that $s_{V_x} = t_{|V_x}$. This implies that s, t agree on the map $P(U) \to \coprod_{x \in U} P(V_x)$. From the sheaf requirements it follows that $t = s$.
2. Onto:
   We want to show that any section $h : U \to \Lambda_p$ is of the form $\eta_U(s) = \dot{s}$ for some $s \in P(U)$. Let us consider a section $h : U \to \Lambda_p$, this will pick for each $x \in U$ an element, say $h(x) = germ_x(s_x)$. Therefore for each $x \in U$ there will exist an open $U_x \ni x$ such that $s_x \in P(U_x)$. By definition $germ_x(s_x) = \dot{s}_x(x)$ where $\dot{s}_x$ is a continuous section, therefore for each open U_x we get $\dot{s}_x(U_x) = \{germ_x(s_x) | \forall x \in U_x\}$, which is open by definition. It follows that for each $x \in U_x$ there will exist some $t, s \in germ_x(s)$, such that $\dot{s}(x) = \dot{t}(x)$. This implies that there exists some open set W_x for which $x \in W_x \subseteq U_x \subseteq U$ and such that $t_{|W_x} = s_{|W_x}$. These open sets W_x form a covering of U, i.e. $U = \coprod_{x \in U_x} W_x$ with $s_{|W_x} \in P(W_x)$ for each $P(W_x)$. Moreover, since $h(x) = germ_x(s_x)$ for $x \in U_x$ it follows that $h = \dot{s}_x$ for each W_x. Now consider two sections $\dot{s}_x$ and $\dot{s}_y$ for $x \in P(W_x)$ and $y \in P(W_y)$, then on the intersection $W_x \cap W_y$, h agrees with both $\dot{s}_x$ and $\dot{s}_y$, therefore the latter agrees in the intersection. This means that $germ_z(s_x) = germ_z(s_y)$ for $z \in W_x \cap W_y$, therefore $s_x|_{W_x \cap W_y} = s_y|_{W_x \cap W_y}$.
   We thus obtain a family of elements s_x for each $x \in U_x$ such that they agree on both maps $P(U_x) \rightrightarrows \coprod_{x \in U_x} P(W_x) \cap P(W_y)$. From the condition of being a sheaf it follows that there exists an $s \in P(U)$, such that $s_{V_x} = s_x$. Then at each $x \in U$ we have $h(x) = germ_x(s_x) = germ_x(s) = \dot{s}(x)$, therefore $h = \dot{s}$.

□

It follows that all sheaves are sheaves of cross sections of some bundle. Moreover it is possible to generalise the above process and define the following pair of functors

$$Sets^{\mathcal{O}(X)^{op}} \xrightarrow{\Lambda} Bund(X) \xrightarrow{\Gamma} Sh(X)\,, \tag{A.4.18}$$

which if we combine together we get the so called *sheafification functor*:

$$\Gamma\Lambda : Sets^{\mathcal{O}(X)^{op}} \rightarrow Sh(X)\,. \tag{A.4.19}$$

Such a functor sends each presheaf P on X to the "best approximation" $\Gamma\Lambda_P$ of P by a sheaf.

In the case of etalé bundles we then obtain the following equivalence of categories:

$$\text{Etalé}(X) \underset{\Gamma}{\overset{\Lambda}{\rightleftarrows}} Sh(X)$$

The pair of functors Γ and Λ are an adjoint pair (see Sect. A.5). Here we have restricted the functors to act on $Sh(X) \subseteq Sets^{\mathcal{O}(X)^{op}}$.

A.5 The Adjoint Pair

As discussed in Chapter 14 of [26], given a map $f : X \rightarrow Y$ between topological spaces X and Y we obtain a geometric morphism, whose inverse and direct image are, respectively,

$$f^* : Sh(Y) \rightarrow Sh(X) \tag{A.5.1}$$
$$f_* : Sh(X) \rightarrow Sh(Y)\,.$$

We also know that $f^* \dashv f_*$, i.e., f^* is the left adjoint of f_*. If f is an etalé map, however, there also exists the left adjoint $f!$ to f^*, namely

$$f! : Sh(X) \rightarrow Sh(Y) \tag{A.5.2}$$

with $f! \dashv f^* \dashv f_*$.

In Theorem A.5.1, below, we will show that

$$f!(p_A : A \rightarrow X) = f \circ p_A : A \rightarrow Y \tag{A.5.3}$$

so that we combine the etalé bundle $p_A : A \to X$ with the etalé map $f : X \to Y$, to give the etalé bundle $f \circ p_A : A \to Y$. Here we have used the fact that sheaves can be defined in terms of etalé bundles. In fact in Chapter 14 of [26] it was shown that there exists an equivalence of categories $Sh(X) \simeq Etale(X)$ for any topological space X.

Theorem A.5.1 *Given the etalé map $f : X \to Y$, the left adjoint functor $f! : Sh(X) \to Sh(Y)$ is defined as follows:*

$$f!(p_A : A \to X) = f \circ p_A : A \to Y \tag{A.5.4}$$

for $p_A : A \to Y$ being an etalé bundle.

Proof In the proof we will first define the functor $f!$ for general presheaf situation, then we will restrict our attention to the case of sheaves ($Sh(X) \subseteq \mathbf{Sets}^{X^{op}}$) and f etalé.

Consider the map $f : X \to Y$, this gives rise to the functor $f! : \mathbf{Sets}^{X^{op}} \to \mathbf{Sets}^{Y^{op}}$. The standard definition of $f!$ is as follows:

$$f! := - \otimes_X ({}_f X^{\bullet}) \tag{A.5.5}$$

such that, for any object $A \in \mathbf{Sets}^{X^{op}}$ we have

$$A \otimes_X ({}_f Y^{\bullet}) . \tag{A.5.6}$$

This is a presheaf in $\mathbf{Sets}^{Y^{op}}$, thus for each element $y \in Y$ we obtain the set

$$(A \otimes_X ({}_f Y^{\bullet}))y := A \otimes_X ({}_f Y^{\bullet})(-, y) \tag{A.5.7}$$

where $({}_f Y^{\bullet})$ is the presheaf

$$({}_f Y^{\bullet}) : X \times Y^{op} \to \mathbf{Sets} . \tag{A.5.8}$$

This presheaf derives from the composition of $f \times id_{Y^{op}} : X \times Y^{op} \to Y \times Y^{op}$ ($(f \times id_{Y^{op}})^* : \mathbf{Sets}^{Y \times Y^{op}} \to \mathbf{Sets}^{X \times Y^{op}}$) with the bi-functor ${}^{\bullet}Y^{\bullet} : Y \times Y^{op} \to \mathbf{Sets}$; $(y, y') \mapsto Hom_Y(y', y)$, i.e.,

$$({}_f Y^{\bullet}) := (f \times id_{Y^{op}})^*({}^{\bullet}Y^{\bullet}) = {}^{\bullet} Y^{\bullet} \circ (f \times id_{Y^{op}}) . \tag{A.5.9}$$

Now coming back to our situation we then have the restricted functor

$$\begin{aligned} ({}_f Y^{\bullet})(-, y) : (X, y) &\to \mathbf{Sets} \\ (x, y) &\mapsto ({}_f Y^{\bullet})(x, y) \end{aligned} \tag{A.5.10}$$

which, from the definition given above is

$$({}_fY^\bullet)(x,y) =^\bullet Y^\bullet \circ (f \times id_{Y^{op}})(x,y) =^\bullet Y^\bullet(f(x),y) = Hom_Y(y,f(x))\,. \tag{A.5.11}$$

Therefore, putting all the results together we have that for each $y \in Y$ we obtain $A \otimes_X ({}_fY^\bullet)(-,y)$, defined for each $x \in X$ as

$$A(-) \otimes_X ({}_fY^\bullet)(x,y) := A(x) \otimes_X Hom_Y(y,f(x))\,. \tag{A.5.12}$$

This represents the presheaf A defined over the element x, plus a collection of maps in Y mapping the original y to the image of x via f.

In particular $A(x) \otimes_X ({}_fX^\bullet) = A(x) \otimes_X Hom_Y(y,f(-))$ represents the following equaliser:

$$\begin{array}{ccc} \coprod_{x,x'} A(x) \times Hom_X(x',x) \times Hom_Y(y,f(x')) & \underset{\theta}{\overset{\tau}{\rightrightarrows}} & \coprod_x A(x) \times Hom_Y(y,f(x)) \\ & & \downarrow \sigma \\ & & A(-) \otimes_X Hom_Y(y,f(-)) \end{array}$$

such that, given a triplet $(a,g,h) \in A(x) \times Hom_X(x',x) \times Hom_Y(y,f(x'))$, we then obtain that

$$\tau(a,g,h) = (ag,h) = \theta(a,g,h) = (a,gh)\,. \tag{A.5.13}$$

Therefore, from the above equivalence conditions, $A(-) \otimes_X Hom_Y(y,f(-))$ is the quotient space of $\coprod_x A(x) \times Hom_Y(y,f(x))$.

We now consider the situation in which A is a sheaf on X, in particular it is an etalé bundle $p_A : A \to X$ and f is an etalé map which means that it is a local homeomorphism, i.e. for each $x \in X$ there is an open set V, such that $x \in V$ and $f_{|V} : V \to f(V)$ is a homeomorphism. It follows that for each $x_i \in V$ there is a unique element y_i such that $f_{|V}(x_i) = y_i$. In particular for each $V \subset X$ then $f_{|V}(V) = U$ for some $U \subset Y$.

Note that, since the condition of being a homeomorphism is only local, it can be the case that $f_{|V_i}(V_i) = f_{|V_j}(V_j)$ even if $V_i \neq V_j$. However in these cases the restricted etalé maps have to agree on the intersections, i.e. $f_{|V_i}(V_i \cap V_j) = f_{|V_j}(V_j \cap V_j)$.

Let us now consider an open set V with local homeomorphism $f_{|V}$. In this setting each element $y_i \in f_{|V}(V)$ will be of the form $f(x_i)$ for a unique x_i. Moreover, if we consider two open sets $V_1, V_2 \subseteq V$, then to each map $V_1 \to V_2$ in X, with associated bundle map $A(V_2) \rightarrow A(V_1)$, there corresponds a map $f_{|V}(V_1) \rightarrow f_{|V}(V_2)$ in Y. Therefore, evaluating $A(-) \otimes_X Hom_Y(-, f(-))$ at the open set $f_{|V}(V) \subset Y$ we get, for each $V_i \subseteq V$, the equivalence classes

$$[A(V_i) \times_X Hom_Y(f_{|V}(V), f_{|V}(V_i))] .$$

The equivalence relation is such that

$$A(V_j) \times_X Hom_Y(f_{|V}(V), f_{|V}(V_j)) \simeq A(V_k) \times_X Hom_Y(f_{|V}(V), f_{|V}(V_k))$$

iff: (1) there exists a map $f_{|V}(V_j) \rightarrow f_{|V}(V_k)$ which combines with $F_{|V} \rightarrow f_{|V}(V_j)$, giving $f_{|V}(V) \rightarrow f_{|V}(V_k)$; and (2) the corresponding bundle map $A(V_k) \rightarrow A(V_j) \rightarrow A(V)$ is given by the map $V \rightarrow V_j \rightarrow V_k$ in X. A moment of thought reveals that such an equivalence class is nothing but $p_A^{-1}(V)$ (the fibre of p_A at V) with associated fibre maps induced by the base maps.

We will now denote such an equivalence class by $[A(V) \times_X Hom_Y(f_{|V}(V), f_{|V}(V))]$ since, obviously, in each equivalence class there will be the element $A(V) \times_X Hom_Y(f_{|V}(V), f_{|V}(V))$.

Applying the same procedure for each open set $V_i \subset X$ we can obtain two cases:

(i) $f_{|V_i}(V_i) = U \neq f_V(V)$. In this case we simply get an independent equivalence class for U.
(ii) If $f_{|V_i}(V_i) = U = f_V(V)$ and there is no map $i : V \rightarrow V_i$ in X then, in this case, for U, we obtain two distinct equivalence classes $[A(V_i) \times_X Hom_Y(f_{|V_i}(V_i), f_{|V_i}(V_i))]$ and $[A(V) \times_X Hom_Y(f_{|V}(V), f_{|V}(V))]$.

Thus the sheaf $A(-) \otimes_X ({}_fY^\bullet)$ is defined for each open set $f_V(V) \subset Y$ as the set

$$[A(V) \times_X Hom_Y(f_{|V}(V), f_{|V}(V))] \simeq A(V))) ,$$

while, for each map $f_{V'}(V') \rightarrow f_V(V)$ in Y (with associated map $V' \rightarrow V$ in X), there is associated the map

$$[A(V) \times_X Hom_Y(f_{|V}(V), f_{|V}(V))] \simeq A(V) \rightarrow [A(V') \times_X Hom_Y(f_{|V'}(V)', f_{|V'}(V'))] \simeq A(V').$$

This is precisely what the etalé bundle $f \circ p_A : A \rightarrow Y$ is. □

Now that we understand the action of $f!$ on sheaves we will try to understand its action on functions. To this end, let us go back to etalé bundles. Given a map $\alpha : A \rightarrow B$ of etalé bundles over X, we obtain the map $f!(\alpha) : f!(A) \rightarrow f!(B)$ which is defined as follows: we start with the collection of fibre maps $\alpha_x : A_x \rightarrow B_x$, $x \in X$, where $A_x := p^{-1}A(\{x\})$. Then, for each $y \in Y$ we want to define the maps $f!(\alpha)_y : f!(A)_y \rightarrow f!(B)_y$, i.e., $f!(\alpha)_y : p^{-1}\big(A(f^{-1}\{y\})\big) \rightarrow p^{-1}\big(B(f^{-1}\{y\})\big)$. This

are defined as

$$f!(\alpha)_y(a) := \alpha_{p_A(a)}(a) \tag{A.5.14}$$

for all $a \in f!(A)_y = p^{-1}\big(A(f^{-1}\{a\})\big)$.

A.6 Lawvere-Tierney Topology and Closure Operator

In this section we will give a very brief review of a Lawvere-Tierney Topology and of the closure operator. Essentially a Lawvere-Tierney topology is a topology on a topos. We will choose the topos to be $\mathbf{Sets}^{\mathcal{V}(\mathcal{H})^{\mathrm{op}}}$ since it is the topos we are interested in.

Definition A.6.1 Given the topos $\mathbf{Sets}^{\mathcal{V}(\mathcal{H})^{\mathrm{op}}}$ with sub-object classifier Ω, a Lawvere-Tierney Topology on $\mathbf{Sets}^{\mathcal{V}(\mathcal{H})^{\mathrm{op}}}$ is a map

$$j : \Omega \to \Omega$$

in $\mathbf{Sets}^{\mathcal{V}(\mathcal{H})^{\mathrm{op}}}$ which satisfies the following properties:

1. $j \circ \text{true} = \text{true}$, i.e. the diagram

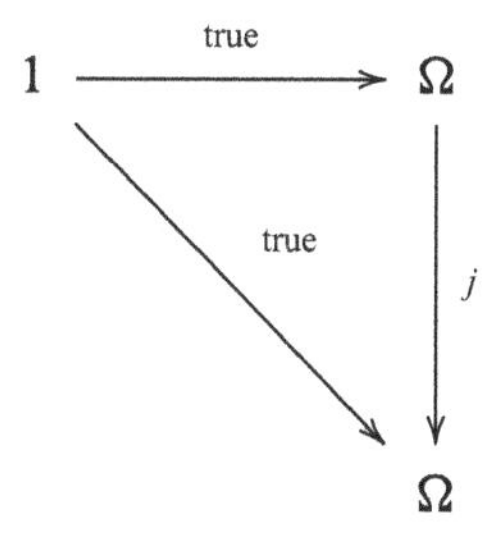

   commutes.
2. $j \circ j = j$, i.e. the diagram

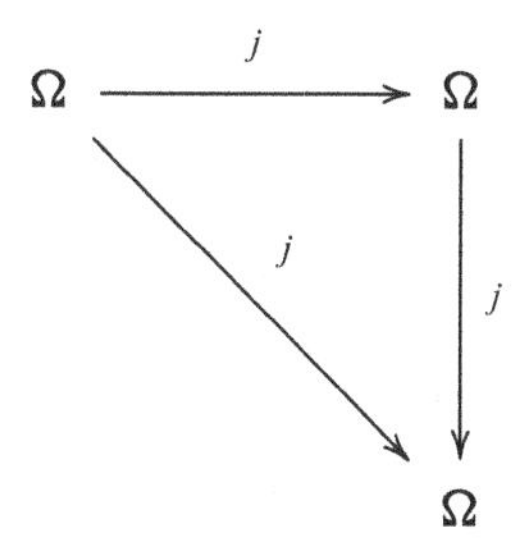

   commutes.

3. $j \circ \wedge = \wedge \circ (j \times j)$, i.e. the diagram

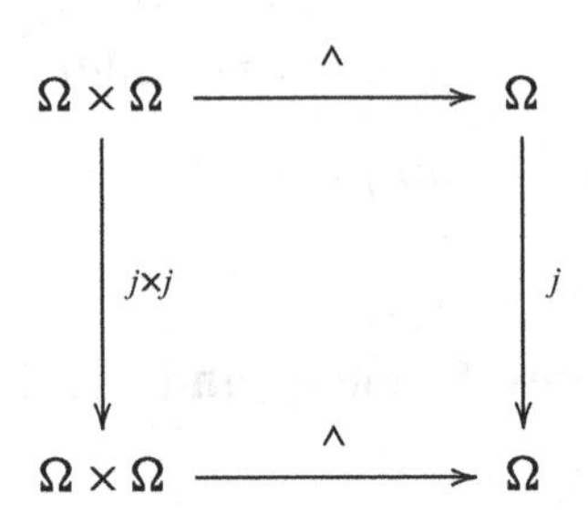

commutes. Here $\wedge : \Omega \times \Omega \to \Omega$ is defined so that for each $V \in \mathcal{V}(\mathcal{H})$ we have

$$\wedge_V : \Omega_V \times \Omega_V \to \Omega_V$$
$$(w_1, w_2) \mapsto \wedge_V(w_1, w_2) := w_1 \cap w_2 .$$

Definition A.6.2 A closure operator $\overline{(\cdot)}$ is a map such that, for every $P \in \mathbf{Sets}^{\mathcal{V}(\mathcal{H})^{\text{op}}}$, it maps a sub-object $S \subseteq P$ ($S \in Sub(P)$) to another sub-object $\overline{S} \subseteq P$. This assignment is such that, given any two sub-objects $S, T \in Sub(Q)$ then the following conditions hold:

$$S \subseteq \overline{S}$$
$$\overline{\overline{S}} = S$$
$$\overline{S \cap T} = \overline{S} \cap \overline{T} .$$

In [55] it was shown that the Lawvere-Tierney topology, the closure operator and the Grothendieck topology are equivalent to each other in the sense that each of them implies the other.

Proof

1. Lawvere-Tierney topology $\Longrightarrow$ the closure operator: let us assume we have a Lawvere-Tierney topology j. We then construct the following pullback

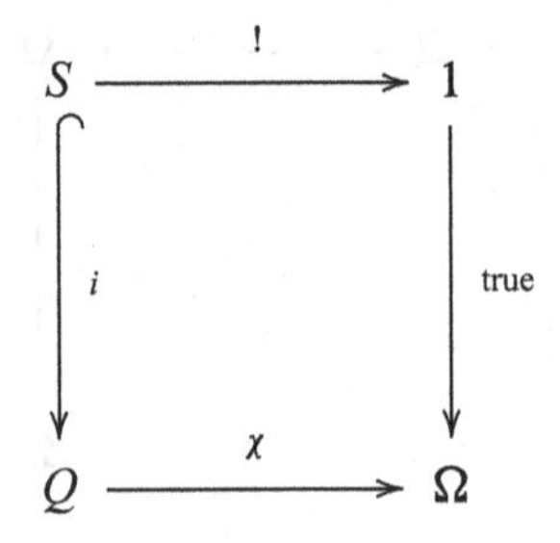

This is utilised to construct the closure $\overline{S}$ of S as the sub-object of Q, whose characteristic morphism is $j \circ \chi$, i.e. such that the outer square is a pullback:

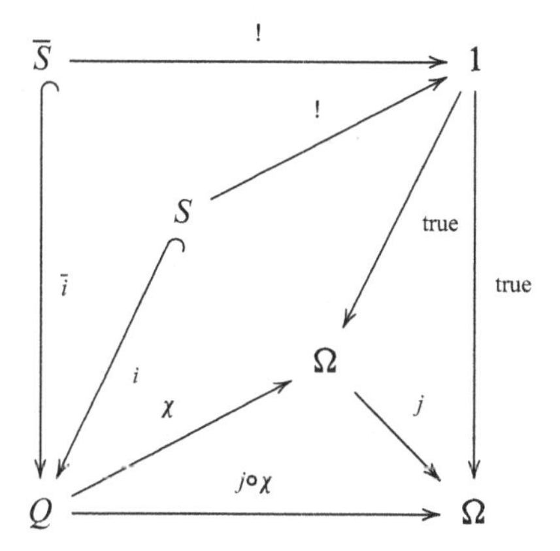

2. Closure operator $\Longrightarrow$ Grothendieck topology: let us assume we have a closure operator $\overline{(\cdot)}$, then the Grothendieck topology is given in terms of the closure of the terminal object 1, i.e.

$$J \hookrightarrow \Omega := \overline{1} \stackrel{\overline{\text{true}}}{\hookrightarrow} \Omega\,.$$

3. Grothendieck topology $\Longrightarrow$ Lawvere-Tierney topology: let us assume we have a Grothendieck topology J, then it is possible to construct a Lawvere-Tierney Topology j in terms of the characteristic morphism of J. In particular we defined j to be the morphism which would make the following diagram a pullback:

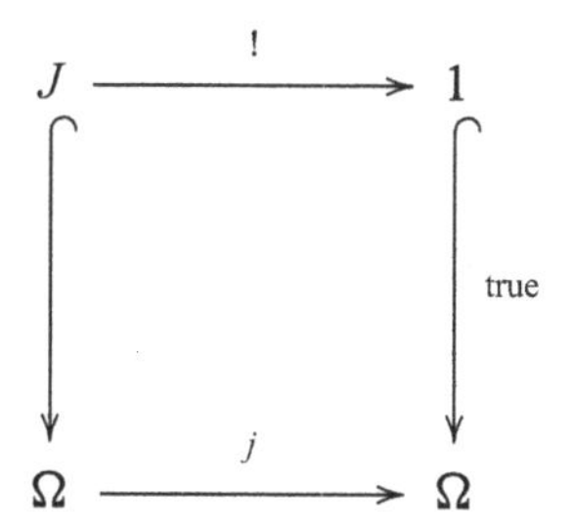

It is straightforward to see that if we reiterate the procedures $1 \to 2 \to 3 \to 1$ we would end up with the Lawvere-Tierney topology we started with. □

A.7 Yoneda Lemma

In this section we will give a very brief review of the Yoneda Lemma since it is used through out the book.

Lemma A.7.1 ***Preliminary***: *if $\mathcal{C}$ is a locally small category,*[4] *then each object A of $\mathcal{C}$ induces a natural contravariant functor from $\mathcal{C}$ to* **Sets** *called a hom-functor* $\mathbf{y}(A) := Hom_{\mathcal{C}}(-, A)$.[5] *Such a functor is defined on objects $C \in \mathcal{C}$ as*

$$\begin{aligned} \mathbf{y}(A) : \mathcal{C} &\to \mathbf{Sets} \\ C &\mapsto Hom_{\mathcal{C}}(C, A) \end{aligned} \tag{A.7.1}$$

on $\mathcal{C}$-morphisms $f : C \to B$ as

$$\begin{aligned} \mathbf{y}(A)(f) : Hom_{\mathcal{C}}(B, A) &\to Hom_{\mathcal{C}}(C, A) \\ g &\mapsto \mathbf{y}(A)(f)(g) := g \circ f \, . \end{aligned} \tag{A.7.2}$$

A very simple graphical example of the above is the following:

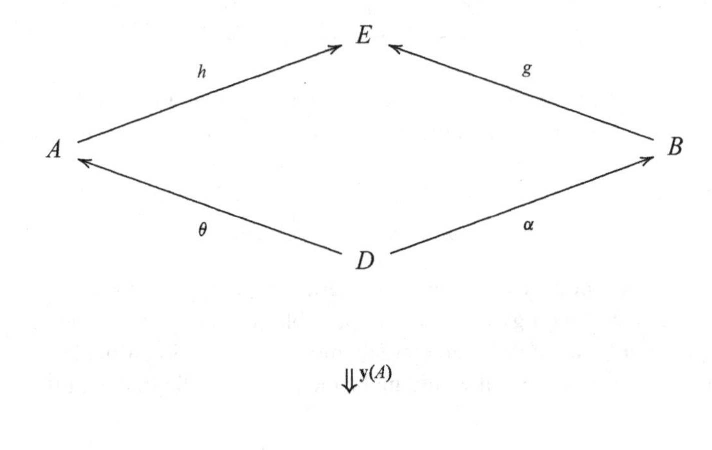

$\Downarrow \mathbf{y}(A)$

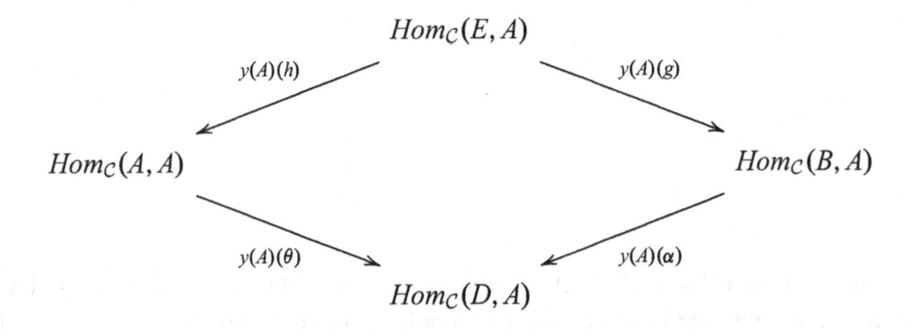

Lemma A.7.2 ***Yoneda lemma***: *Given an arbitrary presheaf P on $\mathcal{C}$ there exists a bijective correspondence between natural transformations $\mathbf{y}(A) \to P$ and elements*

---

[4] A category $\mathcal{C}$ is said to be locally small iff its collection of morphisms form a proper set.

[5] We have already encountered this in Example 5.10 of [51].

of the set $P(A)$ $(A \in \mathcal{C})$ *defined as an arrow*

$$\begin{aligned} \theta : Nat_{\mathcal{C}}(\mathbf{y}(A), P) &\overset{\simeq}{\to} P(A) \\ \left(\alpha : \mathbf{y}(A) \to P\right) &\mapsto \theta(\alpha) = \alpha_A(id_A)\,. \end{aligned} \tag{A.7.3}$$

where $\alpha_A : \mathbf{y}(A)(A) \to P(A)$; $Hom_{\mathcal{C}}(A, A) \to P(A)$.

References

1. E. Alfsen, F. Shultz, *State Spaces of Operator Algebras. Basic Theory, Orientations, and C-Products*. Mathematics: Theory and Applications (Birkhauser, Boston, MA, 2001)
2. J. Ambjorn, J. Jurkiewicz, R. Loll, The universe from scratch. Contemp. Phys. **47**, 103–117 (2006)
3. B. Banaschewski, C.J. Mulvey, The spectral theory of commutative C*-algebras: the constructive Gelfand-Mazur theorem. Quaest. Math. **23**(4), 465–488 (2000)
4. B. Banaschewski, C.J. Mulvey, The spectral theory of commutative C*-algebras: the constructive spectrum. Quaest. Math. **23**(4), 425–464 (2000)
5. B. Banaschewski, C.J. Mulvey, A globalisation of the Gelfand duality theorem. Ann. Pure Appl. Logic **137**, 62–103 (2006)
6. H. Barnum, M.P. Muller, C. Ududec, Higher-order interference and single-system postulates characterizing quantum theory. New J. Phys. **16** (2014)
7. J. Butterfield, C.J. Isham, Space-time and the philosophical challenge of quantum gravity (1999). arXiv:gr-qc/9903072
8. H. Comman, Upper regularization for extended self-adjoint operators. J. Oper. Theory **55**(1), 91–116 (2006)
9. A. Connes, A factor not anti-isomorphic to itself. Ann. Math. (2) **101**, 536–554 (1975)
10. T. Coquand, B. Spitters, Integrals and valuations. J. Logic Anal. **1**(3), 1–22 (2009)
11. L. Crane, What is the mathematical structure of quantum space-time? (2007). arXiv:0706.4452 [gr-qc]
12. H.F. de Groote, Observables IV: the presheaf perspective (2007). arXiv:0708.0677 [math-ph]
13. A. Doering, Flows on generalised Gelfand spectra of non-abelian unital C*-algebras and time evolution of quantum systems (2012). arXiv:1212.4882 [math.OA]
14. A. Doering, Generalised Gelfand spectra of nonabelian unital C*-algebras (2012). arXiv:1212.2613 [math.OA]
15. A. Doering, Topos-based logic for quantum systems and bi-Heyting algebras (2012). arXiv:1202.2750 [quant-ph]
16. A. Doering, Some remarks on the logic of quantum gravity (2013). arXiv:1306.3076 [gr-qc]
17. A. Doering, Topos theory and 'neo-realist' quantum theory. arXiv:0712.4003 [quant-ph]
18. A. Doering, The physical interpretation of daseinisation. arXiv: 1004.3573 [quant-ph]
19. A. Doering, R.S. Barbosa, Unsharp values, domains and topoi, in *Quantum Field Theory and Gravity: Conceptual and Mathematical Advances in the Search for a Unified Framework*, ed. by F. Finster et al. (Birkhäuser, Basel, 2011), pp. 65–96
20. A. Doering, B. Dewitt, Self-adjoint operators as functions I: lattices, Galois connections, and the spectral order (2012). arXiv:1208.4724 [math-ph]

C. Flori, *A Second Course in Topos Quantum Theory*,
Lecture Notes in Physics 944, https://doi.org/10.1007/978-3-319-71108-9

21. A. Doering, B. Dewitt, Self-adjoint operators as functions II: quantum probability. arXiv:1210.5747 [math-ph]
22. A. Doering, C.J. Isham, A topos foundation for theories of physics. II. Daseinisation and the liberation of quantum theory. J. Math. Phys. **49**, 053516 (2008). quant-ph/0703062 [quant-ph]
23. A. Doering, C.J. Isham, Classical and quantum probabilities as truth values (2011). arXiv:1102.2213v1
24. A. Doering, C. Isham, 'What is a thing?': Topos theory in the foundations of physics. arXiv:0803.0417 [quant-ph]
25. H.A. Dye, On the geometry of projections in certain operator algebras. Ann. Math. **61**(1), 73–89 (1955)
26. C. Flori, *A First Course in Topos Quantum Theory*. Lecture Notes in Physics, vol. 868 (Springer, Heidelberg, 2013)
27. C. Flori, Group action in topos quantum physics. J. Math. Phys. **54**, 3 (2013). arXiv:1110.1650 [quant-ph]
28. C. Flori, Concept of quantization in a topos (2017, in preparation)
29. C. Flori, Approaches to quantum gravity. arXiv:0911.2135 [gr-qc]
30. C. Flori, T. Fritz, (Almost) C^*-algebras as sheaves with self-action. J. Noncomm. Geom. **11**(3), 1069–1113 (2017)
31. G.B. Folland, *A Course in Abstract Harmonic Analysis* (CRC Press, Boca Raton, 1995)
32. R. Garner, Remarks on exactness notions pertaining to pushouts. Theory Appl. Categ. **27**(1), 2–9 (2012)
33. R. Goldblatt, *Topoi The Categorial Analysis of Logic* (North-Holland, London, 1984)
34. A. Grinbaum, Reconstruction of quantum theory (2006). philsci-archive.
35. R. Haag, *Local Quantum Physics: Fields, Particles, Algebras*. Texts and Monographs in Physics, 2nd edn. (Springer, Berlin, 1996)
36. J. Hamhalter, *Quantum Measure Theory*. Fundamental Theories of Physics, vol. 134 (Kluwer, Dordrecht, 2003)
37. J. Harding, A. Doering, Abelian subalgebras and the Jordan structure of a von Neumann algebra. arXiv:1009.4945 [math-ph]
38. J. Harding, M. Navara, Subalgebras of orthomodular lattices. Order **28**, 549–563 (2011)
39. L. Hardy, Quantum theory from five reasonable axioms (2001). arXiv:quant-ph/0101012
40. L. Hardy, Reformulating and reconstructing quantum theory (2011). arXiv:1104.2066 [quant-ph]
41. C. Heunen, M.L. Reyes, Active lattices determine AW*-algebras. J. Math. Anal. Appl. **416**(1), 289–313 (2014)
42. C. Heunen, N.P. Landsman, B. Spitters, A topos for algebraic quantum theory. Commun. Math. Phys. **291**(1), 63–110 (2009)
43. C. Heunen, N.P. Landsman, B. Spitters, S. Wolters, The Gelfand spectrum of a non-commutative C*-algebra: a topos-theoretic approach. J. Aust. Math. Soc. **90**, 39 (2011). arXiv:1010.2050 [math-ph]
44. J.R. Isbell, Adequate subcategories. Ill. J. Math. **4**, 541–552 (1960)
45. C.J. Isham, Some reflections on the status of conventional quantum theory when applied to quantum gravity (2002). arXiv:quant-ph/0206090
46. C.J. Isham, J. Butterfield, Some possible roles for topos theory in quantum theory and quantum gravity. Found. Phys. **30**, 1707 (2000). gr-qc/9910005
47. B. Jacobs, New directions in categorical logic, for classical, probabilistic and quantum logic. Log. Methods Comput. Sci. **11**, 3 (2016)
48. P.T. Johnstone, Open locales and exponentiation. Contemp. Math. **30**, 84–116 (1984)
49. P.T. Johnstone, *Stone Space* (Cambridge University Press, Cambridge, 1986)
50. P.T. Johnstone, *Sketches of an Elephant A Topos Theory Compendium I, II* (Oxford Science Publications, Oxford, 2002)
51. S. Kochen, E. Specker, The problem of hidden variables in quantum mechanics. J. Math. Mech. **17**(1), 59–87 (1967)
52. N.P. Landsman, *Algebraic Quantum Mechanics* (Springer, Berlin, 2009), pp. 6–10

53. R.C. Lyndon, P.E. Schupp, *Combinatorial Group Theory*. Classics in Mathematics (Springer, Berlin, 2001). Reprint of the 1977 edition
54. S. MacLane, *Categories for the Working Mathematician* (Springer, London, 1997)
55. S. MacLane, I. Moerdijk, *Sheaves in Geometry and Logic: A First Introduction to Topos Theory* (Springer, London, 1968)
56. S. Mercuri, Introduction to loop quantum gravity (2009). PoS ISFTG:016
57. K. Nakayama, Sheaves in quantum topos induced by quantization. arXiv:1109.1192 [math-ph]
58. M. Nilsen, C^*-bundles and $C_0(X)$-algebras. Indiana Univ. Math. J. **45**(2), 463–477 (1996)
59. nLab, Stuff, structure, property (2014). http://ncatlab.org/nlab/revision/stuff,+structure,+property/38
60. N.C. Phillips, Continuous-trace C*-algebras not isomorphic to their opposite algebras. Int. J. Math. **12**(3), 263–275 (2001)
61. M.L. Reyes, Sheaves that fail to represent matrix rings, in *Ring Theory and Its Applications*. Contemporary Mathematics, vol. 609 (American Mathematical Society, Providence, RI), pp. 285–29 (2012)
62. A. Rosenberg, *Noncommutative 'Spaces' and 'Stacks'*. Selected Papers on Noncommutative Geometry (New Prairie Press, 2014)
63. A. Rosenberg, M. Kontsevich, *Noncommutative Spaces*. Selected Papers on Noncommutative Geometry (New Prairie Press, 2014)
64. C. Rovelli, Relational quantum mechanics. Int. J. Theor. Phys. **35**, 1637 (1996)
65. M. Shulman, Exact completions and small sheaves. Theory Appl. Categ. **27**(7), 97–173 (2012)
66. S. Staton, S. Uijlen, Effect algebras, presheaves, non-locality and contextuality, in *Automata, Languages, and Programming*. Lecture Notes in Computer Science, vol. 9135 (Springer, Berlin, 2015), pp. 401–413
67. E. Størmer, On the Jordan structure of C*-algebras. Trans. Am. Math. Soc. **120**(3), 438–447 (1965)
68. L.N. Stout, Topological space objects in a topos II: ϵ-completeness and ϵ-cocompleteness. Manuscr. Math. **17**(1), 1–14 (1975)
69. L.N. Stout, A topological structure on the structure sheaf, of a topological ring. Commun. Algebra **5**(7), 695–706 (1977)
70. F. Strocchi, *An Introduction to the Mathematical Structure of Quantum Mechanics*. Advanced Series in Mathematical Physics, vol. 28, 2nd edn. (World Scientific, Singapore, 2008)
71. M. Takesaki, *Theory of Operator Algebras I* (Springer, Berlin, 2001)
72. B. van den Berg, C. Heunen, Noncommutativity as a colimit. Appl. Categ. Struct. **20**(4), 393–414 (2012)
73. S. Vickers, *Topology via Logic* (Cambridge University Press, New York, 1989)
74. S.A.M. Wolters, *Quantum Toposophy* UB Nijmegen [host] (2013)
75. S. Wolters, A comparison of two topos-theoretic approaches to quantum theory. arXiv:1010.2031 [math-ph]

Index

C. Flori, *A Second Course in Topos Quantum Theory*,
Lecture Notes in Physics 944, https://doi.org/10.1007/978-3-319-71108-9

Printed in the United States
By Bookmasters